T0399115

Cancer Sensitizing Agents for Chemotherapy
# OVERCOMING DRUG RESISTANCE IN GYNECOLOGIC CANCERS

VOLUME 17

# Cancer Sensitizing Agents for Chemotherapy Series

## Series Editor: Benjamin Bonavida, PhD

# Cancer Sensitizing Agents for Chemotherapy

# OVERCOMING DRUG RESISTANCE IN GYNECOLOGIC CANCERS

## VOLUME 17

*Edited by*

### RIYAZ BASHA, PhD, FAACC, FABAP

*Associate Professor and Vice Chair for Research, Department of Pediatrics and Women's Health, Texas College of Osteopathic Medicine, University of North Texas Health Science Center, Fort Worth, TX, United States*

### SARFRAZ AHMAD, PhD, FAACC, FABAP

*Director of Clinical Research, Gynecologic Oncology, AdventHealth Cancer Institute, Professor of Clinical Sciences, Florida State University, Professor of Medical Education, University of Central Florida, Orlando, FL, United States*

ELSEVIER

**ACADEMIC PRESS**

**An imprint of Elsevier**

Academic Press is an imprint of Elsevier
125 London Wall, London EC2Y 5AS, United Kingdom
525 B Street, Suite 1650, San Diego, CA 92101, United States
50 Hampshire Street, 5th Floor, Cambridge, MA 02139, United States
The Boulevard, Langford Lane, Kidlington, Oxford OX5 1GB, United Kingdom

**Notices**
Knowledge and best practice in this field are constantly changing. As new research and experience broaden our
understanding, changes in research methods, professional practices, or medical treatment may become
necessary.

Practitioners and researchers must always rely on their own experience and knowledge in evaluating and using
any information, methods, compounds, or experiments described herein. In using such information or methods
they should be mindful of their own safety and the safety of others, including parties for whom they have a
professional responsibility.

To the fullest extent of the law, neither the Publisher nor the authors, contributors, or editors, assume any liability
for any injury and/or damage to persons or property as a matter of products liability, negligence or otherwise, or
from any use or operation of any methods, products, instructions, or ideas contained in the material herein.

**Library of Congress Cataloging-in-Publication Data**
A catalog record for this book is available from the Library of Congress

**British Library Cataloguing-in-Publication Data**
A catalogue record for this book is available from the British Library

ISSN: 2468-3183

ISBN: 978-0-12-824299-5

For information on all Academic Press publications
visit our website at https://www.elsevier.com/books-and-journals

Publisher: Stacy Masucci
Senior Acquisitions Editor: Rafael Teixeira
Editorial Project Manager: Samantha Allard
Production Project Manager: Punithavathy Govindaradjane
Cover Designer: Miles Hitchen

Typeset by SPi Global, India

Working together
to grow libraries in
developing countries

www.elsevier.com • www.bookaid.org

# Cover Image Insert

The cover image depicts the female reproductive system, the primary site of origin for the gynecologic malignancies such as the cervical, endometrial/uterine, ovarian/fallopian tubes/peritoneum, vaginal, and vulvar cancers.

# Aims and Scope for Series "Cancer Sensitizing Agents for Chemotherapy"

Current cancer management strategies fail to adequately treat malignancies with chemotherapy, with multivariable dose-restrictive factors such as systemic toxicity and multidrug resistance; hence, limiting therapeutic benefits, quality of life, and complete long-term remission rates. The resistance of cancer cells to anticancer drugs is one of the major reasons for the failure of traditional cancer treatments. Cellular components and dysregulation of signaling pathways contribute to drug resistance. If modulated, such perturbations may restore the drug response and its efficacy. The recent understanding of the molecular mechanisms and targets that are implicated for cancer chemoresistance has paved the way to develop a large battery of small molecules (sensitizing agents) that can target resistant factors and reduce the threshold of resistance, thus allowing their combination with chemotherapeutic drugs to be effective and to reverse the chemoresistance. A large variety of chemotherapy-sensitizing agents has been developed, and several have been shown to be effective in experimental models and in cancer patients.

The main objective of the proposed series "Cancer Sensitizing Agents for Chemotherapy" is to publish individualized and focused volumes, whereby each volume is edited by an invited expert Editor(s). Each volume will dwell on specific chemosensitizing agents with similar targeting activities. The combination treatment of the sensitizing agent with chemotherapy may result in a synergistic/additive activity and the reversal of tumor cells' resistance to drugs.

The Editor(s) will compile nonoverlapping review chapters on reported findings in various cancers, both experimentally and clinically, with particular emphases on underlying biochemical, genetic, and molecular mechanisms of the sensitizing agent and the combination treatment.

The scope of the series is to provide scientists and clinicians with updated basic and clinical information that will be valuable in their quest to investigate, develop, and apply novel combination therapies to reverse drug resistance, thereby prolonging survival and even resulting in a cure in cancer patients.

*Benjamin Bonavida, PhD (Series Editor)*

# About the Series Editor

**Benjamin Bonavida**, PhD, (Series Editor), is currently Distinguished Research Professor at the University of California, Los Angeles (UCLA). He is affiliated with the Department of Microbiology, Immunology and Molecular Genetics, UCLA David Geffen School of Medicine. His research career, thus far, has focused on investigations in the fields of basic immunochemistry and cancer immunobiology. His research investigations have ranged from the biochemical, molecular, and genetic mechanisms of cell-mediated killing and tumor cell resistance to chemoimmuno cytotoxic drugs. The reversal of tumor cell resistance was investigated by the use of various selected sensitizing agents based on molecular mechanisms of resistance. In these investigations, there was the newly characterized dysregulated NF-κB/Snail/YY1/RKIP/PTEN loop in many cancers that was reported to regulate cell survival, proliferation, invasion, metastasis, and resistance. Emphasis was focused on the roles of the tumor suppressor Raf Kinase Inhibitor Protein (RKIP), the tumor promoter Yin Yang 1 (YY1), and the role of nitric oxide as a chemo-immuno-sensitizing factor. Many of the earlier mentioned studies are centered on the clinical challenging features of cancer patients' failure to respond to both conventional and targeted therapies.

The Editor has been active in the organization of regular sequential international miniconferences that are highly focused on the roles of YY1, RKIP, and nitric oxide in cancer and their potential therapeutic applications. Several books edited or coedited by the Editor have been published. In addition, the Editor has been the Series Editor of books (over 23) published by Springer on *Resistance to Anti-Cancer Targeted Therapeutics*. In addition, the Editor is presently the Series Editor of three series published by Elsevier/ Academic Press on *Cancer Sensitizing Agents for Chemotherapy, Sensitizing Agents for Cancer Resistant to Cell Mediated Immunotherapy*, and *Breaking Tolerance to Anti-Cancer Immunotherapy*. Lastly, the Editor is the Editor-in-Chief of the Journal Critical Reviews in Oncogenesis. The Editor has published over 500 research publications and reviews in various scientific journals of high impact.

**Acknowledgments:** The Series Editor acknowledges the Department of Microbiology, Immunology and Molecular Genetics and the UCLA David Geffen School of Medicine for their continuous support. The Series Editor also acknowledges the assistance of Mr. Rafael

Teixeira, Acquisitions Editor for Elsevier/ Academic Press, and the excellent assistance and support of Ms. Samantha Allard, Editorial Project Manager for Elsevier/Academic Press, for their continuous cooperation throughout the development of this book.

# Aims and Scope of the Volume

Gynecologic cancers are the leading causes of cancer-related deaths among women. In recent years, advances in the understanding of gynecologic cancers and their therapeutic strategies have been encouraging. Patients with gynecologic malignancies often develop resistance to the currently used chemotherapy. Overcoming resistance to such therapies poses a decisive challenge in the management of patients with gynecologic cancer that impacts their survival and quality-of-life. The patients who belong to certain ethnicity/race possess differences in developing resistance and exhibit disparities in reacting to the therapy(ies). In this volume, eminent researchers/clinicians from across the globe contributed various chapters covering important topics related to drug resistance, reversing the drug resistance, and associated mechanisms/targets in major gynecologic cancers (i.e., cervical, endometrial, and ovarian cancers). The contributors stipulated consolidated information on these gynecologic cancers and response rates, lifestyle changes, health disparities, genetic polymorphism, and nanotechnology for addressing drug resistance. This volume provides up-to-date information with a broader scope summarized in the same volume for students, researchers, trainees, and clinicians who are interested in the study of gynecologic cancers, covering basic, translational, and applied clinical aspects in reversing the drug resistance. These topics are unique to this volume, which are not generally covered in other peer-reviewed publications, thus representing a valuable scholarly resource for scientists and clinicians across the globe.

*Riyaz Basha, PhD, FAACC, FABAP*
*Sarfraz Ahmad, PhD, FAACC, FABAP*

# About the Volume Editors

**Riyaz Basha, PhD, FAACC, FABAP**

Doctor Basha is an Associate Professor and Vice Chair for Research in the Department of Pediatrics and Women's Health, Texas College of Osteopathic Medicine, The University of North Texas Health Science Center (UNTHSC) at Fort Worth, TX, USA. He graduated with a PhD degree in 1999 from Sri Venkateswara University, Tirupati, India, and received postdoctoral training at the University of Rhode Island, RI, USA. Before joining the UNTHSC in 2013, Doctor Basha worked as an Assistant Professor at the Cancer Research Institute of MD Anderson Cancer Center Orlando (currently named as UF Cancer Center), Orlando, FL, USA.

Doctor Basha's research is focused on improving cancer therapies in children, adolescents, and adults. He has been working closely with physicians and researchers with emphasis on the translational principle of "bench to bedside." He has been investigating drug resistance and novel strategies to improve the response of standard care/treatment, especially chemotherapy for various cancers including ovarian cancer. He has published studies in high impact factor journals such as the *Journal of Biological Chemistry, Cancer Letters, Cancer Medicine, Critical Reviews in Oncology/Hematology,* and *Gynecologic Oncology.* Doctor Basha is a member of several scientific societies, including the Association of Biotechnology and Pharmacy (ABAP), the American Association for Cancer Research, the American Society of Clinical Oncology, and the American Association for Clinical Chemistry (AACC) Academy. He became Fellow of the ABAP and the AACC Academy, in 2014 and 2019, respectively. He is the recipient of numerous competitive grants from the National Institutes of Health, Hyundai Hope On Wheels, nonprofit foundations, and awards including the Young Scientist Travel Award from the Asian Pacific Society for Neurochemistry, four research presentation awards from the Society of Toxicology (USA), and the best presentation award at the International Conference on Drug Discovery and Therapy held in Dubai, UAE. Doctor Basha co-authored more than 100 peer-reviewed publications, served as Guest Editor for two journals, and is serving as peer reviewer for more than 50 scientific journals. His research publications are extensively cited by the peers globally.

**Sarfraz Ahmad, PhD, FAACC, FABAP**

Doctor Ahmad is Director of Clinical Research at the Gynecologic Oncology Department of AdventHealth (formerly Florida Hospital) Cancer Institute (AHCI), Orlando, FL, USA. He earned his PhD degree in Biochemistry from North-Eastern Hill University, Shillong. Before joining AHCI in 2002, he spent 2 years at the Indian Institute of Technology in Delhi as researcher and 10 years in research/teaching at Loyola University of Chicago and University of Illinois at Chicago's Division of Hematology/ Oncology, College of Medicine. Currently, he is Professor of Medical Education at the University of Central Florida College of Medicine and Professor of Clinical Sciences at the Florida State University College of Medicine, Orlando, USA. He has trained numerous fellows, residents, medical students, and graduate/undergraduate students during their scholarly research projects and theses.

Doctor Ahmad's current research focus is on the analyses of clinicopathologic and surgical outcomes of oncology/hematology patients and to better understand the cellular/molecular mechanisms of cancer and related thromboembolic/hematologic disorders. His investigations are also aimed toward the evaluation of novel treatment options (chemo/cellular/immunotherapies) for the better management of hemato-oncologic patients. His past research interests focused on the anticoagulant, antithrombin, antiplatelet, and thrombolytic drug developments for the management of hematologic and cardiovascular patients. In these various areas of biomedical research and international collaborations, Doctor Ahmad has published over 200 peer-reviewed scholarly research articles and book chapters, and nearly 400 scientific abstracts, which are extensively cited globally. He is a reviewer and has editorial responsibilities for several biomedical journals/ books and has received several competitive research grants and national/international awards for his research contributions/accomplishments. In addition to biomedical science endeavors, Doctor Ahmad is passionate about classical Hindi-Urdu literature and poetry and takes pride with active participation in such forums.

# Preface

Gynecologic cancers are one of the leading causes for cancer-related deaths among women. Similar to other cancers, multimodality therapeutic options are used for the treatment of the patients with these malignancies. Patients with gynecologic malignancies often develop resistance to currently available/used therapeutic options, thereby affecting their outcomes in terms of survival and quality of life.

Cancer therapies have revolutionized due to consistent improvements happening over the years for the treatment of cancer patients. Several novel strategies and options have increased the quality of life and survival of patients with gynecologic cancers; however, treating relapsed patients and the patients who develop resistance to the first-line therapy are still complicated and a serious concern. Therefore, it is very important to explore and compile the newer research findings related to the topics on resistance to therapies for gynecologic cancer patients.

This book has 15 chapters covering the topics related to important aspects of cancer therapies, highlighting the issues on resistance to chemotherapy and associated mechanisms. These research chapters elaborate up-to-date information on both clinical and translational aspects of major gynecologic cancers such as cervical, endometrial, and ovarian cancers. The information outlined in these chapters is focusing on the mechanisms of drug resistance, epidemiology, risk factors, lifestyle, genetics, signaling, immunology, health disparities, and targeted therapies. The order of the chapters is arranged in such a way as to cover the information in general (for all gynecologic cancers) and then specific to each major type of gynecologic cancer.

Some of the topics covered provide insights into pathways/mechanisms and are not only helpful to better understand the disease process/development, but also useful in the diagnostic approach, lifestyle changes, and designing precise/personalized treatment options for better management and outcomes. Furthermore, patients belong to certain ethnicity/race possess differences in developing resistance and exhibit disparities in reacting to the therapies. Therefore, additional topics included in this book are on nanotechnology, health disparities, and economic/commercial impacts for tackling drug resistance in gynecologic cancers. The chapters are contributed by eminent researchers and clinicians from across the globe.

We, the Editors, are grateful to the Series Editor, Dr. Benjamin Bonavida, for inviting us to edit this book. We are also grateful to all the authors for their valuable and timely contributions, which form the basis of this important collection. The Editors also thank Mr. Rafael Teixeira and Ms. Samantha Allard from Elsevier for their help and support during the editing, production, and publication processes of the book.

*Riyaz Basha, PhD, FAACC, FABAP*
*Sarfraz Ahmad, PhD, FAACC, FABAP*

# Contents

## 7. Genetic polymorphisms in gynecologic cancers

Ketevani Kankava, Eka Kvaratskhelia, and Elene Abzianidze

## 8. Overcoming drug resistance in cervical cancer: Chemosensitizing agents and targeted therapies

Anum Jalil, James Wert, Aimen Farooq, and Sarfraz Ahmad

## 9. Immune response, inflammation pathway gene polymorphisms, and the risk of cervical cancer

Henu Kumar Verma, Batoul Farran, and Lakkakula V.K.S. Bhaskar

## 10. Overcoming chemotherapy resistance in endometrial cancer

Thomas A. Paterniti, Evan A. Schrader, Aditi Talkad, Kasey Shepp, Jesse Wayson, Alexandra M. Poch, and Sarfraz Ahmad

## 11. Chemoresistance in uterine cancer: Mechanisms of resistance and current therapies

Abeer Arain, Ibrahim N. Muhsen, Ala Abudayyeh, and Maen Abdelrahim

## 12. Ovarian cancer: Targeted therapies and mechanisms of resistance

Deepika Sarvepalli, Mamoon Ur Rashid, Hammad Zafar, Sundas Jehanzeb, Effa Zahid, and Sarfraz Ahmad

## 13. Overcoming drug resistance in ovarian cancer: Chemo-sensitizing agents, targeted therapies

Santoshi Muppala

## 14. Resistance to chemotherapy among ethnic and racial groups: Health disparities perspective in gynecologic cancers

Begum Dariya and Ganji Purnachandra Nagaraju

## 15. Future perspectives and new directions in chemosensitizing activities to reverse drug resistance in gynecologic cancers: Emphasis on challenges and opportunities

Ishna Sharma, Nathan Hannay, Swathi Sridhar, Sarfraz Ahmad, and Riyaz Basha

# Contributors

**Maen Abdelrahim** Houston Methodist Cancer Center, Houston Methodist Hospital, Houston, TX, United States

**Ala Abudayyeh** Section of Nephrology, MD Anderson Cancer Center, Houston, TX, United States

**Elene Abzianidze** Department of Molecular and Medical Genetics, Tbilisi State Medical University, Tbilisi, Georgia

**Luna Acharya** Department of Internal Medicine, University of Iowa, Iowa City, IA, United States

**Sarfraz Ahmad** Gynecologic Oncology, AdventHealth Cancer Institute, Florida State University, University of Central Florida, Orlando, FL, United States

**Saeed Ali** Department of Internal Medicine, University of Iowa, Iowa City, IA, United States

**Abeer Arain** Houston Methodist Cancer Center, Houston Methodist Hospital, Houston, TX, United States

**Riyaz Basha** Department of Pediatrics and Women's Health, Texas College of Osteopathic Medicine, University of North Texas Health Science Center, Fort Worth, TX, United States

**Lakkakula V.K.S. Bhaskar** Department of Zoology, Guru Ghasidas Vishwavidyalaya, Bilaspur, Chhattisgarh, India

**Lorna A. Brudie** Orlando Health Cancer Institute, Gynecologic Oncology Center, Orlando, FL, United States

**Anjali Chandra** Des Moines University College of Podiatric Medicine and Surgery, Des Moines, IA, United States

**Begum Dariya** Department of Bioscience and Biotechnology, Banasthali University, Vanasthali, Rajasthan, India

**Akpedje Dossou** Department of Physiology and Anatomy; Department of Molecular Biology, Immunology and Genetics, Graduate School of Biomedical Sciences, University of North Texas Health Science Center, Fort Worth, TX, United States

**Aimen Farooq** Department of Internal Medicine, AdventHealth, Orlando, FL, United States

**Batoul Farran** Department of Hematology & Medical Oncology, Winship Cancer Institute, Emory University, Atlanta, GA, United States

**Rafal Fudala** Department of Molecular Biology, Immunology and Genetics, Graduate School of Biomedical Sciences, University of North Texas Health Science Center, Fort Worth, TX, United States

**V. Gayathri** Bioengineering Department, SRM Institute of Science & Technology, Chennai, Tamil Nadu, India

**Nnamdi I. Gwacham** AdventHealth Cancer Institute, Gynecologic Oncology Program, Orlando, FL, United States

**Nathan Hannay** Department of Pediatrics and Women's Health, Texas College of Osteopathic Medicine, University of North Texas Health Science Center, Fort Worth, TX, United States

**Sana Hussain** Department of Medicine, Khyber Teaching Hospital, Peshawar, Pakistan

**Anum Jalil** Department of Internal Medicine, AdventHealth, Orlando, FL, United States

**Sundas Jehanzeb** Khyber Medical College, Peshawar, Pakistan

**Ketevani Kankava** Department of Molecular and Medical Genetics, Tbilisi State Medical University, Tbilisi, Georgia

**Deepak Kakara Gift Kumar** Cancer Biology Lab, Department of Biochemistry and Bioinformatics, Institute of Science, Gandhi Institute of Technology and Management (GITAM, Deemed University), Visakhapatnam, Andhra Pradesh, India

**Eka Kvaratskhelia** Department of Molecular and Medical Genetics, Tbilisi State Medical University, Tbilisi, Georgia

**Andras G. Lacko** Department of Physiology and Anatomy; Department of Pediatrics and Women's Health, University of North Texas Health Science Center, Fort Worth, TX, United States

**Rama Rao Malla** Cancer Biology Lab, Department of Biochemistry and Bioinformatics, Institute of Science, Gandhi Institute of Technology and Management (GITAM, Deemed University), Visakhapatnam, Andhra Pradesh, India

**Hariharasudan Mani** Department of Internal Medicine, University of Iowa, Iowa City, IA, United States

**Rakshimitha Marni** Cancer Biology Lab, Department of Biochemistry and Bioinformatics, Institute of Science, Gandhi Institute of Technology and Management (GITAM, Deemed University), Visakhapatnam, Andhra Pradesh, India

**Ezek Mathew** Texas College of Osteopathic Medicine; Department of Molecular Biology, Immunology and Genetics, Graduate School of Biomedical Sciences, University of North Texas Health Science Center, Fort Worth, TX, United States

**Vikas Venkata Mudgapalli** New York Institute of Technology, College of Osteopathic Medicine, Jonesboro, AR, United States

**Ibrahim N. Muhsen** Department of Medicine, Houston Methodist Hospital, Houston, TX, United States

**Santoshi Muppala** Cleveland Clinic, Cleveland, OH, United States

**Bhavani Nagarajan** North Texas Eye Research Institute, University of North Texas Health Science Center, Fort Worth, TX, United States

**Ganji Purnachandra Nagaraju** Department of Hematology and Medical Oncology, Winship Cancer Institute, Emory University, Atlanta, GA, United States

**Maya Nair** Environmental Health & Safety Department, University of North Texas Health Science Center, Fort Worth, TX, United States

**Thomas A. Paterniti** Department of Obstetrics and Gynecology, Augusta University Medical Center, Augusta, GA, United States

**Kiranmayi Patnala** Department of Biotechnology, Institute of Science, Gandhi Institute of Technology and Management (GITAM, Deemed University), Visakhapatnam, Andhra Pradesh, India

**Alexandra M. Poch** Department of Obstetrics and Gynecology, Augusta University Medical Center, Augusta, GA, United States

**Mamoon Ur Rashid** Department of Gastroenterology, Cleveland Clinic, Weston, FL, United States

**Nirupama Sabnis** Department of Physiology and Anatomy, University of North Texas Health Science Center, Fort Worth, TX, United States

**Deepika Sarvepalli** Guntur Medical College, Guntur, Andhra Pradesh, India

**Evan A. Schrader** Department of Obstetrics and Gynecology, Atrium Health Carolinas Medical Center, Charlotte, NC, United States

**Ishna Sharma** Department of Pediatrics and Women's Health, Texas College of Osteopathic Medicine, University of North Texas Health Science Center, Fort Worth, TX, United States

**Kasey Shepp** Medical College of Georgia, Augusta, GA, United States

**Swathi Sridhar** Department of Pediatrics and Women's Health, Texas College of Osteopathic Medicine, University of North Texas Health Science Center, Fort Worth, TX, United States

**Aditi Talkad** Medical College of Georgia, Augusta, GA, United States

**Aman Ullah** Department of Internal Medicine, SoutheastHEALTH, Cape Girardeau, MO, United States

**Henu Kumar Verma** Institute of Endocrinology and Oncology, Naples, Italy

**Jesse Wayson** Department of Obstetrics and Gynecology, Augusta University Medical Center, Augusta, GA, United States

**James Wert** Department of Internal Medicine, AdventHealth, Orlando, FL, United States

**Hammad Zafar** University of Iowa Hospitals & Clinics, Iowa City, IA, United States

**Effa Zahid** Services Institute of Medical Sciences, Lahore, Pakistan

**Amy Zheng** Texas College of Osteopathic Medicine, University of North Texas Health Science Center, Fort Worth, TX, United States

# Introduction to gynecologic cancers: Emphasis on pathogenesis, incidence, and diagnosis

*Nnamdi I. Gwacham[a] and Sarfraz Ahmad[b]*

[a]AdventHealth Cancer Institute, Gynecologic Oncology Program, Orlando, FL, United States
[b]Gynecologic Oncology, AdventHealth Cancer Institute, Florida State University, University of Central Florida, Orlando, FL, United States

## Abstract

Gynecologic cancers often involve the ovaries, uterus, vulva, cervix, fallopian tubes, vagina, or the peritoneum. In the United States, the most commonly diagnosed gynecologic cancer is endometrial cancer, followed by ovarian cancer. Cervical cancer is less common in developed countries because of the wide availability of routine screening with Papanicolaou (Pap) test and with human papillomavirus (HPV) testing. Additionally, the HPV vaccine has become increasingly more acceptable and validated as an added measure to decrease the incidence of preinvasive disease. All women are at risk for gynecologic cancers, and the risk increases with age. Between 2012 and 2019, approximately 94,000 women or more were diagnosed with gynecologic cancer. The incidence rate of gynecologic cancers among women varies by cancer type and race/ethnicity. Uterine cancer occurs at a rate of 26.82 cases per 100,000, whereas the least common cancer, vaginal cancer, occurs at a rate of 0.66 per 100,000. The median age at diagnosis also varies by cancer type and race/ethnicity. Cervical cancer is prevalent at a younger age than the other gynecologic cancers, whereas vaginal and vulvar cancers tend to be diagnosed at an older age. Consequently, cervical cancer is the most common gynecologic cancer among women aged <50 years, while uterine cancer is the most common among women 50 years or older. Each of the gynecologic cancers has distinct pathogenesis behind their development and, thus, has differing clinical presentations. The aim of this chapter is to briefly introduce each of the gynecologic cancers, describe their individual pathogenesis, and discuss their incidence and diagnosis.

## Abbreviations

| | |
|---|---|
| **ACOG** | American College of Obstetrics and Gynecology |
| **AKT1** | protein kinase B |
| **ARID1A** | adenine-thymine-rich interactive domain containing protein 1A |
| **AUB** | abnormal uterine bleeding |

R. Basha, S. Ahmad (eds.)
*Overcoming Drug Resistance in Gynecologic Cancers*
ISBN 978-0-12-824299-5
https://doi.org/10.1016/B978-0-12-824299-5.00006-X

| | |
|---|---|
| **BRCA** | breast cancer gene |
| **CA125** | cancer antigen 125 |
| **CDC** | Centers for Disease Control and Prevention |
| **CIN** | cervical intraepithelial neoplasia |
| **CT** | computed tomography |
| **CTNNB1** | Catenin Beta 1 |
| **D&C** | Dilation and Curettage |
| **DNA** | deoxyribonucleic acid |
| **DSS** | disease-specific survival |
| **EIN** | endometrial intraepithelial hyperplasia |
| **EOC** | epithelial ovarian cancer |
| **ESS** | endometrial stromal sarcoma |
| **FDA** | Food and Drug Administration |
| **FDG-PET** | fluorodeoxyglucose-positron emission tomography |
| **FGFR2** | fibroblast growth factor receptor 2 |
| **FIGO** | International Federation of Gynecology & Obstetrics |
| **HE4** | human epididymis protein 4 |
| **HIV** | human Immunodeficiency Virus |
| **HOX** | homeobox gene |
| **HPV** | human papillomavirus |
| **HSIL** | high-grade intraepithelial lesion |
| **KRAS** | Kirsten rat sarcoma |
| **LMS** | leiomyosarcoma |
| **LSIL** | low-grade intraepithelial lesion |
| **MLH1** | MutL homolog 1 |
| **MMMT** | malignant mixed Mullerian tumor |
| **MRI** | magnetic resonance imaging |
| **MSH2** | MutS homolog 2 |
| **MSH6** | MutS homolog 6 |
| **PAX-2** | paired-box gene 2 |
| **PDV** | Paget disease of the vulva |
| **PIK3CA** | phosphatidylinositol-3-kinase catalytic subunit |
| **PMB** | postmenopausal bleeding |
| **POLE** | polymerase epsilon catalytic subunit |
| **PTEN** | phosphatase and tensin homolog |
| **Rb** | retinoblastoma gene |
| **SCC** | squamous cell carcinoma |
| **SEER** | surveillance, epidemiology and end results |
| **TP53** | tumor protein 53 |
| **TVUS** | transvaginal ultrasound |
| **VAIN** | vaginal intraepithelial neoplasia |
| **VIN** | vulvar intraepithelial neoplasia |

## Conflict of interest

No potential conflicts of interest were disclosed.

## Introduction

Cancer is recognized as a major worldwide public health problem and is the second leading cause of death in the United States. Gynecologic cancers will account for 6% of the anticipated 1.8 million new cancer cases in 2020 [1]. Gynecologic cancers involve the uterine corpus and cervix, the ovaries, vagina, vulva, fallopian tubes, or the peritoneum. Endometrial cancer is the most commonly diagnosed gynecologic cancer in the United States, accounting for >55% of new diagnosis, whereas ovarian cancer is the most lethal, comprising nearly 50% of deaths from gynecologic cancer [1]. The incidence of cervical cancer has declined in developed countries, owing to widely available screening methods, including Pap and HPV test to detect early disease, but continues to plague the developing world. In contrast, due to incremental improvements in care rather than drastic paradigm shifts, reductions in incidence and mortality for ovarian and uterine cancers have been more modest. The statistics for the rarer vulvar and vaginal cancers have remained largely static over the last 30 years. The purpose of this chapter is to briefly discuss each of the gynecologic cancers, with particular emphasis on their incidence, pathogenesis, clinical presentation, and methods of diagnosis.

## Uterine cancer

Accounting for more than one half of all gynecologic cancers in the United States, uterine cancer is the most common malignancy of the female genital tract [2]. According to the CDC, there were greater than 57,000 new cases diagnosed in 2017, and the trend has shown a continued yearly rise [3]. Estimated new cases of uterine cancer in 2020 are expected to be greater than 65,000 [1]. Uterine cancer ranks fourth among the most common cancer in women, after breast, lung, and colon cancers, and it is the sixth leading cause of female death from malignancy [3]. Uterine cancer arises from the uterine corpus and from the glands of the endometrium (thus, endometrial carcinoma). When malignancy arises from the myometrium of the uterus, it is classified as a sarcoma [4]. Endometrial adenocarcinomas account for greater than 80% of all uterine cancers and require a different management approach than uterine sarcomas [4]. Approximately 2%–3% of women will develop uterine cancer during their lifetime [2]. The highest incidences are in the United States and Canada (19.1/100,000) and western (15.6/100,000) and northern Europe (12.9/100,000) [2]. Most uterine cancers are diagnosed in postmenopausal women (i.e., the sixth and seventh decades of life); however, up to 14% are reported in premenopausal and perimenopausal women, with 5% occurring in women under the age of 40 years [5, 6]. The median age at diagnosis is 62 years, with a majority of them diagnosed at an early stage marked by uterine confinement [1].

The increased incidence of endometrial cancer in North America and Europe is related to an increasing prevalence of obesity and metabolic syndrome in these regions, in addition to the aging population [7–9]. Consequently, prolonged estrogen exposure without progesterone opposition serves as the main risk factor for disease. Other risk factors include diabetes mellitus, obesity, early nulliparity, early menarche, late-onset menopause, older age (>55 years), and the use of tamoxifen [10].

### Pathogenesis of disease

In the past 30 years, endometrial cancer has been classically categorized into two broad classifications based on histologic characteristics, hormone reception expression, and grade (type I and type II) [4]. This traditional classification is based on histopathological, endocrine, and clinical features, but these categorizations do not reflect the full heterogeneity of

TABLE 1    Duplex classification of endometrial cancer showing clinical and pathologic features of type I and type II endometrial cancer.

|  | Type I | Type II |
|---|---|---|
| Clinical features | Metabolic syndrome: obesity, unopposed estrogen, hyperglycemia, hyperlipidemia | None |
| Histology | Endometrioid | Nonendometrioid (clear-cell, serous carcinoma) |
| Grade | Low (G1–G2) | High (G3) |
| Hormone receptor expression | Positive | Negative |
| Genomic stability | Diploid, frequent microsatellite instability (approx. 40%) | Aneuploid |
| Prognosis | Favorable (OS 85% at 5 years) | Poor (OS 55% at 5 years) |
| TP53 mutation | No | Yes |
| Clinical course | Nonaggressive | Aggressive |
| Clinical stage at diagnosis (FIGO) | I–II | III–IV |
| Myometrial invasion | Usually <50% | Deep invasion |
| Metastases | Seldom | Frequent |

*Abbreviations:* OS = Overall survival; FIGO = International Federation of Gynecology & Obstetrics.

endometrial cancer. The groupings also differ based on the prognosis, epidemiology, and treatment as summarized in Table 1. The dualistic classification has received support and was therefore incorporated into clinical decision-making algorithms to aid in defining high-risk patients. Its prognostic value, however, remains limited because up to 20% of type I cancers relapse, whereas 50% of type II cancers do not [4].

Type I cancers account for the majority of diagnoses, and they represent low-grade, endometrioid, hormone receptor-positive, and diploid cancers, which tend to have a favorable prognosis [11]. Their favorable prognosis is driven by their low grade and confinement to the uterus at the time of diagnosis [2]. They include the International Federation of Gynecology and Obstetrics (FIGO) grades 1 and 2 cancers. The FIGO classification system assesses the nuclear grade, the architectural pattern, as well as the presence of atypia in the tissue specimen [12, 13]. Grade 1 cancers are well-differentiated, resembling normal tissue, and typically have a favorable prognosis [14]. They also have a <5% solid component. Grade 2 cancers are moderately differentiated with a 6%–50% solid component, and they fall in between grades 1 and 3 in resemblance to normal endometrium. Grade 3 cancers are considered high grade. They are poorly differentiated with a >50% solid component and do not resemble any healthy endometrial tissue. As a result, they are more aggressive with a poorer prognosis. The type I classification is made up of the Grade 1 and 2 tumors which are generally classified as low grade. Grade 3 tumors are classified as high grade and fall into the type II classification [15].

Type II cancers include the serous and clear-cell carcinomas, which are described as aneuploid, high-grade, hormone receptor-negative, nonendometrioid, and *TP-53* mutated tumors that are associated with a high metastasis risk and, therefore, poorer prognosis [16] (Table 1). Furthermore, nearly 20% of endometrioid cancers present as high grade (FIGO grade 3) and are thereby classified as type II. Undifferentiated, dedifferentiated, mixed cell, and neuroendocrine cancers also fall into this category. Carcinosarcomas, also referred to as malignant mixed Mullerian tumors (MMMT), are also classified as type II but are less common than serous and clear-cell histologies [16].

The endometrioid subset develops on a background of prolonged estrogen exposure and is often preceded by a premalignancy stage characterized as endometrial intraepithelial neoplasia (EIN) (or complex atypical hyperplasia) [17, 18]. In the endometrioid pathway, both conditions (precancer and cancer) appear to be related to obesity and anovulatory cycles that expose the endometrium to unopposed estrogen stimulation [19, 20]. The *PIK3CA* pathway is the most frequently altered in type I endometrial cancer: Mutations are found in more than 90% of lesions [21, 22]. Additionally, recent data have reported *KRAS* and *FGFR2* mutations [23].

There is a range of histological subtypes that are included in type II endometrial cancers, each with distinct molecular and genomic features (Table 2) [21–23]. Patients with serous histology endometrial cancer tend to present at an older age, with the antecedent EIN precursor [24]. It is known to share genomic features with high-grade serous ovarian cancer and triple-negative basal-like breast cancer. Clear-cell carcinomas of the uterus resemble its counterpart that arises from the ovary. It is associated with inactivating mutations in the chromatin remodeling gene *Baf250a (ARID1A)* in 20%–40% of cases and to a lesser extent, *p53* mutation [25–27]. Carcinosarcoma (MMMT) is a form of poorly differentiated endometrial carcinoma in which sarcomatous differentiation arises due to epithelial-mesenchymal transformation [28]. They most often present as heterologous elements such as skeletal muscle, or cartilage, but may consist of less differentiated spindle cells thathave lost characteristic epithelial keratins [29].

Molecular analysis, focusing on serous and endometrioid cancer, emphasizes the disease's heterogeneity and confirms that histologic subtypes have divergent pathogenesis. Four distinct molecular subgroups are identified: polymerase epsilon catalytic subunit (*POLE*) ultramutated, mismatch repair deficient (MMRd) or microsatellite instability hypermutated, copy-number-low microsatellite stable, and copy-number-high serous-like [21]. The *POLE* ultramutated subgroup is characterized by abnormalities in the POLE exonuclease and is characterized by a high mutation load and excellent prognosis [30]. Endometrioid tumors frequently demonstrate inactivation of the phosphatase and tensin homolog (PTEN) tumor-suppressor gene, nuclear transcription factor, microsatellite instability, paired-box gene 2 (*PAX-2*), or Kirsten rat sarcoma gene (*KRAS*) inactivation [31]. Generally, 30%–40% of endometrioid cancers demonstrate loss of the DNA mismatch repair proteins:MutS homologs 2 and 6 (*MSH2* and *MSH6*), MutL homolog 1, (*MLH1*), and mismatch repair endonuclease gene 2(*PMS2*) [32]. In contrast, loss of *p53* function (seen as aberrant protein accumulation on immunohistochemistry) is more commonly seen in the serous subtype [28]. The microsatellite-stable subgroup is characterized by a low rate of somatic copy-number alterations, low mutation load, and intermediate prognosis [32]. As gene-specific therapies become more available and sequencing of individual tumors is now routine, primary genomic classification strategies will drive personalized management.

TABLE 2    Genomic and molecular features of endometrial cancer.

| Features | Endometrioid | Serous | Clear cell | Carcinosarcoma |
| --- | --- | --- | --- | --- |
| Subtype | I | II | II | II |
| TP53 mutation | Rare | >90% | 35% | 60%–90% |
| KRAS mutation | 20%–30% | 3% | 0% | 17% |
| FGFR2 amplification or mutation | FGFR2 mutation (12%) | Frequent FGFR1 and FGFR3 amplification FGFR2 mutation (5%) | N/A | FGFR3 amplification (20%) |
| PTEN mutation | 75%–85% | 11% | PTEN loss (80%) | 19% |
| PIK3CA mutation | 50%–60% | 35% | 18% | 35% |
| PIK3CA amplification | N/A | 45% | N/A | 14% |
| ERBB alterations | None | Amplification (25%–30%) | Mutation (12%) Amplification (16%) | Amplification (13%–20%) |
| CTNNB1 mutation | 25% | 3% | N/A | N/A |
| Other | ARID1A mutation (35%–40%) | PPP2R1A mutation (20%) FBXW7 mutation (20%) LRPB1 deletion MYC amplification CCNE1 amplification SOX17 amplification | ARID1A (25%) | ARID1A mutation (25%) SOX17 amplification (25%) CCNE1 amplification (42%) PPP2R1A mutation (28%) FBXW7 mutation (35%–40%) |

*Abbreviations:* TP53, Tumor protein 53; CCNE1, cyclin E1; PTEN, phosphatase and tensin homolog; KRAS, Kirsten rat sarcoma gene; FGFR2, fibroblast growth factor receptor 2; PIK3CA, Phosphatidylinositol-4,5-biphosphate 3-Kinase catalytic subunit alpha; CTNNB1, Catenin Beta 1; ARID1A, AT-rich interaction Domain 1A; ERBB, Erb-B2 Receptor Tyrosine Kinase; PPP2R1A, Protein phosphatase 2 scaffold subunit alpha; FBXW7, F-box and WD repeat domain containing 7; SOX17, SRY-Box transcription factor 17.

## Clinical presentation and diagnosis

Abnormal uterine bleeding (AUB) is the most common presenting symptom, occurring in 90% of patients. This includes postmenopausal bleeding (PMB), abnormal discharge, or pyometra [15, 16]. Other vague complaints include abdominopelvic pain and abdominal distention which can sometimes represent advanced disease [16]. Five percent of endometrial cancer diagnosis is found in asymptomatic women, which is usually associated with an incidental finding of a thickened endometrial lining on ultrasonography [16]. Diagnosis is confirmed with office-based pipelle sampling, which may be infeasible for some patients due to cervical stenosis or unforgiving body habitus. When histologic findings are insufficient to confirm the diagnosis, dilation and curettage (D&C) is recommended, although this requires anesthesia. The gold standard for evaluating the endometrium is a biopsy under hysteroscopic guidance because of its better accuracy than blind D&C [33]. Some patients who have undergone a prior hysteroscopy have been shown to have a higher incidence of malignant peritoneal cytology at the time of hysterectomy; however, no evidence exists that supports an association between hysteroscopy and worse prognosis [34]. Therefore, the proper protocol for the evaluation of AUB is transvaginal ultrasonography, with an endometrial biopsy in cases of abnormal endometrial thickness. When diagnosis is uncertain, a hysteroscopy should be performed [11]. In menopausal women, an endometrial thickness cutoff of 5 mm on transvaginal ultrasonography was demonstrated to have a sensitivity of 90% and specificity of 54% for the diagnosis of malignancy. When the cutoff was reduced to 3 mm, the sensitivity and specificity values were 98% and 35% [35].

## Uterine sarcoma

Sarcomas arise from mesenchymal elements, whereas carcinomas arise from epithelial elements. They are uncommon and are thought to arise primarily from the uterine muscle and endometrial stroma [28, 29]. Sarcomas account for 3%–8% of uterine cancers [1, 3]. The age-adjusted incidences (per 100,000 women age 35 years and older) for all sarcomas in the United States are about 7.02 for black women and 3.58 for white women [2, 3]. Sarcomas tend to behave aggressively. There is also a lack of consensus on risk factors and optimal treatment due to their rarity and histopathological diversity. Consequently, sarcomas are associated with poorer outcomes. Histologically, uterine sarcomas are either pure sarcomas or mixed sarcoma and carcinoma [36]. Pure sarcomas are further subdivided into leiomyosarcomas (LMS), endometrial stromal sarcoma (ESS), or undifferentiated sarcoma [37]. LMS occurs as a result of the malignant transformation of the smooth muscle of the uterus. The mixed sarcoma and carcinoma categories- are comprised of adenosarcoma, which occurs when endometrial mesenchymal tissue undergoes transformation with an accompanying benign-appearing epithelial component [36].

The type and frequency of uterine sarcomas are related to both age and race. LMS has an incidence plateau in middle age and thus presents at an early age with resultant decline thereafter. Most common are LMS, comprising 1%–2% of uterine malignancies, often occurring in women >40 years. Similar to other uterine cancers, women usually present with AUB and, occasionally, a rapidly enlarging uterus causing abdominopelvic pain [16, 29]. The hallmarks of malignant smooth muscle tumors of the uterus are hypercellularity, high mitotic rate exceeding 15 mitotic figures per 10 high-power fields, and severe nuclear atypia [16]. These features lend to the ease of the histopathologic identification and diagnosis of LMS. The diagnostic workup is similar to that for the common uterine neoplasm and involves biopsy or D&C. Imaging studies including computed tomography (CT) and magnetic resonance imaging (MRI) are used perioperatively to also aid in treatment guidance.

## Ovarian cancer

Ovarian cancer accounts for only 3% of cancers in women in the United States, but results in more deaths than any other gynecologic cancer [1, 3]. In 2020 in the United States, 21,750 women are estimated to be diagnosed with ovarian cancer, and 13,940 will unfortunately die from this malignancy [1, 3]. Ovarian cancer has an age-adjusted incidence of 10.2 per 100,000 women, slightly decreased in recent years [1, 2]. The lifetime risk for a woman to develop ovarian cancer is approximately 1 in 70 [2]. The incidence of ovarian cancer is highest among White women, followed by Hispanic, Asian/Pacific Islander, and Black women [3]. Consequently, White women are also more likely to die of ovarian cancer than any other ethnic group. Similar to other cancers affecting women, the incidence of ovarian cancer and associated mortality increase with age, with women over 50 years of age experiencing the highest incidence. However, varying types of ovarian cancer may be diagnosed at any age. There are three main categories of primary ovarian cancer. These are: (i) epithelial ovarian cancers (EOC), which are due to malignant transformation of the ovarian surface epithelium, which neighbors the peritoneal epithelium; (ii) sex cord-stromal tumors, which arise from the ovarian stroma, from sex cord derivatives, or both; and (iii) germ cell tumors, which mainly originate from germ cells [38]. Of the three main groups, epithelial ovarian cancers are the most common, comprising about 60% of all ovarian cancers, and will be the primary focus of this section.

Recent data indicate that the incidence of epithelial ovarian cancer has been decreasing in all age groups as well as in non-Hispanic ethnic groups since the early 1990s [39, 40]. Patients with advanced-stage disease have also noted an improvement in medial survival in the past few decades. According to the Surveillance, Epidemiology, and End Results (SEER) data, there has been an improvement of 8 months in disease-specific survival (DSS) in women with stage III disease [39]. This decrease in incidence in the United States and western Europe and a corresponding rise in eastern and southern Europe are thought to be attributable to increasingly acceptable use of oral contraceptives and to decreased fecundity, respectively [41].

## Pathogenesis

EOC is believed to arise from a single layer of flattened cells covering the ovary. Additionally, these cells also line cysts that are immediately beneath the ovarian surface [42]. They consist of several histotypes that are categorized into a less aggressive type I and more aggressive type II cancers. These histotypes are based on clinical, histological, molecular, and epidemiologic features [43]. Neoplasms with similar behavior and morphology arise from the fallopian tubes, endometriosis, endosalpingiosis, and the peritoneum [42]. These cancers are classified by their appearance into serous (most common), endometrioid, and mucinous subtypes. They are also classified by histopathologic grades 1–3, with less common subtypes being transitional, clear-cell, squamous, mixed, and undifferentiated subtypes. Recently, the fimbria of the fallopian tube has been implicated as the site of origin for as many as 30% of high-grade serous epithelial ovarian cancers, particularly in women with germline mutations of breast cancer (BRCA) susceptibility genes BRCA1 and BRCA2 [44]. Primary peritoneal and fallopian tube cancers share the same embryonic precursor with the ovarian surface epithelium. Therefore, it is thought that primary peritoneal cancers and fallopian tube cancers clinically and morphologically resemble epithelial ovarian cancer [42]. EOC subtypes hold singular transcriptional signatures and molecular aberrations with morphological features that resemble the specialized epithelia of the reproductive tract derived from the Mullerian ducts.

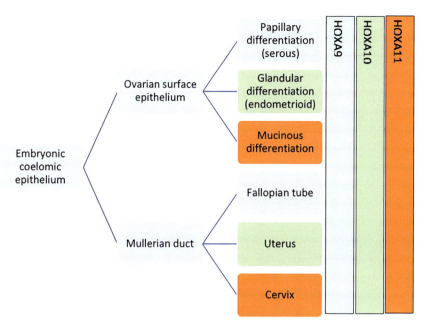

FIG. 1    Homeobox (HOX) gene control of Mullerian differentiation. The embryonic coelomic epithelium gives rise to the ovarian surface epithelium and the Mullerian ducts. HOX gene expression becomes specially restricted as the Mullerian ducts differentiate to form the pelvic structures. These patterns of HOX expression are reiterated in endometrioid, serous, and mucinous epithelial ovarian cancers according to their pattern of Mullerian-like differentiation.

One surface epithelium precursor cell gives rise to all of them, with a specific path of differentiation that is regulated by embryonic pathways that involve the homeobox (*HOX*) genes (Fig. 1). *HOX* genes are highly supported members of the *HOX* superfamily. They have a fundamental role in determining cellular identity [45]. Hematological malignancies were the first to establish a link between cancer and HOX genes and have been involved in the oncogenesis of various other solid tumors. *HOX A* genes have been found to demonstrate evidence of altered expression in ovarian and breast cancers. *HOX B* genes have been implicated in colon cancer, while *HOX C* genes show altered expression in prostate and lung cancers. *HOX D* genes play a role in breast and colon cancer [45]. A variety of the HOX genes demonstrate dysregulation, particularly *HOXA9*, *HOXA10*, and *HOXA11*, and they have been found with wide expression in high-grade serous ovarian cancers as well as normal ovarian surface epithelium [45].

Genomic mutations also play a key role in the pathogenesis of many cancers. Approximately 85%–90% of epithelial ovarian cancers are sporadic and arise in the absence of a family history of the disease, often associated with spontaneous somatic mutations of *TP53* [46]. Other genes implicated include *CTNNB1, PTEN, KRAS, PIK3CA*, and *AKT1* [45, 47, 48]. Germline mutations of *BRCA1* and *BRCA2* are found in the majority of familial ovarian cancers. These cancers have a propensity for earlier occurrence at a younger age than the sporadic tumors and also have a tendency to be high-grade serous carcinomas with *p53* dysfunction [49].

TABLE 3　General symptoms associated with epithelial ovarian cancer.

| **Symptoms** (*in order of clinical significance*) |
| --- |
| Pelvic pain |
| Abdominal bloating, discomfort, or fullness |
| Early satiety, food intolerance, or anorexia |
| Unintentional weight loss |
| Nausea or vomiting |
| Rectal pain |
| Back pain |
| Fatigue, low energy, or general weakness |
| Shortness of breath |
| Diarrhea |
| Irritable bowel syndrome |
| Dysuria |
| Menopausal symptoms or hot flashes |
| Headache |

## Clinical presentation and diagnosis

The symptomatology of epithelial ovarian cancer is nonspecific (Table 3). Pelvicpain is a common symptom in earlystage disease, although it is usually asymptomatic. Patients that present with primarily gastrointestinal symptoms such as nausea, vomiting, diarrhea, and constipation tend to be advanced stage at diagnosis than patients presenting with primarily gynecologic symptoms such as abnormal vaginal bleeding or pain. There are various physical exam findings and typically include a palpable abdominopelvic mass. The formation of ascitic fluid containing mesothelial cells, leukocytes, and a wide fraction of tumor cells is a common finding that is distinct to ovarian cancer [42, 50]. As a result, gross abdominal distention can be the first noticeable symptom of the disease. One of the first imaging modalities to evaluate the pelvis in the setting of suspected ovarian malignancy is transvaginal ultrasonography (TVUS). Certain features that are highly suggestive of cancer on imaging are the findings of a complex ovarian cyst with both cystic and solid components, septations, and internal echoes [16]. Percutaneous biopsy of cysts is ill-advised to avoid spillage of tumor, and operative intervention is generally needed to obtain a diagnosis. Information gained from analysis of ascitic fluid through paracentesis can also aid in making the diagnosis of cancer. A CT scan of the abdomen and pelvis and chest can also be performed to assess for the presence of extrapelvic disease [16].

Serum cancer antigen 125 (CA125) concentration is elevated in more than 80% of patients with advanced-stage disease. However, diagnostic sensitivity and specificity from this assay alone are not sufficient to arrive at a diagnosis because CA125 concentrations can also be raised in several other malignant and benign diseases such as pelvic inflammatory disease and endometriosis [50]. Unfortunately, early detection is limited because CA125 is elevated in only 50% of stage I cancer, and the highest estimates of sensitivity, specificity, and positive predictive values are most applicable to postmenopausal women with a known adnexal mass. The assay is more useful in the setting of evaluation of chemotherapy response and detection of relapse. Human epididymis protein 4 (HE4) is another serum biomarker that, similar to CA125, has obtained approval for surveillance but not for the screening of ovarian malignancy. A combination of CA125 along with HE4 increases the likelihood of serologic detection of EOC.

## Cervical cancer

Worldwide, cervical cancer is the fourth most common female malignancy that continues to represent a major global health challenge [51]. It is the most common cause of death from cancer in developing countries, and thus, the greatest cancer killer of young women [51]. Among the 570,000 new cases of cervical cancer in 2018, the highest incidence rates were in Central and South America, Africa, and Asia [2]. These countries are responsible for almost 90% of the 300,000 deaths from cervical cancer where the mortality is 18 times higher than in developed countries [52]. In the United States, 13,800 new cases will be diagnosed with 4290 deaths from cervical cancer in 2020 [1]. Recently, trends across 38 countries have shown a decrease in age-standardized incidence rates in higher-income countries, while those rates have either stayed the same or increased in the lower-income nations [53]. Unsurprisingly, the disparities in incidence and stage at time of diagnosis in the United States are identifiable among different ethnic groups—Asian Americans, African Americans, Latin Americans, and European Americans. The recent decreases in the incidence of cervical cancer observed in the developing countries have largely been due to opportunistic screening [54, 55].

In 2012, cervical cancer was ranked as the11th most common female malignancy and the 9th most common cause of cancer mortality in high-income countries; however in middle- and low-income countries, it was the second most common type of cancer and the third most common cause of cancer death [56]. In Latin America and Africa, it is the principal cause of cancer-specific mortality in women [52]. The lifetime risk of developing cervical cancer is 0.9% for women in high-resource countries and 1.6% for women in middle- and low-income countries. The risk of death due to cervical cancer is 0.9% and 0.3% in low- to middle-income and high-income countries, respectively [52]. Risk factors include history of sexually transmitted infections, early sexual debut, multiple sexual partners, cigarette smoking, low socioeconomic status, and immunosuppression, as evidenced by its high infectivity with Human Immunodeficiency Virus (HIV) [16].

## Pathogenesis

Cervical cancer usually arises from the cervical transformation zone—a ring of mucosa located on the cervix [57]. The epithelial cells undergo viral transformation by the high-risk type of the human papilloma virus (HPV). Therefore, it is the only gynecologic cancer that can be prevented by routine and regular screening via cytology and HPV cotesting [58–60]. Persistent HPV infections give rise to cancers mainly at the transformation zone between different kinds of epithelium (e.g., anus, cervix, and oropharynx). Carcinogenic HPV infections are equivalently common in vaginal and cervical specimens; however, vaginal cancer is exceedingly rare. This indicates the importance of the transformation zone. The transformation zone position is dynamic and shifts gradually over time toward and into the endocervical canal. In this process, the mucus-producing glandular epithelium is replaced by stratified squamous epithelium. The virus's capability of transforming human epithelium has been associated with the expression of two viral gene products, E6 and E7, which interact with *p53* and retinoblastoma protein (*Rb*), respectively, and affect the control mechanism of the cell cycle [57].

Cervical cancer arises through a series of four steps and is illustrated in Fig. 2—(i) HPV transmission, which occurs mainly by mucosal or direct skin contact. The likelihood of infection per each sexual act is unknown but is probably high. Additionally, there is no established difference between the HPV types. HPV types have a common transmission route and will likely be transmitted together. As a result, a higher proportion of concurrent infections with various types is demonstrated in the general population. Because men are also infected with various HPV types concurrently, an individual sexual act is capable of transmitting several HPV types. Thusly, it is likely that most women in the world probably harbor at least one or more types of HPV during their sexual life [61]. Cell-mediated immunity is responsible for clearing or suppressing most cervical HPV infections within 1–2 years of exposure [62]. (ii) Due to persistence of HPV of any given subtype, the likelihood of resultant clearance decreases, giving rise to precancer. The median time to clearance of detected HPV infections

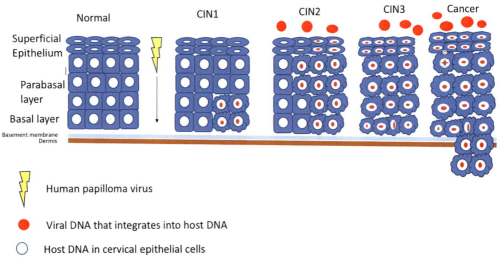

FIG. 2   The pathogenesis of cervical cancer. Distribution of normal and HPV-infected epithelial cells in normal, precancerous lesions (mild, moderate, and severe dysplasia; CIN 1, CIN 2, and CIN 3, respectively), and cancer of the cervix. The integration of viral HPV and host factors marks the initial stage of carcinogenesis. HPV gains entrance to the basal epithelial cells through a microscopic wound. Its genome is then integrated into the host genome via breaks in the nuclear envelope. HPV gains control of the host genome upon entrance into the nucleus. It further self-replicates and spreads throughout the epithelium. The host cells begin to grow irregularly in a disordered fashion compared to normal cells. Further transmission is then facilitated as the virions become sloughed off with dead cells of the host epithelium. *Abbreviations:* CIN=cervical intraepithelial neoplasia, HPV=human papillomavirus.

at time of screening is 6–18 months; however, the same HPV type can reappear even after clearance [55]. The mechanism by which the infections resolve is not completely understood. They may resolve by maintaining a latent state in the basal-cell epithelium whereby the virus may continue to replicate at very low levels or may completely resolve by viral clearance. (iii) Persistence of the virus may inevitably result in progression of a clone of persistently infected cells to precancer. In the setting of precancer, undifferentiated cells that have fixed genetic abnormalities have supplanted almost the full thickness of the normal epithelium of the cervix. Cervical intraepithelial neoplasia (CIN) is regarded as the precursor lesion to carcinoma and is characterized (grades 1–3) based on probability of developing carcinoma in situ. CIN1 is not precancer and, thus, is an insensitive sign of HPV infection; therefore, these lesions demonstrate a low risk of progression to cervical cancer. CIN2 lesions can be produced by noncarcinogenic types of HPV and are heterogeneous. Their potential for causing cancer is equivocal. CIN3, however, is a reliable morphological diagnosis that is associated with severe dysplasia or dyskeratosis, and carcinoma in situ [58]. (iv) The incidence of progression to invasive cervical cancer occurs earlier in unscreened populations than for most adult cancers, peaking from about 35 to 55 years of age [3, 52]. This is largely in part due to sexual transmission in late adolescence and early adulthood.

The average time between HPV infection and precancer is shorter when compared to the average time of precancer growth that leads to invasion [16]. Consequently, there are more precancers than cancers, which suggests that only a lesser proportion of precancers result in invasion. Due to ethical reasons, the significance and timing and risk of invasion in the event that precancers were left untreated will remain unknown due to the mandated treatment of precancer.

## Clinical presentation and diagnosis

Cellular and molecular biology has allowed the development of viral-like particles using the L1 protein of the HPV virus [59]. The vaccines are based on the self-assembly of recombinant L1 protein into noninfectious capsids without any genetic material and are approved for use in boys/girls and women/men between the ages of 9 and 26 years. In 2018, the Food and Drug Administration (FDA) expanded its approval to include women and men between the ages of 27 and 45 years old. Vaccine awareness has increased, but vaccination rates for US adolescents are only at 48.6% vaccinated as of 2017 [63]. The current standard screening guidelines as proposed by the American Society for Colposcopy and Cervical Pathology (ASCCP) for prevention of cervical cancer are presented in Table 4. Colposcopic examination with guided biopsies of the exocervix, and in most cases, the endocervix, is indicated in women with abnormal cytology. This determines the type of follow-up and/or therapy that is required.

Historically, the clinical reference standard to distinguish between CIN grades 1, 2, and 3 or even making the diagnosis of precancer has relied upon colposcopically directed biopsies. But the choice of biopsy site and the resultant diagnosis are subjective and thus variable among clinicians. Though the technology has not markedly improved, colposcopy is utilized by clinicians to establish the presence or absence of lesions. Screening sensitivity has improved, but the combination of both results in weak correlations between disease severity

TABLE 4   Cervical cytologic screening guidelines as recommended by the American Society for Colposcopy and Cervical Pathology.

| Population | Recommendation for cytologic screening | Comments |
| --- | --- | --- |
| <21 years | Avoid screening | Do not use HPV testing for screening in this group. Do not use HPV testing to guide management of ASC-US in this group. |
| 21–29 years | Screen every 3 years with cytology | |
| 30–65 years | HPV and cytology every 5 years (preferred) | |
| | Cytology alone every 3 years (acceptable) | |
| >65 years | No screening following adequate negative prior screening | Continue routine screening in women with a history of CIN2 or more for at least 20 years |
| After hysterectomy | No screening recommended | Applies to women without a cervix and without history of CIN2 or more in the past 20 years, or cervical cancer ever |

*Abbreviations:* HPV, Human papillomavirus; ASCUS, atypical squamous cells of undetermined significance; CIN, cervical intraepithelial neoplasia.

and visual changes, thereby inhibiting reproducibility among clinicians. Obtaining more than one nonrandom biopsy will increase sensitivity, however, irrespective of expertise or training [53].

The initial workup of cervical cancer depends on whether a lesion is visible. In its early stages, it is often asymptomatic and diagnosed through routine screening methods. Women may present with symptoms of malodorous discharge, flank pain, or postcoital abnormal vaginal bleeding. Patients without a visible lesion require a cone biopsy to diagnose early-stage disease [58]. Patients with a visible cervical lesion may require examination under anesthesia for biopsy, cystoscopy, and proctosigmoidoscopy to evaluate local involvement of nearby organs. Advanced imaging modalities like CT, MRI, and fluorodeoxyglucose-positron emission tomography (FDG-PET) are useful for the evaluation of nodal and other metastatic sites [64]. PET-CT is the most sensitive modality in identifying nodal metastasis or metastatic disease assisting with radiotherapy planning, and evaluation of metabolic response to therapy [64].

# Vulvar cancer

Vulvar cancer is an uncommon gynecologic malignancy that is predominantly a disease of postmenopausal women in the seventh and eight decades of life, with a median age at diagnosis of 68 years [3, 16]. Age-specific incidence rates in the United States steeply increase at age 70 years and continue to increase beyond 80 years of age [3]. The incidence of vulvar cancer has increased by an average of 4.6% every 5 years, whereas relative survival appears to be decreasing [1, 3]. Interestingly, this increase has largely been driven by a growing incidence among younger women. Vulvar cancer-associated deaths reached 1200 in the year 2018, and it is estimated that 6190 cases will occur each year, accounting for 0.7% of all cancers in women, 5.6% of gynecologic malignancies, 0.42% of cancer deaths in women, and 3.6% of gynecologic cancer deaths [1]. The paucity of the disease has made it difficult to evaluate new treatment strategies in prospective randomized-controlled trials. Several decades ago, the incidence of vulvar cancer in women <50 years was 2% [65]. Currently greater than 20% of cases occur in women younger than 50 years [3, 65]. Until the year 2004, the 5-year disease-specific survival was unchanged at 75%; however, since then, death rates have increased 0.7% per year, and the 5-year disease-specific survival has dropped to 68% [65]. The most common vulvar malignancy is squamous cell carcinoma (SCC), accounting for greater than 90% of all vulvar cancers [66, 67]. These are usually preceded by dysplastic lesions [68, 69]. Melanoma of the vulva is the second most common vulvar malignancy, and other less commonly encountered histologies include Bartholin gland carcinoma, basal-cell carcinoma, verrucous carcinoma, sarcoma, and Paget disease of the vulva (PDV) [65, 66, 70].

## Pathogenesis and diagnosis

Vulvar cancer incorporates both invasive SCC and its precursor lesions. Precursor lesions were characterized as vulvar intraepithelial neoplasia (VIN) in the older nomenclature and were subdivided into two categories: differentiated VIN (dVIN) and usual-type VIN [66, 68]. The coexistence of VIN or invasive SCC with in situ or epidermoid carcinoma of the cervix has long been established. Twenty percent of women with primary vulvar lesions may also demonstrate evidence of synchronous or sequential cervical lesions [71]. This observation suggests a shared etiology in at least some patients. VIN of the usual type is driven by the oncogenic types of HPV, whereas differentiated VIN is not typically attributable to HPV, but rather is more commonly related to chronic inflammatory vulvar dermatoses, such as lichen sclerosis [71]. Lichen sclerosis is a chronic inflammatory, mucocutaneous lesion that is most commonly found on the genital skin, presenting with pruritus, pain, viding dysfunction, sexual dysfunction, and bleeding. It commonly arises on the hood of the clitoris and then progresses to the labia minora, vestibule, perineal, and perianal regions to form the characteristic figure-of-eight distribution [66, 69].

In HPV-positive SCC, oncoproteins E6 and E7 are major driving factors leading to tumor-suppressor protein inactivation. The inactivation of *p53* and *Rb*, respectively, by E6 and E7 results in the overexpression of the cell-cycle protein *p16ink4a* [67]. Alternatively, *p16ink4a* is not overexpressed in HPV-negative SCC, but *p53* is expressed [67]. The terminology was modified in the year 2015 to coincide with other HPV-associated lesions of the lower genital tract [69]. Under the current classification, vulvar high-grade squamous intraepithelial lesions (HSIL) are associated with HPV, and non-HPV-associated precursors remain dVIN [68]. Low-grade squamous intraepithelial lesions (LSIL) are defined as flat lesions associated with basal atypia and koilocytic changes. Grossly, invasive SCC and its precursors can present as a raised, flat, ulcerated, polypoid, warty, or plaque-like mass on the vulva. Histologically, it is characterized by full-thickness or near-full-thickness atypia [71]. Typical findings include nuclear enlargement, nuclear hyperchromasia, apoptosis, and increased mitotic activity, with mitotic figures present in the upper portion of the epithelium [16]. Many patients are without symptoms at diagnosis, but others present with burning pain, pruritus, or vaginal bleeding. HSIL tends to be slow growing and multifocal, with a potential for spontaneous regression [71].

SCC is classified into three primary histologic subtypes: warty, basaloid, and keratinizing [16]. The basaloid and warty variants are the subtypes that are often HPV-mediated, accounting for the minority of cases, and are primarily noted in younger women. Basaloid cancers have a histologic appearance similar to that of HSIL and are often referred to as the intraepithelial type of invasive SCC, whereas the warty variant has surface koilocytic atypia [71]. Keratinizing SCC is found in older patients, arises from the HPV pathway, and accounts for 65%–80% of SCC subtypes [68, 71].

## Melanoma

The second most common vulvar malignancy is vulvar melanoma. Vulvar melanoma accounts for up to 10% of all vulvar malignancies [72]. The median age at diagnosis of vulvar melanoma is 68 years, and it is more prevalent among Whites than non-Whites [65, 70, 72]. The incidence of vulvar melanoma is <0.2 cases per 100,000 women [72]. Around 50% of patients will present with localized disease, whereas 8.4% will present with advanced disease [70]. Vulvar melanoma is significantly rarer than cutaneous melanoma. It is also diagnosed at a much older age and frequently at advanced stages, resulting in significantly worse survival [65]. The vulva represents less than 1% of total body surface area, but 2% of all melanomas occurring in women arise from the vulva. This means that the vulva is almost 3 times more likely to develop melanoma than any other location sharing a similar surface area [72]. Classical presentation of vulvar melanoma is an asymmetric black papule, nodule, or macule, with jagged borders and coloration, extending >7 mm in diameter. These lesions arise commonly on the clitoral hood, labia minora, and labia majora [65, 70]. Genetic and histopathologic studies have suggested that vulvar melanoma is more comparable to acral lentiginous melanoma than to cutaneous melanoma [70]. Vulvar melanomas also have a distinctly different molecular profile from that of cutaneous melanomas.

## Paget disease of the vulva

Paget disease of the vulva (PDV) is an extremely rare gynecologic malignancy that represents only 1%–2% of vulvar malignancies [73]. The true incidence and prevalence are not entirely known due to its rarity. It does primarily affect postmenopausal women with a median age of 72 years. It originates from the intraepidermal element of apocrine sweat glands. In approximately 9% of cases, it is associated with an underlying adenocarcinoma of dermal apocrine glands [66]. PDV is classified by the origin of the neoplastic cells; primary lesions arise from within the vulvar epithelium itself, and secondary lesions result from the spread of a noncutaneous internal malignancy [71]. The pathogenesis of primary disease is not yet fully understood, but theories suggest that Paget cells originate from mammary-like glands between the labia majora, minora, perineum, and perianal area [65]. PDV has a classic characterization marked by a painful, sharply demarcated, and erythematous rash over the labia majora. The visual borders of the disease generally underestimate the histopathologic extent of the disease. It is believed that this feature accounts for the higher rates of recurrence after primary resection [66] (Fig. 3).

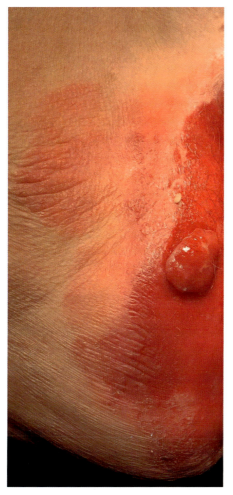

FIG. 3    Paget's disease of the vulva. PDV has a classic characterization marked by a painful, sharply demarcated, and erythematous rash over the labia. The visual borders on physical exam generally underestimate the histopathologic extent of the disease. This feature likely accounts for the higher rates of recurrence after primary resection. On this hemivulva, there is also a superimposed SCC of the vulva as evidenced by the raised lesion. *Abbreviations:* PDV = Paget's disease of the vulva.

## Vaginal cancer

Primary invasive cancer of the vagina is rare. Most vaginal malignancies are usually metastatic, originating from the cervix or the endometrium. In 2020, it is estimated that 6230 cases will be diagnosed in the United States, resulting in 1450 deaths [1]. Vaginal cancer comprises of about 0.56% of all cancers in women, 4.5% of gynecologic malignancies, 0.44% of cancer deaths in women, and 3.9% of gynecologic cancer deaths [1, 74]. Women aged 60 years or older make up approximately 70% of detected primary vaginal cancer. Of these patients, 60% of diseases actually arise in women that are 70 years of age and older. Primary vaginal cancer must involve the vagina, with no external cervical or vulvar involvement to be diagnosed as such. If this occurs, the malignancy is classified as cervical or vulvar in origin. Of note, invasive SCC that occurs in the vagina within 5 years of a prior invasive SCC of the cervix is considered a recurrence of the cervical primary. SCC of the vagina is the most common type of vaginal cancer, accounting for up to 90% of primary vaginal cancers [3, 75, 76].

The precursor lesion, vaginal intraepithelial neoplasia (VAIN), is also uncommon and occurs less frequently than comparable lesions in the vulvar and cervix [77]. Intraepithelial neoplasia of the vagina has a similar histologic appearance to that of the cervix and is classified correspondingly. VAIN-1 characterizes mild dysplasia. VAIN-2 signifies moderate dysplasia, and severe dysplasia to carcinoma in situ is defined as VAIN-3 [75]. VAIN-1 is classified as a low-grade squamous intraepithelial lesion (LSIL), while VAIN-2 and VAIN-3 make up the aggregate group characterized as high-grade squamous intraepithelial lesions (HSIL) [78].

Most cases are likely mediated by HPV infection, as with cervical cancer, but the natural history of VAIN is not well understood, and the potential of VAIN progression to invasive SCC is not precisely known [75, 78]. It is estimated to be low, approximately 9% [78]. Women with high-risk strains of HPV, those who are immunocompromised, and those with VAIN-3 are thought to be at highest risk of progression. Histologically, vaginal LSIL parallels that of the cervix and includes flat and exophytic condyloma [78]. Vaginal HSIL is characterized by nuclear atypia in all levels of the epithelium, with a varying degree of surface maturation [77].

Delay in the diagnosis of vaginal cancers occurs frequently, in part due to their rarity. The most common symptom at presentation is abnormal bleeding or discharge. In advanced disease, involvement of adjacent structures may lead to urinary obstruction, and constipation or tenesmus from posterior vaginal wall involvement. They primarily involve the upper two-thirds of the vagina, where direct extension can involve the adjacent parametria, paracolpos, bladder, or rectum [76]. Grossly, it may appear as a polypoid, fungating, or ulcerating mass. A secondary infection may be accompanied by a foul discharge or odor. Microscopically, the tumor demonstrates the classic findings of an invasive SCC infiltrating the vaginal epithelium [76].

There is strong evidence that implicates HPV in the pathogenesis of most SCC of the vagina. Consequently, use of HPV vaccines in phase III trials has suggested that significant reductions in the incidence of VAIN as well as invasive SCC of the vagina can be anticipated with the implementation of prophylactic immunization [59]. Most cases of VAIN spontaneously regress without intervention, but some do not. Therefore, a policy for longitudinal surveillance and effective treatment for patients with persistent VAIN is warranted. In doing so, the reduction of the risk of developing invasive cancer is therefore decreased, especially in women in whom high-grade VAIN is a precursor lesion. This underscores the importance of continued vaginal cytology even after total hysterectomy for dysplastic conditions of the cervix. The American College of Obstetrics & Gynecology (ACOG) recommends screening for these women for at least 20 years after the initial treatment for grades CIN2 and CIN3 [60].

## Conclusions

Gynecologic cancers originate from the female reproductive system and genitalia, consisting of cervical, uterine, ovarian, vulvar, and vaginal cancers. Multiple histologic types are associated with malignancies in each of the disease sites; therefore, the collective group of gynecologic cancers is large and heterogeneous. Mutually, these cancers will affect 113,000 US women in 2020 and will be responsible for nearly 33,000 deaths. Accounting for >55% of new diagnoses per year, the most commonly diagnosed gynecologic cancer is uterine cancer, while the most lethal gynecologic cancer is ovarian cancer, comprising nearly 50% of deaths. The incidence and mortality rates for cervical cancer have declined by about 50% owing to newer methods and technology that enable early detection and prevention. In contrast, reductions in mortality for uterine and ovarian cancer have been much more limited, due in large part to gradual improvements in care, rather than drastic shifts in treatment paradigms. In contrast, the incidence and mortality rates of the rarer gynecologic cancers, vulvar and vaginal cancer, have remained largely stagnant over the last 30 years. As a whole, the collective group of malignancies continues to impact women worldwide with a predilection for different ethnic groups, and socioeconomic status, based on the type of cancer. As advances in treatment continue to improve and include targeted therapies, including earlier screening based on genetics, we may have an opportunity in the future to decrease the impact of gynecologic malignancy in society.

# References

[1] Siegel RL, Miller KD, Jemal A. Cancer statistics, 2020. CA Cancer J Clin 2020;70:6–30.

[2] Ferlay J, Soerjomataram I, Dikshit R, Eser S, Mathers C, Rebelo M, Parkin DM, Forman D, Bray F. Cancer incidence and mortality worldwide: sources, methods and major patterns in GLOBOCAN 2012. Int J Cancer 2015;136:E359–86.

[3] Centers for Disease Control and Prevention. 2020. United States cancer statistics. Available at https://gis.cdc.gov/Cancer/USCS/DataViz.html. (Accessed 18 September 2020).

[4] Bokhman JV. Two pathogenetic types of endometrial carcinoma. Gynecol Oncol 1983;15:10–7.

[5] Duska LR, Garrett A, Rueda BR, Haas J, Chang Y, Fuller AF. Endometrial cancer in women 40 years old or younger. Gynecol Oncol 2001;83:388–93.

[6] Lee TS, Jung JY, Kim JW, Park NH, Song YS, Kang SB, Lee HP. Feasibility of ovarian preservation in patients with early stage endometrial carcinoma. Gynecol Oncol 2007;104:52–7.

[7] Lacey JV, Chia VM, Rush BB, Carreon DJ, Richesson DA, Ioffe OB, Ronnett BM, Chatterjee N, Langholz B, Sherman ME, Glass AG. Incidence rates of endometrial hyperplasia, endometrial cancer and hysterectomy from 1980 to 2003 within a large prepaid health plan. Int J Cancer 2012;131:1921–9.

[8] Sheikh MA, Althouse AD, Freese KE, Soisson S, Edwards RP, Welburn S, Sukumvanich P, Comerci J, Kelley J, LaPorte RE, Linkov F. USA endometrial cancer projections to 2030: should we be concerned? Future Oncol 2014;10:2561–8.

[9] Trabert B, Wentzensen N, Felix AS, Yang HP, Sherman ME, Brinton LA. Metabolic syndrome and risk of endometrial cancer in the United States: a study in the SEER-medicare linked database. Cancer Epidemiol Biomarkers Prev 2015;24:261–7.

[10] Renehan AG, Tyson M, Egger M, Heller RF, Zwahlen M. Body-mass index and incidence of cancer: a systematic review and meta-analysis of prospective observational studies. Lancet 2008;371:569–78.

[11] Burke WM, Orr J, Leitao M, Salom E, Gehrig P, Olawaiye AB, Brewer M, Boruta D, Herzog T, Abu SF. Endometrial cancer: a review and current management strategies: part I. Gynecol Oncol 2014;134:385–92.

[12] National Cancer Institute. 2020. Surveillance, epidemiology, and end results program. Cancer Stat Facts: Uterine Cancer. Available at: https://seer.cancer.gov/statfacts/html/corp.html. (Accessed 19 September 2020).

[13] National Cancer Institute. 2020. Endometrial cancer treatment (PDQÒ): health professional version. Available at: https://www.cancer.gov/types/uterine/hp/endometrial-treatment-pdq/. (Accessed 19 September 2020).

[14] Kurman RJ, Cancangiu ML, Herrington CS, Young RH. WHO classification of tumours of female reproductive organs. 4th ed. Geneva, Switzerland: World Health Organization; 2014.

[15] The American College of Obstetricians and Gynecologists. Committee opinion no. 601: tamoxifen and uterine cancer. Obstet Gynecol 2014;123:1394–7.

[16] Barakat RR, Berchuck A, Markman M, Randall M. Principles and practice of gynecologic oncology. 6th ed. Philadelphia, PA: Lippincott Williams & Wilkins; 2013.

[17] Key TJ, Pike MC. The dose-effect relationship between 'unopposed' oestrogens and endometrial mitotic rate: its central role in explaining and predicting endometrial cancer risk. Br J Cancer 1988;57:205–12.

[18] Pike MC, Peters RK, Cozen W, Probst-Hensch NM, Felix JC, Wan PC, Mack TM. Estrogen-progestin replacement therapy and endometrial cancer. J Natl Cancer Inst 1997;89:1110–6.

[19] Kaaks R, Lukanova A, Kurzer MS. Obesity, endogenous hormones and endometrial cancer risk: a synthetic review. Cancer Epidemiol Biomarkers Prev 2002;11:1531–43.

[20] Purdie DM, Green AC. Epidemiology of endometrial cancer. Best Pract Res Clin Obstet Gynaecol 2001;15:341–54.

[21] Kandoth C, Schultz N, Cherniack AD, Akbani R, Liu Y, Shen H, Robertson AG, Pashtan I, Shen R, Benz CC, Yau C, Laird PW, Ding L, Zhang W, Mills GB, Kucherlapati R, Mardis ER, Levine DA. Integrated genomic characterization of endometrial carcinoma. Nature 2013;497:67–73.

[22] Kuhn E, Ayhan A, Bahadirli-Talbot A, Zhao C, Shih IM. Molecular characterization of undifferentiated carcinoma associated with endometrioid carcinoma. Am J Surg Pathol 2014;38:660–5.

[23] Byron SA, Pollock PM. FGFR2 as a molecular target in endometrial cancer. Future Oncol 2009;5:27–32.

[24] Brinton LA, Felix AS, McMeekin DS, Creasman WT, Sherman ME, Mutch D, Cohn DE, Walker JL, Moore RG, Downs LS, Soslow RA, Zaino R. Etiologic heterogeneity in endometrial cancer: evidence from a gynecologic oncology group trial. Gynecol Oncol 2013;129:277–84.

[25] Zhang ZM, Xiao S, Sun GY, Liu YP, Zhang FH, Yang HF, Li J, Qiu HB, Liu Y, Zhang C, Kang S, Shan BE. The clinicopathologic significance of the loss of BAF250a (ARID1A) expression in endometrial carcinoma. Int J Gynecol Cancer 2014;24:534–40.

[26] Fadare O, Gwin K, Desouki MM, Crispens MA, Jones III HW, Khabele D, Liang SX, Zheng W, Mohammed K, Hecht JL, Parkash V. The clinicopathologic significance of p53 and BAF-250a (ARID1A) expression in clear cell carcinoma of the endometrium. Mod Pathol 2013;26:1101–10.

[27] Hoang LN, Han G, McConechy M, Lau S, Chow C, Gilks CB, Huntsman DG, Köbel M, Lee CH. Immunohistochemical characterization of prototypical endometrial clear cell carcinoma—diagnostic utility of HNF-1β and oestrogen receptor. Histopathology 2014;64:585–96.

[28] Castilla MA, Moreno-Bueno G, Romero-Pérez L, Van De Vijver K, Biscuola M, López-García MÁ, Prat J, Matías-Guiu X, Cano A, Oliva E, Palacios J. Micro-RNA signature of the epithelial-mesenchymal transition in endometrial carcinosarcoma. J Pathol 2011;223:72–80.

[29] Cherniack AD, Shen H, Walter V, Stewart C, Murray BA, Bowlby R, Hu X, Ling S, Soslow RA, Broaddus RR, Zuna RE, Robertson G, Laird PW, Kucherlapati R, Mills GB, Cancer Genome Atlas Research Network, Weinstein JN, Zhang J, Akbani R, Levine DA. Integrated molecular characterization of uterine carcinosarcoma. Cancer Cell 2017;31:411–23.

[30] Hussein YR, Weigelt B, Levine DA, Schoolmeester K, Dao LN, Balzer BL, Liles G, Karlan B, Kobel M, Lee CH, Soslow RA. Clinicopathological analysis of endometrial carcinomas harboring somatic POLE exonuclease domain mutations. Mod Pathol 2015;28:505–14.

[31] Monte NM, Webster KA, Neuberg D, Dressler GR, Mutter GL. Joint loss of PAX2 and PTEN expression in endometrial precancers and cancer. Cancer Res 2010;70:6225–32.

[32] Modica I, Soslow RA, Black D, Tornos C, Kauff N, Shia J. Utility of immunohistochemistry in predicting microsatellite instability in endometrial carcinoma. Am J Surg Pathol 2007;31:744–51.

[33] Touboul C, Piel B, Koskas M, Gonthier C, Ballester M, Cortez A, Darai E. Factors predictive of endometrial carcinoma in patients with atypical endometrial hyperplasia on preoperative histology. Anticancer Res 2014;34:5671–6.

[34] Chang YN, Zhang Y, Want YJ, Wang LP, Duan H. Effect of hysteroscopy on the peritoneal dissemination of endometrial cancer cells: a meta-analysis. Fertil Steril 2011;96:957–61.

[35] Timmermans A, Opmeer BC, Khan KS, Bachmann LM, Epstein E, Clark TJ, Gupta JK, Bakour SH, van den Bosch T, van Doorn HC, Cameron ST, Giusa MG, Dessole S, Dijkhuizen FP, TerRiet G, Mol BW. Endometrial thickness measurement for detecting endometrial cancer in women with postmenopausal bleeding: a systematic review and meta-analysis. Obstet Gynecol 2010;116:160–7.

[36] McCluggage WG. Mullerian adenosarcoma of the female genital tract. Adv Anat Pathol 2010;17:122–9.

[37] Chang KL, Crabtree GS, Lim-Tan SK, Kempson RL, Hendrickson MR. Primary uterine endometrial stromal neoplasms. A clinicopathologic study of 117 cases. Am J Surg Pathol 1990;14:415–38.

[38] Cannistra SA. Cancer of the ovary. N Engl J Med 2004;351:2519–29.

[39] Hayat MJ, Howlander N, Reichman ME, Edwards BK. Cancer statistics, trends, and multiple primary cancer analyses from the surveillance, epidemiology, and end results (SEER) program. Oncologist 2007;12:20–37.

[40] Morris CR, Rodrigue AO, Epstein J, Cress RD. Declining trends of epithelial ovarian cancer in California. Gynecol Oncol 2008;108:207–13.

[41] Brewster WR. Temporal trends in ovarian cancer: incidence and mortality across Europe. Nat Clin Pract Oncol 2005;2:286–7.

[42] Bast Jr RC, Mills G. Molecular pathogenesis of epithelial ovarian cancer. In: Mendelsohn J, Howley P, Israel M, Gray JW, Thompson CB, editors. The molecular basis of cancer. 3rd ed. Philadelphia, PA: Saunders-Elsevier; 2008. p. 441–54.

[43] Pavlik EJ, Smith C, Dennis TS, Harvey E, Huang B, Chen Q, Piecoro DW, Burgess BT, McDowell A, Gorski J, Baldwin LA, Miller RW, DeSimone CP, Dietrich 3rd C, Gallion HH, Ueland FR, van Nagell Jr JR. Disease-specific survival of type I and type II epithelial ovarian cancers-stage challenges categorical assignments of indolence & aggressiveness. Diagnostics (Basel) 2020;10:56–69.

[44] Crum CP, Drapkin R, Kindelberger D, Medeiros F, Miron A, Lee Y. Lessons from BRCA: the tubal fimbria emerges as an origin for pelvic serous cancer. Clin Med Res 2007;5:35–44.

[45] Hennessy BT, Mills GB. Ovarian cancer: homebox genes, autocrine/paracrine growth, and kinase signaling. Int J Biochem Cell Biol 2006;38:1450–6.

[46] Kohler MF, Marks JR, Wiseman RW, Jacobs IJ, Davidoff AM, Clarke-Pearson DL, Soper JT, Bast Jr RC, Berchuck A. Spectrum of mutation and frequency of allelic deletion of the p53 gene in ovarian cancer. J Natl Cancer Inst 1993;85:1513–9.

[47] Campbell IG, Russell SE, Choong DY, Montgomery KG, Ciavarella ML, Hooi CS, Cristiano BE, Pearson RB, Phillips WA. Mutation of the PIK3CA gene in ovarian and breast cancer. Cancer Res 2004;64:7678–81.

[48] Carpten JD, Faber AL, Horn C, Donoho GP, Briggs SL, Robbins CM, Hostetter G, Boguslawski S, Moses TY, Savage S, Uhlik M, Lin A, Du J, Qian YW, Zeckner DJ, Tucker-Kellogg G, Touchman J, Patel K, Mousses S, Bittner M, Schevitz R, Lai MH, Blanchard KL, Thomas JE. A transforming mutation in the pleckstrin homology domain of AKT1 in cancer. Nature 2007;448:439–44.

[49] Shaw PA, Zweemer RP, McLaughlin J, Sarod SA, Risch H, Jacobs IJ. Characteristics of genetically determined ovarian cancer. Mod Pathol 1999;12:124.

[50] Merritt MA, Cramer DW. Molecular pathogenesis of endometrial and ovarian cancer. Cancer Biomark 2010;9:233–62.

[51] Bray F, Ferlay J, Soerjomataram I, Siegel RL, Torre LA, Jemal A. Global cancer statistics 2018: GLOBOCAN estimates of incidence and mortality worldwide for 36 cancers in 185 countries. CA Cancer J Clin 2018;68:394–424.

[52] WHO. Cervical cancer. Geneva: World Health Organization; 2018. http://www.who.int/cancer/prevention/diagnosis-screening/cervical-cancer/en/. [Accessed 20 September 2020].

[53] Vaccarella S, Lortet-Tieulent J, Plummer M, Franceschi S, Bray F. Worldwide trends in cervical cancer incidence: impact of screening against changes in disease risk factors. Eur J Cancer 2013;49:3262–673.

[54] Sriplung H, Singkham P, Iamsirithaworn S, Jiraphongsa C, Bilheem S. Success of a cervical cancer screening program: trends in incidence in Songkhla, Southern Thailand, 1989–2010, and prediction of future incidences to 2030. Asian Pac J Cancer Prev 2014;15:10003–8.

[55] Olorunfemi G, Ndlovu N, Masukume G, Chikandiwa A, Pisa PT, Singh E. Temporal trends in the epidemiology of cervical cancer in South Africa (1994–2012). Int J Cancer 2018;143:2238–49.

[56] Torre LA, Bray F, Siegel RL, Ferlay J, Lortet-Tieulent J, Jemal A. Global cancer statistics, 2012. CA Cancer J Clin 2015;65:87–108.

[57] Waggoner SE. Cervical cancer. Lancet 2003;361:2217–25.

[58] Stapley S, Hamilton W. Gynaecological symptoms reported by young women: examining the potential for earlier diagnosis of cervical cancer. Fam Pract 2011;28:592–8.

[59] Schiller JT, Castellsague X, Garland SM. A review of clinical trials of human papillomavirus prophylactic vaccines. Vaccine 2012;30:F123–38.

[60] American College of Obstetricians and Gynecologists' Committee on Practice Bulletins—Gynecology. Practice bulletin No. 168: cervical cancer screening and prevention. Obstet Gynecol 2016;128:111–30.

[61] Crosbie EJ, Einstein MH, Franceschi S, Kitchener HC. Human papillomavirus and cervical cancer. Lancet 2013;382:889–99.

[62] Dunne EF, Unger ER, Sternberg M, McQuillan G, Swan DC, Patel SS, Markowitz LE. Prevalence of HPV infection among females in the United States. JAMA 2007;297:813–9.

[63] Hirth J. Disparities in HPV vaccination rates and HPV prevalence in the United States: a review of the literature. Hum Vaccin Immunother 2019;15:146–55.

[64] Havrilesky LK, Kulasingam SL, Matchar DB, Myers ER. FDG-PET for management of cervical and ovarian cancer. Gynecol Oncol 2005;97:183–91.

[65] Mert I, Semaan A, Winer I, Morris RT, Ali-Fehmi R. Vulvar/vaginal melanoma: an updated surveillance epidemiology and end results database review, comparison with cutaneous melanoma and significance of racial disparities. Int J Gynecol Cancer 2013;23:1118–25.

[66] Sand FL, Thomsen SF. Clinician's update on the benign, premalignant, and malignant skin tumours of the vulva: the dermatologist's view. Int Sch Res Notices 2017;25:1155–65.

[67] Hinten F, Molijn A, Eckhardt L, Massuger LFAG, Quint W, Bult P, Bulten J, Melchers WJG, de Hullu JA. Vulvar cancer: two pathways with different localization and prognosis. Gynecol Oncol 2018;149:310–7.

[68] Hoang LN, Park KJ, Soslow RA, Murali R. Squamous precursor lesions of the vulva: current classification and diagnostic challenges. Pathology 2016;48:291–302.

[69] American College of Obstetricians and Gynecologists' Committee on Gynecologic Practice, American Society for Colposcopy and Cervical Pathology. Committee opinion No. 675: management of vulvar intraepithelial neoplasia. Obstet Gynecol 2016;128:e178–82.

[70] Murzaku EC, Penn LA, Hale CS, Pomeranz MK, Polsky D. Vulvar nevi, melanosis, and melanoma: an epidemiologic, clinical, and histopathologic review. J Am Acad Dermatol 2014;71:1241–9.

[71] Terlou A, Blok LJ, Helmerhorst TJ, van Beurden M. Premalignant epithelial disorders of the vulva: squamous vulvar intraepithelial neoplasia, vulvar Paget's disease and melanoma in situ. Acta Obstet Gynecol Scand 2010;89:741–8.

[72] Piura B. Management of primary melanoma of the female urogenital tract. Lancet Oncol 2008;9:973–81.

[73] Parker LP, Parker JR, Bodurka-Bevers D, Deavers M, Bevers MW, Shen-Gunther J, Gershenson DM. Paget's disease of the vulva: pathology, pattern of involvement, and prognosis. Gynecol Oncol 2000;77:183–9.

[74] American Cancer Society. Cancer facts & figures 2018. Atlanta; 2018.

[75] Aho M, Vesterinen E, Meyer B, Purola E, Paavonen J. Natural history of vaginal intraepithelial neoplasia. Cancer 1991;68:195–7.

[76] Davis KP, Stanhope CR, Garton GR, Atkinson EJ, O'Brien PC. Invasive vaginal carcinoma: analysis of early-stage disease. Gynecol Oncol 1991;42:131–6.

[77] Audet-Lapointe P, Body G, Vauclair R, Drouin P, Ayoub J. Vaginal intraepithelial neoplasia. Gynecol Oncol 1990;36:232–9.

[78] Srodon M, Stoler MH, Baber GB, Kurman RJ. The distribution of low and high-risk HPV types in vulvar and vaginal intraepithelial neoplasia (VIN and VaIN). Am J Surg Pathol 2006;30:1513–8.

## Further reading

Levert M, Clavel C, Graesslin O, Masure M, Birembaut P, Quereux C, Gabriel R. Human papillomavirus typing in routine cervical smears. Results from a series of 3778 patients. Gynecol Obstet Fertil 2000;28:722–8.

# Lifestyle, nutrition, and risk of gynecologic cancers

*Evan A. Schrader[a], Thomas A. Paterniti[b], and Sarfraz Ahmad[c]*

[a]Department of Obstetrics and Gynecology, Atrium Health Carolinas Medical Center, Charlotte, NC, United States [b]Department of Obstetrics and Gynecology, Augusta University Medical Center, Augusta, GA, United States [c]Gynecologic Oncology, AdventHealth Cancer Institute, Florida State University, University of Central Florida, Orlando, FL, United States

## Abstract

Gynecologic cancers are an increasingly widespread and well-studied malignancy in today's world of female reproductive health. The pathophysiologic mechanisms of developing cervical cancer, endometrial (uterine) cancer, and ovarian cancer are unique and come with their list of risk factors. As we discuss these risk factors in this chapter, we also review the modifiable lifestyle changes that can reduce patients' risk of developing those cancers. Some of the interventions come through the diet, and in all cases are improved by exercise plans, two foundational components to essential lifestyle change. In every step of this chapter, we review evidence-based interventions that have been shown to have a positive impact. We also provide an extensive review of the various lifestyle, genetic, and reproductive risk factors that play functional roles in the development of gynecologic cancers. This chapter is ultimately a provider's guide to the important and necessary discussions with patients about what risk factors a patient has and what we know can impact that individual patient's outcomes and quality of life.

## Abbreviations

| | |
|---|---|
| AARP | American Association of Retired Persons |
| ACSM | American College of Sports Medicine |
| BMI | body mass index |
| CIN | cervical intraepithelial neoplasia |
| EC | endometrial cancer |
| EOC | epithelial ovarian cancer |
| ER | estrogen receptor |
| GOG | Gynecologic Oncology Group |
| HER2 | human epidermal growth factor receptor 2 |
| HPV | human papilloma virus |

R. Basha, S. Ahmad (eds.)
*Overcoming Drug Resistance in Gynecologic Cancers*
ISBN 978-0-12-824299-5
https://doi.org/10.1016/B978-0-12-824299-5.00019-8

**HRT**    hormone replacement therapy
**MSI**    microsatellite instability
**NSAID**    nonsteroidal antiinflammatory drugs.
**OCPs**    oral contraceptives
**PEPI**    postmenopausal estrogen/progestin intervention
**ROS**    reactive oxygen species
**SERM**    selective estrogen-receptor modifier
**SHBG**    sex hormone-binding globulin
**UCC**    uterine-cervix cancer
**UPSC**    uterine papillary serous carcinoma

## Conflict of interest

No potential conflicts of interest were disclosed.

## Introduction

Understanding the pathophysiology surrounding the development of gynecologic cancers allows providers to educate patients and advocate for the prevention of disease by reducing known risk factors. At the core of risk factor reduction is the general principle of living a wholly healthy lifestyle, but what exactly does that look like? The concept of living a healthy lifestyle incorporates many factors—diet and nutrition, exercise, and even social factors like psychology and support.

Delving into the realm of what it means to live a healthy lifestyle, we must look at diet and nutrition, the fuel sources that supply a person's energy. As patients and their families come into the clinic with new diagnoses of gynecologic cancers, what can we tell them about diet that will improve outcomes? In this chapter, we sought to discuss the dietary components that have evidence-based data on decreasing the risk of gynecologic cancers, which ultimately empowers physicians to better educate their patients. We know that in many cases, one of the greatest risk factors for common gynecologic cancers is obesity and an increased BMI [1], so what can we tell patients to do to reduce this risk? Depending on each patient's unique diagnosis, can we take that advice one step further and individualize dietary counseling for each patient?

In a broader sense than just the diet, many lifestyle factors also influence the risk for various types of gynecologic cancer. Many studies have investigated different lifestyle factors throughout a large sampling of female cohorts and found aspects that have an impact on rates of female pelvic cancers. The main and most heavily researched target is exercise. What do we know about the potential benefits of exercise in evading the development of gynecologic cancer? In our patients with cancer or who are undergoing strict chemo/radiation therapy protocols, how does exercise impact their treatment?

Finally, providers will forever ask: how can we best identify risk factors for the development of gynecologic cancers in our patients? Depending on the tissue type, there are very well understood pathophysiologic mechanisms behind the inception of cancerous growth. Obesity and smoking are some of the more obvious associated risk factors, but even social factors like race and socioeconomic status have also been proven to confer risk upon patients. Obstetric history, parity, and exogenous hormone administration are other factors to be considered when evaluating a patient's potential risk factors for female pelvic cancers.

In this chapter, we delineate various protective effects of nutritional components on patients with gynecologic cancers and identify what parts of a person's diet can reduce the risk of future gynecologic cancer development. We broaden this investigation to lifestyle factors as a whole to show what proven lifestyle changes we can advocate for in our patients to make a difference. Finally, this chapter outlines various risk factors for developing gynecologic cancers to help us hone in on patients who might be at the highest risk during their lifetimes.

## Nutrition

A targeted nutritional investigation comes first from a foundational understanding of how diet can predispose the body to the development of cancer. One of the greatest risk factors for common gynecologic cancers is obesity. At all ages and stages of life, endometrial cancer is strongly positively linked to higher body-mass index (BMI) and overweight body habitus [2]. The underlying mechanism here is found in the increased proportion of adipose tissue, which uses intrinsic aromatase enzymes to increase the circulating levels of estrogen. Estrogen exerts its proliferative effects on the uterine endometrium and thus is a cancer-promoting factor [3]. Thus, by surveying a population of overweight and obese women, providers will find a higher incidence of endometrial cancer and a target population for appropriate dietary change [4]. It is not just endometrial cancer, but an increased BMI confers a higher risk of all gynecologic cancer subtypes [1].

### Endometrial cancer

Several large studies have investigated the link between diet and endometrial cancer (EC) risk reduction. One such study found that diets with high glycemic loads, meaning foods densely packed with caloric content to raise the blood glucose level after ingestion, increased EC risk by up to 20% [5]. The glycemic index classifies meal quality by its ability to increase postprandial blood glucose levels, whereas the glycemic load accounts for not only the glycemic index but the total carbohydrate content as well [5]. This would lead you to believe a diet low in fat and rich in fruits, vegetables, and grains would prove beneficial in reducing EC risk, but one study found no difference in EC risk after even 8-years of follow-up [6]. However, another peer-reviewed publication contradicts these results, reporting that the consumption of whole grains, fresh fruit, and fresh vegetables decreases EC risk [7]. While nutritional research remains crucial to understanding the effects of diet on female pelvic cancers, the literature can wholly support the notion that the Mediterranean diet is superior to any other nutritional combination. Multiple case-control studies have shown that the Mediterranean diet is associated with a decreased risk of EC [8] (Table 1; Fig. 1).

Looking more closely at the link between endometrial cancer and caloric intake, we know that fat intake confers the most risk. A large metaanalysis of overall energy intake that studied caloric breakdown via specific macronutrient calorie analysis found that associated EC risk increased with higher fat energy intake [9]. Demonstrating a numerical association, Zhao et al. showed that with every 10% kcal increase in total fat intake was a 5% increased EC risk [10]. Saturated fat intake was even more costly, showing a 17% increase in EC risk for every 10% increase in the proportion of the total fat intake [10]. The fat-derived source has also shown associated risk to EC. Animal-derived nutrients and polyunsaturated fatty acids were both

TABLE 1    Principles of the Mediterranean diet (Mayo clinic: "Mediterranean diet: a heart-healthy eating plan").

**The Mediterranean diet**

- Daily consumption of fruits, vegetables, whole grains, and healthy fats
- Weekly intake of fish, poultry, beans, and eggs
- Moderate portions of dairy with limited red meat intake

*Emphasis on plant-based*
- Foundation of vegetables, fruits, herbs, nuts, beans, and whole grains

*Healthy fats*
- Olive oil provides monounsaturated fat, different from less healthy fats such as saturated or trans fats
- Fish are rich in omega-3 fatty acids

*Wine*
- Intake in moderation

associated with a significantly higher EC risk among the cohort of obese women [11]. As fatty macronutrients are the most densely packed caloric units, it is clear that fat intake reductions are essential to reducing gynecologic cancer risk.

Other dietary components like red meats, phytoestrogens, and vitamin D have been studied to unveil their links to gynecologic cancers as well. Red meat has been linked to carcinogenic potential in various cancer subtypes, including increased EC risk [12]. Phytoestrogens are another studied entity of various foods that contain the potential to exert effects on human tissue. These nonsteroidal compounds are structurally similar to endogenous estrogens and can thus exert both estrogenic and antiestrogenic properties [13]. Primarily found in soy

FIG. 1    Representative elements of the Mediterranean diet.

products, beans, and some grains [14], phytoestrogen studies regarding EC risk are contradictory. The majority of studies, however, show no increased risk of EC with phytoestrogen consumption [14]. Studies with various vitamin D supplementation protocols, thought to alter EC risk, have also shown no benefit to EC risk reduction [15]. Overall, these more minute dietary components have not been as thoroughly studied, warranting further research.

Two commonly consumed beverages, caffeine and alcohol, and their effects on endometrial cancer risk have been thoroughly studied as well. Of all caffeine-related beverages, green tea has been shown to reduce EC risk, but black tea has not shown any association [16]. Increased coffee consumption has also been linked to decreased EC risk. One study found that drinking more than four cups of coffee per day conferred a 25% EC risk reduction compared to individuals who drank less than one cup per day [17]. The metaanalysis found a similar dose-responsive relationship and overall reduced risk of EC with coffee consumption [18]. One study found liquor intake to confer a slightly increased risk of EC, but no association with beer or wine [19]. However, a large metaanalysis found no association between alcohol consumption and EC risk [19], so the true effects of alcohol and EC risk remain to be studied further. An Italian study found that in women with endometrial cancer, the specific proanthocyanidin component in wine contributed to a reduced EC risk, particularly in normal-weight women [20].

Lastly, chemicals related to the food processing industry have been under investigation for their potential roles in cancer forming mechanisms as well. One such compound is acrylamide, commonly found in potato products and cookies. Acrylamide forms in food products that rely on high-temperature cooking, providing the necessary ingredients for the Maillard-Browning reaction [21]. Animal studies show that acrylamide is a probable Class 2A human carcinogen in hormone-sensitive organs, like the uterine endometrium [3,22]. A few large-cohort studies have currently shown no association between dietary acrylamide and *overall* EC risk, but statistically, the significant risk was shown for nonusers of oral contraceptives (OCPs) and never-smokers and their risk of EC [23, 24]. With further research into this dietary compound, it likely plays a contributing biochemical role in the development of EC.

Many important dietary factors play a role in endometrial cancer risk reduction, most of which revolve around preventing the development of obesity. While nutritionists emphasize the introduction of fresh fruits and vegetables and whole grains into the diet, the Mediterranean diet is evidence-based and shows proven EC risk reduction. When planning meals, patients must prioritize foods with both a low glycemic index and glycemic load, as these have also been proven to increase EC risk. The avoidance of high-fat foods, foods with a particularly high saturated fat content, and animal-derived fats should also be avoided to decrease the risk of EC. Red meat intake should be decreased, if not eliminated from the diet completely, and products containing acrylamide should also be avoided out of an abundance of precaution. This literature review on various dietary components that play roles in EC and risk reduction should provide a foundation on which providers counsel their patients.

## Cervical cancer

The impact of patient nutrition has been most thoroughly investigated in cervical cancer patients. Cervical cancer manifests through persistent infection with specific high-risk human papillomavirus (HPV) genotypes. These HPV genotypes lead to the development, maintenance, and progression of cervical intraepithelial neoplasia (CIN). Sixty percent of preinvasive intraepithelial lesions from HPV infection regress spontaneously [25] due to factors in the patient's environment, immune system, and overall lifestyle. For patients with HPV exposure or lesions on the CIN spectrum, it is important to know what dietary changes have proven impact on better outcomes.

Important nutritional studies have focused on diet and its effects on HPV risk and subsequent development of CIN. The traditional Western diet, characterized by a high intake of red and processed meats, sauces, snacks, and low amounts of olive oil was associated with higher HPV infection rates [26]. A specific investigation found that oleic acid, commonly found in animal fats, exerts a stimulatory effect on cervical cancer growth and thus should be avoided [27]. These components lead to increased inflammation, decreased infection control, and increased risk of developing autoimmune diseases [28]. In contrast, individuals who adhere to the Mediterranean diet high in vegetables, legumes, fruits, nuts, cereals, fish, and higher levels of unsaturated versus saturated fats have shown a decreased risk of HPV infection [26]. Particular investigation into high vegetable consumption in one study was associated with a 54% reduction in risk of HPV persistence [29] and increased fruit intake showed an inverse association to intraepithelial squamous cancer [30].

It is the fundamental molecules in the Mediterranean diet that confer the nutritional benefit at the cellular level. For example, antioxidant vitamins act as scavengers of free radicals to prevent DNA damage [31]. Vitamins A, C, and E have also been shown to prevent the proliferation of cancer cells [32,33], stabilize the p53 protein involved in many cancer development cascades [26, 34]. Interestingly, papaya consumption was studied and found to reduce the incidence of HPV persistence. Through the carotenoids, or pro-vitamin A compounds, and vitamin C levels in this fruit, consumption decreased HPV persistence in females [35]. Papaya showed a reduced risk of squamous intraepithelial lesion and carotenoids showed a marginally positive protective impact against these lesions as well [36, 37]. Another study reported carotenoids to be associated with a 50%–63% reduced risk of HPV persistence [38] and an overall reduction in cervical cancer risk [39, 40]. Studies on vitamin A deficiency confirm its importance and showed an increased risk of developing CIN 1[36]. Vitamin A and C compounds are important protective factors in reducing the risk of cervical cancer. Further investigation into other members of this tropical, citrus fruit family could yield similar, valuable information.

CIN 2 and 3 are more advanced precancerous lesions that mark the tipping point between abnormal tissue and invasive cancer, and nutritional interventions are equally as important in this population. Studies of women who had a lower intake of fruits and vegetables reported a higher HPV viral load and higher risk of developing CIN 2 and 3 [41]. This stresses the need for promoting particular dietary changes in at-risk populations. Another such intervention is daily multivitamin administration. Women who took a multivitamin every day have shown a decreased risk for developing higher risk CIN [42]. Case studies have also identified low folate levels as a risk factor for developing all types of CIN [43]. Folate is an important micronutrient found in multivitamins that is essential for DNA synthesis, repair, methylation, and ultimately cell proliferation [44]. The use of multivitamins is supported in other research work as well. One study found that the risk of developing CIN 3 is reduced with adequate serum levels of vitamins A and E [45, 46]. It is important to note that these studies support the importance of multivitamins in preventing high-grade CIN, but compounding this data with the previously reviewed studies suggests multivitamin use is important throughout the prevention of the entire HPV-CIN pathway.

Another realm for nutritional discovery lies in the effects of diet on actively diagnosed patients and patients undergoing various chemotherapy and radiation treatments. One study focusing on the effects of green tea consumption found that properties within the tea were capable of enhancing the therapeutic effects of bleomycin [47]. Another study found that green tea is capable of chemo-sensitizing cervical cancer cells to cisplatin, a commonly used cervical cancer agent. Compounds in green tea are said to enhance cytotoxicity and induce apoptosis of cancer cells as an additive effect with cisplatin administration. This is exciting research and begs for a greater emphasis on exploring the dietary impact of various nutritional components on currently accepted treatment regimens.

## Ovarian cancer

Ovarian cancer risk is increased by many of the same nutritional factors as endometrial and cervical cancers. Similar to the EC risk described earlier, studies found that a high glycemic load was also associated with increased ovarian cancer risk [5]. Specific diets have not been reported to have particular protective effects against ovarian cancer, but multiple dietary components have been identified as impacting ovarian cancer risk. As in other female pelvic cancers, red meat and processed meats are associated with higher rates of ovarian cancer as well [48–50]. The American Association of Retired Persons (AARP) Diet and Health study reported an increased risk of ovarian cancer from increased dietary fat intake, particularly from animal sources [51]. This notion is supported in other studies that showed increasing ovarian cancer risk with increased saturated fat intake [52]. High-fat diets, especially animal-derived fats, are considered pro-inflammatory diets which have been studied across Italian and US populations, showing increased ovarian cancer risk [53, 54]. In females who prioritized poultry and fish, traditionally low-fat animal protein sources, they had a 25%–75% reduction in ovarian cancer risk [49, 50, 55]. These studies show the importance of avoiding high-fat diets and choosing fish and lean poultry as dietary protein sources to reduce ovarian cancer risk.

A variety of other foods and micronutrients have been shown to have effects on ovarian cancer risk and survival. Interestingly, a Danish study found that the consumption of milk and yogurt was associated with a higher risk of developing ovarian cancer, and an American study concluded supporting evidence by showing how whole milk and lactose consumption were also associated with a higher risk of developing ovarian cancer [56]. As dairy products are our primary source of calcium, it is important to note that increased calcium intake is associated with reduced ovarian cancer risk [57], so calcium supplementation remains crucial if various yogurt and milk products are removed from one's diet (Fig. 2). Dietary calcium sources are listed in Table 2. In this study, vitamin D supplementation was not associated with decreased ovarian cancer risk, but another study did show that females who had longer periods of sun exposure during the summer months were associated with a decreased ovarian cancer risk, likely due to vitamin D activation [57, 58]. The greatest risk reduction here was found to be with serous borderline and mucinous tumors. Multivitamins and individual

FIG. 2   Representative calcium (Ca)-rich foods.

TABLE 2   Dietary calcium sources (U.S. Department of Agriculture, Agricultural Research Service: "Food Data Central, 2019"; Cleveland Clinic Health Library Online, 2019: "Increasing Calcium in Your Diet"; NIH Office of Dietary Supplements online: "Calcium: Fact sheet for Health Professionals").

| Food group | Foods |
| --- | --- |
| Dairy | Yogurt, milk, cheese, soymilk, cottage cheese |
| Protein | Sardines, canned fish with bones |
| Vegetables and fruits | Kale, broccoli, mustard greens, collard and turnip greens, spinach, orange juice, tofu |
| Nuts, seeds, and legumes | Almonds, chia seeds, sesame seeds, beans |
| Grains | Fortified breads, fortified cereals |

vitamin level analyses including B vitamins and folate have not been shown to decrease ovarian cancer risk [59, 60].

While phytoestrogens and dietary acrylamide showed slightly positive risk association with cervical cancer patients, more research is needed to determine the risk with ovarian cancer. The only study between dietary acrylamide and ovarian cancer showed no increased risk with consumption [61]. There is also scant research around phytoestrogen and its potential effects on ovarian cancer risk. One small study found an overall reduced risk of ovarian cancer in patients who consumed foods containing phytoestrogens, but the results were deemed statistically insignificant [62] while others found no association [61, 63] (Fig. 2).

Various dietary components have also been studied in patients with active ovarian cancer diagnoses, patients enrolled in treatment, and patients with recurrent cancers. Interestingly, a survivorship study for patients with active ovarian cancer made connections to dietary intake. This Australian study reported a reduced risk of dying for women who were in the highest tertile of vegetable intake. They concluded no association between protein, carbohydrate, or fat intake analysis and increased survival. Supporting previous data on dairy intake and increased ovarian cancer risk, this study did see an increased risk of dying for women in the highest tertile of dairy intake, suggesting that reductions in dairy intake could be important in patients with active ovarian cancer [63–65]. The consumption of green tea was studied across the spectrum of ovarian cancer patients and found to decrease the risk of ovarian cancer recurrence [4]. The polyphenol compounds in tea were then investigated in actively treated patients and one study found that tea can also enhance the therapeutic properties of bleomycin [47]. Much research has been done to create preventive dietary strategies and these discoveries reveal the potential impact of these studies on dietary interventions for currently diagnosed patients as well.

## Risk factors

Another important approach to gynecologic cancers includes accounting for risk factors for the disease. Some of these risk factors take shape in the form of genetic predisposition through complex factors like race and other physiologic changes specific to a person. Many risk factors are modifiable components to patients' lives, though there are even things we can do for patients with an unchangeable genetic predisposition. Obesity is one of the most well-known risk factors for endometrial cancer, so addressing a patient's comorbid obesity is arguably as important to address in consultation. Smoking is another dangerous habit that confers risk to women for gynecologic cancers. In this section, we explore various lifestyle factors that have been shown to add or reduce the risk of cancers of the female reproductive organs.

### Obesity

While unhealthy foods and sedentary lifestyles become more common in today's society, obesity has become increasingly prevalent in today's modern world. Obesity increases stress on many different systems of human physiology and is ultimately linked to higher rates of many different types of cancers. One study showed that over the last decade, over 25% of all cancer incidences across the population could have been prevented if the rate of obesity had not increased [66]. The Centers for Disease Control (CDC) reported that between 2017 and 2018, the incidence rate of obesity in America was 42.4% [67], a quite astonishing yet equally alarming statistic about the United States' population. Obesity also happens to be the strongest known risk factor for endometrial cancer [68]. One study reported that almost half of all new EC diagnoses today are attributable to obesity alone [69]. Compared to their nonobese counterparts, obese women are at 2.4–2.5 times the risk of being diagnosed with EC [70]. Even studies that have adjusted risk scores for other known risk factors of EC have still found an obese woman with a BMI greater than $40 \text{kg/m}^2$ to have seven times the odds of developing EC compared to a normal weight woman [71]. This risk is not just associated with older age, either, as some studies have found that women in their 20s with an increased BMI greater than $25 \text{kg/m}^2$ are at an increased risk of EC as well [72].

So why is obesity so dangerous to females concerning EC? It is well known to today's gynecologists that excess adiposity is a breeding ground for continuous androgen conversion to estradiol, which stimulates endometrial proliferation. This excess estradiol in the setting of endometrial proliferation leads to hyperplasia, a precursor lesion in the cancer cascade, setting the stage for eventual cancer development [73]. Obesity and low levels of physical exercise are also associated with constant, low-grade inflammation, which adds to cancer risk in all tissues of the body [74].

Obesity is also clearly linked to an associated diagnosis of diabetes mellitus, a comorbid condition that increases the risk of obesity and thus, cancer. Diabetes and insulin resistance are both highly associated with obesity and have independently been implicated in EC pathogenesis [75]. In type II diabetes, insulin resistance leads to elevated basal insulin levels, which decreases serum levels of sex hormone-binding globulin (SHBG). This decreased SHBG level decreases the capacity for binding excess estradiol, thus increasing circulating estradiol levels even further and decreasing the body's natural balancing mechanism to account for the excess estrogen produced by adipose aromatase [76–78]. At the cellular level, insulin receptors are present on each type of endometrial cancer cell, so it has also been postulated that insulin itself can increase endometrial proliferation by augmenting the effects of insulin-like growth factors and elevated circulating insulin levels [7, 79]. There exists data to support this conclusion as well. Two large epidemiologic studies on diabetic females showed a two- to threefold increased risk of EC compared to their nondiabetic counterparts [80, 81]. As health experts magnified diabetic and concurrently obese females in these studies, these women were at a sixfold increased risk of EC, and a ninefold increased risk for these women who were not physically active [82]. One study even found a significant positive association between EC risk in type I diabetic patients [81].

Obesity is a clear and prevalent risk factor that providers must target with their patients to combat EC risk. Obesity likely puts patients at an increased risk for other types of gynecologic cancers as well, but these associations have not been as clearly revealed. In the typical cervical cancer patient, other risk factors seem to be the major players in cancer development. One study of ovarian cancer epidemiology reported a modest risk and an increased risk of dying associated with obesity for the 5-years before ovarian cancer diagnosis [83]. This study also acknowledged the careful consideration of chemotherapy dosing in obese patients and how standard-dosing protocols may lead to suboptimal chemotherapy dosing in larger patients.

## Smoking

Smoking has known carcinogenic activity in a multitude of organs, especially in those in direct contact with these inhaled tobacco products, like the oral cavity and the lungs. Smoking's harmful effects reach all organ systems of the body, including the gynecologic tract. Smoking is most well-known for its association in the early stages of cervical cancer development. Smoking plays a role in carcinogenesis from the acquisition of HPV infection to the development of cancer in the cervix. The risk of cervical cancer in smokers is twofold higher compared to women who do not smoke [84]. Smoking in the setting of ovarian cancer has conflicting data. One study showed that smoking at the time of ovarian cancer diagnosis or within 1-year afterward was associated with a higher risk of dying [85], but a Swedish study did not find any significant associations [60].

Interestingly, cigarette smoking has a positive effect on EC cancer risk. Cigarette smoking exerts antiestrogenic effects on the body through various mechanisms. Smoking decreases appetite and increases weight loss. Daily smoking habits lead to an earlier induction of menopause, among other hormonal alterations, leading to decreased circulating levels of estradiol [86]. Large metaanalyses have shown a reduced risk of EC in women who had ever smoked in their lifetime both independent of BMI and age of menopause [87]. Studies also showed that smoking 20 cigarettes per day was associated with a 16% reduced risk of uterine-cervix cancer (UCC) [86].

## Excess estrogen

Increased levels of serum estradiol and its different derivatives are a very well-known risk factor for EC. This well-studied mechanism, as described earlier, where prolonged estrogen exposure stimulates endometrial cell proliferation, ultimately increases the occurrence and accumulation of cellular mutations [88]. Thus, pathophysiologically elevated estradiol levels become problematic, but iatrogenic circumstances that result in increased serum estradiol also increase the risk for EC. Epidemiologic data looking at users of estrogen-only OCPs and estrogen-containing hormone replacement therapy showed a significantly increased risk of EC [89, 90]. This epidemiologic data is backed by in vitro lab studies as well, where endometrial cells were maximally stimulated by estrogen and showed significant hyperplasia and genetic mutations [91].

We see increased endogenous estrogen in other medical conditions as well. Premenopausal women with anovulatory disorders or with progesterone deficiency, postmenopausal women with estrogen excess, and conditions of androgen excess leading to peripheral aromatization to estrogen are all various scenarios where patients are at increased risk of EC from increased estrogen exposure [92, 93]. Anovulation removes the cyclic production of progesterone and appropriate endometrial shedding and places a woman in a setting of prolonged estrogen exposure [94]. Polycystic ovarian syndrome is one of those conditions where anovulation combines with increased aromatization of androgens to estradiol in excess adipose tissue, thereby increasing a patient's risk of developing EC via endogenous estrogen excess [95]. Another interesting case of endogenous estrogen excess is a sex cord-stromal tumor, specifically a granulosa or theca cell tumor. These estrogen and androgen-secreting tumors lead to prolonged exposure to tumor-derived estradiol, also resulting in hyperplasia and EC risk [96, 97].

It makes sense also that exogenous estrogen, or estrogen coming from an outside source, also increases the risk of developing EC. Historically and even today, postmenopausal women often undergo an unopposed estrogen therapy regimen to treat symptoms of menopause in the form of hormone replacement therapy (HRT). In these cases, endometrial hyperplasia naturally occurs. The postmenopausal estrogen/progestin interventions (PEPI) trial was the foundational study that demonstrated the deleterious effects of unopposed estrogen as women in this study assigned to estrogen-only OCPs were more likely than patients in the placebo group to develop hyperplasia of all grades [98]. One study showed that some 20% of women taking HRT without progesterone developed endometrial hyperplasia compared to a mere 1% who developed endometrial hyperplasia when HRT therapy included progesterone [99]. Peer-reviewed case studies quote the relative risk as high as 12 times for women who have ever used unopposed estrogen therapy and a relative risk of up to 15-fold in long-term HRT users [100,101]. With the addition of progesterone, one large randomized control trial found that the use of combined HRT with progesterone did not increase the rate of hyperplasia or EC [102]. For this reason, the addition of progesterone is extremely important in protecting the endometrium against EC.

## Fertility associated factors: Age at menarche, age at menopause, parity, OCPs

The unopposed estrogen theory predicts that any factor that prolongs the duration of estrogen exposure in a patient confers a carcinogenic effect. Building off of this theory and the reviewed outcomes of HRT, factors that prolong a woman's menstrual lifecycle should confer risk. In cases when menarche comes significantly earlier than compared to a patient's counterparts, a woman's ovulatory timespan is extended. Two large studies support this notion that EC risk is increased by early-onset menstruation and found that EC occurrence was associated with early menarche and late menopause. In the same way, one study looked at women who experienced menarche later and described decreased EC risk in those patients [103, 104]. There is inconsistent data to show whether or not age at menarche affects ovarian cancer risk. A Norwegian study found a worse prognosis in female patients who reports a later average age at menarche around 14–15 years, but a population-based control study also found a worse prognosis in women with a younger age at menarche [60, 105–107].

Applying the theory of unopposed estrogen in reverse, factors that decrease exposure to cyclic estrogen rise monthly should be protective against the development of EC. One protective factor is parity, which represents the number of pregnancies a woman has carried. Increased parity is protective against EC development due to the effects of elevated progesterone and extended periods where the endometrium is not cyclically stimulated. This is proven by a large metaanalysis of 171 publications that reported a 40% reduction in EC incidence among parous compared to nulliparous women [108, 109]. In addition, the later age of final pregnancy is also protective, showing decreased rates of EC. Large randomized-control trials and pooled analyses studied women who birthed a child after the age of 40 years and showed that these women were at a 44%–49% decreased risk of developing EC compared to those who birthed their last child before the age of 29 years [103, 108, 110]. Increasing parity also increases periods during a woman's life where she is breastfeeding. Breastfeeding suppresses ovulation and decreases estrogen levels, and these benefits offer a risk reduction of up to 11% for the development of EC in those who have ever breastfed [111].

The use of OCPs is an important modulator in female gynecologic cancers as well. Since their inception in the 1960s, over 300 million women have used OCPs, with 100–150 million women currently using them [112,113]. With systemic, exogenous hormone administration via OCPs, the hypothalamic-pituitary-ovarian axis is diverted and regulated at a lower serum concentration of hormone levels. In addition, OCPs increase SHBGs, which further decreases serum hormone circulation of free estrogen. In this way, combined OCPs confer longer-term protection against EC. Studies have shown risk reduction after 10-years of use from 30% to 80% and were shown to have reduced the incidence rate by 10% [113]. As far as oral or injectable progesterone-only contraceptives, little is known about the impact on EC risk [114]. In a study published in 2015, OCPs were estimated to have prevented over 400,000 cases of EC, including 100,000 deaths in just the past 50 years [115, 116]. This epidemiologic data also addresses the protective effect of OCPs on ovarian cancer risk.

These same fertility-associated factors have been studied in ovarian cancer risk as well. With regards to OCP use and related ovarian cancer risk, most studies have found a small, nonsignificant decrease in the risk of death compared to nonusers of OCPs. These studies showed that OCP users were more likely to get tumors with a better prognosis and an overall decreased risk of aggressiveness. However, a European study showed that OCP uses greater than 10-years increased the patients' risk of dying from ovarian cancer, so the association may be contingent upon the length of OCP use [115]. When looking at age and parity, the results stand conflicted. Some studies show a worse survival in ovarian cancer for women who have their first child at an older age [105, 117], while others show no difference [60, 107]. Breastfeeding history in patients with ovarian cancer was also unclear, as one study found an association with longer survival [107], while others did not.

## Other pharmacotherapy

Some drugs have the potential to impact gynecologic cancer risk. The most well-known of these include the selective estrogen-receptor modifiers (SERMs), which are nonsteroidal drugs with estrogen agonist and antagonist properties in different tissues. Tamoxifen is primarily used in estrogen receptor (ER)-positive breast cancer to inhibit growth at the estrogen receptors in breast tissue. Raloxifene is used to prevent osteoporosis as it has agonistic properties at the estrogen receptors in bone tissue. Some of these SERM drugs, however, have agonistic properties at the endometrium, namely Tamoxifen. These drugs have demonstrated significant increases in the risk of endometrial hyperplasia and carcinoma with their use. Notably, raloxifene is the safest option as it exerts antagonistic properties on the estrogen receptor in the endometrial tissue, and studies have confirmed that it reduces the risk of EC compared to Tamoxifen or placebo [118, 119]. Aromatase inhibitors are another class of drugs used in breast cancer patients. These drugs are responsible for blocking the aromatic transformation of androgens into estrogen, which can be used to treat breast cancer, endometriosis, ovulatory dysfunction, and other estrogen-modulating conditions. These drugs also show a decreased risk of EC compared to tamoxifen. However, as they prevent the conversion of estrogen in all tissues, they can exert other harmful effects like decreased bone density [120].

Nonsteroidal antiinflammatory drugs (NSAIDs) were also studied in EC risk. As NSAID drugs mediate chronic inflammation from carcinogenic properties, it was theorized that antiinflammatory medications might play a therapeutic role in cytokine release and subsequent DNA-damaging reactive oxygen species (ROS)[121]. One metaanalysis found that daily aspirin administration was only found to decrease EC risk in obese women, and no other NSAID medication was found to have a positive impact [122]. This small body of peer-reviewed literature represents an area where more high-quality research must be done to recommend NSAIDs as a therapeutic agent.

## Race and socioeconomic status

Race and socioeconomic status have been studied extensively in the setting of female gynecologic cancers and play a large, complex role in conferring risk factors to patients who identify with certain racial and ethnic groups. African American women are adversely affected by uterine cancer when compared to their white counterparts. African American women with EC have an 80% higher mortality rate than white women [123], and a significantly lower survival rate as well. One study reports that white women have an 86% survival rate at the 5-year interval, whereas black women have only a 64% survival rate [124]. The reasons behind this are likely multifactorial and all related to the unequal distribution of resources to people in lower socioeconomic status, which largely includes people of color, as well as factors like diversity and trust in the general community of medical providers. This cohort of women is also more likely to be overweight or obese [125], which increases the risk of diabetes and impaired quality-of-life compared to their white counterparts [126].

Black women have a considerably lower survival rate with ovarian cancer as well. These striking statistics probe the question of inherent biological differences or whether there are socioeconomic factors that confer this increased risk [127]. Multiple studies have addressed this question by adjusting for socioeconomic status and still show increased rates of EC and ovarian cancer for black patients [83]. Although EC has a slightly lower incidence in black women, it has a significantly higher mortality rate than that in whites (9.0 versus 4.6 deaths per 100,000), and survival is substantially lower at every stage of diagnosis [128]. For any person with economic hardship, studies also have outlined lower ovarian cancer survival rates [127].

Black women are also twice as likely to die from EC as women from any other racial/ethnic group. Multiple reasons exist including a higher incidence of aggressive histologic subtypes, idiosyncratic patterns of gene expression, failure to access standard of care healthcare services, and increased incidence of comorbidities [128]. From a treatment perspective, some Gynecologic Oncology Group (GOG) trials found that blacks were less likely to experience complete or partial responses to chemotherapy [125]. Some even venture as far as to question if they are less represented in federally funded clinical trials, a reality that could account for reduced effectiveness of standard treatment regimens [129]. Specific genetic studies have found that pathologic human epidermal growth factor receptor 2 (HER-2)/neu gene amplification in uterine papillary serous carcinoma (UPSC) occurs more often in blacks than whites [130], and p53 mutations are also seen 2–3 times more in blacks compared to whites [131]. However, beyond genetic factors, black women are generally worse surgical candidates due to increased comorbidities like an increased BMI or Type II diabetes [132]. These women then become less likely to undergo cytoreductive or therapeutic surgeries, and factors like reduced access to care and potential for discriminatory practices by surgeons further that instance of suboptimal care [133]. This issue of race and socioeconomic status will always be an area of investigation and one that should be at the forefront of providers' minds when treating diverse gynecologic oncology patients.

## Genetic syndromes

Many well-studied genes and genetic mutations affect various gynecologic cancer risks and have been described in the peer-reviewed literature. Genetic syndromes such as Lynch Syndrome, Hereditary Nonpolyposis Colorectal Cancer, and Cowen Syndrome are known risk factors that predispose patients to gynecologic cancers. Lynch Syndrome is an autosomal dominant condition characterized by an increased susceptibility to colorectal, endometrial, and several other cancers, including ovarian as well [134]. Patients with Lynch Syndrome have defects in the mismatch repair genes *ML1, MSH2, MSH6,* and *PMS2* which result in unrepaired nucleotide repeats throughout the genome, also known as microsatellite instability [134, 135]. However, the risk of EC is 3% in the general female population, one with Lynch Syndrome has a 40%–60% cumulative lifetime risk of EC [134]. Cowden syndrome is another genetically determined condition that increases the risk of EC, among other cancers as well. Cowden Syndrome has an autosomal dominant inheritance pattern and is characterized mainly by hamartomatous growths [136]. The lifetime risk of EC is 12.5%–19% for patients with Cowden syndrome [137].

The *BRCA1* and *BRCA2* genes have also been clearly outlined in the peer-reviewed literature to affect both EC and ovarian cancer risk. The *BRCA1* and *BRCA2* genes are widely known to confer a greatly increased risk of breast and ovarian cancers but have also been associated with various types of EC, most notably the papillary-serous histologic subtype [138]. A large randomized control trial reported a significantly elevated risk of EC in women with BRCA1 mutations specifically, but this report notes that further investigation is warranted to further clarify the association between EC and *BRCA* genes mutation [139], as other studies report conflicting evidence between *BRCA* mutations and EC risk [140, 141].

## Lifestyle

For many patients, the concept of a healthy lifestyle is where nutrition, exercise, and risk factors merge into one ultimate plan that generally directs life choices. Based on a predetermined genetic profile and current daily routine, what lifestyle changes can patients make to positively impact their future health? Research shows that many patients are motivated to make lifestyle changes after their cancer diagnosis, and likely many are before as well [142]. This makes sense because a diagnosis of gynecologic cancer is life-changing and warrants life-changing lifestyle adjustments. Research also shows that lifestyle changes and pharmacologic interventions reduce cancer incidence [143]. When we combine these two forces, we can take the foundational research from this chapter and apply it to promote healthy weight maintenance, exercise habits, and lifestyle choices. Engaging in behaviors such as physical activity, avoiding tobacco use, eating a healthy diet, and learning stress-reduction techniques offer control over one's health to lessen disease recurrence and progression [144].

Exercise alone is a powerful lifestyle factor that can improve a woman's risk for female gynecologic cancers. From a biochemical perspective, exercise modulates the response to inflammation and the immune response and makes hormonal alterations that produce antitumorigenic results [144, 145]. Additionally, myokines, or molecules released during muscle contraction during physical activity, have also been shown to regulate immune cells and play roles in cell migration and proliferation [146]. Applying this biochemical understanding with the known pathogenesis in EC, studies have shown a 20%–30% reduction in EC risk in women who reported moderate- or high-intensity exercise [147]. This makes sense as exercise addresses elevated BMI and obesity, a strong precursor in the pathogenesis of EC. Another study supports this notion and found that women who reported high-intensity workouts found a 35% EC risk reduction [148]. It is not only vigorous, high-intensity exercise that provides benefit, but even daily walking reduces EC risk [148] and the 2018 World Cancer Research Fund Network's Continuous Update Project found risk reduction irrespective of the type of physical activity [2]. When these studies looked across the spectrum of women enrolled in the studies, they also found that the greatest risk reduction comes for women who are in the highest tertiles of obesity, meaning the women at highest risk of EC may glean the most benefit from exercise [149].

Exercise is truly an area for great lifestyle change for these patients. A study in 2016 showed that only 19% of women met national physical activity guidelines, meaning roughly 80% of the female population is missing out on a key risk-reducing lifestyle change impacting gynecologic cancer risk [150]. While the well-known association between obesity and EC risk makes endometrial cancer patients the most probable beneficiaries of this lifestyle change, decreased risk of cervical cancer has been found as well. A Korean study found a decreased risk of cervical cancer in patients who reported regular exercise [151] but lacked evidence for patients already with a cervical cancer diagnosis. Exercise seems to also confer benefit for women with ovarian cancer as well, both in obese and nonobese women ovarian cancer patients [152]. Another study showed a higher quality-of-life reported in ovarian cancer patients who adhered to an exercise program, the majority of which were just brisk walking [153]. At the end of this data review, the patients were also found to have a higher overall survival with their ovarian cancer. This report shows how important the integration of exercise into the daily life of all female patients is in the prevention of gynecologic malignancies.

Exercise has also been shown to provide benefits to patients undergoing cancer treatment. Physical activity has a beneficial effect on physical fitness and quality-of-life metrics in patients with gynecologic cancers who are actively undergoing treatment, and studies show that exercise is well-tolerated in these patients [154]. Studies show that exercise is not only safe, but it reduces symptoms of treatment and disease. Fatigue is one of the most frequently reported side effects of various chemotherapy, radiation therapy, and postsurgical therapy schemes, and exercise is proven to reduce fatigue and the burden of fatigue-like symptoms [155]. Unfortunately, however, as few as 20% of oncologists make recommendations to patients to increase exercise and physical activity to prevent disease and lessen side-effect profiles. This is a great area for improvement for oncologists across the globe, especially as patients have shown that they increase their physical activity in response to the recommendations of their oncologists [156]. When patients who increased their level of physical exercise after a diagnosis of gynecologic cancer reported greater perceived quality of life [157], it becomes evident how important it is for providers to advocate for the necessary lifestyle changes for patients. All of the evidence supports a positive relationship between physical activity and gynecologic cancer in the prevention and treatment of gynecologic malignancy [158].

So, what can providers who are not well versed in exercise recommendations provide to patients? In 2010, the American College of Sports Medicine (ACSM) made physical activity and exercise guidelines with specific considerations for cancer survivors and with particular addendums for people with chronic conditions [159], with the ultimate goal of avoiding inactivity altogether (see Table 3). Table 3 was primarily produced by the ACSM and offers exercise prescriptions that are evidence-based in targeting various medical problems encountered with oncology patients. Referral to a primary care provider for evaluation and medical assessment of significant side effects from chemotherapy, radiation therapy, and postsurgical changes may be necessary. Patients may have significant neuropathies, musculoskeletal abnormalities, new fracture risk from bony metastases, and new cardiac conditions, which legitimize a primary care appointment before the initiation of an exercise plan. Here, a primary care provider may work with a rehabilitation specialty team including physical therapists and occupational therapists who will best guide and formulate a safe exercise regimen for cancer patients and survivors.

TABLE 3   Exercise guidelines for patients with cancer, as originated adapted/endorsed by the American College of Sports Medicine (ASCM) [159] and https://carleysangels.ca/sweat-and-cancer/.

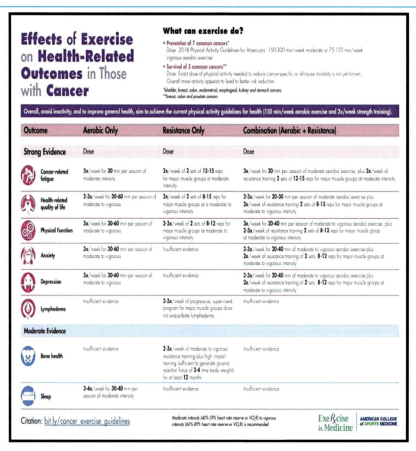

Not only does exercise offer a myriad of health benefits to patients regardless of their current health status, but it plays a vital role in the prevention, treatment, and survivorship of patients with gynecologic cancers. It is so vitally important that healthcare providers are aware of these benefits and advocate for their patients. It is important to encourage physical activity and patient buy-in in various exercise plans. The review of a patient's exercise plan should be a regular part of any gynecologic cancer patient's regular surveillance as it has been shown to impact risk, survival, outcomes, and overall quality of life during and after potentially harsh oncology treatment regimens.

## Conclusions and prospects

A patient's lifestyle is an intricate collaboration of dietary patterns, exercise habits, and particular individual risk factors. As a gynecologic oncologist, this relationship between a patient's diet, exercise, and risk factors becomes a personalized calculation of many different factors that pertain to cervical, uterine, and ovarian cancer risk in that patient. As we described in this chapter, the Mediterranean diet is evidence-based to have a positive impact on a patient's risk for developing gynecologic risk factors. We broke down the diet into individual components as well to assess each food category's risk of developing certain gynecologic cancers. Cervical cancer has a unique pathophysiologic development that has been addressed by multiple vitamins found in citrus fruits with high nutritional value. Endometrial cancer and its densely studied relationship with obesity were reviewed to stress the importance of a healthy diet and regular exercise. Various dietary elements with proposed carcinogenic properties were also addressed, like acrylamide, red meats, and alcohol, as well as overall caloric intake and macronutrient balance. This review of nutrition and lifestyle exercise interventions as it pertains to gynecologic oncology patients is important for providers to more accurately counsel patients on what happens next in their cancer diagnoses, and what patients can do to help decrease cancer risk.

This chapter also reviewed the myriad of various risk factors for gynecologic cancers. Some well-studied associations with medical conditions were reviewed, like diabetes and estrogen-related conditions. Conditions like diabetes and PCOS elevate serum estradiol levels and therefore increase EC risk. Other risk factors like a patient's parity, age at menopause, and patterns of OCP use of OCPs were also reviewed with their role in cancer risk. Another especially important area of investigation is the role of socioeconomic status and race on gynecologic cancer risk. Here further research and attention are warranted regarding health equity, access to care, and resource distribution for underrepresented communities. Finally, we reviewed the well-known genetic risk factors for gynecologic cancers, like Lynch Syndrome and BRCA testing.

After addressing nutritional habits and risk factors for gynecologic cancer, we reviewed the importance of lifestyle modifications—lifelong changes in one's daily patterns—and how those can alter the course of gynecologic cancer development. We probed topics like weight management, exercise regimens, and lifestyle choices, which can alter disease risk and progression. Lifestyle choices like the use of tobacco products, dietary choices, and stress management all play roles in a person's life. We reviewed the evidence-based literature which explains not only the preventative and protective effects of daily exercise but also the benefits of exercise during chemoradiation treatment protocols as well. Research shows that incorporating an exercise routine into a patient's life increases treatment efficacy and decreases disease burden. Exercising for gynecologic oncology patients even increases their quality-of-life ability to mitigate treatment side effects. We conclude this chapter by stressing the important responsibility gynecologic oncologists have in being able to advocate for and prescribe exercise plans for patients to ultimately improve patient outcomes.

# References

[1] Zeng X, Zhang Y, Kwong JS, Zhang C, Li S, Sun F, Niu Y, Du L. The methodological quality assessment tools for preclinical and clinical studies, systematic review and meta-analysis, and clinical practice guideline: a systematic review. J Evid Based Med 2015;8:2–10.

[2] Abu-Abid S, Szold A, Klausner J. Obesity and cancer. J Med 2002;33:73–86.

[3] Koshiyama M. The effects of the dietary and nutrient intake on gynecologic cancers. Healthcare (Basel) 2019;7(3):88.

[4] Acmaz G, Aksoy H, Albayrak E, Baser M, Ozyurt S, Aksoy U, Unal D. Evaluation of endometrial precancerous lesions in postmenopausal obese women—a high risk group? Asian Pac J Cancer Prev 2014;15:195–8.

[5] Mulholland HG, Murray LJ, Cardwell CR, Cantwell MM. Dietary glycaemic index, glycaemic load and endometrial and ovarian cancer risk: a systematic review and meta-analysis. Br J Cancer 2008;99:434–41.

[6] Neuhouser ML, Howard B, Lu J, Tinker LF, Van Horn L, Caan B, Rohan T, Stefanick ML, Thomson CA. A low-fat dietary pattern and risk of metabolic syndrome in postmenopausal women: the Women's Health Initiative. Metabolism 2012;61:1572–81.

[7] Hale GE, Hughes CL, Cline JM. Endometrial cancer: hormonal factors, the perimenopausal "window of risk," and isoflavones. J Clin Endocrinol Metab 2002;87:3–15.

[8] Filomeno M, Bosetti C, Bidoli E, Levi F, Serraino D, Montella M, La Vecchia C, Tavani A. Mediterranean diet and risk of endometrial cancer: a pooled analysis of three Italian case-control studies. Br J Cancer 2015; 112:1816–21.

[9] Chu KT, Song Y, Zhou JH. No effect of energy intake overall on risk of endometrial cancers: a meta-analysis. Asian Pac J Cancer Prev 2014;15:10293–8.

[10] Zhao J, Lyu C, Gao J, Du L, Shan B, Zhang H, Wang HY, Gao Y. Dietary fat intake and endometrial cancer risk: a dose response meta-analysis. Medicine (Baltimore) 2016;95, e4121.

[11] Bravi F, Bertuccio P, Turati F, Serraino D, Edefonti V, Dal Maso L, Decarli A, Montella M, Zucchetto A, La Vecchia C, Bosetti C, Ferraroni M. Nutrient-based dietary patterns and endometrial cancer risk: an Italian case-control study. Cancer Epidemiol 2015;39:66–72.

[12] Bandera EV, Williams MG, Sima C, Bayuga S, Pulick K, Wilcox H, Soslow R, Zauber AG, Olson SH. Phytoestrogen consumption and endometrial cancer risk: a population-based case-control study in New Jersey. Cancer Causes Control 2009;20:1117–27.

[13] Unfer V, Casini ML, Costabile L, Mignosa M, Gerli S, Di Renzo GC. Endometrial effects of long-term treatment with phytoestrogens: a randomized, double-blind, placebo-controlled study. Fertil Steril 2004;82:145–8 [quiz 265].

[14] Bandera EV, Kushi LH, Moore DF, Gifkins DM, McCullough ML. Consumption of animal foods and endometrial cancer risk: a systematic literature review and meta-analysis. Cancer Causes Control 2007;18:967–88.

[15] McCullough ML, Bandera EV, Moore DF, Kushi LH. Vitamin D and calcium intake in relation to risk of endometrial cancer: a systematic review of the literature. Prev Med 2008;46:298–302.

[16] Zhou Q, Li H, Zhou JG, Ma Y, Wu T, Ma H. Green tea, black tea consumption and risk of endometrial cancer: a systematic review and meta-analysis. Arch Gynecol Obstet 2016;293:143–55.

[17] Je Y, Hankinson SE, Tworoger SS, De Vivo I, Giovannucci E. A prospective cohort study of coffee consumption and risk of endometrial cancer over a 26-year follow-up. Cancer Epidemiol Biomarkers Prev 2011;20:2487–95.

[18] Zhou Q, Luo ML, Li H, Li M, Zhou JG. Coffee consumption and risk of endometrial cancer: a dose-response meta-analysis of prospective cohort studies. Sci Rep 2015;5:13410.

[19] Sun Q, Xu L, Zhou B, Wang Y, Jing Y, Wang B. Alcohol consumption and the risk of endometrial cancer: a meta-analysis. Asia Pac J Clin Nutr 2011;20:125–33.

[20] Rossi M, Edefonti V, Parpinel M, Lagiou P, Franchi M, Ferraroni M, Decarli A, Zucchetto A, Serraino D, Dal Maso L, Negri E, La Vecchia C. Proanthocyanidins and other flavonoids in relation to endometrial cancer risk: a case-control study in Italy. Br J Cancer 2013;109:1914–20.

[21] Stadler RH, Blank I, Varga N, Robert F, Hau J, Guy PA, Robert MC, Riediker S. Acrylamide from Maillard reaction products. Nature 2002;419:449–50.

[22] Friedman MA, Dulak LH, Stedham MA. A lifetime oncogenicity study in rats with acrylamide. Fundam Appl Toxicol 1995;27:95–105.

[23] Je Y. Dietary acrylamide intake and risk of endometrial cancer in prospective cohort studies. Arch Gynecol Obstet 2015;291:1395–401.

[24] Obon-Santacana M, Kaaks R, Slimani N, Lujan-Barroso L, Freisling H, Ferrari P, Dossus L, Chabbert-Buffet N, Baglietto L, Fortner RT, Boeing H, Tjonneland A, Olsen A, Overvad K, Menendez V, Molina-Montes E, Larranaga N, Chirlaque MD, Ardanaz E, Khaw KT, Wareham N, Travis RC, Lu Y, Merritt MA, Trichopoulou A, Benetou V, Trichopoulos D, Saieva C, Sieri S, Tumino R, Sacerdote C, Galasso R, Bueno-de-Mesquita HB, Wirfalt E, Ericson U, Idahl A, Ohlson N, Skeie G, Gram IT, Weiderpass E, Onland-Moret NC, Riboli E, Duell EJ. Dietary intake of acrylamide and endometrial cancer risk in the European prospective investigation into cancer and nutrition cohort. Br J Cancer 2014;111:987–97.

[25] Schiffman M, Kjaer SK. Chapter 2: natural history of anogenital human papillomavirus infection and neoplasia. J Natl Cancer Inst Monogr 2003;14–9.

[26] Barchitta M, Maugeri A, Quattrocchi A, Agrifoglio O, Scalisi A, Agodi A. The association of dietary patterns with high-risk human papillomavirus infection and cervical cancer: a cross-sectional study in Italy. Nutrients 2018;10(4):469.

[27] Yang P, Su C, Luo X, Zeng H, Zhao L, Wei L, Zhang X, Varghese Z, Moorhead JF, Chen Y, Ruan XZ. Dietary oleic acid-induced CD36 promotes cervical cancer cell growth and metastasis via up-regulation Src/ERK pathway. Cancer Lett 2018;438:76–85.

[28] Myles IA. Fast food fever: reviewing the impacts of the Western diet on immunity. Nutr J 2014;13:61.

[29] Sedjo RL, Roe DJ, Abrahamsen M, Harris RB, Craft N, Baldwin S, Giuliano AR. Vitamin A, carotenoids, and risk of persistent oncogenic human papillomavirus infection. Cancer Epidemiol Biomarkers Prev 2002;11:876–84.

[30] Gonzalez CA, Travier N, Lujan-Barroso L, Castellsague X, Bosch FX, Roura E, Bueno-de-Mesquita HB, Palli D, Boeing H, Pala V, Sacerdote C, Tumino R, Panico S, Manjer J, Dillner J, Hallmans G, Kjellberg L, Sanchez MJ, Altzibar JM, Barricarte A, Navarro C, Rodriguez L, Allen N, Key TJ, Kaaks R, Rohrmann S, Overvad K, Olsen A, Tjonneland A, Munk C, Kjaer SK, Peeters PH, van Duijnhoven FJ, Clavel-Chapelon F, Boutron-Ruault MC, Trichopoulou A, Benetou V, Naska A, Lund E, Engeset D, Skeie G, Franceschi S, Slimani N, Rinaldi S, Riboli E. Dietary factors and in situ and invasive cervical cancer risk in the European prospective investigation into cancer and nutrition study. Int J Cancer 2011;129:449–59.

[31] Salganik RI. The benefits and hazards of antioxidants: controlling apoptosis and other protective mechanisms in cancer patients and the human population. J Am Coll Nutr 2001;20:464S–72S [discussion 473S–475S].

[32] Borutinskaite VV, Navakauskiene R, Magnusson KE. Retinoic acid and histone deacetylase inhibitor BML-210 inhibit proliferation of human cervical cancer HeLa cells. Ann N Y Acad Sci 2006;1091:346–55.

[33] Garcia-Closas R, Castellsague X, Bosch X, Gonzalez CA. The role of diet and nutrition in cervical carcinogenesis: a review of recent evidence. Int J Cancer 2005;117:629–37.

[34] Reddy L, Odhav B, Bhoola KD. Natural products for cancer prevention: a global perspective. Pharmacol Ther 2003;99:1–13.

[35] Giuliano AR, Siegel EM, Roe DJ, Ferreira S, Baggio ML, Galan L, Duarte-Franco E, Villa LL, Rohan TE, Marshall JR, Franco EL, Ludwig-McGill HPVNHS. Dietary intake and risk of persistent human papillomavirus (HPV) infection: the Ludwig-McGill HPV Natural History Study. J Infect Dis 2003;188:1508–16.

[36] Yeo AS, Schiff MA, Montoya G, Masuk M, van Asselt-King L, Becker TM. Serum micronutrients and cervical dysplasia in Southwestern American Indian women. Nutr Cancer 2000;38:141–50.

[37] Siegel EM, Salemi JL, Villa LL, Ferenczy A, Franco EL, Giuliano AR. Dietary consumption of antioxidant nutrients and risk of incident cervical intraepithelial neoplasia. Gynecol Oncol 2010;118:289–94.

[38] Giuliano AR, Sedjo RL, Roe DJ, Harri R, Baldwi S, Papenfuss MR, Abrahamsen M, Inserra P. Clearance of oncogenic human papillomavirus (HPV) infection: effect of smoking (United States). Cancer Causes Control 2002;13:839–46.

[39] Zhang X, Dai B, Zhang B, Wang Z. Vitamin A and risk of cervical cancer: a meta-analysis. Gynecol Oncol 2012;124:366–73.

[40] Myung SK, Ju W, Kim SC, Kim H, Korean Meta-analysis (KORMA) Study Group. Vitamin or antioxidant intake (or serum level) and risk of cervical neoplasm: a meta-analysis. BJOG 2011;118:1285–91.

[41] Hwang JH, Kim MK, Lee JK. Dietary supplements reduce the risk of cervical intraepithelial neoplasia. Int J Gynecol Cancer 2010;20:398–403.

[42] Hwang JH, Lee JK, Kim TJ, Kim MK. The association between fruit and vegetable consumption and HPV viral load in high-risk HPV-positive women with cervical intraepithelial neoplasia. Cancer Causes Control 2010;21:51–9.

[43] Kwanbunjan K, Saengkar P, Cheeramakara C, Thanomsak W, Benjachai W, Lais P. Low folate status as a risk factor for cervical dysplasia in Thai women. Nutr Res 2005;25:641–54.

[44] Piyathilake CJ, Henao OL, Macaluso M, Cornwell PE, Meleth S, Heimburger DC, Partridge EE. Folate is associated with the natural history of high-risk human papillomaviruses. Cancer Res 2004;64:8788–93.

[45] Chih HJ, Lee AH, Colville L, Binns CW, Xu D. A review of dietary prevention of human papillomavirus-related infection of the cervix and cervical intraepithelial neoplasia. Nutr Cancer 2013;65:317–28.

[46] Tomita LY, Roteli-Martins CM, Villa LL, Franco EL, Cardoso MA, Team BS. Associations of dietary dark-green and deep-yellow vegetables and fruits with cervical intraepithelial neoplasia: modification by smoking. Br J Nutr 2011;105:928–37.

[47] Alshatwi AA, Periasamy VS, Athinarayanan J, Elango R. Synergistic anticancer activity of dietary tea polyphenols and bleomycin hydrochloride in human cervical cancer cell: caspase-dependent and independent apoptotic pathways. Chem Biol Interact 2016;247:1–10.

[48] Tavani A, La Vecchia C, Gallus S, Lagiou P, Trichopoulos D, Levi F, Negri E. Red meat intake and cancer risk: a study in Italy. Int J Cancer 2000;86:425–8.

[49] Bosetti C, Negri E, Franceschi S, Pelucchi C, Talamini R, Montella M, Conti E, La Vecchia C. Diet and ovarian cancer risk: a case-control study in Italy. Int J Cancer 2001;93:911–5.

[50] Kolahdooz F, van der Pols JC, Bain CJ, Marks GC, Hughes MC, Whiteman DC, Webb PM, Australian Cancer Study (Ovarian Cancer) and the Australian Ovarian Cancer Study Group. Meat, fish, and ovarian cancer risk: results from 2 Australian case-control studies, a systematic review, and meta-analysis. Am J Clin Nutr 2010;91:1752–63.

[51] Blank MM, Wentzensen N, Murphy MA, Hollenbeck A, Park Y. Dietary fat intake and risk of ovarian cancer in the NIH-AARP diet and health study. Br J Cancer 2012;106:596–602.

[52] Merritt MA, Tzoulaki I, van den Brandt PA, Schouten LJ, Tsilidis KK, Weiderpass E, Patel CJ, Tjonneland A, Hansen L, Overvad K, His M, Dartois L, Boutron-Ruault MC, Fortner RT, Kaaks R, Aleksandrova K, Boeing H, Trichopoulou A, Lagiou P, Bamia C, Palli D, Krogh V, Tumino R, Ricceri F, Mattiello A, Bueno-de-Mesquita HB, Onland-Moret NC, Peeters PH, Skeie G, Jareid M, Quiros JR, Obon-Santacana M, Sanchez MJ, Chamosa S, Huerta JM, Barricarte A, Dias JA, Sonestedt E, Idahl A, Lundin E, Wareham NJ, Khaw KT, Travis RC, Ferrari P, Riboli E, Gunter MJ. Nutrient-wide association study of 57 foods/nutrients and epithelial ovarian cancer in the European prospective investigation into cancer and nutrition study and the Netherlands cohort study. Am J Clin Nutr 2016;103:161–7.

[53] Shivappa N, Hebert JR, Paddock LE, Rodriguez-Rodriguez L, Olson SH, Bandera EV. Dietary inflammatory index and ovarian cancer risk in a New Jersey case-control study. Nutrition 2018;46:78–82.

[54] Shivappa N, Hebert JR, Rosato V, Rossi M, Montella M, Serraino D, La Vecchia C. Dietary inflammatory index and ovarian cancer risk in a large Italian case-control study. Cancer Causes Control 2016;27:897–906.

[55] McCann SE, Freudenheim JL, Marshall JR, Graham S. Risk of human ovarian cancer is related to dietary intake of selected nutrients, phytochemicals and food groups. J Nutr 2003;133:1937–42.

[56] Faber MT, Jensen A, Sogaard M, Hogdall E, Hogdall C, Blaakaer J, Kjaer SK. Use of dairy products, lactose, and calcium and risk of ovarian cancer—results from a Danish case-control study. Acta Oncol 2012;51:454–64.

[57] Qin B, Moorman PG, Alberg AJ, Barnholtz-Sloan JS, Bondy M, Cote ML, Funkhouser E, Peters ES, Schwartz AG, Terry P, Schildkraut JM, Bandera EV. Dairy, calcium, vitamin D and ovarian cancer risk in African-American women. Br J Cancer 2016;115:1122–30.

[58] Merritt MA, Cramer DW, Vitonis AF, Titus LJ, Terry KL. Dairy foods and nutrients in relation to risk of ovarian cancer and major histological subtypes. Int J Cancer 2013;132:1114–24.

[59] Dixon SC, Ibiebele TI, Protani MM, Beesley J, de Fazio A, Crandon AJ, Gard GB, Rome RM, Webb PM, Nagle CM, Australian Ovarian Cancer Study Group. Dietary folate and related micronutrients, folate-metabolising genes, and ovarian cancer survival. Gynecol Oncol 2014;132:566–72.

[60] Yang L, Klint A, Lambe M, Bellocco R, Riman T, Bergfeldt K, Persson I, Weiderpass E. Predictors of ovarian cancer survival: a population-based prospective study in Sweden. Int J Cancer 2008;123:672–9.

[61] Hedelin M, Lof M, Andersson TM, Adlercreutz H, Weiderpass E. Dietary phytoestrogens and the risk of ovarian cancer in the women's lifestyle and health cohort study. Cancer Epidemiol Biomarkers Prev 2011;20:308–17.

[62] Bandera EV, King M, Chandran U, Paddock LE, Rodriguez-Rodriguez L, Olson SH. Phytoestrogen consumption from foods and supplements and epithelial ovarian cancer risk: a population-based case control study. BMC Womens Health 2011;11:40.

[63] Neill AS, Ibiebele TI, Lahmann PH, Hughes MC, Nagle CM, Webb PM, Australian Ovarian Cancer Study Group and Australian National Endometrial Cancer Study Group. Dietary phyto-oestrogens and the risk of ovarian and endometrial cancers: findings from two Australian case-control studies. Br J Nutr 2014;111:1430–40.

[64] Nagle CM, Purdie DM, Webb PM, Green A, Harvey PW, Bain CJ. Dietary influences on survival after ovarian cancer. Int J Cancer 2003;106:264–9.

[65] Thomson CA, Crane TE, Wertheim BC, Neuhouser ML, Li W, Snetselaar LG, Basen-Engquist KM, Zhou Y, Irwin ML. Diet quality and survival after ovarian cancer: results from the Women's Health Initiative. J Natl Cancer Inst 2014;196(11):dju314.

[66] Ryan AM, Cushen S, Schellekens H, Bhuachalla EN, Burns L, Kenny U, Power DG. Poor awareness of risk factors for cancer in Irish adults: results of a large survey and review of the literature. Oncologist 2015;20:372–8.

[67] Hales CM, Carroll MD, Fryar CD, Ogden CL. Prevalence of obesity and severe obesity among adults: United States, 2017-2018. NCHS Data Brief Feb 2020;360:1–8.

[68] Beavis AL, Smith AJ, Fader AN. Lifestyle changes and the risk of developing endometrial and ovarian cancers: opportunities for prevention and management. Int J Womens Health 2016;8:151–67.

[69] Reeves GK, Pirie K, Beral V, Green J, Spencer E, Bull D, Million Women Study Collaboration. Cancer incidence and mortality in relation to body mass index in the Million Women Study: cohort study. BMJ 2007;335:1134.

[70] Zhang Y, Liu H, Yang S, Zhang J, Qian L, Chen X. Overweight, obesity and endometrial cancer risk: results from a systematic review and meta-analysis. Int J Biol Markers 2014;29:e21–9.

[71] Setiawan VW, Yang HP, Pike MC, McCann SE, Yu H, Xiang YB, Wolk A, Wentzensen N, Weiss NS, Webb PM, van den Brandt PA, van de Vijver K, Thompson PJ, Australian National Endometrial Cancer Study Group, Strom BL, Spurdle AB, Soslow RA, Shu XO, Schairer C, Sacerdote C, Rohan TE, Robien K, Risch HA, Ricceri F, Rebbeck TR, Rastogi R, Prescott J, Polidoro S, Park Y, Olson SH, Moysich KB, Miller AB, McCullough ML, Matsuno RK, Magliocco AM, Lurie G, Lu L, Lissowska J, Liang X, Lacey Jr JV, Kolonel LN, Henderson BE, Hankinson SE, Hakansson N, Goodman MT, Gaudet MM, Garcia-Closas M, Friedenreich CM, Freudenheim JL, Doherty J, De Vivo I, Courneya KS, Cook LS, Chen C, Cerhan JR, Cai H, Brinton LA, Bernstein L, Anderson KE, Anton-Culver H, Schouten LJ, Horn-Ross PL. Type I and II endometrial cancers: have they different risk factors? J Clin Oncol 2013;31:2607–18.

[72] Schouten LJ, Goldbohm RA, van den Brandt PA. Anthropometry, physical activity, and endometrial cancer risk: results from the Netherlands cohort study. Int J Gynecol Cancer 2006;16(Suppl 2):492.

[73] Schmandt RE, Iglesias DA, Co NN, Lu KH. Understanding obesity and endometrial cancer risk: opportunities for prevention. Am J Obstet Gynecol 2011;205:518–25.

[74] Feng Y-H. The association between obesity and gynecological cancer. Gynaecol Minim Invasive Ther 2015;4:102–5.

[75] Talavera F, Reynolds RK, Roberts JA, Menon KM. Insulin-like growth factor I receptors in normal and neoplastic human endometrium. Cancer Res 1990;50:3019–24.

[76] Zeleniuch-Jacquotte A, Akhmedkhanov A, Kato I, Koenig KL, Shore RE, Kim MY, Levitz M, Mittal KR, Raju U, Banerjee S, Toniolo P. Postmenopausal endogenous oestrogens and risk of endometrial cancer: results of a prospective study. Br J Cancer 2001;84:975–81.

[77] Amant F, Moerman P, Neven P, Timmerman D, Van Limbergen E, Vergote I. Endometrial cancer. Lancet 2005;366:491–505.

[78] Fader AN, Arriba LN, Frasure HE, von Gruenigen VE. Endometrial cancer and obesity: epidemiology, biomarkers, prevention and survivorship. Gynecol Oncol 2009;114:121–7.

[79] Nagamani M, Stuart CA. Specific binding and growth-promoting activity of insulin in endometrial cancer cells in culture. Am J Obstet Gynecol 1998;179:6–12.

[80] Friberg E, Mantzoros CS, Wolk A. Diabetes and risk of endometrial cancer: a population-based prospective cohort study. Cancer Epidemiol Biomarkers Prev 2007;16:276–80.

[81] Lindemann K, Vatten LJ, Ellstrom-Engh M, Eskild A. Body mass, diabetes and smoking, and endometrial cancer risk: a follow-up study. Br J Cancer 2008;98:1582–5.

[82] Friberg E, Orsini N, Mantzoros CS, Wolk A. Diabetes mellitus and risk of endometrial cancer: a meta-analysis. Diabetologia 2007;50:1365–74.

[83] Poole EM, Konstantinopoulos PA, Terry KL. Prognostic implications of reproductive and lifestyle factors in ovarian cancer. Gynecol Oncol 2016;142:574–87.

[84] Plummer M, Herrero R, Franceschi S, Meijer CJ, Snijders P, Bosch FX, de Sanjose S, Munoz N, IARC Multicentre Cervical Cancer Study Group. Smoking and cervical cancer: pooled analysis of the IARC multi-centric case—control study. Cancer Causes Control 2003;14:805–14.

[85] Kjaerbye-Thygesen A, Frederiksen K, Hogdall EV, Glud E, Christensen L, Hogdall CK, Blaakaer J, Kjaer SK. Smoking and overweight: negative prognostic factors in stage III epithelial ovarian cancer. Cancer Epidemiol Biomarkers Prev 2006;15:798–803.

[86] Zhou B, Yang L, Sun Q, Cong R, Gu H, Tang N, Zhu H, Wang B. Cigarette smoking and the risk of endometrial cancer: a meta-analysis. Am J Med 2008;121:501–8 [e503].

[87] Loerbroks A, Schouten LJ, Goldbohm RA, van den Brandt PA. Alcohol consumption, cigarette smoking, and endometrial cancer risk: results from the Netherlands Cohort Study. Cancer Causes Control 2007;18:551–60.

[88] Siiteri PK. Steroid hormones and endometrial cancer. Cancer Res 1978;38:4360–6.

[89] Herrinton LJ, Weiss NS. Postmenopausal unopposed estrogens. Characteristics of use in relation to the risk of endometrial carcinoma. Ann Epidemiol 1993;3:308–18.

[90] Weiss NS, Sayvetz TA. Incidence of endometrial cancer in relation to the use of oral contraceptives. N Engl J Med 1980;302:551–4.

[91] Key TJ, Pike MC. The dose-effect relationship between 'unopposed' oestrogens and endometrial mitotic rate: its central role in explaining and predicting endometrial cancer risk. Br J Cancer 1988;57:205–12.

[92] Lukanova A, Lundin E, Micheli A, Arslan A, Ferrari P, Rinaldi S, Krogh V, Lenner P, Shore RE, Biessy C, Muti P, Riboli E, Koenig KL, Levitz M, Stattin P, Berrino F, Hallmans G, Kaaks R, Toniolo P, Zeleniuch-Jacquotte A. Circulating levels of sex steroid hormones and risk of endometrial cancer in postmenopausal women. Int J Cancer 2004;108:425–32.

[93] Potischman N, Hoover RN, Brinton LA, Siiteri P, Dorgan JF, Swanson CA, Berman ML, Mortel R, Twiggs LB, Barrett RJ, Wilbanks GD, Persky V, Lurain JR. Case-control study of endogenous steroid hormones and endometrial cancer. J Natl Cancer Inst 1996;88:1127–35.

[94] Barry JA, Azizia MM, Hardiman PJ. Risk of endometrial, ovarian and breast cancer in women with polycystic ovary syndrome: a systematic review and meta-analysis. Hum Reprod Update 2014;20:748–58.

[95] Ding DC, Chen W, Wang JH, Lin SZ. Association between polycystic ovarian syndrome and endometrial, ovarian, and breast cancer: a population-based cohort study in Taiwan. Medicine (Baltimore) 2018;97, e12608.

[96] Podfigurna-Stopa A, Czyzyk A, Katulski K, Moszynski R, Sajdak S, Genazzani AR, Meczekalski B. Recurrent endometrial hyperplasia as a presentation of estrogen-secreting thecoma—case report and minireview of the literature. Gynecol Endocrinol 2016;32:184–7.

[97] Schumer ST, Cannistra SA. Granulosa cell tumor of the ovary. J Clin Oncol 2003;21:1180–9.

[98] XX. Effects of hormone replacement therapy on endometrial histology in postmenopausal women. The postmenopausal estrogen/progestin interventions (PEPI) trial. The writing group for the PEPI Trial. JAMA 1996;275:370–5.

[99] Schiff I, Sela HK, Cramer D, Tulchinsky D, Ryan KJ. Endometrial hyperplasia in women on cyclic or continuous estrogen regimens. Fertil Steril 1982;37:79–82.

[100] Henderson BE. The cancer question: an overview of recent epidemiologic and retrospective data. Am J Obstet Gynecol 1989;161:1859–64.

[101] Persson I, Adami HO, Bergkvist L, Lindgren A, Pettersson B, Hoover R, Schairer C. Risk of endometrial cancer after treatment with oestrogens alone or in conjunction with progestogens: results of a prospective study. BMJ 1989;298:147–51.

[102] Anderson GL, Judd HL, Kaunitz AM, Barad DH, Beresford SA, Pettinger M, Liu J, McNeeley SG, Lopez AM, Women's Health Initiative Investigators. Effects of estrogen plus progestin on gynecologic cancers and associated diagnostic procedures: the Women's Health Initiative randomized trial. JAMA 2003;290:1739–48.

[103] Karageorgi S, Hankinson SE, Kraft P, De Vivo I. Reproductive factors and postmenopausal hormone use in relation to endometrial cancer risk in the Nurses' Health Study cohort 1976-2004. Int J Cancer 2010;126:208–16.

[104] McPherson CP, Sellers TA, Potter JD, Bostick RM, Folsom AR. Reproductive factors and risk of endometrial cancer. The Iowa Women's Health Study. Am J Epidemiol 1996;143:1195–202.

[105] Besevic J, Gunter MJ, Fortner RT, Tsilidis KK, Weiderpass E, Onland-Moret NC, Dossus L, Tjonneland A, Hansen L, Overvad K, Mesrine S, Baglietto L, Clavel-Chapelon F, Kaaks R, Aleksandrova K, Boeing H, Trichopoulou A, Lagiou P, Bamia C, Masala G, Agnoli C, Tumino R, Ricceri F, Panico S, Bueno-de-Mesquita HB, Peeters PH, Jareid M, Quiros JR, Duell EJ, Sanchez MJ, Larranaga N, Chirlaque MD, Barricarte A, Dias JA, Sonestedt E, Idahl A, Lundin E, Wareham NJ, Khaw KT, Travis RC, Rinaldi S, Romieu I, Riboli E, Merritt MA. Reproductive factors and epithelial ovarian cancer survival in the EPIC cohort study. Br J Cancer 2015;113:1622–31.

[106] Mascarenhas C, Lambe M, Bellocco R, Bergfeldt K, Riman T, Persson I, Weiderpass E. Use of hormone replacement therapy before and after ovarian cancer diagnosis and ovarian cancer survival. Int J Cancer 2006;119:2907–15.

[107] Nagle CM, Bain CJ, Green AC, Webb PM. The influence of reproductive and hormonal factors on ovarian cancer survival. Int J Gynecol Cancer 2008;18:407–13.

[108] Raglan O, Kalliala I, Markozannes G, Cividini S, Gunter MJ, Nautiyal J, Gabra H, Paraskevaidis E, Martin-Hirsch P, Tsilidis KK, Kyrgiou M. Risk factors for endometrial cancer: an umbrella review of the literature. Int J Cancer 2019;145:1719–30.

[109] Wu QJ, Li YY, Tu C, Zhu J, Qian KQ, Feng TB, Li C, Wu L, Ma XX. Parity and endometrial cancer risk: a meta-analysis of epidemiological studies. Sci Rep 2015;5:14243.

[110] Setiawan VW, Pike MC, Karageorgi S, Deming SL, Anderson K, Bernstein L, Brinton LA, Cai H, Cerhan JR, Cozen W, Chen C, Doherty J, Freudenheim JL, Goodman MT, Hankinson SE, Lacey Jr JV, Liang X, Lissowska J, Lu L, Lurie G, Mack T, Matsuno RK, McCann S, Moysich KB, Olson SH, Rastogi R, Rebbeck TR, Risch H, Robien K, Schairer C, Shu XO, Spurdle AB, Strom BL, Thompson PJ, Ursin G, Webb PM, Weiss NS, Wentzensen N, Xiang YB, Yang HP, Yu H, Horn-Ross PL, De Vivo I, Australian National Endometrial Cancer Study Group. Age at last birth in relation to risk of endometrial cancer: pooled analysis in the epidemiology of endometrial cancer consortium. Am J Epidemiol 2012;176:269–78.

[111] Jordan SJ, Na R, Johnatty SE, Wise LA, Adami HO, Brinton LA, Chen C, Cook LS, Maso LD, De Vivo I, Freudenheim JL, Friedenreich CM, La Vecchia C, McCann SE, Moysich KB, Lu L, Olson SH, Palmer JR, Petruzella S, Pike MC, Rebbeck TR, Ricceri F, Risch HA, Sacerdote C, Setiawan VW, Sponholtz TR, Shu XO, Spurdle AB, Weiderpass E, Wentzensen N, Yang HP, Yu H, Webb PM. Breastfeeding and endometrial cancer risk: an analysis from the epidemiology of endometrial cancer consortium. Obstet Gynecol 2017;129:1059–67.

[112] Hannaford PC, Selvaraj S, Elliott AM, Angus V, Iversen L, Lee AJ. Cancer risk among users of oral contraceptives: cohort data from the Royal College of General Practitioner's oral contraception study. BMJ 2007;335:651.

[113] Iversen L, Sivasubramaniam S, Lee AJ, Fielding S, Hannaford PC. Lifetime cancer risk and combined oral contraceptives: the Royal College of General Practitioners' Oral Contraception Study. Am J Obstet Gynecol 2017;216:580 e581–9.

[114] Mueck AO, Seeger H, Rabe T. Hormonal contraception and risk of endometrial cancer: a systematic review. Endocr Relat Cancer 2010;17:R263–71.

[115] Collaborative Group on Epidemiological Studies of Ovarian Cancer, Beral V, Doll R, Hermon C, Peto R, Reeves G. Ovarian cancer and oral contraceptives: collaborative reanalysis of data from 45 epidemiological studies including 23,257 women with ovarian cancer and 87,303 controls. Lancet 2008;371:303–14.

[116] Collaborative Group on Epidemiological Studies on Endometrial Cancer. Endometrial cancer and oral contraceptives: an individual participant meta-analysis of 27 276 women with endometrial cancer from 36 epidemiological studies. Lancet Oncol 2015;16:1061–70.

[117] Jacobsen BK, Vollset SE, Kvale G. Reproductive factors and survival from ovarian cancer. Int J Cancer 1993;54:904–6.

[118] Cummings SR, Eckert S, Krueger KA, Grady D, Powles TJ, Cauley JA, Norton L, Nickelsen T, Bjarnason NH, Morrow M, Lippman ME, Black D, Glusman JE, Costa A, Jordan VC. The effect of raloxifene on risk of breast

cancer in postmenopausal women: results from the MORE randomized trial. Multiple outcomes of Raloxifene evaluation. JAMA 1999;281:2189–97.

[119] DeMichele A, Troxel AB, Berlin JA, Weber AL, Bunin GR, Turzo E, Schinnar R, Burgh D, Berlin M, Rubin SC, Rebbeck TR, Strom BL. Impact of raloxifene or tamoxifen use on endometrial cancer risk: a population-based case-control study. J Clin Oncol 2008;26:4151–9.

[120] Committee Opinion No. 738. Aromatase inhibitors in gynecologic practice: correction. Obstet Gynecol 2018;132:786.

[121] Matsuo K, Cahoon SS, Yoshihara K, Shida M, Kakuda M, Adachi S, Moeini A, Machida H, Garcia-Sayre J, Ueda Y, Enomoto T, Mikami M, Roman LD, Sood AK. Association of low-dose aspirin and survival of women with endometrial cancer. Obstet Gynecol 2016;128:127–37.

[122] Neill AS, Nagle CM, Protani MM, Obermair A, Spurdle AB, Webb PM, Australian National Endometrial Cancer Study Group. Aspirin, nonsteroidal anti-inflammatory drugs, paracetamol and risk of endometrial cancer: a case-control study, systematic review and meta-analysis. Int J Cancer 2013;132:1146–55.

[123] Allard JE, Maxwell GL. Race disparities between black and white women in the incidence, treatment, and prognosis of endometrial cancer. Cancer Control 2009;16:53–6.

[124] Long B, Liu FW, Bristow RE. Disparities in uterine cancer epidemiology, treatment, and survival among African Americans in the United States. Gynecol Oncol 2013;130:652–9.

[125] Maxwell GL, Tian C, Risinger J, Brown CL, Rose GS, Thigpen JT, Fleming GF, Gallion HH, Brewster WR, Gynecologic Oncology Group study. Racial disparity in survival among patients with advanced/recurrent endometrial adenocarcinoma: a gynecologic oncology group study. Cancer 2006;107:2197–205.

[126] Rossi A, Moadel-Robblee A, Garber CE, Kuo D, Goldberg G, Einstein M, Nevadunsky N. Physical activity for an ethnically diverse sample of endometrial cancer survivors: a needs assessment and pilot intervention. J Gynecol Oncol 2015;26:141–7.

[127] Peterson CE, Rauscher GH, Johnson TP, Kirschner CV, Freels S, Barrett RE, Kim S, Fitzgibbon ML, Joslin CE, Davis FG. The effect of neighborhood disadvantage on the racial disparity in ovarian cancer-specific survival in a large hospital-based study in cook county, Illinois. Front Public Health 2015;3:8.

[128] Henley SJ, Miller JW, Dowling NF, Benard VB, Richardson LC. Uterine cancer incidence and mortality—United States, 1999-2016. MMWR Morb Mortal Wkly Rep 2018;67:1333–8.

[129] Baskovic M, Lichtensztajn DY, Nguyen T, Karam A, English DP. Racial disparities in outcomes for high-grade uterine cancer: a California cancer registry study. Cancer Med 2018;7:4485–95.

[130] Santin AD, Bellone S, Van Stedum S, Bushen W, Palmieri M, Siegel ER, De Las Casas LE, Roman JJ, Burnett A, Pecorelli S. Amplification of c-erbB2 oncogene: a major prognostic indicator in uterine serous papillary carcinoma. Cancer 2005;104:1391–7.

[131] Ferguson SE, Olshen AB, Levine DA, Viale A, Barakat RR, Boyd J. Molecular profiling of endometrial cancers from African-American and Caucasian women. Gynecol Oncol 2006;101:209–13.

[132] Ko EM, Walter P, Clark L, Jackson A, Franasiak J, Bolac C, Havrilesky L, Secord AA, Moore DT, Gehrig PA, Bae-Jump VL. The complex triad of obesity, diabetes and race in type I and II endometrial cancers: prevalence and prognostic significance. Gynecol Oncol 2014;133:28–32.

[133] Randall TC, Armstrong K. Differences in treatment and outcome between African-American and white women with endometrial cancer. J Clin Oncol 2003;21:4200–6.

[134] Berends MJ, Wu Y, Sijmons RH, van der Sluis T, Ek WB, Ligtenberg MJ, Arts NJ, ten Hoor KA, Kleibeuker JH, de Vries EG, Mourits MJ, Hollema H, Buys CH, Hofstra RM, van der Zee AG. Toward new strategies to select young endometrial cancer patients for mismatch repair gene mutation analysis. J Clin Oncol 2003;21:4364–70.

[135] Meyer LA, Broaddus RR, Lu KH. Endometrial cancer and Lynch syndrome: clinical and pathologic considerations. Cancer Control 2009;16:14–22.

[136] Riegert-Johnson DL, Gleeson FC, Roberts M, Tholen K, Youngborg L, Bullock M, Boardman LA. Cancer and Lhermitte-Duclos disease are common in Cowden syndrome patients. Hered Cancer Clin Pract 2010;8:6.

[137] Heald B, Mester J, Rybicki L, Orloff MS, Burke CA, Eng C. Frequent gastrointestinal polyps and colorectal adenocarcinomas in a prospective series of PTEN mutation carriers. Gastroenterology 2010;139:1927–33.

[138] Levine DA, Lin O, Barakat RR, Robson ME, McDermott D, Cohen L, Satagopan J, Offit K, Boyd J. Risk of endometrial carcinoma associated with BRCA mutation. Gynecol Oncol 2001;80:395–8.

[139] Thompson D, Easton DF, Breast Cancer Linkage Consortium. Cancer incidence in BRCA1 mutation carriers. J Natl Cancer Inst 2002;94:1358–65.

[140] Goshen R, Chu W, Elit L, Pal T, Hakimi J, Ackerman I, Fyles A, Mitchell M, Narod SA. Is uterine papillary serous adenocarcinoma a manifestation of the hereditary breast-ovarian cancer syndrome? Gynecol Oncol 2000;79:477–81.

[141] Hornreich G, Beller U, Lavie O, Renbaum P, Cohen Y, Levy-Lahad E. Is uterine serous papillary carcinoma a BRCA1-related disease? Case report and review of the literature. Gynecol Oncol 1999;75:300–4.

[142] Clark LH, Ko EM, Kernodle A, Harris A, Moore DT, Gehrig PA, Bae-Jump V. Endometrial cancer survivors' perceptions of provider obesity counseling and attempted behavior change: are we seizing the moment? Int J Gynecol Cancer 2016;26:318–24.

[143] Toledo E, Salas-Salvado J, Donat-Vargas C, Buil-Cosiales P, Estruch R, Ros E, Corella D, Fito M, Hu FB, Aros F, Gomez-Gracia E, Romaguera D, Ortega-Calvo M, Serra-Majem L, Pinto X, Schroder H, Basora J, Sorli JV, Bullo M, Serra-Mir M, Martinez-Gonzalez MA. Mediterranean diet and invasive breast cancer risk among women at high cardiovascular risk in the PREDIMED trial: a randomized clinical trial. JAMA Intern Med 2015;175:1752–60.

[144] Cella D. The functional assessment of cancer therapy-anemia (FACT-An) scale: a new tool for the assessment of outcomes in cancer anemia and fatigue. Semin Hematol 1997;34:13–9.

[145] Hojman P. Exercise protects from cancer through regulation of immune function and inflammation. Biochem Soc Trans 2017;45:905–11.

[146] Coughlin SS, Yoo W, Whitehead MS, Smith SA. Advancing breast cancer survivorship among African-American women. Breast Cancer Res Treat 2015;153:253–61.

[147] Voskuil DW, Monninkhof EM, Elias SG, Vlems FA, van Leeuwen FE, Task Force Physical Activity and Cancer. Physical activity and endometrial cancer risk, a systematic review of current evidence. Cancer Epidemiol Biomarkers Prev 2007;16:639–48.

[148] Du M, Kraft P, Eliassen AH, Giovannucci E, Hankinson SE, De Vivo I. Physical activity and risk of endometrial adenocarcinoma in the Nurses' Health Study. Int J Cancer 2014;134:2707–16.

[149] Patel AV, Feigelson HS, Talbot JT, McCullough ML, Rodriguez C, Patel RC, Thun MJ, Calle EE. The role of body weight in the relationship between physical activity and endometrial cancer: results from a large cohort of US women. Int J Cancer 2008;123:1877–82.

[150] Centers for Disease Control and Prevention. Prevalence of physical activity, including lifestyle activities among adults—United States, 2000-2001. MMWR Morb Mortal Wkly Rep 2003;52:764–9.

[151] Lee JK, So KA, Piyathilake CJ, Kim MK. Mild obesity, physical activity, calorie intake, and the risks of cervical intraepithelial neoplasia and cervical cancer. PLoS One 2013;8, e66555.

[152] Moorman PG, Jones LW, Akushevich L, Schildkraut JM. Recreational physical activity and ovarian cancer risk and survival. Ann Epidemiol 2011;21:178–87.

[153] Zhou Y, Cartmel B, Gottlieb L, Ercolano EA, Li F, Harrigan M, McCorkle R, Ligibel JA, von Gruenigen VE, Gogoi R, Schwartz PE, Risch HA, Irwin ML. Randomized trial of exercise on quality of life in women with ovarian cancer: women's activity and lifestyle study in Connecticut (WALC). J Natl Cancer Inst 2017;109(12):djx072.

[154] Tucker K, Staley SA, Clark LH, Soper JT. Physical activity: impact on survival in gynecologic cancer. Obstet Gynecol Surv 2019;74:679–92.

[155] Donnelly CM, Blaney JM, Lowe-Strong A, Rankin JP, Campbell A, McCrum-Gardner E, Gracey JH. A randomised controlled trial testing the feasibility and efficacy of a physical activity behavioural change intervention in managing fatigue with gynaecological cancer survivors. Gynecol Oncol 2011;122:618–24.

[156] Demark-Wahnefried W, Aziz NM, Rowland JH, Pinto BM. Riding the crest of the teachable moment: promoting long-term health after the diagnosis of cancer. J Clin Oncol 2005;23:5814–30.

[157] Stevinson C, Faught W, Steed H, Tonkin K, Ladha AB, Vallance JK, Capstick V, Schepansky A, Courneya KS. Associations between physical activity and quality of life in ovarian cancer survivors. Gynecol Oncol 2007;106:244–50.

[158] Schmitz KH, Courneya KS, Matthews C, Demark-Wahnefried W, Galvao DA, Pinto BM, Irwin ML, Wolin KY, Segal RJ, Lucia A, Schneider CM, von Gruenigen VE, Schwartz AL, American College of Sports Medicine. American College of Sports Medicine roundtable on exercise guidelines for cancer survivors. Med Sci Sports Exerc 2010;42:1409–26.

[159] Campbell KL, Winters-Stone KM, Wiskemann J, May AM, Schwartz AL, Courneya KS, Zucker DS, Matthews CE, Ligibel JA, Gerber LH, Morris GS, Patel AV, Hue TF, Perna FM, Schmitz KH. Exercise guidelines for cancer survivors: consensus statement from international multidisciplinary roundtable. Med Sci Sports Exerc 2019;51:2375–90.

# Drug resistance in gynecologic cancers: Findings and underlying mechanisms

*Luna Acharya[a], Hariharasudan Mani[a], Aman Ullah[b],*
*Sana Hussain[c], Saeed Ali[a], and Sarfraz Ahmad[d]*

[a]Department of Internal Medicine, University of Iowa, Iowa City, IA, United States [b]Department of Internal Medicine, SoutheastHEALTH, Cape Girardeau, MO, United States [c]Department of Medicine, Khyber Teaching Hospital, Peshawar, Pakistan [d]Gynecologic Oncology, AdventHealth Cancer Institute, Florida State University, University of Central Florida, Orlando, FL, United States

## Abstract

Drug resistance is a serious issue in treating common and lethal gynecologic cancers like cervical, endometrial, and ovarian cancers. It causes the failure of cancer treatment and tumor relapse, at the cellular and genetic levels. The commonly used FDA-approved chemotherapy treatments are taxanes (paclitaxel) and platinums (carboplatin and cisplatin) for ovarian cancer; cisplatin, carboplatin, gemcitabine, paclitaxel, and bevacizumab for advanced-stage cervical cancer; and platinum, taxanes, and doxorubicin for endometrial cancer. Decrease of intracellular accumulation of chemotherapy agents by acting on drug efflux and influx mechanisms is an important mechanism for developing resistance to chemotherapy. Signaling pathways like MAPK, PI3K, and mTOR help in regulating survival of cell, proliferation, and growth, contributing to resistance of drugs. The treatment of gynecologic cancers is affected by affecting the membrane transporter, cell death, and apoptosis. The roles of oncogenes, tumor-suppressor genes, and some proteins such as vimentin, connexin-32 (Cx-32), and tumor necrosis factor-associated protein 1 (TRAP-1) are being investigated to look for an involvement in drug resistance mechanisms. Other mechanisms leading to drug resistance are DNA repair mechanisms, cancer stem cell resistance, epithelial-mesenchymal transition activation, microRNAs (miRNAs), and circular RNAs (circRNAs). There are possibilities of other mechanisms of resistance, which is yet to be completely understood. This understanding is crucial in identifying possible biomarkers and monitoring the pattern of resistance and development of novel and effective targeting therapies to ultimately prevent failure of treatment.

## Abbreviations

| | |
|---|---|
| **ABCG2** | ATP-binding cassette super-family G member 2 |
| **AKT** | protein kinase B |
| **ANXA** | Annexin A |

R. Basha, S. Ahmad (eds.)
*Overcoming Drug Resistance in Gynecologic Cancers*
ISBN 978-0-12-824299-5
https://doi.org/10.1016/B978-0-12-824299-5.00007-1

| | |
|---|---|
| **APC** | adenomatous polyposis coli |
| **APOA1** | apolipoprotein A-1 |
| **ATP7A** | transmembrane protein |
| **BAX** | Bcl-2-associated X protein |
| **Bcl-2** | B-cell lymphoma |
| **BIRC2** | baculoviral IAP repeat containing 2 |
| **BIRC3** | baculoviral IAP repeat containing 3 |
| **CD99** | cluster of differentiation 99 |
| **COL1A1** | collagen type I alpha 1 |
| **Cx32** | Connexin-32 |
| **CYP450** | cytochrome P450 |
| **DAPK** | death-associated protein kinase |
| **DDR2** | discoidin domain receptor 2 |
| **EFTU** | elongation factor thermo unstable |
| **EGFR** | epidermal growth factor receptor |
| **EMT** | epithelial-mesenchymal-transition |
| **ENOA** | α-Enolase |
| **EPHA7** | ephrin type-A receptor 7 |
| **ERCC1** | excision repair 1 |
| **FOXM1** | forkhead box protein M1 |
| **FOXO3a** | forkhead box O3 |
| **G3P** | glyceraldehyde-3-phosphate dehydrogenase |
| **GOLPH3L** | golgi phosphoprotein 3 like |
| **GRP78** | glucose-regulated protein 78 |
| **HNF1** | homeobox B (HNF1B)-hepatocyte nuclear factor-1 beta (HNF-1β) |
| **IFI16** | interferon gamma inducible protein 16 |
| **IFIH1** | interferon induced with helicase C domain 1-M |
| **JNK** | c-Jun N-terminal kinase |
| **MAPK** | mitogen activated protein kinase |
| **MCTP1** | multiple C2 and transmembrane domain containing 1 |
| **MMR** | mismatch repair |
| **MRP-2** | multidrug associated protein |
| **MYOT** | myofibrillar titin-like protein |
| **NF-kB** | nuclear factor kappa-light-chain-enhancer of activated B cells |
| **Nrf2** | nuclear factor, erythroid 2 like 2 |
| **OATP** | organic ion transporting protein |
| **Par4** | prostate apoptosis response4 |
| **PI3K** | phosphoinositide 3-kinase |
| **PLAGL1** | pleomorphic adenoma gene-like 1 |
| **PRDX2** | peroxiredoxin |
| **PTEN** | phosphatase and tensin homolog |
| **PTPRK** | receptor-type tyrosine-protein phosphatase kappa |
| **S100A3** | S100 calcium-binding protein A3 |
| **SAMD4A** | sterile alpha motif domain containing 4A |
| **SHH** | SONIC hedgehog |
| **SPP1** | secreted phosphoprotein 1 |
| **TGF-β1** | transforming growth factor beta 1 |
| **TP63** | tumor protein 63 |
| **TRAP1** | tumor necrosis factor-associated protein 1 |
| **TSG** | tumor suppressor gene |
| **UBC** | ubiquitin C |
| **XIAP** | X-linked inhibitor of apoptosis protein |
| **XPA** | xeroderma pigmentosum |

## Conflict of interest

No potential conflicts of interest were disclosed.

# Introduction

Drug resistance is cell tolerance to one or multiple drugs ultimately leading to treatment failure. This causes failing of targeted treatment of underlying cancer and other type of chemotherapy [1]. It can occur in various stages, before starting cancer treatment, called intrinsic resistance, or later in beginning of response to drug itself, called acquired resistance [2].

Drug resistance has been widely reported and studied for antibiotic-resistant infections, and it has now been reported in cancer as major causes of failure of cancer treatment [3]. It is becoming difficult to overcome multidrug resistance to cancer treatment, which has paved a path for newer modalities of combining ongoing chemotherapy with drug targeting cancer stem cell, microRNA (miRNA), epigenetics, etc. [4]. Drug resistance can occur at various levels like cell membrane, cell cycle, cell metabolism, or gene-level expression [3, 5]. Tumor or cancer cells can develop resistance simultaneously or in subset of cells, prior used or unexposed drugs, via mechanisms either interrelated or unrelated, by creating a drug-resistant phenotype [6].

A few mechanisms are genetic, epigenetic, posttranslational, cellular, or pharmacokinetic causing drug resistance [7]. Cancer cells can acquire genetic modifications that make these cells insensitive to therapeutic agents [8]. These tumor modifications cause change inexpression patterns of gene, causing easy adaptation to the targeted drug therapy and ultimately causing acquired drug resistance [9]. DNA methylation, histone modification, and noncoding RNA expression are a few epigenetic changes that are known to cause cancer growth in ovarian, endometrial, and cervical cancers [10]. Another mechanism is derangement of cell signaling pathways by posttranslational activations that are known to affect the gene expression patterns of oncogenes causing acquired drug resistance [7]. The tumor cells, that express drug resistance comprises of heterogeneous subpopulation of cancer stem cells that has the ability to multiply and self-renew like a normal stem cells [7, 11]. A few well-studied drug resistance mechanisms occur at cellular level via drug efflux [12], overexpression of antiapoptotic molecules [13], epithelial-mesenchymal transition [14, 15], and autophagy [16, 17].

The in-depth and thorough understanding of the various ways of development of drug resistance will not only help to improve the treatment outcome but also lead to discovery of biomarkers. These biomarkers can itself be used to formulate newer targeted therapy as well as to monitor and diagnose drug-resistant cancer [5, 18].

# Drug resistance in endometrial (uterine) cancer

Endometrial cancer represents most frequent gynecological malignancy. The National Comprehensive Cancer Network (NCCN) has highlighted the role of chemotherapy during the treatment of high-risk and advanced-stage disease [19]. The standard available chemotherapeutic regimens consist of doxorubicin, platinum compounds such as cisplatin or carboplatin, and taxanes like paclitaxel [20–22]. However, chemoresistance is a major hurdle to overcome and responsible for failure to treatment, resulting in mortality of more than 90% of patients with advanced-stage disease [23, 24]. The better understanding of various mechanisms of development of drug resistance will help in development of targeted therapy to eradicate cancer cells selectively [25].

Multiple proposed mechanisms of drug resistance include increased efflux pump to reject drugs, decreased drug influx, mechanisms of DNA repair, tumor suppressors (Par-4 and P53), and survival pathways (MAPK, PI3K/AKT, estrogen signaling, EGFR, and mTOR) [25] (see Fig. 1).

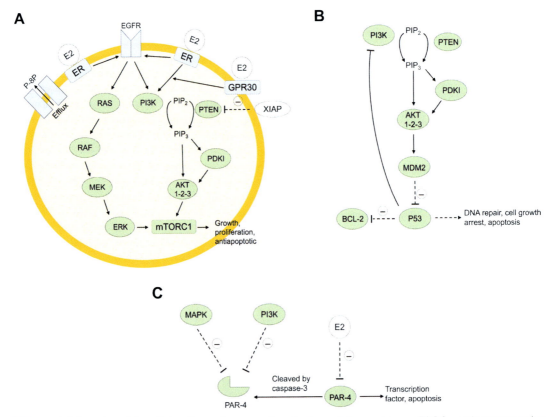

FIG. 1    (A) Overview of the cell signaling pathways and mechanisms of drug resistance. (B) Schematic representation of P53 mechanisms. (C) Schematic representation of Par-4 mechanisms. *Abbreviations*: E2—Estradiol, ER—estrogen receptors, EFGR—Epidermal growth factor receptors, MAPK—Mitogen-activated protein kinase, Par-4—Prostate apoptosis response-4, P-gp—P-glycoprotein, PI3K—Phosphatidylinositol 3-kinase, PTEN—Phosphatase and tensin homolog, XIAP—X-Linked Inhibitor of Apoptosis Protein, Bcl-2—B-Cell Lymphoma-2, ERK—Extracellular-signal-regulated kinase.

## Role of membrane transporters

### Efflux pumps

The overexpression of the multidrug resistant gene MDR1 causes increased expression of P-glycoprotein (P-gp) in expelling the chemotherapy drugs like paclitaxel from cancer cell [26, 27]. Copper transporter ATP7B overexpression works similarly for cisplatin resistance [28].

## DNA repair mechanisms

### Mismatch repair (MMR)

MMR helps in recognizing a mismatched DNA base, and then repairing, followed by re-assembling the DNA in a precise manner [29]. Exposure to chemotherapy disables the MMR process, leading to apoptosis [30]. Deficiency of MMR in endometrial cancer frequently impacts the efficacy of platinum compounds leading to resistance [31, 32].

## Survival pathways

Several survival pathways are involved in mechanisms of drug resistance during treatment of endometrial cancers. One of them is the PI3K/AKT pathway, also known as phosphatidylinositol 3'-kinase (PI3K/AKT), which is the most important signaling pathway that is hyperactivated or mutated at different stages [23, 33–36]. PI3K is a type of kinase located on the cell membrane which causes phosphorylation of PIP2 to PIP3 and activating downstream targets such as PDK1 kinases and AKT involved in cell growth and protein synthesis [37, 38]. AKT isoforms (AKT1, 2, 3) have been responsible for chemoresistance against taxanes, doxorubicin, and platinum drugs, during treatment of gynecologic cancers like endometrial and ovarian cancers [38–42]. AKT1 and AKT2 isoforms increase resistance to drugs like paclitaxel and cisplatin, while all three isoforms of AKT are responsible for causing doxorubicin resistance in endometrial cancer cells [41].

### Phosphatase and tensin homolog

Phosphatase and tensin homolog (PTEN) is a tumor-suppressor lipid phosphatase that negatively impacts the PI3K pathway and controls the activity of PI3K by its ability to dephosphorylate PIP3 to PIP2. Mutation in endometrium cells causes down-regulation or inactivity of PTEN and leads to more resistance against platinum compounds [34, 37, 43, 44]. X-linked inhibitor of apoptosis protein (XIAP) inhibits apoptosis of targets of PI3K/AKT pathway and protects cells by promoting AKT activity by interacting with PTEN, thus negatively regulating PTEN protein levels [45, 46]. XIAP is also involved in developing doxorubicin and cisplatin resistance in endometrial cancer [40, 46–49].

### MAPK pathway

The mitogen-activated protein kinase (MAPK) pathway includes cascades of protein kinases which are activated by genotoxic stress and growth factors, including chemotherapeutic compounds. MAPK pathway activation stimulates the RAS oncogene, which stimulates RAF, consequently leading to ERK1/ERK2 cascade. MAPK pathway stimulation is associated with cell survival, cell division, and apoptosis. The inability of cisplatin to stimulate the

MAPK pathway to induce apoptosis causes chemotherapy resistance in ovarian cancer [50, 51]. MAPK cell signaling pathway and chemoresistance have not been studied well in endometrial cancer, highlighting the need for more studies to understand the drug resistance mechanism in endometrial cancer.

## Cell death inhibition

### Prostate apoptosis response-4

Prostate Apoptosis Response (Par-4) is a protein that acts selectively by inducing apoptosis in various cancer cells [52, 53]. It causes drug resistance of compounds like platinum agents, tamoxifen, and taxane [54–58]. Par-4 can increase apoptosis to paclitaxel during treatment of ovarian cancer [59]. In drug-resistant gynecologic cancers, a cleaved form of Par-4 is either reduced or absent. Par-4 is regulated by the MAPK and PI3K pathways, which play a major role in various mechanisms of drug resistance [60].

## Role of oncogene and tumor-suppressor genes

### P53

P53 is an extensively studied tumor-suppressor protein that protects against cancer by arrest of cell growth, apoptosis, and repair of DNA. P53 is regulated by MDM2, activated, or stabilized by various stimuli such as DNA damage, hypoxia, and oncogene stimulation. P53 is highly mutated and overexpressed in endometrial cancer [61–63]. P53 is involved in drug resistance by alterations in the PI3K pathway [64–66]. P53 inactivation leads to upregulation of the BCL-2 antiapoptotic protein responsible for resistance of platinum compounds [67, 68].

## Role of receptors

### HER family

The epidermal growth factor receptors EGFR (HER-1) and ErbB2 (HER-2) are cell receptors that are overly expressed in advanced stages of endometrial cancers and act as markers of poor prognosis [69–72]. EGFR and ErbB2 overexpression, along with stimulation of the MAPK and PI3K cell signaling pathways, increases drug resistance to paclitaxel and cisplatin in gynecological cancers [73–76]. The effects of ErbB2 and EGFR on chemoresistance are controversial in the literature, and more research work is required.

### Estrogen receptors

Estrogen along with estrogen receptors can also directly bind to G-coupled protein such as GPR30 receptor [71, 77]. Nongenomic binding of estrogen with GPR30 receptor activates the PI3K and MAPK pathways [77, 78]. Estrogen can also positively stimulateGRP78 and prevent apoptosis and is responsible for chemoresistance to both cisplatin and paclitaxel [79]. A brief summary of the chemoresistance mechanisms in endometrial cancer is shown in Table 1.

TABLE 1  Chemoresistance mechanisms in endometrial cancers.

| Mechanism(s) | Chemo drugs | Comments | References |
|---|---|---|---|
| *Efflux pumps* | | | |
| ↑ P-glycoprotein | Taxanes and doxorubicin | Increased level of the efflux pump P-glycoprotein (Pgp) eliminates chemotherapeutic drugs from cancer cells. | [26, 27] |
| ↑ ATP7A/ATP7B | Platinum | | [28] |
| *DNA repair mechanisms* | | | |
| ↓ MMR | Platinum | ↓ MMR allows the cells to continue proliferating, leading failure to enter into apoptosis after exposure to chemotherapy, thus enabling chemoresistance. | [29, 31, 32] |
| *Signaling pathways* | | | |
| ↑ AKT | Platinum, taxanes, and doxorubicin | ↑ AKT leads to increase protein synthesis and cell growth, thus ↑ the chemoresistance. | [28, 38–42] |
| ↓ PTEN | Platinum | ↓ PTEN fails to impact negatively on the PI3K pathway and leads to an increase of resistance against platinum compounds. | [45, 46, 49] |
| ↑ XIAP | Platinum, taxanes, and doxorubicin | XIAP is involved in chemoresistance against cisplatin (endometrial and ovarian), taxane (ovarian),and doxorubicin (endometrial). | [45, 46, 49] |
| ↑ ErbB2 | Platinum and taxanes | ErbB2 via activation of PI3K and MAPK signaling pathways increases resistance to cisplatin and paclitaxel | [69–72] |
| ↑ GRP78 | Platinum and taxanes | ↑ GRP78 prevents apoptosis and provides chemoresistance to both paclitaxel and cisplatin | [79] |
| ↓/mut P53 | Platinum, taxanes | ↓/mut P53 is associated with an increase of the mitochondrial BCL-2, an antiapoptotic protein, leading to acquired chemoresistance to platinum compounds. | [67, 68] |
| ↓ PAR-4 | Taxane | ↓ PAR-4 reduces apoptosis in cancer cells. This mechanism is regulated by PI3K and MAPK pathways, which are involved in chemoresistance mechanisms | [60] |

# Drug resistance in ovarian cancer

Ovarian cancer has caused worldwide cancer-related mortality and is known to be aggressive gynecologic cancer, with drug resistance to chemotherapy causing major limitations [80, 81]. It has less than 40% of 5-year survival rate, likely due to treatment failure from acquired resistance of ovarian tumors to standard chemotherapy like platinum drugs [82]. It has about 50% of drug resistance-related relapse in last 5 years [83–85].

Drug resistance in ovarian cancer can occur anytime, after prolonged cycles of treatment or even before exposure to drugs via intrinsic pathways, that is either de novo or acquired [86]. Acquired drug resistance in ovarian cancer can occur via enhanced drug export, mutations or changes in the target of drug, miRNAs, activation or inhibition of various cell signaling pathways, or molecular alterations in tubulin isotype composition, to name a few [87].

Platinum-resistant ovarian cancer is defined as recurrence in 6 months after the completion of first-line platinum therapy and or patient who has progression of disease. It has poor and unpredictable outcome, with low response rate and variable median survival of less than 12 months [88]. The mechanism of resistance to cisplatin occurs at various levels by limiting cisplatin to enter cells (also called pretarget mechanism); efficient repair mechanisms of DNA (also called on-target mechanisms); limiting apoptosis (also called posttarget mechanisms); and increasing resistance via other compensatory ways (also known as off-target mechanisms). miRNAs play a major role in causing cisplatin resistance [82]. The multidrug resistance in ovarian cancer is often related with each other, like the drug efflux phenomenon and antiapoptotic pathways [89].

## Drug metabolism

Drug is metabolized by various enzyme systems like the cytochrome P450 (CYP450) system, glutathione-S-transferase (GST) system, and uridine diphospho-glucuronosyltransferase (UGT) system. It helps the drug to be metabolized, activated, or detoxified by the help of various drug-metabolizing enzymes [3, 5, 90].

### Cytochrome P450 (CYP450) system

CYP450 helps in metabolism of taxanes and is usually metabolized in the liver. A few solid tumors like ovarian cancer cells express CYP450 enzyme activity that increases the ability to metabolize docetaxel. The ratio of CYP4A5 to ABCB1 gene expression is correlated to pharmacokinetics of taxanes, explaining important mechanism of chemotherapy resistance. Epithelial ovarian tumor cells have shown CYP2C8, CYP3A4/A5, and the ABC transporter expression, knowing which can help in development of selective agents and potential target of chemotherapy [91, 92].

### Glutathione S-transferase system

The increased expression of glutathione S-transferase (GST) causes inactivation of drugs or increased resistance to cell death by inhibiting MAP kinase pathway, through JNK and p38 [93]. Glutathione S Transferase P1 (GSTP1) is upregulated in ovarian cancer showing correlation with platinum (cisplatin and carboplatin) sensitivity [94].

### Uridine diphospho-glucuronosyl transferase

This family of enzymes helps in drug metabolism by glycosylation of lipophilic compounds and glucuronidation. The increased expression of UGT1A1 causes increased glucuronidation, leading to epigenetic changes resulting in resistance to irinotecan, which is used to treat malignant ovarian cancer [95, 96].

## Role of membrane transporters

Drug resistance in cancer cells commonly occurs either by decreasing the influx or by increasing the efflux of drugs by various cell membrane transporters, which ultimately leads to decreased intracellular drug accumulation [24, 97]. Genetic polymorphisms in membrane transporters affect drug pharmacokinetics and pharmacodynamics by affecting gene expressions [98].

### Efflux pumps

ATP-binding cassette (ABC) transporter belongs to ABC protein superfamily, which is a type of well-known efflux transporters causing multiple chemotherapy drug resistance. P-glycoprotein (P-gp), also known as ABCB1/MDR1, belongs to this superfamily that helps in efflux of drugs. It is highly expressed in ovarian cancer cells resistant to taxanes and olaparib [18, 99]. Fletcher et al. reported that the overexpression and upregulation of this ATP-driven efflux transporter have major correlation with development of multidrug resistance in ovarian cancer. It is an important cause for resistance of docetaxel therapy in ovarian cancer to recur [100]. Hedditch et al. reported poor outcomes in primary serous epithelial ovarian cancer with the upregulation and increased expression of ABCA1 [101]. ABC transporters are upregulated by activation of FOXO3a, NF-κB, PI3K/Akt pathways, extracellular signal-regulated kinase (ERK), and HSP27 depletion pathways, and are downregulated by AKT/ERK and JAK2/STAT3 pathways [5, 100].

### Influx pumps

Solute carrier transporters are ion-coupled transporters or exchangers that act as influx pumps usually located in cytoplasmic and organelle membranes functioning in cellular uptake of anticancer drugs [85, 102, 103]. OATP1B3 is a type of organic anion transporting protein (OATP) that is upregulated in ovarian cancer cells. Cancer cells with decreased expression are correlated with decreased sensitivity to platinum chemotherapy [102]. Copper influx transporter is highly correlated with platinum resistance to ovarian cancer, and the downregulation of CRT1 causes decreased uptake of cisplatin leading to drug resistance [80, 104–106]. The downregulation of CTR2 increases uptake and sensitivity to cisplatin, but ovarian cancer cells have shown upregulation of CTR2 resulting in poor outcomes [104–106].

## Cellular metabolism

Cancer cells derive energy for increased proliferation via cell metabolisms like fatty acid synthesis, glutaminolysis, and glycolysis. For the cancer cell survival, energy is required, suggesting possible correlation between cell metabolism and drug resistance [5, 107]. Ovarian cells which are sensitive to platinum chemotherapy have upregulation of glucose and glutamine metabolism with platinum chemotherapy [108]. Ovarian cancer cells that are resistant to chemotherapy show increased expression of glutamine transporter like ASCT2 and glutaminase, causing significant dependency on availability of glutamine. Development of drugs targeting the various targets or outcomes of glutamine metabolism can help to overcome emerging resistance to platinum chemotherapy in ovarian cancer.

## Cell death

Autophagy is the process by which cytoplasmic materials get destroyed or degraded by the help of lysosomal enzymes present in lysosomes [17]. Various stimuli such as hypoxia, hyperthermia, lack of nutrients, oxidative stress as response to radiation, and drugs and protein aggregates can activate autophagy [17]. Cisplatin activates the ERK pathway, which promotes autophagy, leading to cisplatin resistance [109]. Chemokine ligand 5 (CCL5), also known as RANTES, is a chemotactic for T cell which is released from cancer-associated fibroblasts that regulate STAT3 and PI3K/Akt signal pathways causing cisplatin resistance [110]. Hypoxia-inducible factor 1α (HIF-1α), type of oncogene, which is highly overexpressed in ovarian cancer, causes autophagy via PI3K/AKT/mTOR signaling pathway [111].

## Epigenetics

Epigenetic plays a major role in development of resistance to drugs during the treatment of ovarian cancer by acting on the process of methylation of DNA, histone modification, and miRNAs [5, 112, 113]. It causes resistance to platinum compounds in ovarian cancer [114]. Lund et al. reported association of cisplatin resistance with loss of hypermethylation site in a genome [115]. Transcriptomic changes were found in cisplatin-resistant cells. These resistant cells had induction of IL6, whose expression increased further with response to cisplatin. Other components of IL-6 signaling were also affected, explaining the role and importance on development of malignant changes along with development of drug resistance.

### microRNAs

The microRNAs (miRNAs) are noncoding RNA, 22 nucleotides in length, which act as posttranscriptional regulator. In epithelial ovarian cancer, it acts as oncogenes or tumor-suppressor gene associated with poor prognosis phenotypes [116, 117]. They act on various levels of genetic pathways and induce resistance to cisplatin in ovarian cancer cells [82]. Exo-miRNAs are type of miRNAs that lie within exosomes and that help to regulate gene expression both locally and systematically [99, 116]. miR-199a-3p and miR-29a/b/c play a major role in treatment of ovarian cancer by causing resistance to cisplatin [118]. They are downregulated and have low expression in cells causing cisplatin drug resistance, respectively. Along with impairment in the cell signaling pathway, the inhibition of miR-29a/b/c causes upregulation of collagen type I alpha 1 (COL1A1), which in turn helps cancer cells to avoid cisplatin-induced cell death. The downregulation of miR-29 causes the active form of caspase-9 and caspase-3 to be decreased, promoting development of resistance to drug regimen in treatment of gynecologic cancers [119]. Other miRNAs that can lead to ovarian cancer drug resistance are listed in Table 2 [117].

## Epithelial-mesenchymal transition

The epithelial-mesenchymal transition (EMT) plays an important role in development of resistance to treatment by various drugs in ovarian cancer. It acts on various steps of cell invasion, migration, or metastasis along with contributing to aggressiveness of cancer [120, 121]. The key transcriptional factor called Nanog is highly expressed in resistant epithelial ovarian

TABLE 2 MicroRNA causing drug resistance in ovarian cancer cell.

| Cancer cells | Upregulation | Downregulation |
|---|---|---|
| Ovarian cancer cells | miR-141, miR-106a, miR-200c, miR-96, and miR-378 [118] | miR-411, miR-432, miR-494, miR-409-3p, and miR-655 [118] |
| Cisplatin-resistant ovarian cancer cells | | miR-199a-3p [117], miR-29a/b/c [116] |
| Paclitaxel-resistant ovarian cancer cells | MiR-181a [116], miR-663, miR-622, and HS_188 [118] | |
| Paclitaxel-sensitive ovarian cancer cells | | miR-497, miR-187, miR-195, and miR-107 [118] |

cancer cells compared to normal epithelial ovarian cancer cells, leading to overall poor prognosis [122]. The exhibition of a biphasic pattern has direct correlation with the development of resistance to drug treatment in ovarian cancer.

## Role of oncogenes and tumor-suppressor genes

Oncogenes like v-akt murine thymoma viral oncogene (AKT) and B-cell CLL/lymphoma 2-associated X protein (BAX) playa major role in development of resistance to drug treatment by causing either apoptosis/cell death or phosphorylation along with miRNAs. About 11 microRNAs are identified as targets of 7 out of the 25 oncogenes, suggesting that miRNAs act as regulator of oncogenes [123]. Increased expression of ABCG2, HERC5, IFIH1, MYOT, S100A3, SAMD4A, SPP1, and TGFBI and decreased expression levels of MCTP1 and PTPRK are seen in resistant cell lines as shown in Table 3 [124]. This involvement of genes plays an important role in development of drug resistance in ovarian cancer.

The loss of function of tumor-suppressor genes (TSGs) leads to development of cancer. About 15 TSGs contribute to development of resistance during drug treatment of ovarian cancer. It can affect various pathways such as apoptosis, cell cycle, DNA binding, DNA damage, or DNA methylation. Adenomatous polyposis coli gene (APC), death-associated protein kinase gene (DAPK), pleomorphic adenoma gene-like 1 (PLAGL1), a gene encoding an apoptosis-associated speck-like protein (PYCARD/ASC), retinoblastoma susceptibility gene (RB1), and tumor protein 63 (TP63) are few tumor-suppressor genes along with Ubiquitin C (UBC) that are involved in development of resistance to drug treatment in ovarian cancer [81].

TABLE 3 Different genes have varied levels of expression in the drug-resistant ovarian cancer cells.

| Drug-resistant ovarian cell line | Upregulated genes | Downregulated genes |
|---|---|---|
| Cisplatin-resistant cell line | – | ABCC6, BST2, ERAP2, and MCTP1 |
| Paclitaxel-resistant cell line | ABCB1, EPHA7, and RUNDC3B | LIPG, MCTP1, NSBP1, PCDH9, PTPRK, and SEMA3A |
| Docetaxel-resistant cell line | ABCB1, ABCB4, and IFI16 | – |

## Cancer stem cells

The cancer stem cells (CSCs) are highly correlated with drug resistance and tumor relapse by resisting drug treatment, self-renewing, and repopulating the tumor with various undifferentiated and differentiated cells [125]. It can readily adapt to environmental, immunologic, and pharmacologic factors [126]. Various signaling pathways such as NF-kB, NOTCH, PI3K/PTEN, SONIC Hedgehog (SHH), and WNT exhibit properties of stem cell, suggesting possible correlation of cell signaling pathways with the survival of cancer stem cell [125, 127].

## Miscellaneous proteins and genes

Proteins such as α-enolase (ENOA), annexin A (ANXA), apolipoprotein A-1 (APOA1), elongation factor Tu, glyceraldehyde-3-phosphate dehydrogenase (G3P), mitochondrial (EFTU), mitochondrial (GRP75), peroxiredoxin (PRDX2), and stress-70 protein have been investigated as potential biomarker for drug-resistant ovarian cancer [128]. It signifies the underlying role in the development on resistance to drug treatment in ovarian cancer.

The molecular chaperone protein, also known as TRAP1 (tumor necrosis factor-associated protein 1), plays an important role in epithelial-to-mesenchymal transition and development of resistance to cisplatin in ovarian cancer. It acts as a key junctional molecule and metabolic switch causing oxidative phosphorylation which triggers cytokines secretion leading to remodeling of gene expression remodeling [129].

Vimentin is a cytoskeletal protein present in mesenchymal cells. It functions to anchor and support various organelles and helps to maintain the integrity of the cell [130]. The downregulation of expression leads to prolonged G2 arrest and increased exocytosis, causing cisplatin resistance in ovarian cancer. Thus, vimentin can be used as potential target for treatment of drug-resistant ovarian cancer [131].

Connexin-32 (Cx32) is a type of protein, overexpressed in ovarian cells, that induces cisplatin resistance by acting in opposite manner to the expression of efflux transporters like ATP7A and MRP-2. It activates the EGFR-protein kinase B-cell signaling pathway and causes insensitivity to apoptosis induced by cisplatin in ovarian cells [80].

Qin et al. reported that the level of p-cofilin level is higher in cisplatin-resistant ovarian cancer cells than that of cisplatin-sensitive ovarian cancer cells [116]. Therefore, the higher level of expression of p-cofilin is directly correlated to resistance of cisplatin treatment in ovarian cancer.

Nrf2 (nuclear factor, erythroid 2 like 2) promotes CD99 (CD99 molecule) expression at the level of transcriptional and plays an important role in cisplatin resistance in ovarian cancer [132]. Drugs targeting CD99 induced by Nrf2 can effectively reverse cisplatin resistance.

Lu et al. reported that FOXO3a plays an important role in development of platinum resistance in ovarian cancer. They reported that the upregulated expression of FOXO3a is inversely related to the platinum resistance [133].

The decreased expression of HNF1 homeobox B (HNF1B) via four pathways such as apoptosis, ErbB signaling, focal adhesion, and p53 signaling causes resistance to drug treatment in ovarian cancer [134].

NEK2 [NIMA (never in mitosis gene A)-related expression kinase 2], serine/threonine centrosomal kinase, is a type of oncogene. It plays an important role in mitosis via increased expression in S and G2 phases of cell cycle [135, 136]. The interaction with various genes, miRNA, proteins, and other biological processes is associated with development of drug resistance in ovarian cancer [137].

Chiu and colleagues reported that the FOXM1 expression level is directly correlated with cisplatin resistance in ovarian cancer [138]. The increased level of expression of FOXM1 led to increased nuclear accumulation and accelerated the activity of β-catenin in drug-resistant ovarian cancer cells, whereas decreased FOXM1 expression led to opposite. The correlation of FOXM1 and cisplatin resistance in ovarian cancer had developed FOXM1 as potential biomarker for disease progression. The high level of FOXM1 is associated with short progression-free interval and high cancer progression. Thus, FOXM1 inhibitors could be the potential drug treatment to overcome cisplatin resistance in ovarian cancer.

Glucose-regulated protein 78 (GRP78) is a type of protein that plays a major role in development of drug resistance in ovarian cancer. The high expression causes paclitaxel resistance; therefore, the siRNA of GRP78 can be useful in tumor-specific gene therapy [139].

Collagen type XI alpha 1 (COL11A1) is a gene that encodes a protein collagen of the extracellular matrix. It functions by binding to another protein called integrin α1β1 and discoidin domain receptor 2 (DDR2). This activates downstream signaling pathways such as Src-PI3K/Akt-NF-kB, which in turn activates to induce the expression of three inhibitor apoptosis proteins (IAPs), such as BIRC2, BIRC3, and XIAP. This causes inhibition of cisplatin-induced apoptosis. Thus, COL11A1 plays an important role in development of cisplatin resistance in ovarian cancer. It is upregulated in cisplatin-resistant ovarian cells and shows high level of expression in recurrent ovarian cancer [140].

Parl et al. reported that clusterin was significantly upregulated in small group of short-term ovarian cancer survivors [84]. The increased levels of expression of clusterin cause higher level of physical binding to paclitaxel, thus preventing paclitaxel from interacting with microtubules to induce apoptosis leading to overall paclitaxel resistance in ovarian cancer. It is directly correlated with poor survival outcome. Thus, to overcome emerging resistance to drug treatment in ovarian cancer, it is vital to target patient with increased expression of clusterin.

Ni et al. found that the increased expression of long intergenic noncoding RNA 00515 (LINC00515) directly correlated with the increased platinum resistance in high-grade serous ovarian. It can be utilized as a potential predictive biomarker for prognosis of RFS and platinum resistance in ovarian cancer [141]. He and colleagues found that expression of GOLPH3L was remarkably upregulated in ovarian cancer cells, and increased expression was associated with poor prognosis and more aggressive phenotype [85].

## DNA repair mechanisms

DNA damage response is a natural process by which cells maintain genome integrity and prevent harmful mutations [142, 143]. Platinum-based drugs can act by inducing direct or indirect DNA damage, causing inhibition during the progression of cell-cycle progression. This causes cell death with the help of cell signaling pathways [5]. The ability to repair damaged DNA predicts the efficiency of anticancer drugs [24, 144]. Apoptosis pathways are activated by severe DNA damage. Cancer cells can avoid apoptosis induced by DNA damage by various mechanisms. The aberrations of DNA mismatch repair can lead to resistance to platinum, which acts by directly or indirectly damaging the DNA. The resistance to cisplatin in ovarian cancer is associated with overexpression of ERCC1 and XPA [145]. It is shown that drug-resistant cancer cells are highly efficient in repairing DNA damage than sensitive cancer cells [146].

## Drug resistance in cervical cancer

Cervical cancer is reported as the third most common gynecological cancer in United States [147], and HPV infection is associated with 99.7% of cases [148]. Surgery is the first modality of treatment for early stage cervical cancer (stages 1A, 1B1, and 1B2) [149, 150], whereas cisplatin along with radiation therapy is for the treatment for advanced-stage cervical cancer [151]. Second-line chemotherapeutic agents are carboplatin, gemcitabine, paclitaxel, and bevacizumab [152–154].

Both intrinsic and acquired resistances to chemotherapeutic agents are the main barriers in determining patient outcomes. Cisplatin and carboplatin exert action by forming covalent bonds with DNA and inhibiting DNA replication [155]. The mechanism underlying resistance to various drug is thought to be multifactorial (see Fig. 2).

### Reduction of intracellular accumulation

Decreased intracellular accumulation is an important factor among all other mechanisms [156].

#### Decreased influx

Intracellular uptake of cisplatin occurs by passive diffusion dependent on the copper transmembrane protein (CTR1). Some cervical cancer cells seem to downregulate CTR1 transmembrane protein, causing decreased cisplatin uptake (Fig. 2) [157].

#### Increased efflux

Certain transmembrane proteins are also involved with increased efflux of cisplatin, causing resistance. These include multidrug resistance proteins (MRP1,2,3,5) [158], ATP7A and ATP7B (copper transporting ATPase) [157], organic cation transporter 3 (OCT3) [159], and lysosome-associated protein transmembrane 4ß-35 (LAPTM4B-35) [160]. Additionally, P-glycoprotein (P-gp) expression increases cisplatin resistance by causing efflux of platinum-based chemotherapeutic agents [161].

#### Intracellular inactivation

Proteins such as glutathione (GSH), metallothionein, and methionine are thiol containing proteins which helps to regulate the intracellular oxidative stress and deplete available cisplatin [162, 163]. Similarly, high levels of GSH conjugation enzymes such as GSH-S-transferase (GST), gamma-glutamyl cysteine synthase, and gamma-glutamyl transferase also cause platinum-based chemotherapy agent depletion [162].

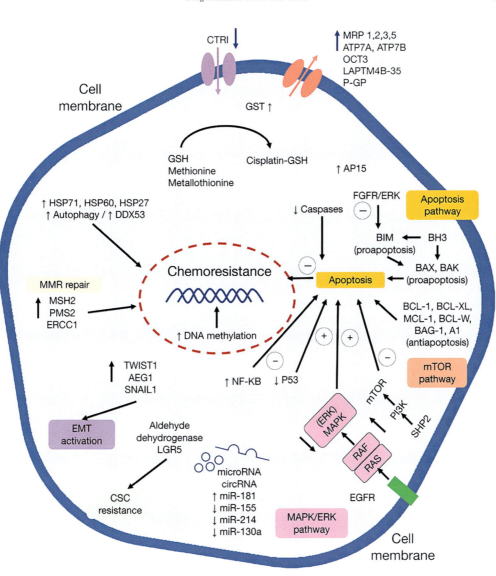

FIG. 2 Pathways involved in chemotherapy resistance in cervical cancer. *Abbreviations*: CTR1—Copper transmembrane protein, MRP—Multidrug resistance protein, ATP7A—ATP7B-copper transporting ATPase, OCT3—Organic cation transporter 3, LAPTM4B-35—Lysosome-associated protein transmembrane 4ß-35, P-gp—P-glycoprotein, GSH—Glutathione, GST—GSH-S-transferase, MSH—MutS homolog 2 protein, PMS2—Postmeiotic segregation 2, ERCC1—Excision repair cross-complement 1, FGFR/ERK—Fibroblast growth factor/Extracellular signal-regulated kinase, MAPK—Mitogen-activated protein kinase, EMT—Epithelial-mesenchymal transition, CSC—Cancer stem cell, HSP—Heat shock protein, DDX53—DEAD-box helicase 53.

## DNA damage repair

DNA MMR rectifies DNA mismatches generated during DNA replication processes. MutS homolog 2 (MSH2), an MMR protein, is found to be linked with cisplatin resistance [164]. Over expression of one of the major MMR system proteins, postmeiotic segregation 2 (PMS2), is also shown to be linked with cisplatin resistance [165]. A 15-kb human nucleotide excision repair gene, also called excision repair cross-complement 1(ERCC1), is associated with platinum resistance by altering protein function or mRNA in gynecologic cancer such as cervical and ovarian cancer as well as other type of solid cancer such as NSCLC (nonsmall cell lung cancer) and colorectal cancer [166].

## Apoptosis inactivation

### Bcl-2 protein family

The Bcl-2 protein group comprises of three distinct protein components: (1) antiapoptotic proteins such as Bcl-2, Bcl-xL, Mcl-1, Bcl-w, Bag-1, and A1; (2) proapoptotic and antiapoptotic BH3-only proteins; and (3) proapoptotic effector proteins, bax and Bak. Multidrug-resistant cervical cancer cells exhibit suppression of proapoptotic proteins and show higher expression of antiapoptotic factors [167, 168]. The over expression of AP15 through the fibroblast growth factor/extracellular signal-regulated kinase (FGFR/ERK)/Bim axis decreases the proapoptotic molecule Bim, which is seen in acquired resistance after multiple exposures to cisplatin [169].

### Caspase activation

The proapoptotic factors are released through the membrane for mitochondria which activates caspases that plays an important role in cellular apoptosis [170]. Cisplatin has been shown to decrease the activity of caspase-3, caspase-8, and caspase-9, causing resistance to apoptosis [168].

### MAPK/ERK pathway

Three mitogen-activated protein kinase (MAPK) pathways play a major role in stress-induced gene expression, activating cell signaling pathways via stress-activated protein kinase/c-Jun-N-terminal kinase (SAPK/JNK), extracellular signal-regulated kinase (ERK), and p38 kinase [171]. It plays a major role in cisplatin-induced cell death. The inhibition or blockade of the MAPK pathway is linked to resistance to cisplatin in cervical cancer [171, 172].

### PI3K/AKT/mTOR pathway

The phosphatidylinositol 3-kinase (PI3K) pathway is related to cellular function and longevity. It is associated with paclitaxel resistance in cervical cancer [173]. PI3K inhibitors, along with paclitaxel, have been shown to increase apoptosis and sensitivity.

### P53 signaling pathway

Cisplatin induces apoptosis through the P53-BAX-mediated pathway. Loss of P53 can cause resistance to drug, impairing apoptosis [168]. Conversely, improved response to cisplatin chemotherapy has been associated with wild-type P53 and higher proportions of P53 [174].

### NF-kB pathway

NF-kB regulates pathways of inflammation, apoptosis, immune regulation, and carcinogenesis [175]. NF-kB inhibition sensitizes cisplatin-induced cell death, and NF-kB activation causes cisplatin resistance in cervical cancer [176].

### Tumor necrosis factor

Tumor-necrosis-factor-[alpha]-induced protein 8 (TNFAIP8) is an inhibitor of apoptosis, and higher levels of TNFAIP8 cause chemotherapy-induced resistance in cervical cancer [177].

## Epithelial-mesenchymal transition (EMT) activation

The EMT is a complex process of transformation of epithelial cells into cells with mesenchymal features, and it is directly associated with cisplatin resistance [178]. The highly conserved transcription factor TWIST1 plays an important role in EMT activation [179]. Similarly, other factors such as SNAIL1 and E-cadherin are associated with activation of EMT and resistance to cisplatin therapy at high doses [180]. Also, astrocyte-elevated gene-1 (AEG-1) induced by human immunodeficiency virus-1 (HIV-1) is associated with EMT activation and cisplatin resistance [181].

## DNA methylation

DNA methylation blocks the expression of genes necessary for drug response. The combination of platinum-based agents and demethylation drugs such as 5-aza cytidine seems to have a role in sensitizing cervical tumor cells [182, 183]. Epigenetic changes, namely promoter methylation and histone methylation, play particular parts in gemcitabine resistance of cervical cancer cells. In one study, resistant cells showed downregulation of hENT1 and dCK expression [184]. Hydralazine, a weak demethylating agent, seems to reverse chemoresistance to gemcitabine in cervical cancer cells [185].

## Cancer stem cell resistance

A very small subsets among the heterogeneous population of cancer cells in tumors are cancer stem cells (CSC). Cancer stem cells have multidrug resistance transporters, are highly chemo-resistant, and have enhanced self-renewal capabilities [161]. CSCs have been reported to harbor aldehyde dehydrogenase as a marker, and high levels have been associated with chemotherapy resistance [186, 187]. LGR5 is recently reported as a potential marker for cervical CSCs and chemoresistance in cervical tumors [188].

## MicroRNA and circRNAs

miRNAs are noncoding RNAs and are 22 nucleotides in length. They can bind to complementary sequences within target mRNAs and generally lead to their degradation [189]. miRNAs regulate multiple cellular pathways, and some, such as miR-181, increase cisplatin resistance by inhibiting apoptosis [190], while others, such as miR-155, suppress EMT [191]. miR-214 increases Bax and caspases [192], thus sensitizing cisplatin to cervical cancer cells. miR-130a was found to regulate CTR1; thus, inhibition of miR-130a could induce higher platinum-based chemotherapy resistance [193]. Hypoxic conditions occurring during cervical cancer tumorigenesis cause increased miR-100 and regulate ubiquitin-specific protease-15 (USP-15), representing a mechanistic pathway for paclitaxel-induced chemoresistance in cervical tumors [194].

CircRNAs are closed loop, novel, noncoding RNAs involved in various cellular functions. CircMTO1 (mitochondrial translation optimization 1 homolog) promotes resistance to drug in cervical cancer via sponging miR-6893and then regulating autophagy and S100A1 [189]. Another circRNA called hsa_circ_0023404 has been recently found to be a promoting factor for cervical tumor chemoresistance mediated through miR-5047 and vascular endothelial growth factor-A (VEGFA), eventually signaling autophagy [195].

## Heat shock proteins and autophagy

Heat shock proteins (HSP) protect cells from oxidative stress damage. Heat shock cognate protein 71 (HSP71) and HSP60 are overexpressed in cisplatin-resistant cervical cancer cells [196]. HSP27 inhibits cisplatin-induced activation of ASK1/p38 and Akt, thus impairing cytotoxic response and subsequently causing cisplatin resistance [197]. Taxol treatment has been associated with chemoresistance through increased autophagy flux through increased expression of cancer/testis antigen DDX53 (DEAD-box helicase 53) and overexpression of P-gp multidrug resistance protein (P-gp, MDR1) [198]. Increased autophagy in cells with paclitaxel resistance has shown a switch from cellular metabolism to glycolytic pathways from the tricarboxylic acid cycle (Warburg effect) [199].

## Miscellaneous mechanisms

Calcium-binding proteins such as S100A8 and S100A9 are increased in patients with cisplatin-sensitive cervical cell lines and could be a new prognostic marker [200]. Annexin A2 (receptor for some S100A proteins) overexpression was associated with paclitaxel-resistant HPV-18-positive cervical carcinoma, and lower levels were associated with sensitivity to chemotherapy in cervical cancer cells [200]. Increased lymph nodal metastasis and human papilloma viral (HPV) load is associated with Src homology phosphotyrosine phosphatase 2(Shp2) via Akt signaling. Inhibition of Shp2 is associated with increased cisplatin sensitivity [155].

Based on multifactorial mechanisms, drugs targeting one mechanism could be overridden by other mechanisms. Also, differences in response to chemotherapeutic agents could be due to the interplay between different mechanisms active in the particular cervical tumor cell line.

## Conclusions and future perspectives

To understand the mechanisms of resistance to drug therapy in gynecological malignancy, significant advances have been made in recent years. Still, we have only scratched the surface when it comes to exploring the complete pathways involved, focusing on the need for more work to be done in understanding multidrug resistance in both intrinsic and acquired chemoresistance. Multiple tools have been discovered, designed, and implemented such as gene therapy, cancer stem cell therapy, immunotherapy, and targeting angiogenesis. The principles of targeted therapy are to utilize acquired changes responsible for chemoresistance by developing drugs specifically targeting them. This potential targeted therapy will help to overcome chemoresistance and lead to fewer treatment failures and better patient outcomes.

Though many studies are done and a few are ongoing, more studies are to be done to overcome the ongoing problem of drug resistance in gynecologic cancers.

## Acknowledgments

The authors are grateful to Ms. Kristina K. Greiner for expert editing and graphics assistance.

## References

[1] Camidge DR, Pao W, Sequist LV. Acquired resistance to TKIs in solid tumours: learning from lung cancer. Nat Rev Clin Oncol 2014;11(8):473–81.

[2] Persidis A. Cancer multidrug resistance. Nat Biotechnol 1999;17(1):94–5.

[3] Housman G, Byler S, Heerboth S, Lapinska K, Longacre M, Snyder N, et al. Drug resistance in cancer: an overview. Cancers (Basel) 2014;6(3):1769–92.

[4] Wu Q, Yang Z, Nie Y, Shi Y, Fan D. Multi-drug resistance in cancer chemotherapeutics: mechanisms and lab approaches. Cancer Lett 2014;347(2):159–66.

[5] Norouzi-Barough L, Sarookhani MR, Sharifi M, Moghbelinejad S, Jangjoo S, Salehi R. Molecular mechanisms of drug resistance in ovarian cancer. J Cell Physiol 2018;233(6):4546–62.

[6] Mattern J. Drug resistance in cancer: a multifactorial problem. Anticancer Res 2003;23(2C):1769–72.

[7] Sun X, Hu B. Mathematical modeling and computational prediction of cancer drug resistance. Brief Bioinform 2018;19(6):1382–99.

[8] Hu X, Zhang Z. Understanding the genetic mechanisms of cancer drug resistance using genomic approaches. Trends Genet 2016;32(2):127–37.

[9] Brown R, Curry E, Magnani L, Wilhelm-Benartzi CS, Borley J. Poised epigenetic states and acquired drug resistance in cancer. Nat Rev Cancer 2014;14(11):747–53.

[10] Eskander RN. The epigenetic landscape in the treatment of gynecologic malignancies. Am Soc Clin Oncol Educ Book 2018;38:480–7.

[11] Beck B, Blanpain C. Unravelling cancer stem cell potential. Nat Rev Cancer 2013;13(10):727–38.

[12] Golebiewska A, Brons NH, Bjerkvig R, Niclou SP. Critical appraisal of the side population assay in stem cell and cancer stem cell research. Cell Stem Cell 2011;8(2):136–47.

[13] Yang YP, Chien Y, Chiou GY, Cherng JY, Wang ML, Lo WL, et al. Inhibition of cancer stem cell-like properties and reduced chemoradioresistance of glioblastoma using microRNA145 with cationic polyurethane-short branch PEI. Biomaterials 2012;33(5):1462–76.

[14] Ma JL, Zeng S, Zhang Y, Deng GL, Shen H. Epithelial-mesenchymal transition plays a critical role in drug resistance of hepatocellular carcinoma cells to oxaliplatin. Tumour Biol 2016;37(5):6177–84.

[15] Shang Y, Cai X, Fan D. Roles of epithelial-mesenchymal transition in cancer drug resistance. Curr Cancer Drug Targets 2013;13(9):915–29.

[16] Chen S, Rehman SK, Zhang W, Wen A, Yao L, Zhang J. Autophagy is a therapeutic target in anticancer drug resistance. Biochim Biophys Acta 2010;1806(2):220–9.

[17] Lin L, Baehrecke EH. Autophagy, cell death, and cancer. Mol Cell Oncol 2015;2(3):e985913.

[18] Vaidyanathan A, Sawers L, Gannon AL, Chakravarty P, Scott AL, Bray SE, et al. ABCB1 (MDR1) induction defines a common resistance mechanism in paclitaxel- and olaparib-resistant ovarian cancer cells. Br J Cancer 2016;115(4):431–41.

[19] Wang J, Zhang L, Jiang W, Zhang R, Zhang B, Silayiding A, et al. MicroRNA-135a promotes proliferation, migration, invasion and induces chemoresistance of endometrial cancer cells. Eur J Obstet Gynecol Reprod Biol X 2020;5:100–3.

[20] Lalwani N, Prasad SR, Vikram R, Shanbhogue AK, Huettner PC, Fasih N. Histologic, molecular, and cytogenetic features of ovarian cancers: implications for diagnosis and treatment. Radiographics 2011;31(3):625–46.

[21] Plataniotis G, Castiglione M, ESMO Guidelines Working Group. Endometrial cancer: ESMO Clinical practice guidelines for diagnosis, treatment and follow-up. Ann Oncol 2010;21(suppl_5) v41-v5.

[22] Casciato DA, Territo MC. Manual of clinical oncology. Lippincott Williams & Wilkins; 2009.

[23] Agarwal R, Kaye SB. Ovarian cancer: strategies for overcoming resistance to chemotherapy. Nat Rev Cancer 2003;3(7):502–16.

[24] Longley D, Johnston P. Molecular mechanisms of drug resistance. J Pathol 2005;205(2):275–92.

[25] Brasseur K, Gévry N, Asselin E. Chemoresistance and targeted therapies in ovarian and endometrial cancers. Oncotarget 2017;8(3):4008.

[26] Esteller M, Martinez-Palones JM, García A, Xercavins J, Reventós J. High rate of MDR-1 and heterogeneous pattern of MRP expression without gene amplification in endometrial cancer. Int J Cancer 1995;63(6):798–803.

[27] Kamazawa S, Kigawa J, Kanamori Y, Itamochi H, Sato S, Iba T, et al. Multidrug resistance gene-1 is a useful predictor of paclitaxel-based chemotherapy for patients with ovarian cancer. Gynecol Oncol 2002;86(2):171–6.

[28] Aida T, Takebayashi Y, Shimizu T, Okamura C, Higasimoto M, Kanzaki A, et al. Expression of copper-transporting P-type adenosine triphosphatase (ATP7B) as a prognostic factor in human endometrial carcinoma. Gynecol Oncol 2005;97(1):41–5.

[29] Kunkel TA, Erie DA. DNA mismatch repair. Annu Rev Biochem 2005;74:681–710.

[30] Martin LP, Hamilton TC, Schilder RJ. Platinum resistance: the role of DNA repair pathways. Clin Cancer Res 2008;14(5):1291–5.

[31] Gurin CC, Federici MG, Kang L, Boyd J. Causes and consequences of microsatellite instability in endometrial carcinoma. Cancer Res 1999;59(2):462–6.

[32] Masuda K, Banno K, Yanokura M, Kobayashi Y, Kisu I, Ueki A, et al. Relationship between DNA mismatch repair deficiency and endometrial cancer. Mol Biol Int 2011;2011: 256063. https://doi.org/10.4061/2011/256063.

[33] Cheung LW, Hennessy BT, Li J, Yu S, Myers AP, Djordjevic B, et al. High frequency of PIK3R1 and PIK3R2 mutations in endometrial cancer elucidates a novel mechanism for regulation of PTEN protein stability. Cancer Discov 2011;1(2):170–85.

[34] Oda K, Stokoe D, Taketani Y, McCormick F. High frequency of coexistent mutations of PIK3CA and PTEN genes in endometrial carcinoma. Cancer Res 2005;65(23):10669–73.

[35] Rudd ML, Price JC, Fogoros S, Godwin AK, Sgroi DC, Merino MJ, et al. A unique spectrum of somatic PIK3CA (p110α) mutations within primary endometrial carcinomas. Clin Cancer Res 2011;17(6):1331–40.

[36] Urick ME, Rudd ML, Godwin AK, Sgroi D, Merino M, Bell DW. PIK3R1 (p85α) is somatically mutated at high frequency in primary endometrial cancer. Cancer Res 2011;71(12):4061–7.

[37] Lee S, Choi E-J, Jin C, Kim D-H. Activation of PI3K/Akt pathway by PTEN reduction and PIK3CA mRNA amplification contributes to cisplatin resistance in an ovarian cancer cell line. Gynecol Oncol 2005;97(1):26–34.

[38] Mitsuuchi Y, Johnson SW, Selvakumaran M, Williams SJ, Hamilton TC, Testa JR. The phosphatidylinositol 3-kinase/AKT signal transduction pathway plays a critical role in the expression of p21WAF1/CIP1/SDI1 induced by cisplatin and paclitaxel. Cancer Res 2000;60(19):5390–4.

[39] Gagnon V, Mathieu I, Sexton É, Leblanc K, Asselin E. AKT involvement in cisplatin chemoresistance of human uterine cancer cells. Gynecol Oncol 2004;94(3):785–95.

[40] Gagnon V, Van Themsche C, Turner S, Leblanc V, Asselin E. Akt and XIAP regulate the sensitivity of human uterine cancer cells to cisplatin, doxorubicin and taxol. Apoptosis 2008;13(2):259–71.

[41] Girouard J, Lafleur M-J, Parent S, Leblanc V, Asselin E. Involvement of Akt isoforms in chemoresistance of endometrial carcinoma cells. Gynecol Oncol 2013;128(2):335–43.

[42] Peng D-J, Wang J, Zhou J-Y, Wu GS. Role of the Akt/mTOR survival pathway in cisplatin resistance in ovarian cancer cells. Biochem Biophys Res Commun 2010;394(3):600–5.

[43] Tashiro H, Blazes MS, Wu R, Cho KR, Bose S, Wang SI, et al. Mutations in PTEN are frequent in endometrial carcinoma but rare in other common gynecological malignancies. Cancer Res 1997;57(18):3935–40.

[44] Wu H, Cao Y, Weng D, Xing H, Song X, Zhou J, et al. Effect of tumor suppressor gene PTEN on the resistance to cisplatin in human ovarian cancer cell lines and related mechanisms. Cancer Lett 2008;271(2):260–71.

[45] Van Themsche C, Leblanc V, Parent S, Asselin E. X-linked inhibitor of apoptosis protein (XIAP) regulates PTEN ubiquitination, content, and compartmentalization. J Biol Chem 2009;284(31):20462–6.

[46] Asselin E, Wang Y, Tsang BK. X-linked inhibitor of apoptosis protein activates the phosphatidylinositol 3-kinase/Akt pathway in rat granulosa cells during follicular development. Endocrinology 2001;142(6): 2451–7.

[47] Dan HC, Sun M, Kaneko S, Feldman RI, Nicosia SV, Wang H-G, et al. Akt phosphorylation and stabilization of X-linked inhibitor of apoptosis protein (XIAP). J Biol Chem 2004;279(7):5405–12.

[48] Li J, Sasaki H, Sheng YL, Schneiderman D, Xiao CW, Kotsuji F, et al. Apoptosis and chemoresistance in human ovarian cancer: is Xiap a determinant? Neurosignals 2000;9(2):122–30.

[49] Ma J-j, B-l C, X-y X. XIAP gene downregulation by small interfering RNA inhibits proliferation, induces apoptosis, and reverses the cisplatin resistance of ovarian carcinoma. Eur J Obstet Gynecol Reprod Biol 2009;146 (2):222–6.

[50] Mansouri A, Ridgway LD, Korapati AL, Zhang Q, Tian L, Wang Y, et al. Sustained activation of JNK/p38 MAPK pathways in response to cisplatin leads to Fas ligand induction and cell death in ovarian carcinoma cells. J Biol Chem 2003;278(21):19245–56.

[51] Villedieu M, Deslandes E, Duval M, Héron J-F, Gauduchon P, Poulain L. Acquisition of chemoresistance following discontinuous exposures to cisplatin is associated in ovarian carcinoma cells with progressive alteration of FAK, ERK and p38 activation in response to treatment. Gynecol Oncol 2006;101(3):507–19.

[52] El-Guendy N, Zhao Y, Gurumurthy S, Burikhanov R, Rangnekar VM. Identification of a unique core domain of par-4 sufficient for selective apoptosis induction in cancer cells. Mol Cell Biol 2003;23(16):5516–25.

[53] Shrestha-Bhattarai T, Rangnekar VM. Cancer-selective apoptotic effects of extracellular and intracellular Par-4. Oncogene 2010;29(27):3873–80.

[54] Boehrer S, Chow KU, Beske F, Kukoc-Zivojnov N, Puccetti E, Ruthardt M, et al. In lymphatic cells par-4 sensitizes to apoptosis by down-regulating bcl-2 and promoting disruption of mitochondrial membrane potential and caspase activation. Cancer Res 2002;62(6):1768–75.

[55] Jagtap JC, Dawood P, Shah RD, Chandrika G, Natesh K, Shiras A, et al. Expression and regulation of prostate apoptosis response-4 (Par-4) in human glioma stem cells in drug-induced apoptosis. PLoS One 2014;9(2): e88505.

[56] Jagtap JC, Parveen D, Shah RD, Desai A, Bhosale D, Chugh A, et al. Secretory prostate apoptosis response (Par)-4 sensitizes multicellular spheroids (MCS) of glioblastoma multiforme cells to tamoxifen-induced cell death. FEBS Open Bio 2015;5:8–19.

[57] Pereira MC, de Bessa-Garcia SA, Burikhanov R, Pavanelli AC, Antunes L, Rangnekar VM, et al. Prostate apoptosis response-4 is involved in the apoptosis response to docetaxel in MCF-7 breast cancer cells. Int J Oncol 2013;43(2):531–8.

[58] Tan J, You Y, Xu T, Yu P, Wu D, Deng H, et al. Par-4 downregulation confers cisplatin resistance in pancreatic cancer cells via PI3K/Akt pathway-dependent EMT. Toxicol Lett 2014;224(1):7–15.

[59] Meynier S, Kramer M, Ribaux P, Tille J-C, Delie F, Petignat P, et al. Role of PAR-4 in ovarian cancer. Oncotarget 2015;6(26):22641.

[60] Brasseur K, Fabi F, Adam P, Parent S, Lessard L, Asselin E. Post-translational regulation of the cleaved fragment of Par-4 in ovarian and endometrial cancer cells. Oncotarget 2016;7(24):36971.

[61] Berchuck A, Kohler MF, Marks JR, Wiseman R, Boyd J, Bast Jr. RC. The p53 tumor suppressor gene frequently is altered in gynecologic cancers. Am J Obstet Gynecol 1994;170(1):246–52.

[62] Bieging KT, Mello SS, Attardi LD. Unravelling mechanisms of p53-mediated tumour suppression. Nat Rev Cancer 2014;14(5):359–70.

[63] Cerami E, Gao J, Dogrusoz U, Gross BE, Sumer SO, Aksoy BA, et al. The cBio cancer genomics portal: an open platform for exploring multidimensional cancer genomics data. Cancer Discov 2012;2(5):401–4.

[64] Abedini M, Muller E, Bergeron R, Gray D, Tsang B. Akt promotes chemoresistance in human ovarian cancer cells by modulating cisplatin-induced, p53-dependent ubiquitination of FLICE-like inhibitory protein. Oncogene 2010;29(1):11–25.

[65] Fraser M, Bai T, Tsang BK. Akt promotes cisplatin resistance in human ovarian cancer cells through inhibition of p53 phosphorylation and nuclear function. Int J Cancer 2008;122(3):534–46.

[66] Yang X, Fraser M, Moll UM, Basak A, Tsang BK. Akt-mediated cisplatin resistance in ovarian cancer: modulation of p53 action on caspase-dependent mitochondrial death pathway. Cancer Res 2006;66(6):3126–36.

[67] Perego P, Righetti S, Supino R, Delia D, Caserini C, Carenini N, et al. Role of apoptosis and apoptosis-related proteins in the cisplatin-resistant phenotype of human tumor cell lines. Apoptosis 1997;2(6):540–8.

[68] Rouette A, Parent S, Girouard J, Leblanc V, Asselin E. Cisplatin increases B-cell-lymphoma-2 expression via activation of protein kinase C and Akt2 in endometrial cancer cells. Int J Cancer 2012;130(8):1755–67.

[69] Reyes HD, Thiel KW, Carlson MJ, Meng X, Yang S, Stephan JM, et al. Comprehensive profiling of EGFR/HER receptors for personalized treatment of gynecologic cancers. Mol Diagn Ther 2014;18(2):137–51.

[70] Sheng Q, Liu J. The therapeutic potential of targeting the EGFR family in epithelial ovarian cancer. Br J Cancer 2011;104(8):1241–5.

[71] Konecny GE, Santos L, Winterhoff B, Hatmal M, Keeney GL, Mariani A, et al. HER2 gene amplification and EGFR expression in a large cohort of surgically staged patients with nonendometrioid (type II) endometrial cancer. Br J Cancer 2009;100(1):89–95.

[72] Villella JA, Cohen S, Smith DH, Hibshoosh H, Hershman D. HER-2/neu overexpression in uterine papillary serous cancers and its possible therapeutic implications. Int J Gynecol Cancer 2006;16(5):1897–902.

[73] Marth C, Widschwendter M, Kaern J, Jørgensen NP, Windbichler G, Zeimet AG, et al. Cisplatin resistance is associated with reduced interferon-gamma-sensitivity and increased HER-2 expression in cultured ovarian carcinoma cells. Br J Cancer 1997;76(10):1328–32.

[74] Pegram MD, Finn RS, Arzoo K, Beryt M, Pietras RJ, Slamon DJ. The effect of HER-2/neu overexpression on chemotherapeutic drug sensitivity in human breast and ovarian cancer cells. Oncogene 1997;15(5):537–47.

[75] Mori N, Kyo S, Nakamura M, Hashimoto M, Maida Y, Mizumoto Y, et al. Expression of HER-2 affects patient survival and paclitaxel sensitivity in endometrial cancer. Br J Cancer 2010;103(6):889–98.

[76] Sasaki N, Kudoh K, Kita T, Tsuda H, Furuya K, Kikuchi Y. Effect of HER-2/neu overexpression on chemoresistance and prognosis in ovarian carcinoma. J Obstet Gynaecol Res 2007;33(1):17–23.

[77] Prossnitz ER, Arterburn JB, Sklar LA. GPR30: A G protein-coupled receptor for estrogen. Mol Cell Endocrinol 2007;265–266:138–42.

[78] Bjornstrom L, Sjoberg M. Mechanisms of estrogen receptor signaling: convergence of genomic and nongenomic actions on target genes. Mol Endocrinol 2005;19(4):833–42.

[79] Luvsandagva B, Nakamura K, Kitahara Y, Aoki H, Murata T, Ikeda S, et al. GRP78 induced by estrogen plays a role in the chemosensitivity of endometrial cancer. Gynecol Oncol 2012;126(1):132–9.

[80] Zhang Y, Tao L, Fan LX, Huang K, Luo HM, Ge H, et al. Cx32 mediates cisplatin resistance in human ovarian cancer cells by affecting drug efflux transporter expression and activating the EGFRAkt pathway. Mol Med Rep 2019;19(3):2287–96.

[81] Yin F, Liu X, Li D, Wang Q, Zhang W, Li L. Tumor suppressor genes associated with drug resistance in ovarian cancer (review). Oncol Rep 2013;30(1):3–10.

[82] Samuel P, Pink RC, Brooks SA, Carter DR. miRNAs and ovarian cancer: a miRiad of mechanisms to induce cisplatin drug resistance. Expert Rev Anticancer Ther 2016;16(1):57–70.

[83] Alharbi M, Zuñiga F, Elfeky O, Guanzon D, Lai A, Rice GE, et al. The potential role of miRNAs and exosomes in chemotherapy in ovarian cancer. Endocr Relat Cancer 2018;25(12):R663–85.

[84] Park DC, Yeo SG, Wilson MR, Yerbury JJ, Kwong J, Welch WR, et al. Clusterin interacts with paclitaxel and confer paclitaxel resistance in ovarian cancer. Neoplasia 2008;10(9):964–NaN.

[85] He S, Niu G, Shang J, Deng Y, Wan Z, Zhang C, et al. The oncogenic Golgi phosphoprotein 3 like overexpression is associated with cisplatin resistance in ovarian carcinoma and activating the NF-κB signaling pathway. J Exp Clin Cancer Res 2017;36(1):137.

[86] Zheng H-C. The molecular mechanisms of chemoresistance in cancers. Oncotarget 2017;8(35):59950.

[87] Vergara D, Tinelli A, Iannone A, Maffia M. The impact of proteomics in the understanding of the molecular basis of paclitaxel-resistance in ovarian tumors. Curr Cancer Drug Targets 2012;12(8):987–97.

[88] Davis A, Tinker AV, Friedlander M. "Platinum resistant" ovarian cancer: what is it, who to treat and how to measure benefit? Gynecol Oncol 2014;133(3):624–31.

[89] Risnayanti C, Jang Y-S, Lee J, Ahn HJ. PLGA nanoparticles co-delivering MDR1 and BCL2 siRNA for overcoming resistance of paclitaxel and cisplatin in recurrent or advanced ovarian cancer. Sci Rep 2018;8(1):1–12.

[90] Michael M, Doherty MM. Tumoral drug metabolism: overview and its implications for cancer therapy. J Clin Oncol 2005;23(1):205–29.

[91] DeLoia JA, Zamboni WC, Jones JM, Strychor S, Kelley JL, Gallion HH. Expression and activity of taxane-metobolizing enzymes in ovarian tumors. Gynecol Oncol 2008;108(2):355–60.

[92]   van Eijk M, Boosman RJ, Schinkel AH, Huitema AD, Beijnen JH. Cytochrome P450 3A4, 3A5, and 2C8 expression in breast, prostate, lung, endometrial, and ovarian tumors: relevance for resistance to taxanes. Cancer Chemother Pharmacol 2019;1–13.

[93]   Townsend DM, Tew KD. The role of glutathione-S-transferase in anti-cancer drug resistance. Oncogene 2003;22 (47):7369–75.

[94]   Sawers L, Ferguson M, Ihrig B, Young H, Chakravarty P, Wolf C, et al. Glutathione S-transferase P1 (GSTP1) directly influences platinum drug chemosensitivity in ovarian tumour cell lines. Br J Cancer 2014;111(6):1150–8.

[95]   Hirasawa A, Zama T, Akahane T, Nomura H, Kataoka F, Saito K, et al. Polymorphisms in the UGT1A1 gene predict adverse effects of irinotecan in the treatment of gynecologic cancer in Japanese patients. J Hum Genet 2013;58(12):794–8.

[96]   Wilson T, Longley D, Johnston P. Chemoresistance in solid tumours. Ann Oncol 2006;17:315–24.

[97]   Kilari D, Guancial E, Kim ES. Role of copper transporters in platinum resistance. World J Clin Oncol 2016;7 (1):106.

[98]   Sissung TM, Baum CE, Kirkland CT, Gao R, Gardner ER, Figg WD. Pharmacogenetics of membrane transporters: an update on current approaches. Mol Biotechnol 2010;44(2):152–67.

[99]   Guo W, Dong W, Li M, Shen Y. Mitochondria P-glycoprotein confers paclitaxel resistance on ovarian cancer cells. OncoTargets Ther 2019;12:3881.

[100]  Fletcher JI, Williams RT, Henderson MJ, Norris MD, Haber M. ABC transporters as mediators of drug resistance and contributors to cancer cell biology. Drug Resist Updat 2016;26:1–9.

[101]  Hedditch EL, Gao B, Russell AJ, Lu Y, Emmanuel C, Beesley J, et al. ABCA transporter gene expression and poor outcome in epithelial ovarian cancer. J Natl Compr Inst 2014;106(7).

[102]  Lancaster CS, Sprowl JA, Walker AL, Hu S, Gibson AA, Sparreboom A. Modulation of OATP1B-type transporter function alters cellular uptake and disposition of platinum chemotherapeutics. Mol Cancer Ther 2013;12(8):1537–44.

[103]  Pressler H, Sissung TM, Venzon D, Price DK, Figg WD. Expression of OATP family members in hormone-related cancers: potential markers of progression. PLoS One 2011;6(5):e20372.

[104]  Blair BG, Larson CA, Adams PL, Abada PB, Safaei R, Howell SB. Regulation of copper transporter 2 expression by copper and cisplatin in human ovarian carcinoma cells. Mol Pharmacol 2010;77(6):912–21.

[105]  Hsu K-F, Shen M-R, Huang Y-F, Cheng Y-M, Lin S-H, Chow N-H, et al. Overexpression of the RNA-binding proteins Lin28B and IGF2BP3 (IMP3) is associated with chemoresistance and poor disease outcome in ovarian cancer. Br J Cancer 2015;113(3):414–24.

[106]  Huang CP, Fofana M, Chan J, Chang CJ, Howell SB. Copper transporter 2 regulates intracellular copper and sensitivity to cisplatin. Metallomics 2014;6(3):654–61.

[107]  Zhao Y, Butler EB, Tan M. Targeting cellular metabolism to improve cancer therapeutics. Cell Death Dis 2013;4 (3):e532.

[108]  Hudson CD, Savadelis A, Nagaraj AB, Joseph P, Avril S, DiFeo A, et al. Altered glutamine metabolism in platinum resistant ovarian cancer. Oncotarget 2016;7(27):41637–49.

[109]  Wang J, Wu GS. Role of autophagy in cisplatin resistance in ovarian cancer cells. J Biol Chem 2014;289 (24):17163–73.

[110]  Zhou B, Sun C, Li N, Shan W, Lu H, Guo L, et al. Cisplatin-induced CCL5 secretion from CAFs promotes cisplatin-resistance in ovarian cancer via regulation of the STAT3 and PI3K/Akt signaling pathways. Int J Oncol 2016;48(5):2087–97.

[111]  Huang J, Gao L, Li B, Liu C, Hong S, Min J, et al. Knockdown of hypoxia-inducible factor 1α (HIF-1α) promotes autophagy and inhibits phosphatidylinositol 3-kinase (PI3K)/AKT/mammalian target of rapamycin (mTOR) signaling pathway in ovarian cancer cells. Med Sci Monit 2019;25:4250.

[112]  Borley J, Brown R. Epigenetic mechanisms and therapeutic targets of chemotherapy resistance in epithelial ovarian cancer. Ann Med 2015;47(5):359–69.

[113]  Wilting RH, Dannenberg J-H. Epigenetic mechanisms in tumorigenesis, tumor cell heterogeneity and drug resistance. Drug Resist Updat 2012;15(1–2):21–38.

[114]  de Leon M, Cardenas H, Vieth E, Emerson R, Segar M, Liu Y, et al. Transmembrane protein 88 (TMEM88) promoter hypomethylation is associated with platinum resistance in ovarian cancer. Gynecol Oncol 2016;142 (3):539–47.

[115]  Lund RJ, Huhtinen K, Salmi J, Rantala J, Nguyen EV, Moulder R, et al. DNA methylation and transcriptome changes associated with cisplatin resistance in ovarian cancer. Sci Rep 2017;7(1):1–11.

[116] Qin Y, Li W, Long Y, Zhan Z. Relationship between p-cofilin and cisplatin resistance in patients with ovarian cancer and the role of p-cofilin in prognosis. Cancer Biomark 2019;24(4):469–75.

[117] Kim YW, Kim EY, Jeon D, Liu JL, Kim HS, Choi JW, et al. Differential microRNA expression signatures and cell type-specific association with Taxol resistance in ovarian cancer cells. Drug Des Devel Ther 2014;8:293–314.

[118] Cui Y, Wu F, Tian D, Wang T, Lu T, Huang X, et al. miR-199a-3p enhances cisplatin sensitivity of ovarian cancer cells by targeting ITGB8. Oncol Rep 2018;39(4):1649–57.

[119] Yu PN, Yan MD, Lai HC, Huang RL, Chou YC, Lin WC, et al. Downregulation of miR-29 contributes to cisplatin resistance of ovarian cancer cells. Int J Cancer 2014;134(3):542–51.

[120] Yamaguchi N, Mimoto R, Yanaihara N, Imawari Y, Hirooka S, Okamoto A, et al. DYRK2 regulates epithelial-mesenchymal-transition and chemosensitivity through snail degradation in ovarian serous adenocarcinoma. Tumor Biol 2015;36(8):5913–23.

[121] Yun-Ju Huang R, Yee Chung V, Paul TJ. Targeting pathways contributing to epithelial-mesenchymal transition (EMT) in epithelial ovarian cancer. Curr Drug Targets 2012;13(13):1649–53.

[122] Qin S, Li Y, Cao X, Du J, Huang X. NANOG regulates epithelial–mesenchymal transition and chemoresistance in ovarian cancer. Biosci Rep 2017;37(1).

[123] Liu X, Gao Y, Lu Y, Zhang J, Li L, Yin F. Oncogenes associated with drug resistance in ovarian cancer. J Cancer Res Clin Oncol 2015;141(3):381–95.

[124] Januchowski R, Sterzyńska K, Zawierucha P, Ruciński M, Świerczewska M, Partyka M, et al. Microarray-based detection and expression analysis of new genes associated with drug resistance in ovarian cancer cell lines. Oncotarget 2017;8(30):49944.

[125] Keyvani V, Farshchian M, Esmaeili S-A, Yari H, Moghbeli M, Nezhad S-RK, et al. Ovarian cancer stem cells and targeted therapy. J Ovarian Res 2019;12(1):120.

[126] Al-Alem LF, Pandya UM, Baker AT, Bellio C, Zarrella BD, Clark J, et al. Ovarian cancer stem cells: what progress have we made? Int J Biochem Cell Biol 2019;107:92–103.

[127] Reya T, Morrison SJ, Clarke MF, Weissman IL. Stem cells, cancer, and cancer stem cells. Nature 2001;414 (6859):105–11.

[128] Cruz IN, Coley HM, Kramer HB, Madhuri TK, Safuwan NA, Angelino AR, et al. Proteomics analysis of ovarian cancer cell lines and tissues reveals drug resistance-associated proteins. Cancer Genomics Proteomics 2017;14 (1):35–51.

[129] Amoroso MR, Matassa DS, Agliarulo I, Avolio R, Maddalena F, Condelli V, et al. Stress-adaptive response in ovarian cancer drug resistance: role of TRAP1 in oxidative metabolism-driven inflammation. In: Advances in protein chemistry and structural biology. vol. 108. Elsevier; 2017. p. 163–98.

[130] Satelli A, Li S. Vimentin in cancer and its potential as a molecular target for cancer therapy. Cell Mol Life Sci 2011;68(18):3033–46.

[131] Huo Y, Zheng Z, Chen Y, Wang Q, Zhang Z, Deng H. Downregulation of vimentin expression increased drug resistance in ovarian cancer cells. Oncotarget 2016;7(29):45876.

[132] Wu J, Zhang L, Li H, Wu S, Liu Z. Nrf2 induced cisplatin resistance in ovarian cancer by promoting CD99 expression. Biochem Biophys Res Commun 2019;518(4):698–705.

[133] Lu M, Chen X, Xiao J, Xiang J, Yang L, Chen D. FOXO3a reverses the cisplatin resistance in ovarian cancer. Arch Med Res 2018;49(2):84–8.

[134] Li J, Zhang Y, Gao Y, Cui Y, Liu H, Li M, et al. Downregulation of HNF1 homeobox B is associated with drug resistance in ovarian cancer. Oncol Rep 2014;32(3):979–88.

[135] Fry AM, Schultz SJ, Bartek J, Nigg EA. Substrate specificity and cell cycle regulation of the Nek2 protein kinase, a potential human homolog of the mitotic regulator NIMA of Aspergillus nidulans. J Biol Chem 1995;270 (21):12899–905.

[136] Lee J, Gollahon L. Nek2-targeted ASO or siRNA pretreatment enhances anticancer drug sensitivity in triple-negative breast cancer cells. Int J Oncol 2013;42(3):839–47.

[137] Liu X, Gao Y, Lu Y, Zhang J, Li L, Yin F. Upregulation of NEK2 is associated with drug resistance in ovarian cancer. Oncol Rep 2014;31(2):745–54.

[138] Chiu W-T, Huang Y-F, Tsai H-Y, Chen C-C, Chang C-H, Huang S-C, et al. FOXM1 confers to epithelial-mesenchymal transition, stemness and chemoresistance in epithelial ovarian carcinoma cells. Oncotarget 2015;6(4):2349.

[139] Zhang LY, Li PL, Xu A, Zhang XC. Involvement of GRP78 in the resistance of ovarian carcinoma cells to paclitaxel. Asian Pac J Cancer Prev 2015;16(8):3517–22.

[140] Rada M, Nallanthighal S, Cha J, Ryan K, Sage J, Eldred C, et al. Inhibitor of apoptosis proteins (IAPs) mediate collagen type XI alpha 1-driven cisplatin resistance in ovarian cancer. Oncogene 2018;37(35):4809–20.

[141] Ni M-W, Zhou J, Zhang Y-L, Zhou G-M, Zhang S-J, Feng J-G, et al. Downregulation of LINC00515 in high-grade serous ovarian cancer and its relationship with platinum resistance. Biomark Med 2019;13(07):535–43.

[142] Polo SE, Jackson SP. Dynamics of DNA damage response proteins at DNA breaks: a focus on protein modifications. Genes Dev 2011;25(5):409–33.

[143] Özeş AR, Miller DF, Özeş ON, Fang F, Liu Y, Matei D, et al. NF-κB-HOTAIR axis links DNA damage response, chemoresistance and cellular senescence in ovarian cancer. Oncogene 2016;35(41):5350–61.

[144] Wang L, Mosel AJ, Oakley GG, Peng A. Deficient DNA damage signaling leads to chemoresistance to cisplatin in oral cancer. Mol Cancer Ther 2012;11(11):2401–9.

[145] Hassan M, Watari H, AbuAlmaaty A, Ohba Y, Sakuragi N. Apoptosis and molecular targeting therapy in cancer. Biomed Res Int 2014;2014:.

[146] Stefanou DT, Bamias A, Episkopou H, Kyrtopoulos SA, Likka M, Kalampokas T, et al. Aberrant DNA damage response pathways may predict the outcome of platinum chemotherapy in ovarian cancer. PLoS One 2015;10 (2):e0117654.

[147] Siegel RL, Miller KD, Jemal A. Cancer statistics, 2020. CA Cancer J Clin 2020;70(1):7–30.

[148] Walboomers JM, Jacobs MV, Manos MM, Bosch FX, Kummer JA, Shah KV, et al. Human papillomavirus is a necessary cause of invasive cervical cancer worldwide. J Pathol 1999;189(1):12–9.

[149] Landoni F, Maneo A, Colombo A, Placa F, Milani R, Perego P, et al. Randomised study of radical surgery versus radiotherapy for stage Ib-IIa cervical cancer. Lancet 1997;350(9077):535–40.

[150] Bansal N, Herzog TJ, Shaw RE, Burke WM, Deutsch I, Wright JD. Primary therapy for early-stage cervical cancer: radical hysterectomy vs radiation. Am J Obstet Gynecol 2009;201(5):485 [e1–9].

[151] Kim YS, Shin SS, Nam JH, Kim YT, Kim YM, Kim JH, et al. Prospective randomized comparison of monthly fluorouracil and cisplatin versus weekly cisplatin concurrent with pelvic radiotherapy and high-dose rate brachytherapy for locally advanced cervical cancer. Gynecol Oncol 2008;108(1):195–200.

[152] Tewari KS, Sill MW, Long 3rd HJ, Penson RT, Huang H, Ramondetta LM, et al. Improved survival with bevacizumab in advanced cervical cancer. N Engl J Med 2014;370(8):734–43.

[153] Cetina L, Rivera L, Candelaria M, de la Garza J, Dueñas-González A. Chemoradiation with gemcitabine for cervical cancer in patients with renal failure. Anticancer Drugs 2004;15(8):761–6.

[154] Cerrotta A, Gardan G, Cavina R, Raspagliesi F, Stefanon B, Garassino I, et al. Concurrent radiotherapy and weekly paclitaxel for locally advanced or recurrent squamous cell carcinoma of the uterine cervix. A pilot study with intensification of dose. Eur J Gynaecol Oncol 2002;23(2):115–9.

[155] Cao M, Gao D, Zhang N, Duan Y, Wang Y, Mujtaba H. Shp2 expression is upregulated in cervical cancer, and Shp2 contributes to cell growth and migration and reduces sensitivity to cisplatin in cervical cancer cells. Pathol Res Pract 2019;215(11):152621.

[156] Zhu H, Luo H, Zhang W, Shen Z, Hu X, Zhu X. Molecular mechanisms of cisplatin resistance in cervical cancer. Drug Des Devel Ther 2016;10:1885–95.

[157] Zisowsky J, Koegel S, Leyers S, Devarakonda K, Kassack MU, Osmak M, et al. Relevance of drug uptake and efflux for cisplatin sensitivity of tumor cells. Biochem Pharmacol 2007;73(2):298–307.

[158] Borst P, Evers R, Kool M, Wijnholds J. A family of drug transporters: the multidrug resistance-associated proteins. J Natl Cancer Inst 2000;92(16):1295–302.

[159] Li Q, Peng X, Yang H, Rodriguez J-A, Shu Y. Contribution of organic cation transporter 3 to cisplatin cytotoxicity in human cervical cancer cells. J Pharm Sci 2012;101(1):394–404.

[160] Meng F, Tan S, Liu T, Song H, Lou G. Predictive significance of combined LAPTM4B and VEGF expression in patients with cervical cancer. Tumour Biol 2016;37(4):4849–55.

[161] Sakaeda T, Nakamura T, Hirai M, Kimura T, Wada A, Yagami T, et al. MDR1 pp-regulated by apoptotic stimuli suppresses apoptotic signaling. Pharm Res 2002;19(9):1323–9.

[162] Galluzzi L, Senovilla L, Vitale I, Michels J, Martins I, Kepp O, et al. Molecular mechanisms of cisplatin resistance. Oncogene 2012;31(15):1869–83.

[163] Mellish KJ, Kelland LR, Harrap KR. In vitro platinum drug chemosensitivity of human cervical squamous cell carcinoma cell lines with intrinsic and acquired resistance to cisplatin. Br J Cancer 1993;68(2):240–50.

[164] Martelli L, Ragazzi E, Di Mario F, Basato M, Martelli M. Cisplatin and oxaliplatin cytotoxic effects in sensitive and cisplatin-resistant human cervical tumor cells: time and mode of application dependency. Anticancer Res 2009;29(10):3931–7.

[165] Zhang Y, Shu YM, Wang SF, Da BH, Wang ZH, Li HB. Stabilization of mismatch repair gene PMS2 by glycogen synthase kinase 3beta is implicated in the treatment of cervical carcinoma. BMC Cancer 2010;10:58.

[166] Karageorgopoulou S, Kostakis ID, Gazouli M, Markaki S, Papadimitriou M, Bournakis E, et al. Prognostic and predictive factors in patients with metastatic or recurrent cervical cancer treated with platinum-based chemotherapy. BMC Cancer 2017;17(1):451.

[167] Ding Z, Yang X, Pater A, Tang SC. Resistance to apoptosis is correlated with the reduced caspase-3 activation and enhanced expression of antiapoptotic proteins in human cervical multidrug-resistant cells. Biochem Biophys Res Commun 2000;270(2):415–20.

[168] Siddik ZH. Cisplatin: mode of cytotoxic action and molecular basis of resistance. Oncogene 2003;22 (47):7265–79.

[169] Jang HS, Woo SR, Song KH, Cho H, Chay DB, Hong SO, et al. API5 induces cisplatin resistance through FGFR signaling in human cancer cells. Exp Mol Med 2017;49(9):e374.

[170] Brozovic A, Osmak M. Activation of mitogen-activated protein kinases by cisplatin and their role in cisplatin-resistance. Cancer Lett 2007;251(1):1–16.

[171] Brozovic A, Fritz G, Christmann M, Zisowsky J, Jaehde U, Osmak M, et al. Long-term activation of SAPK/JNK, p38 kinase and fas-L expression by cisplatin is attenuated in human carcinoma cells that acquired drug resistance. Int J Cancer 2004;112(6):974–85.

[172] Wang X, Martindale JL, Holbrook NJ. Requirement for ERK activation in cisplatin-induced apoptosis. J Biol Chem 2000;275(50):39435–43.

[173] Liu JJ, Ho JY, Lee HW, Baik MW, Kim O, Choi YJ, et al. Inhibition of phosphatidylinositol 3-kinase (PI3K) signaling synergistically potentiates antitumor efficacy of paclitaxel and overcomes paclitaxel-mediated resistance in cervical cancer. Int J Mol Sci 2019;20(14):3383. https://doi.org/10.3390/ijms20143383.

[174] Garzetti GG, Ciavattini A, Lucarini G, Goteri G, De Nictolis M, Romanini C, et al. Modulation of expression of p53 and cell proliferation in locally advanced cervical carcinoma after neoadjuvant combination chemotherapy. Eur J Obstet Gynecol Reprod Biol 1995;63(1):31–6.

[175] Karin M. Nuclear factor-kappaB in cancer development and progression. Nature 2006;441(7092):431–6.

[176] Zheng X, Lv J, Shen Q, Chen Y, Zhou Q, Zhang W, et al. Synergistic effect of pyrrolidine dithiocarbamate and cisplatin in human cervical carcinoma. Reprod Sci 2014;21(10):1319–25.

[177] Shi TY, Cheng X, Yu KD, Sun MH, Shao ZM, Wang MY, et al. Functional variants in TNFAIP8 associated with cervical cancer susceptibility and clinical outcomes. Carcinogenesis 2013;34(4):770–8.

[178] Jing L, Bo W, Yourong F, Tian W, Shixuan W, Mingfu W. Sema4C mediates EMT inducing chemotherapeutic resistance of miR-31-3p in cervical cancer cells. Sci Rep 2019;9(1):17727.

[179] Zhu K, Chen L, Han X, Wang J. Short hairpin RNA targeting Twist1 suppresses cell proliferation and improves chemosensitivity to cisplatin in HeLa human cervical cancer cells. Oncol Rep 2012;27(4):1027–34.

[180] Shen DW, Pouliot LM, Hall MD, Gottesman MM. Cisplatin resistance: a cellular self-defense mechanism resulting from multiple epigenetic and genetic changes. Pharmacol Rev 2012;64(3):706–21.

[181] Liu X, Wang D, Liu H, Feng Y, Zhu T, Zhang L, et al. Knockdown of astrocyte elevated gene-1 (AEG-1) in cervical cancer cells decreases their invasiveness, epithelial to mesenchymal transition, and chemoresistance. Cell Cycle 2014;13(11):1702–7.

[182] Bai T, Tanaka T, Yukawa K, Umesaki N. A novel mechanism for acquired cisplatin-resistance: suppressed translation of death-associated protein kinase mRNA is insensitive to 5-aza-2′-deoxycitidine and trichostatin in cisplatin-resistant cervical squamous cancer cells. Int J Oncol 2006;28(2):497–508.

[183] Chen CC, Lee KD, Pai MY, Chu PY, Hsu CC, Chiu CC, et al. Changes in DNA methylation are associated with the development of drug resistance in cervical cancer cells. Cancer Cell Int 2015;15:98.

[184] Candelaria M, Edl C-H, Gonzalez-Fierro A, Perez-Cardenas E, Trejo-Becerril C, Taja-Chayeb L, et al. Epigenetic changes in nucleoside transporter hENT1 and dCK, as mechanism for gemcitabine-aquired resistance in cervical cancer cell lines. J Clin Oncol 2010;28(15_suppl):e13633.

[185] Candelaria M, de la Cruz-Hernandez E, Taja-Chayeb L, Perez-Cardenas E, Trejo-Becerril C, Gonzalez-Fierro A, et al. DNA methylation-independent reversion of gemcitabine resistance by hydralazine in cervical cancer cells. PLoS One 2012;7(3):e29181.

[186] Liu SY, Zheng PS. High aldehyde dehydrogenase activity identifies cancer stem cells in human cervical cancer. Oncotarget 2013;4(12):2462–75.

[187] Dean M, Fojo T, Bates S. Tumour stem cells and drug resistance. Nat Rev Cancer 2005;5(4):275–84.

[188] Cao HZ, Liu XF, Yang WT, Chen Q, Zheng PS. LGR5 promotes cancer stem cell traits and chemoresistance in cervical cancer. Cell Death Dis 2017;8(9):e3039.

[189] Chen M, Ai G, Zhou J, Mao W, Li H, Guo J. circMTO1 promotes tumorigenesis and chemoresistance of cervical cancer via regulating miR-6893. Biomed Pharmacother 2019;117:109064.

[190] Chen Y, Ke G, Han D, Liang S, Yang G, Wu X. MicroRNA-181a enhances the chemoresistance of human cervical squamous cell carcinoma to cisplatin by targeting PRKCD. Exp Cell Res 2014;320(1):12–20.

[191] Lei C, Wang Y, Huang Y, Yu H, Wu L, Huang L. Up-regulated miR155 reverses the epithelial-mesenchymal transition induced by EGF and increases chemo-sensitivity to cisplatin in human Caski cervical cancer cells. PLoS One 2012;7(12):e52310.

[192] Wang F, Liu M, Li X, Tang H. MiR-214 reduces cell survival and enhances cisplatin-induced cytotoxicity via down-regulation of Bcl2l2 in cervical cancer cells. FEBS Lett 2013;587(5):488–95.

[193] Feng C, Ma F, Hu C, Ma JA, Wang J, Zhang Y, et al. SOX9/miR-130a/CTR1 axis modulates DDP-resistance of cervical cancer cell. Cell Cycle 2018;17(4):448–58.

[194] Kato R, Nishi H, Nagamitsu Y, Sasaki T, Isaka K. Abstract 4378: miR-100 mediates resistance to paclitaxel in cervical cancer cells. Cancer Res 2014;74(Suppl. 19):4378.

[195] Guo J, Chen M, Ai G, Mao W, Li H, Zhou J. Hsa_circ_0023404 enhances cervical cancer metastasis and chemoresistance through VEGFA and autophagy signaling by sponging miR-5047. Biomed Pharmacother 2019;115:108957.

[196] Castagna A, Antonioli P, Astner H, Hamdan M, Righetti SC, Perego P, et al. A proteomic approach to cisplatin resistance in the cervix squamous cell carcinoma cell line A431. Proteomics 2004;4(10):3246–67.

[197] Zhang Y, Shen X. Heat shock protein 27 protects L929 cells from cisplatin-induced apoptosis by enhancing Akt activation and abating suppression of thioredoxin reductase activity. Clin Cancer Res 2007;13(10):2855–64.

[198] Park SY, Wj K, Jh B, Lee JJ, Jeoung D, Park ST, et al. Role of DDX53 in taxol-resistance of cervix cancer cells in vitro. Biochem Biophys Res Commun 2018;506(3):641–7.

[199] Peng X, Gong F, Chen Y, Jiang Y, Liu J, Yu M, et al. Autophagy promotes paclitaxel resistance of cervical cancer cells: involvement of Warburg effect activated hypoxia-induced factor 1-α-mediated signaling. Cell Death Dis 2014;5(8):e1367.

[200] Jin L, Shen Q, Ding S, Jiang W, Jiang L, Zhu X. Immunohistochemical expression of Annexin A2 and S100A proteins in patients with bulky stage IB-IIA cervical cancer treated with neoadjuvant chemotherapy. Gynecol Oncol 2012;126(1):140–6.

# Mechanisms of chemoresistance and approaches to overcome its impact in gynecologic cancers

*Nirupama Sabnis[a], Ezek Mathew[b,c], Akpedje Dossou[a,c], Amy Zheng[b], Bhavani Nagarajan[d], Rafal Fudala[c], and Andras G. Lacko[a,e]*

[a]Department of Physiology and Anatomy, University of North Texas Health Science Center, Fort Worth, TX, United States [b]Texas College of Osteopathic Medicine, University of North Texas Health Science Center, Fort Worth, TX, United States [c]Department of Molecular Biology, Immunology and Genetics, Graduate School of Biomedical Sciences, University of North Texas Health Science Center, Fort Worth, TX, United States [d]North Texas Eye Research Institute, University of North Texas Health Science Center, Fort Worth, TX, United States [e]Department of Pediatrics and Women's Health, University of North Texas Health Science Center, Fort Worth, TX, United States

## Abstract

Gynecologic cancers (GC) affect millions of women of all ages globally each year. Apart from a crucial economic impact of these cancers on a society, it affects the emotional, physical, material, and social well-being of survivors and their kinfolks as well. GC diagnostics is often at advanced stages, restricting the options of therapeutics. In addition, patients with GCs undergoing chemotherapy often show resistance to chemotherapeutic drugs and recurrence. In fact, the largest hurdle in GC therapeutics is multidrug (chemo) resistance (MDR). Comprehending the key mechanisms involved in chemoresistance, innovative research in the area of targeted drug discoveries and identifying unique and effective drug targets/inhibitorsare necessary for the clinical management of GC patients efficiently. In this chapter, classical as well as novel molecular mechanisms of drug resistance in GCs, particularly ovarian, cervical, and endometrial cancers from relevant peer-reviewed literature, have been investigated. Finally, targeted drug deliveries to combat MDR in GCs are discussed in detail. The knowledge gained could assist design meaningful preclinical and translational studies to overcome MDR in GCs.

R. Basha, S. Ahmad (eds.)
*Overcoming Drug Resistance in Gynecologic Cancers*
ISBN 978-0-12-824299-5
https://doi.org/10.1016/B978-0-12-824299-5.00008-3

# Abbreviations

| | |
|---|---|
| **3-MA** | 3-methyl adenine |
| **4-HPR** | (dihydro) ceramide desaturase |
| **AA** | arachidonic acid |
| **ABC** | ATP-binding cassette |
| **ABCB1** | ATP-binding cassette sub-family B member 1 |
| **Alix** | ALG-2 interacting protein X |
| **ALK5** | activin like kinase 5 |
| **AMBRA1** | activating molecule in Beclin-1-regulated autophagy |
| **AMPK** | 5′ AMP-activated protein kinase |
| **ANGPTL4** | angiopoietin-like 4 |
| **ATG** | autophagy-related genes |
| **ATP7A/7B** | ATPase (copper-transporting) |
| **BBC3** | Bcl-2-binding component 3 |
| **BCL-2** | B-cell lymphoma-2 |
| **Bcl-xL** | B-cell lymphoma-extra large |
| **CAFs** | cancer-associated fibroblasts |
| **Cav1** | caveolin-1 |
| **CCR2** | C-C chemokine receptor type 2 |
| **CLU** | clusterin |
| **CNT** | concentrative nucleoside transporters |
| **CRCs** | colorectal cancer cells |
| **CSCs** | cancer stem cells |
| **CSF-1R** | colony stimulating factor 1 receptor |
| **CTLA-4** | cytotoxic T-lymphocyte-associated protein 4 |
| **CYP** | cytochrome P450 |
| **DAPK** | death-associated protein kinase |
| **DDR** | DNA damage repair |
| **DOPC** | 1,2-DiOleoyl-sn-glycero-3-PhosphoCholine |
| **EIG121** | estrogen-induced gene 121 |
| **EMT** | epithelial-to-mesenchymal transition |
| **ENT1 and 2** | equilibrative nucleoside transporters 1 & 2 |
| **EOC** | epithelial ovarian cancer |
| **ERB** | tyrosine protein kinase |
| **ERK1/2** | extracellular-signal-regulated kinase 1/2 |
| **FAK** | focal adhesion kinase |
| **FAO** | fatty acid oxidation |
| **FASN** | fatty acid synthase |
| **FOXO3** | forkhead Box O3 |
| **GST** | glutathione-S-transferase |
| **HDACs** | histone deacetyltransferases |
| **HGSOC** | high-grade serous ovarian cancer |
| **Hh** | hedgehog |
| **HIF-α** | hypoxia inducible factor 1 α |
| **HMGB1** | high-mobility Group Box 1 |
| **HOTAIR** | HOX transcript antisense RNA |
| **HPV** | human papilloma virus |
| **HR** | homologous recombination |
| **HSP27** | heat shock protein 27 |
| **IAPs** | inhibitor of apoptotic proteins |
| **IL** | interleukin |
| **INFγ** | interferon gamma |

| | |
|---|---|
| **JAK** | Janus kinase |
| **LDL** | low-density lipoproteins |
| **LncRNA** | long noncoding RNA |
| **mAb** | monoclonal antibody |
| **MAPK** | mitogen activated protein kinase |
| **MATE1** | multidrug and toxin extrusion 1 |
| **Mcl-1** | myeloid cell leukemia 1 |
| **MCT** | mono-carboxylate transporters |
| **MDR** | multidrug resistance |
| **MEK** | mitogen-activated protein kinase |
| **MET** | mesenchymal to epithelial transition |
| **miRNA or miR** | micro RibonNucleic acid |
| **MMR** | mismatch repair |
| **MRP2** | multidrug resistance-associated protein 2 |
| **MSCs** | mesenchymal stem cells |
| **mTOR** | mammalian target of tapamycin |
| **mTORC1** | mammalian target of rapamycin complex 1 |
| **NF-κB** | nuclear factor kappa-light-chain-enhancer of activated B cells |
| **NO** | nitric oxide |
| **Nrf2** | erythroid transcription factor NF-E2 |
| **OATPs** | organic anion T=transporting proteins OATPs |
| **OC** | ovarian cancer |
| **OCTs** | organic cation transporters |
| **PARP** | poly ADP-ribose polymerase |
| **PD-L1** | programmed death-ligand 1 |
| **PEPT** | peptide transporters |
| **P-gp** | P-glycoprotein |
| **PI3K** | phosphoinositide-3-kinase complex |
| **PPARs** | peroxisome proliferator-activated receptors |
| **PTEN** | phosphatase and tensin homolog |
| **rHDL** | reconstituted high-density lipoprotein |
| **RNS** | reactive nitrogen species |
| **ROS** | reactive oxygen species |
| **s&lncRNAs** | small & long noncoding RNAs |
| **S1P** | sphingosine 1-phosphate |
| **SCT** | solute career transporter |
| **Sema4C** | Semaphorin-4C |
| **shRNA** | short hairpin RNA |
| **SIRP1α** | signal regulatory protein 1α |
| **SIRT1** | Sirutunin1 |
| **SLC** | solute career |
| **SNAi2** | snail family transcriptional repressor 2 |
| **SOX4** | SRY-related high mobility group box 4 |
| **SphK** | sphingosine kinase |
| **Src** | Sarc (Sarcoma) |
| **STAT3** | signal transducer and activator of transcription 3 |
| **TAMs** | tumor-associated macrophages |
| **TGF-β1** | transforming growth factor-beta 1 |
| **TME** | tumor micro-environment |
| **TNFα** | tumor-necrosis factor α |
| **TRP14** | thioredoxinrelated protein of 14 kDa |
| **TRPM-2** | testosterone-repressed prostate message-2 |
| **UGT** | uridine diphospho-glucuronosyl transferase |
| **Wnt** | wingless related integration site |

## Conflict of interest

No potential conflicts of interest were disclosed.

# Introduction

In 2019, gynecologic cancers accounted for approximately 12% of all new cancer diagnoses and approximately 12% of cancer mortality in women in the United States [1]. These cancers carry a significant economic burden and diminish the well-being and quality of life of survivors and their families. In all cases of gynecologic cancers, early diagnosis is associated with better survival and cure rates. Recent rapid advancements in research, in the area of cancer biology and the discovery of novel immunotherapies, have enhanced the progress toward modern, personalized approaches for managing all cancers in general and women's cancers in particular.

Each of the five gynecologic cancers (viz., cervical, ovarian, endometrial, vaginal, and vulvar) presents with distinct risk factors, prevalence, survival rate, and response to treatment. For instance, the most common gynecologic cancer globally, epithelial ovarian cancer (EOC), is associated with a 5-year survival rate of about 49% compared to 66% for cervical cancer [2], while the death rates for all gynecologic cancers have decreased since the year 2009, including for cervical cancer, due to increased screening, and administration of Human papilloma virus (HPV) vaccines [2, 3].

Endometrial cancer (EC) originates from the inner epithelial lining of the uterus. It is the fourth most common cancer in women in the United States, after breast, lung, and colorectal cancers [4]. Although the survival rate (5 years) is high (81%) compared to other gynecologic cancers, the rising incidence of EC is expected to have 65,000 new cases and 12,500 related deaths in 2020 augmented by its association with obesity (BMI $\geq 30 \, \text{kg/m}^2$) perhaps due to the impact of estrogens and their mitogenic role in endometrial tissue growth [5, 6]. Adipose tissues also contain the enzyme aromatase, which converts the androgen androstenedione to estrone, an estrogen. Excess adipose tissue results in high levels of estrogen synthesis and secretion into the blood. This ultimately inhibits ovulation and signals the endometrium to proliferate [7].

Cervical cancer (CC) is the fourth frequently observed cancer in women around the world [8]. In the United States, it is the third most common GC in women after endometrial and ovarian cancers [9]. The main factor for the risk of cervical cancer is persistent HPV infection, a common sexually transmitted virus [10]. There are over 200 strains of HPV, and 14 of which cause human cancer [11]. Although the majority of those infected, experience spontaneous resolution, HPV is the major cause of almost 99% of the cases [12]. From the year 2000 to 2017, cervical cancer incidence declined from 9.6 to 7.5 cases per 100,000 women [13]. Increased routine HPV screening and Pap tests have reduced the deaths from cervical cancer by 80% [14, 15]. The Pap test involves collecting cells at the transformation zone of the cervix to identify cellular changes that could be caused by HPV [15]. The HPV vaccines, which became available in the year 2006, are safe and highly effective [9, 16].

Ovarian cancer (OC) is the number one killer and fifth most frequently observed cancer among women [1, 17]. The 20,000 new cases diagnosed yearly are accompanied by a survival rate of only 49% in 5 years, accounting for 2.3% of all cancer deaths [3, 14]. From the year 2000 to 2017, cases of OC decreased slightly from 14.4 to 10.2 per 100,000 women per year, indicating a slight reduction in the death rate [18]. OC is considered to be the most typical malignancy among postmenopausal women, and its highest risk factor is increased age. The poor prognosis of OC is driven in part by the lack of effective screening tools (diagnostic tests) resulting in most patients having already advanced disease at diagnosis. Unlike cervical cancer, where HPV is the primary cause, the cause of OC is not well known. The known risk factors include increasing age, more frequent lifetime ovulation, early menarche, late menopause, nulliparity, and BReast CAncer (BRCA)1 or BRCA2 gene mutations [9].

OC is also the top cause of morbidity among patients with gynecologic malignancies in the United States. A recent study conducted by the USInstitute of Medicine came to the overarching consensus that "… *ovarian cancer is not a single disease as it is represented by a number of tumor types with characteristic histologic features, distinctive molecular signatures, and disease trajectories. Moreover, these OC tumors are heterogeneous, and they can arise from different tissues of the female reproductive tract*" [19]. Among the subtypes of OC, high-grade serous type (HGSC) is currently the most fatal among gynecologic malignancies [20].

Currently, the majority of OC patients exhibit favorable response to initial therapy of surgery combined with subsequent platinum/taxane-based chemotherapy [19]. Recent progress in the treatment of OC is limited to empirical optimization [21] of chemotherapeutic agents (or their combinations) and their enhanced delivery. Although, these efforts have led to some improvements in the survival of OC patients [19], approximately 70% with advanced-stage disease develop cancer recurrence, typically due to multidrug resistance (MDR) and eventually these patients succumb to the recurrent disease. In order to improve the prognosis of OC patients, the focus for developing new strategies has recently shifted to concentrating on single subtypes of OC (instead of a broader/all-encompassing approach). In this regard, HGSC is of special interest as it is responsible for the maximum deaths from OC [20] and patient outcomes have only shown limited improvements in the last several decades. Approximately, 50% of the HGSC tumors showdefects in the homologous recombination (HR) DNA repair pathway, a major determinant of platinum sensitivity [21]. Despite the excellent response to poly (ADP-ribose) polymerase (PARP) inhibitors in several ongoing clinical trials of patients with homologous recombination (HR) deficient tumors, the majority of patients do not respond or develop resistance [22].

## Key biological mechanisms of drug resistance

Multidrug- or chemoresistance has been recognized as an utmost challenge for cancer biologists in the past three decades as it continues to be a key limiting factor for cancer therapy. Chemoresistance against drugs has been characterized as either "intrinsic" or "acquired." The detoxification mechanisms overexpressed, prior to the initiation of chemotherapy is considered "intrinsic" or "primary" chemoresistance. The "acquired" form of drug resistance on the other hand develops after the chemotherapy regimen is started or in response to the secondary therapy leading to tumor relapse [23, 24]. The mechanisms underlying the limited success of cancer chemotherapy (due to resistance) has been explored via clinical studies and attempts to elucidate the mechanisms of molecular biological processes, associated with drug resistance [25, 26]. Insufficient distribution of the chemotherapeutics to the "refuge sites" is one of the mechanisms responsible for the limited effectiveness of chemotherapy. This mechanism has been attributed, at least in part, to the intricate and complex nature of the tumor microenvironment (TME). The TME is comprised of a variety of components including leaky vasculature, dense extracellular matrix, hypoxia, and acidosis [27]. Matsumoto et al. demonstrated that antiangiogenic agents such as sunitinib has a potential to render the characteristics of TME vasculature back to its original form, thereby ameliorating the administration of chemotherapeutic drugs to their respective target sites [28].

It has also been observed that the suboptimal dosing of the chemotherapeutic regimens often results in the failure of chemotherapy via pharmacological mechanisms [25, 29]. Once the MDR is acquired by the cancer cells, the impact of anticancer drugs on the development and metastatic growth of malignant lesions decreases precipitously. MDR has been considered as the most detrimental factor limiting the efficiency of cancer chemotherapeutics and thus contributing to cancer metastasis and recurrence [30]. The potential mechanisms of MDR currently reported are shown in Fig. 1, including the membrane transporters, cell death induction, altered drug metabolism, disruption of redox homeostasis, regulation of miRNA, induction of hypoxia, epigenetic regulation, DNA damage and repair, long noncoding RNAs, and oncogenic regulation.

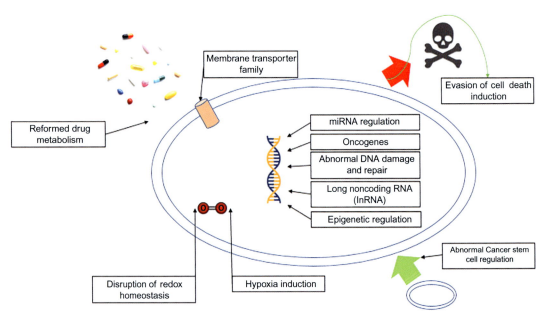

FIG. 1    Mechanisms of drug resistance in cancer cells.

# Membrane transporters

Decrease in cellular accumulation of chemotherapeutics is often considered as one of the most universal mechanisms in the chemoresistance in oncogenic cells. This usually occurs via elevated efflux or reduced influx of drugs via distinct membrane transporters [30, 31]. The balance between the efflux and influx transporters as well as the nature of heterogeneity in genes encoding them are the likely contributors of the altered pharmacodynamics of the therapeutic agents [32–34]. The most pivotal efflux and influx transporters that contribute to drug resistance in gynecologic cancers are discussed in the following section.

## *Efflux transporters*

ATP-binding cassette (ABC) transporters are members of the protein superfamily. To date, these transporters are known to consist of 49 proteins that mediate diverse ATP-driven transport processes [35, 36]. Phylogenetic analysis distinguishes these transporters into seven well-defined subfamilies (ABCAABCG) [37, 38]. These transporters act as efflux pumps to deliver a large number of anticancer drugs. The most widely studied among these are the ABC transporters:ABCC1 (MRP1), ABCB1 (P-glycoprotein or P-gp), and ABCG2 (BCRP). These are the chief transporters that are upregulatedas well as main contributors to the development of drug resistance in different malignancies. Various signaling pathways are known to be activated in upregulation of ABC transporters. These include extracellular signal-regulated kinase (ERK), Forkhead Box O3 (FOXO3a), phosphoInositide-3-kinase complex (PI3K/Akt), nuclear factor kappa-light-chain-enhancer of activated B cells (NF-κB), and heat shock protein 27 (HSP27) depletion pathways. Some studies have revealed that hyperactivation of Hedgehog (Hh) pathway is directly associated with a chemotherapy-resistant phenotype in OC [39]. A significant number of ABC transporters' inhibitors have been identified. However, more screening of unique nontoxic inhibitors that selectively target cancer cells and exert minimal side effects are needed in the clinical setting [30, 40–42].

## *Influx transporters*

Several influx transporters have been identified and characterized. These include the Organic Anion Transporting Proteins (OATPs), Organic Cation Transporters (OCTs), Concentrative Nucleoside Transporters (CNT), Peptide Transporters (PEPT), Monocarboxylate Transporters (MCT), multidrug and toxin extrusion 1 (MATE1), and Equilibrative Nucleoside Transporters (ENT1 and ENT2) [32, 34, 38]. These receptors/pump systems are located in the membranes of cytoplasm and organelles. They have gained attention in recent studies and belong to the solute carrier transporter (SCT) family. These encompasspassive, ion-coupled, and exchanger transporters. These carriers are involved in the uptake of multiple substrates into the cells [34, 43–45]. OATPs are one of the most studied solute carrier family which are expressed in the hepatocytes [43]. Various human cancer tissues, including OC, as well as a wide range of cancer cell lines have demonstrated upregulation of a member of this cluster, OATP1B3. A rigorous study of solute carrier (SLC) and ABC transporters detection and expression analysis using microarray in chemo-resistant OC cell lines demonstrated drastic changes in the corresponding gene expression levels in drug (methotrexate, cisplatin, doxorubicin, vincristine, topotecan, and paclitaxel)-resistant strain of W1 OC cell line. It was shown that among the 38 SLC genes studied, upregulation was observed in 18 genes, downregulation in 16 genes, and 4 were either up- or downregulated based on the type of cell line studied [45]. In another study, several genetic screenings were performed in haploid human cell lines to decipher the relationship between SLC transporter and 60 cytotoxic compounds. SLC-focused CRISPR-Cas9 library was utilized to identify the transporters, which were key for inducing drug resistance. These mechanisms demonstrated therole of SLCs transporters for artemisinin derivatives and cisplatin in the uptake and cellular activity of cytotoxic drugs. These studies validated the resistance observed to a number of therapeutic agents to their association with SLCs [46].

## Reformed drug metabolism

Metabolic pathway alterations are often associated with MDR. Among the reformed metabolic and signaling pathways affected in GCs, dysregulation of enzymes involved in hormone biosynthesis also contribute to chemoresistance along with drug-metabolizing enzymes [47–49]. The intracellular concentrations of drugs are often decreased by overexpressed metabolizing enzymes prior to exerting their detrimental effects. These are categorized based on their effect on the phase of metabolism [38]. Among these, the key players are: cytochrome P450, glutathione-S-transferase (GST), and uridine diphospho-glucuronosyltransferase (UGT) super families [49–53]. Dysregulation of these metabolizing enzymes are further categorized as phase I and phase II affecting enzymes.

### *Phase I affecting enzymes*

Cytochrome P450 (also known as P450s or CYP) enzymes are the leading participants in phase I drug metabolism and comprised of hemoprotein superfamily. To date, more than 57 CYP genes and 58 pseudogenes have been identified in human genome. Studies show that more than 90% of the chemotherapeutic drugs used currently in clinical setting are metabolized by these enzymes [54–56]. Change in the drug efficacy as a result of altered drug metabolism could be facilitated by mutation-induced alterations in CYP activity and/or other types of modifications in tumor cells [50]. This could be observed when overexpression of P450s, in particular CYP1B1, results in increased sequestration and inactivation of anticancer drugs [57, 58]. It has recently been shown that docetaxel and other anticancer drug resistance is affected by overexpression of CYP1B [59]. Furthermore, patients diagnosed with advanced stages of OC may exhibit resistance to platinum-containing drugs. This resistance may occur via detoxifying drug. Although, there are several factors that influence bioactivation of prodrugs among GC patients, CYP superfamily elements are known to play a major role [50, 55, 56]. Hence, P450 profiling could be an additional tool to correlate the key P450 variants that could affect the patient's response to therapeutics. This technique would be a major advance toward selecting for the effectiveness of a specific chemotherapeutic agent to treat patients via "personalized therapy" [60].

### *Phase II affecting enzymes*

The GST enzymes play a significant part in protecting tissues from oxidative damage. Encoded by *GST* genes, the chief function of these enzymes is to detoxify chemicals and metabolites present in the milieu. Pharmacological variability in certain types of cancers is often associated with the genetic polymorphism involved in the detoxification of xenobiotics. Studies have demonstrated an increased risk of cervical cancer in women with *GST* gene polymorphism indicating a close association between *GST* gene polymorphism and cervical cancer [61–63]. GSTs are also shown to be engaged in the inhibition of mitogen-activated protein kinase (MAPK) pathway. Yang et al. (2017) showed a GSTP1-mediated chemoresistance conferred by high expression of CLDN6 in breast cancer [64]. A role of GSTP1 in cisplatin and carboplatin metabolism in OC cells is well characterized. Differential expressions of *GSTP1* may be a diagnostic tool to determine whether platinum-based chemotherapy is suitable for particular OC patients [65].

The UGTs are another important phase II conjugating enzymes responsible for glycosylation of endobiotics and xenobiotics [66–68]. Several biomolecules, identified as substrates for UGTs, include numerous xenobiotics, chemotherapeutic agents, environmental carcinogens, as well as fat-soluble vitamins, bilirubin, bile acids, steroid, and thyroid hormones. An exhaustive analysis of 17 UGTs suggests that UGT1A and UGT2B variants (1A1, 1A3, 1A8, 1A9, 1A10, and 2B7) are the major components involved in glucuronidation process [69–72]. Analysis of UGTs in the endometrium of healthy postmenopausal women demonstrated expression of UGT variants 1A1, 1A8, 1A9, 2B7, and 2B15 [73]. In another study, rise in glucuronidation was observed due to overexpression of UGT1A1 in Irinotecan-treated GC patients. UGT gene polymorphism was observed to be associated with resistance to irinotecan and other drugs as well as with irinotecan off target toxicity [74, 75].

### Other alterations

Among other alterations in drug metabolism leading to drug resistance are Wnt/β signaling pathway components that target genes-regulating cell growth and apoptosis. Involvement of Wnt/β-catenin signaling pathway has been demonstrated in epithelial-to-mesenchymal transition (EMT). Alterations in Wnt pathway proteins have been shown to influence the progression of OC [47, 76]. Another well-studied enzyme that is likely to be involved in drug resistance is aromatase, which is a rate-limiting enzyme in estrogen biosynthesis. Although inhibitors of aromatase have been efficiently used to treat breast cancer in postmenopausal women, their severe side effects such as drastic estrogen deprivation, bone loss, and abnormal lipid metabolism have restricted their use in other GCs. Instead, aromatase promoter regulation in malignant tissues was manipulated to target breast and endometrial cancers. These studies indicated a potential opportunity to inhibit aromatase activity specifically in breast and endometrial tissue without the earlier side effects observed in other organs, including bone [77].

## Dysregulation of cellular metabolism

During rapid proliferation, cancer cells experience intricate metabolic rearrangements characterized by alterations in signaling and metabolic pathways. The dysregulation of cellular metabolism has been recognized as an emerging endorsement of both treatment-naive and treatment-resistant GCs. Among the metabolic processes affected, glycolysis, lipid metabolism (fatty acid synthesis), and amino acid metabolism are the ones studied in depth. Studies have demonstrated that dysregulated metabolism in tumor cells is directly associated with drug resistance [78, 79]. Because one of the mechanisms whereby cells become resistant to therapy is the rewiring of their metabolism these mechanisms thus may serve as potential targets for therapeutic intervention. Examples for paradigmatic dysregulations observed in gynecologic cancers are discussed in the following section.

### Glucose metabolism

Since the early 20th century, Warburg, a physiologist Nobel Laureate, observed the dependency of tumor cells on glycolysis for energy generation, although there is an adequate quantity of oxygen readily available in the surrounding TME. This phenomenon later known as the Warburg effect of cancer cells has been extensively discussed in the peer-reviewed literature [80–83] including those components that could be the potential therapeutic targets for the gynecological cancers [84–86]. The key factors that may serve in this capacity are pyruvate kinase, lactate dehydrogenase, tumor suppressor gene P53, and hypoxia-inducible factor 1 [87–94].

Studies on factors attributing to the elevation in prevalence of endometrial cancer show that metabolic syndrome has a pivotal role in triggering high estrogen and high insulin levels in the blood. Metformin, a proposed drug for an effective therapeutic regimen for endometrial and cervical cancers, is currently prescribed to treat type-2 diabetes because it suppresses gluconeogenesis, thereby lowering blood glucose levels. In the reverse Warburg effect model, this property of metformin is exploited to arrest cancer cell metabolism by directly interfering with aerobic respiration [95, 96]. In another study, using transgenic mouse model, statin drugs were shown to reduce the development of serous tubal intraepithelial carcinomas in human OC. These studies suggested the possibility of using statin drugs via interfering with the Warburg effect as a novel strategy to prevent and manage OC and other GCs [97, 98].

### Lipid metabolism

Reprogramming of lipid metabolism in gynecologic cancers has been reported in the peer-reviewed literature [99–101]. Studies by Cruz et al. have discussed the cholesterol biosynthesis in malignant and nonmalignant cells [102]. A systematic analysis of the collective omics data suggested that the predominantly influenced lipid metabolism pathways include but not limited to fatty acid metabolism, arachidonic acid metabolism, cholesterol metabolism, and peroxisome proliferator-activated receptors (such as PPAR) signaling. In this study, the gene expression profiles of fatty acid metabolism revealed the similarities and differences among different types of cancers. The alterations observed in cholesterol and arachidonic acid metabolism were attributed to the tissue origin of the effect. These studies suggested that alteration in lipid metabolism in different types of malignancies could be "tissue specific" [103].

Low-density lipoproteins (LDL) metabolism in cancer cells derived from various types of GCs with multiple degrees of differentiation (epidermoid vaginal carcinoma, epidermoid cervical carcinoma, and endometrial adenocarcinoma) suggested that these cells metabolized LDL at 20 times higher rate than the noncancerous cervical fibroblasts and epithelial cells originated from endometrial glands [99]. In other studies, the increased expression and activity of lipogenic enzymes was shown to be largely responsible for increased lipid synthesis. Dysregulation of a sphingolipid metabolism pathway has been shown to affect drug resistance [104]. During exposure of cells to pro-apoptotic signals, such as chemotherapeutic agents, ceramide acts as an important mediator of apoptosis and growth inhibitor. Ceramide production is downregulated when some enzymes involved in sphingolipid metabolism are dysregulated. Decreased levels of ceramide are shown to be associated with the development of MDR [105]. Attempts to coadminister ceramide and paclitaxel in OC cellswere shown to revert the susceptibility to chemotherapeutic agents leading to apoptosis, compared to paclitaxel alone [104]. In contrast, activation of sphingosine kinase (SphK) was shown to upregulate sphingosine 1-phosphate (S1P) in OC cells that resulted in resistance to (dihydro) ceramide desaturase (4-HPR), a sphingolipid inhibitor (also a chemotherapeutic agent). These studies emphasize the importance of ceramide and S1P in the development of chemoresistance [104, 105]. Other genes of lipid metabolism involved in drug-resistant studies using elaborate gene expression analysis and other related studies include *HMGCS2*, *GPX2,*and *CD36*. These genes are shown to be associated with poor prognosis as a result of drug resistance [103].

### Amino acid metabolism

Amino acids are major nutrients for the cell survival, especially for malignant cells. The amino acids primarily function as the building blocks for protein synthesis; however, some amino acids have multiple roles including specific biological (signaling) functions. In proliferating cancer cells, amino acids are involved in multiple oncogenic pathways to fulfill the bioenergetic and biosynthetic needs for rapid tumor growth. Leucine, glutamine, and arginine have a pivotal role as signaling molecules and activation of mTOR, whereas glutamine, glycine, and aspartate are contributors to nucleotide biosynthesis. Serine plays a dual role as a one-carbon source essential for nucleotide synthesis as well as DNA methylation. Locasale and Phang et al. have shown that tumor cells have altered metabolic pathways for specific amino acids. These include: threonine, proline, serine, and glutamine [106, 107]. Due to their amphipathic nature, amino acids need assistance from selective transport proteins to cross the plasma membrane.

Mammalian cells exhibit several amino acid transporters. These are expressed differentially in different malignancies either by a tissue-specific and/or by a development-specific manner [108, 109]. Because the signaling pathways responsible for tumor progressiondrastically differ among tumor types, it is likely that different amino acid transporters are employed to acquire amino acids for their growth. This differential regulation of selective amino acid transporters in different cancers presents an opportunity for targeted and personalized cancer therapy.

Glutamine is a critical factor in protein synthesis. It functions as a nitrogen donor for the synthesis of nonessential amino acids, as well as nucleotides and hexosamines. Many cancer cells have been observed to be "addicted" to glutamine, a process which is often exhibited in cells performing oxidative as well as in glycolytic metabolism [108, 109]. In addition, glutamine also serves as akey precursor of the biosynthesis process of glutathione and a carbon source in the tricarboxylic acid (TCA) cycle [110, 111]. This concept expands the theory that glucose is the only source of energy for tumor cell proliferation and suggests that glutamine can be a major alternative for glucose in the TCA cycle [112].

In a recent study, targeting glutamine metabolism combined with platinum-based chemotherapy has shown the potential for effectively treating drug-resistant ovarian cancer [113]. In this study, the investigators demonstrated that in platinum-resistant OC cell, TCA cycle is a predominant mechanism for glutamine metabolism via an upregulated transcription factor c-Myc. Additionally, they have observed that overexpression of a mitochondrial enzyme glutaminase rendersOC cells resistant to platinum treatment. Furthermore, treatment of OC cells with a combination of cisplatin and a glutaminase inhibitor, BPTES, was able to exhibit platinum sensitivity back to cells. In another study, using inhibitors of focal adhesion kinase (FAK), the effect on glutamine metabolism was studied in ovarian clear cell carcinomas [114]. Inhibition of glutaminolysis resulted in apoptosis activation as well as sensitization ofOC cells to one of the mTOR inhibitors(PP242). Also, phosphorylation of STAT3 was considerably decreased by inhibition of GLS1 in OC cells. These studies show a novel therapeutic approach where chemoresistance developed due to drugs that include PI3K/Akt/mTOR inhibitors can be overcome by manipulating "glutamine addiction" in OC. These studies could be extended to other GC therapeutics as well [115].

## Disruption of redox homeostasis

Redox homeostasis is referred to as an equilibrium between cellular oxidants and antioxidants in living cells. The disturbance of this equilibrium promotes cancer progression and development. Studies demonstrate that malignant cells exhibit elevated levels of reactive oxygen or nitrogen species (ROS or RNS) as compared to normal cells [116, 117]. In fact, induction of tumor cell deathdue to excess ROS concentrations has been reported in literature [118]. In vivo, inflammatory cells, and "xanthine/xanthine oxidase system" have been identified as sources for generation of free radicals. The ROS and RNS are also formed during metabolism of a number of anticancer drugs and xenobiotics [116, 119]. In the oxidative stress state, abnormally high levels of ROS damage the building blocks of cells (nucleic acids, proteins, and lipids), producing a cascade effect impacting key cell functions, including proliferation, angiogenesis, genetic instability, as well as drug resistance in malignant cells, via disruption of key signaling pathways [116, 118, 120].

Studies show that chemoresistance in cancer cells and in cancer stem cells has decreased ROS levels. Their antioxidant levels however are increased compared to sensitive cancer cells [121, 122]. In another study, apoptosis and autophagy mechanism mediated via ROS/RNS stress was identified to induce OC cells to undergo cell death after treatment with JS-K (anitric oxide [NO] donor that reacts with glutathione to generate NO) in vitro. Moreover, the ovarian tumor growth in vivowas shown to be inhibited by JS-K [123]. Chen et al. utilized isotypes of dihydrodiol dehydrogenases (DDH), novel detoxifyingagents to regulate ROS expression, in cisplatin-sensitive and-resistant OC responsible cells [124]. These results provide supporting evidence that the increased levels of DDH can inversely affect for the expression of ROS and could potentially be translated to therapeutics for platinum resistance in OC.

Different signaling mechanisms of ROS-induced chemoresistance have been studied. Extended exposure to oxidative stress may inhibit caspase activation and consequently may blockthe apoptosis process [119, 121]. Erythroid transcription factor NF-E2 (Nrf2) and its internal inhibitor, Keap1, play a major role in preventing oxidative stress. Wang et al. have reported that a continued overexpression of Nrf2 is responsible for upregulation of genes involved in pharmacodynamics of drugs. This overexpression of Nrf2 has been linked to resistance to chemotherapeutic agents, including cisplatin, doxorubicin, and etoposide in gynecological cancers [125]. In many cancer cells, progression of chemoresistance is correlated with the antioxidant defense system mediated via Nrf2. Drug resistance in cervical cancer and a role of Nrf2have been investigated [126]. The knockdown of endogenous Nrf2 gene showed a downregulation of Nrf2-regulated genes. Using cervical carcinoma model (CaSki cells), this decrease in Nrf2 expression was shown totrigger drug-induced apoptotic death. Additionally, a synergistic effect of cisplatin and Nrf2 knockdown drastically suppressed tumor growth in vivo. Understanding different mechanisms of redox homeostasis, the relevant signaling pathways, and the role of ROS/RNS/antioxidant system in the development of chemoresistance in GCs is essential for designing new therapeutic strategies to overcome drug resistance [116].

## Cell death inhibition

Dysregulation of signaling pathways can result in the death of cells, categorized as apoptosis, autophagy, necrosis, mitotic catastrophe, and senescence [127, 128]. In a number of types of cancers, "programmed" cell death (apoptosis and autophagy), which induce degradation of proteins and organelles or cell death upon cellular stress, are significant in the pathophysiology. Since the autophagy and apoptosis share common signaling pathways, understanding the mechanistic connection of them on drug resistance becomes complicated. In addition, crosstalk of proteins regulating both autophagy and apoptosis has been reported in different malignancies [129, 130].

### *Autophagy*

Autophagyis a conserved metabolic operation in which double-membraned vesicles (autophagosomes) are transported to lysosomes and degenerated by lysosomal enzymes [131, 132]. Autophagy process operates at the posttranslational level. Autophagy-related genes (ATG) respond to different stimuli, including endoplasmic reticulum (ER) stress, hypoxia, hyperthermia, as well as nutrient deprivation via the ATF4 and MIT/TFE transcription factors [133, 134]. Autophagosome formation takes place in three major steps: i) preinitiation process, involves the serine kinase activity; ii) initiation process, demands the lipid kinase activity; and iii) final process, requires the ligase activity to bring the (ATG5/ATG12/ATG16) complex to nascent phagophores [135]. The molecular mechanism and the corresponding signaling pathways of apoptosis are yet to beexplored extensively. Tumor progression and heterogeneity to various therapy-resistant malignancies in human has been attributed to the characteristics of cancer stem cells (CSCs). These cells have the ability to induce heterogeneity in tumorsas well as to generate cell death. These functions are attributed to the resistant genes, anomalous expression of drug transporters, andtheir latent state [136, 137]. Maintenance of "stem" properties in both normal tissue stem cells and CSCs has been considered as a characteristic of autophagy [138, 139].

The key autophagy pathways that have been demonstrated in GCs are the PI3K/AKT/mTOR pathway, the LKB1/AMPK pathway, and the Ras/Raf/ERK pathway [140, 141]. In order to constitutively maintain cellular homeostasis, extracellular factors such as levels of energy, nutrients, and growth factors are known to stimulate these pathways. During exposure to prolonged stress, however, the cells are capable of progressing to autophagic type II programmed cell death. Autophagy-related proteins in cervical, ovarian, and endometrial cancers and their connection with clinico-pathological characteristics have been discussed by Orfenelli et al. 2013 [142]. In cervical cancer, Beclin 1, LC3, and ATAD3A are the three proteins that have been reported to play a vital role in autophagy. Downregulation of Beclin1 in malignant cervical cancer was demonstrated as a risk factor affecting overall survival and an independent predictor of progression-free survival. Endometrial cancer was demonstrated to have LC3, Beclin1, and EIG121 proteins as major players. Autophagy has been shown as a survival mechanism in response to external stress in endometrial cancer [142]. Autophagy in ovarian cancer has been extensively studied. Regulation of several proteins has been demonstrated to have impact on the response to extracellular stress. These proteins include ARHI, IL-6, PEA-15, TNF-a, PTEN, mTOR, AKT, p54, Beclin 1, LC3, and DRAM2 [142]. Chemoresistance has been shown to be induced by upregulation of autophagy which results in a downregulation of apoptosis in a variety of cancers, including OC [143].

### *Apoptosis*

Apoptosis is a programmed cell death mechanism, involving distinct biochemical and genetic processes that play significant roles in eukaryotic cells for their development and the maintenance of the equilibrium between live and dead cells. This equilibrium is mediated by various pro- and antiapoptotic biomolecules [144, 145]. Although there is a broad spectrum of physiological and pathological conditions that trigger apoptosis, in cancer, apoptosis is initiated by two main pathways: the intrinsic (mitochondrial-mediated) pathway and extrinsic (the death-receptor) pathway [146, 147]. These two pathways may work in synchronization to start apoptosis. The intrinsic pathway is induced by sensing various external stresses such as radiation, DNA-damage, oncogenic factors, chemotherapeutic agents, and viral infections [148, 149]. The intrinsic apoptotic pathway involves B-cell lymphoma-2 (BCL-2) family and inhibitors of apoptotic proteins (IAPs) that serve as an apoptotic switch that regulates the permeabilization of mitochondrial membrane. The extrinsic pathway is initiated by the association of cell surface receptors, a subset of tumor necrosis factor (TNF) with their respective activating cytokine ligands from TNF-superfamily of proteins.

Key apoptotic regulators in OC have been extensively reported in literature, including the direct apoptotic targets, the Bcl-2 family of proteins, and the IAPs [150, 151]. Major indirect targets studied include processes related to microRNA, p53 mutation along with homologous recombination DNA repair. Other pathways that are disrupted by apoptosis inducers may include, wingless-related integration site (Wnt)/β-Catenin, Janus Kinase (JAK), mesenchymal-epithelial transition factor (MET)/hepatocyte growth factor (HGF), phosphatidylinositol 3-kinase (PI3K)/v-AKT, signal transducer and activator of transcription 3 (STAT3), mitogen-activated protein kinase (MAPK)/extracellular signal-regulated kinase (ERK), and murine thymoma viral oncogene homolog (AKT)/mammalian target of rapamycin (mTOR) pathways. Taxane- and platinum-based chemotherapy resistance studies show a strong association with overexpression of B-cell lymphoma 2 (Bcl-2), B-cell lymphoma-extra-large (Bcl-xL), A1, myeloid cell leukemia-1(Mcl-1), and survivin or downregulation or mutation of pro-apoptotic factors, such as Bcl-2-associated X protein (Bax) and caspases [146, 149, 152, 153]. A genomic study using sequencing of tumor and germline DNA samples from 92 ovarian cancer patients with primary refractory, resistant, sensitive, and matched acquired resistant disease, scientists have shown that genetic alteration is responsible for inactivating the tumor suppressors *RB1*, *NF1*, *RAD51B*, and *PTEN* in HGSC. These alterations have influencedthe acquired chemotherapy resistance in OC patients [154]. In another study, overexpression of IAP has been shown to play a key role in development of drug resistance in some cancers [127]. In a study using a chemo-resistant OC model, overexpression of Xiap and downregulation of p53 in OC cells are shown to be associated with resistance to cisplatin. This was followed by elevated ubiquitination anddecreased regulation of pro-apoptotic proteins, such as Bax and Fas [151]. Chemoresistance due to upregulation of HSPs, particularly HSP27, HSP70, and HSP90, mediated via blocking of procaspase-9 activation leading to restriction of apoptosome formation has been reported [155].

## DNA damage repair

DNA damage repair (DDR) is a complex phenomenon in mammalian cells in response to exogenous and endogenous stress. This process is essential to maintain vital genomic functions by preventing mutations and thereby helps cell survival [156, 157]. Advances in cancer biology coupled with increased understanding of the human genome, including the elucidation of six major interactive pathways involved in DNA damage and repair, have offered novel opportunities to combat outcomes for patients with GCs [158]. These parallel and intercalating pathways are described as mismatch repair (MMR), base excision repair (BER), nonhomologous end-joining (NHEJ), homologous recombination repair (HRR), nucleotide excision repair (NER), and translational joining (TLJ). Lethal DNA lesions generated during DNA repair in cancer cells also contribute to intrinsic drug resistance [159]. A dysfunction of one DNA repair pathway may be damaging to a function of another interconnected response pathway, which may eventually produce the resistance to chemotherapy targeting DNA-damage mechanism, including platinum-based drugs [158, 160].

HRR is a major conservative repair mechanism by which the original double-stranded DNA is cleaved due to stress and preserved during the cell cycle phases, S and G2/M. The main constituents in the HRR process are the *BRCA1/2* genes and the other corresponding genes in the signaling pathway. In case of defective *BRCA1* or *BRCA2* genes, due to malfunctioning homologous recombination, double-strand break repair is accomplished via alternative repair mechanisms such as NHEJ and single-strand repair pathway [161–163]. Moreover, NHEJ is normally prevalent in G0 or G1 cell cycle phases and is characterized by amplifying errors rather than their repair [164]. Most chemotherapeutic agents, including platinum-based drugs (cisplatin, carboplatin, and oxaliplatin) as well as radiation therapy, are used to treat GCs via inducing DNA damage, either directly or indirectly.

It has been observed by several researchers that majority of DNA repair mechanisms contribute in the removal of platinum compounds in GCs [161, 164, 165]. Initially, the process involves recognition of DNA damage by proteins (XPA, XPC, RPA) followed by the uncoiling of DNA by helicases XPD and XPB. The cleavage of the targeted damaged site is performed by nucleases XPG 30 and XPF-ERCC1 50. The DNA sequence on the complementary strand is later resynthesized by DNA polymerase. The major players in combating the cisplatin-induced apoptosis are the defects in DNA MMR genes, for example, the hMLH1 promoter [38]. As compared to sensitive OC cell lines, reduced level of intrinsic DNA damage was reported in platinum-resistant OC cells [166]. In another study, the expression of several vital DNA repair genes, including BARD1 (BRCA1-associated RING domain protein 1), H2AFX (H2A histone family, member X), NBN (Nibrin), NTHL1 (nth endonuclease III-like 1), and

SIRT1 (Sirtuin 1) were shown to be downregulated by targeting the Endoglin gene (ENG), a part of TGF beta receptor complex [160].

The role of defective DNA repair in other GCs is not yet well characterized. In a study with cervical cancer, HPV infection and its association with expression of the onco-viral proteins E6 and E7 were shown to inactivate the tumor-suppressor genes, p53 and pRB leading to altered DNA repair capacity [167]. Using a limited number of patients with squamous cervical cancer, a correlation between patients' response to therapy with DNA damaging agents and activation of DNA repair pathways was studied. The nonresponsive patients showed increased expression of BRCA1, BRIP1, FANCD2, and RAD51 as compared with responsive patients [168].

While DDR with platinum drugs has been extensively studied in OC, more studies are required to elucidate the detailed mechanism in other GCs. Recent studies suggest that application of DNA repair pathways for GC therapeutics could be useful. A number of ongoing clinical trials in patients with advanced-stage cervical cancer incorporate DNA repair regulatorsalong with standard chemo or radiotherapy (NCT01281852, NCT02223923, NCT02595879, NCT02466971) [168].

## Epigenetics

Alterations of signaling cascades, due to genetic and/or epigenetic changes, may lead to carcinogenesis. The study of epigenetic markers and the mechanisms involved in chemoresistance in GCs seem essential for the design of unique therapeutic agents to circumvent drug resistance [169, 170]. The epigenetic markers most studied so far include DNA methylation, histone modification, and microRNAs.

### DNA methylation

In eukaryotes, gene expression by cells is governed by DNA methylation. It is a universal epigenetic signal operated by cells to switch off genes when they are not needed. DNA methylation is an operation where an enzyme, DNA methyl transferase attaches a methyl group to d-cytosine moiety of DNA. This affects many biogenetic processes that conserve events such as genomic imprinting, chromosomestability, and X-chromosome inactivation [171]. Hence, errors in DNA methylation lead to a number of downstream consequences including drug resistance in cancer cells.

DNA methylation in OC drug resistance has been studied in detail [172–174]. Using demethylating agents in isogenic, cisplatin-sensitive and -resistant OC cells, Zeller et al. studied relationship between chemoresistance and DNA methylation [175]. Using DNA methylation profiling technique of the entire genome, they identified several hypermethylated loci in genes of OC cells exhibiting chemoresistance as compared to the wild type (sensitive) cell line. They further associated cisplatin resistance to 13 key genes that showed methylation of their CpG sites. [175]. Papakonstantinou et al. have reviewed patterns of DNA methylation inepithelial OC, and have emphasized its futuristic role in prediction, diagnosis, and screening in chemoresistance [176]. In another study, breast and ovarian cancer cells exposed to platinum-based chemotherapy were demonstrated to have hypermethylated BRCA1 promoter when compared to the wild type [177]. Several candidate epigenetic biomarkers of endometrial and ovarian cancers have been identified and discussed in detail by Balch et al.

[172]. Atypical DNA methylation is common in ovarian and endometrial tumors, including global hypo- and hypermethylation that were found to induce silencing of oncogenes. Many of these abnormally methylated genes have been associated with chemoresistance and metastasis [172, 178]. These findings suggest that the shift of epigenetic silencing in specific genes have a potential for combating drug resistance. However, it needs further investigation to gainmoreknowledge of epigenetic mechanisms involved in the treatment of GCs. Moreover, if used as an adjuvant therapy along with conventional therapeutics, therapies based on epigenetics may suggest an effective approach for rendering drug-resistant ovarian tumors respond to chemotherapy.

### Histone modification

In response to cellular signals, chromatin template modulation depends mainly on the post-translational modification of histones. These alterations include methylation, phosphorylation, sumoylation, acetylation, ADP-ribosylation, deimination, ubiquitylation, proline isomerization, and propionylation [179, 180]. When DNA damage occurs, one or more of these histone modification processes are observed on the candidate histone residues. In cells, these modifications are mostly reversible and can result in chromatin modulation, which primarily act via the "writer" and "eraser" enzymes [181, 182]. Eminent role played by histone modifications in gene expression, DDR, and other cellular processes that may lead to malignancy protects the integrity of genetic information, cellular homeostasis, and chemoresistance [179, 180].

In ovarian cancer, histone deacetyltransferases (HDACs) have been reported to facilitate histone alterations. Several histone modifying enzymes are associated with processes underlined in chemoresistance in GCs [183, 184]. For example, inhibition of HDAC enzymes in chemo-resistant and -sensitive cells were shown to enhance an expression of a gene coding for Regulatory Protein for G Signaling (RGS10) and cisplatin resistance [183]. Overexpression of Sirtuin 1 (SIRT1), a NAD+-dependent enzyme family member and deregulation of EZH2, G9a, DOT1L, and SETD2 (Histone modifying transferases) in malignant epithelial OC promoter region were directly involved with chemoresistance in OC cells [183–185]. Although most of the research in this area is in ovarian cancer model, the findings could be extended to other gynecologic cancers. These findings suggest that targeting histone modifications or enzymes regulating may provide novel approaches for gynecological cancer treatment.

### microRNAs

Also referred as miRNAs or miRs are fragments of RNA molecules that are mostly non-coding and short sequences. These have gained importance as a novel class of bioregulators. These are evolutionarily conserved agents that function via modulation of posttranscriptional modification of multiple target miRs by direct cleavage or repression of translation [186, 187]. The National Cancer Institute panel (NCI-60) profiling for miR expression showed a strong association between the expression patterns of miRs and the potency patterns of anticancer agents [188, 189]. Depending on the compound class and cancer type, the change in the magnitude ofspecific cellular miRs (miR-16, miR-21, and let-7i) was shown to have large effect on the efficacy of several anticancer drugs studied [190]. Dysregulation of miR expression in malignant cells has been widely observed and can be attributed to several mechanisms. Most widely studied mechanisms include the epigenetic modulation of miR genes, miRs at fragile sites of cancer-associated genomic locations, and abnormalities in miR-processing genes and proteins [191]. However, additional mechanisms, including genetic aberrations and transcriptional regulation, may also contribute in miR deregulation. In the past decade, a number of miRs were suggested to impact invasion, proliferation, metastasis, chemoresistance, and the recurrence in gynecologic cancers. miRs contributing to malignancies are classified either as oncogenes or tumor suppressors based on their function and expression pattern. In GCs, miR signatures have been suggested as predictors of how therapy will respond as well as early detection biomarkers. Table 1 is a compilation of the miRs in GCs-regulated upstream and downstream, in the peer-reviewed published studies [192, 193]. Several studies have reported distinct miR signatures using miR profiling from serum, plasma, and tissues from patients with gynecologic cancers. The first differential expression of miRs in OCs was demonstrated by Zhang et al. [192]. A role of miRs in chemoresistance is extensively reviewed by Srivastava et al. (2018) and have suggested as one of the key players in chemoresistance of gynecologic cancers [193].

## Long noncoding RNA

Current literature suggests that more than 97%of the total transcriptional residues of human genome are short and long noncoding RNAs (s&lncRNAs) [194]. These lncRNAs have been shown to lack protein coding activity and are known to constitute more than 200 nucleotides. Due to their multifaceted role in various cellular and genomic functions, lncRNAs

TABLE 1   A List of anomalously expressed miRNAs in major gynecologic cancers.

| Type of cancer | miRNAs |
| --- | --- |
| *Downregulated miRNAs* | |
| Cervical cancer | Let-7a, Let-7b, Let-7c, miR-26a, miR-27b, miR-30d, miR-34a[b], miR-100, miR-103b, miR-125a, miR-125b, miR-126, miR-141[b], miR-143[a], miR-145[a,b], miR-149, miR-200b, miR-200c, miR-203[a,b], miR-204, miR-375, miR-451, miR-3185, miR-3196, miR-3960, miR-4324, miR-4467, miR-4488, miR-4525 |
| Endometrial cancer | miR-99b[a], miR-143, miR-145, miR-193b, miR-204[a] |
| Ovarian cancer | miR-1, miR-99b, miR-127, miR-130b, miR-133a, miR-148a, miR-148b, miR-155, miR-375b, miR-451 |
| *Upregulated miRNAs* | |
| Cervical cancer | miR-9, miR-10a, miR-15b, miR-20a[a], miR-20b, miR-21[a], miR-27a, miR-34a[b], miR-127, miR-133a, miR-133b, miR-141[b], miR-145[b], miR-155, miR-196a, miR-199a*, miR-199a, miR-199b, miR-199-s, miR-200a, miR-203[b], miR-214, miR-221, miR-224, miR-500, miR-505, miR-711, miR-888, miR-892b miR-1246, miR-1290, miR-2392, miR-3147, miR-3162 and miR-4484 |
| Endometrial cancer | miR-9, miR-92a, miR-141, miR-182, miR-183, miR-186, miR-200a, miR-205[a], miR-222, miR-223, miR-410, miR-429, miR-449, miR-1228 |
| Ovarian cancer | miR-16, miR-17, miR-20a[a], miR-21[a], miR-23a, miR-23b, miR-27a, miR-29a, miR-30a, miR-92a, miR-93[a], miR-106b, miR-126[a], miR-141[a], miR-144, miR-150, miR-182, miR196a, miR-196b, miR-199a, miR-200a[a], miR-200b, miR-200c[a], miR-375[b], miR-486, miR-449a, miR-499-5p, miR-3613 |

[a] *miRNAs that have been identified in multiple studies.*
[b] *miRNAs that have been reported up- and/or downregulated within the same gynecologic cancer [192, 193].*

have attracted attention in recent years in the domain of cancer research and chemoresistance. Their genomic involvement includes processes such as splicing, transcription, and epigenetics, whereas their role in cellular processes is mainly in different phases of cell cycle as well as pluripotency [195]. These express uniquely in some cancer types, and their role in tumorigenesis and chemoresistance has been studied in past decade [196–198].

During tumorigenesis, there is a distinct difference in the regulation and expression of signature lncRNAs compared to the normal tissues in the vicinity. These lncRNAs have a potential to function as either tumor enhancers (oncogenes) or suppressors [199]. A classic example is a correlation of overexpression of HOX transcript antisense RNA (HOTAIR) lncRNA with aggressiveness in GCs (breast, ovarian, cervical, and endometrial) and other tumors [200–204]. On the other hand, there are some maternally expressed imprinted genes (MEG3) lncRNAs that may act as tumor suppressors in a variety of gynecologic cancers [205]. lncRNA involved in metastasis-associated adenocarcinoma transcript 1 (MALAT1) is shown to be upregulated in GCs like ovarian and cervical resulting in enhanced invasion and migration [206].

Scientists are still examining different possible mechanisms of action of lncRNAs. Their role in GCs is currently in its infancy, although an increasing number of publications on this topic has appeared in the last 5 years. Several specific lncRNAs for ovarian and cervical cancers have been shown to be clinically relevant. Their role in drug resistance via different mechanisms is under study that is anticipated to facilitate the design of new biomarkers for improved diagnosis and better patient outcomes.

## Oncogenes

Oncogenes are eukaryotic genes that have potential to participate in tumorigenesis. They are having pivotal functions in normal cells but undergo mutations to become proto-oncogenes that code for key components of signal transduction pathways that regulate vital cellular processes. The oncogenes encode receptors of membrane growth factors, cytoplasmic signaling molecules, and nuclear transcription factors. Because oncogenes contribute to cancer initiation, tumor growth, as well as chemoresistance, they are considered as potential targets for cancer therapeutics [207, 208].

How the oncogenes contribute to drug resistance is not yet fully understood. However, drug resistance is known to be directly or indirectly affected by oncogenes and signaling pathways [209]. The available literature to date reveals chemoresistance in OC has been linked to more than 25 oncogenes. This phenomenon involves regulation of different cellular and signaling pathways comprising of apoptosis and cell cycle regulators as well as growth and/or transcription factors [210, 211]. Using comprehensive bioinformatics analysis, 21 of these oncogenes were directly associated with signaling pathways, including Bcl-2-associated-X (BAX), PI3K/AKT, ErbB, and MAPK. The two oncogenes controlling the AKT- and BAX-mediated signaling were found to be responsible for chemoresistance in OC. Furthermore, data analysis indicated apoptosis and phosphorylation as the chief biological roadways through which oncogenes are likely to contribute to drug resistance in OC. Furthermore, the bioinformatic analysis employing microarray data retrieved from online databases revealed a link between development of drug resistance and the oncogene NEK2 (NIMA-like serine-threonine kinase family) via cell cycle and microtubules regulation in OC [211].

In another study, chemoresistance in OC was studied to elucidate the drivers of clinical phenotypes using whole-genome sequencing of tumor and germline DNA samples from 92 patients with various stages and grades of disease [212]. The damage to the gene was shown to be associated with inactivation of the tumor suppressors PTEN, NF1, RB1, and RAD51B. This phenomenon was shown to contribute the acquired chemoresistance. In primary resistant and refractory disease, frequent amplification of a cyclin E1 gene (CCNE1) was reported. Additionally, another oncogene S100A10 was confirmed to play a major role in the progression of OC [212]. Functional analyses demonstrated that S100A10 suppression significantly reduced key cellular malignancy characteristic processes in OC cells. Regulation of cleaved-Caspase3 and cleaved-PARP could be responsible for markedly increased carboplatin-induced apoptosis, in SKOV3 and A2780 ovarian cancer cells as well as tumor growth inhibition *in vivo*. The role of oncogenes in other gynecological cancers is yet to be explored.

## Targeted therapies to combat multidrug resistance

For OC patients, the median time to relapse is 18 to 24 months. For patients with earlier staged OC, the relapse rate is about 20%; for those with advanced-stage OC, the recurrence rate is about 70% [213]. This relapse pattern is known to affect the efficacy of the first-line treatment (a combination of paclitaxel and cisplatin) for OC [214]. In addition to exploring the mechanisms by which cancer cells develop chemoresistance, it is also important to examine methods by which this resistance may be limited or prevented all together. A number of effective therapies and preventive approached are discussed in the following section.

### Lipoprotein-based therapies

An important factor mediating drug resistance is P-glycoprotein (P-gp), alternatively known as ABCB1 and multidrug resistance protein 1 (MDR1). It has been shown that P-gp promotes efflux of toxins from cells, thereby promoting drug resistance of cancer cells utilizing a similar pump mechanism.

### The relationship between lipoproteins and resistance to chemotherapy

In a study based on a human uterine sarcoma cell line adhering doxorubicin-resistance, the influence of lipoproteins on P-gp expression was examined. P-gp expression was seen to increase with lower concentrations of high-density lipoprotein (HDL), whereas higher HDL levels (above 0.85 mg HDL protein/ml) reduced P-gp expression [215]. Accordingly, it seems that increased blood or cellular HDL levels may prevent the cell from developing drug resistance. The increased HDL levels, however, may have emerged due to the slower progression of the disease as most cancer cells and tumors are known to be high-level consumers of HDLcholesterol [102]. Regarding LDL, however, the results are relatively less clear. While there appeared to be a dose-related increase in P-gp expression in response to increased LDL concentrations in uterine culture, other studies lead to differing findings. In a vinblastine-resistant leukemia cell lines, LDL was shown to decrease P-gp activity while the timing incubations of LDL with cells varied and the types of cells used were also different in this study [216]. Regarding very low density lipoproteins (VLDL), treatment with VLDL significantly reduced the expression of P-gp, in breast cancer cell studies, an isolated finding that is difficult to explain in the absence of additional supporting data [217].

Apolipoproteins are the surface proteins that bind lipid molecules and thus stabilize lipoprotein structure [218]. Researchers have described the role of apolipoprotein expression as a prognostic factor in cancers, related to metastatic potential, patient survival, and invasiveness [219]. In certain cancers, quantification of apolipoprotein expression levels may reveal the tendency of tumor cells to exhibit resistance toward a particular drug.

Although multiple apolipoproteins have shown positive effects in the treatment of cancers, chemotherapy resistance is another potential application. For example, there has been research utilizing Apo A1 to counter resistance in breast cancer. Although other apolipoproteins, including Apo D and Apo E, have been investigated to combat multidrug resistance, Apo J, also known as Clusterin (CLU) or Testosterone-repressed prostate message-2 (TRPM-2), sulfated glycoprotein-2, has been studied to alleviate the drug resistance observed with gynecologic tumors [220, 221].

### What were the outcomes of these studies and what are the proposed mechanisms involved?

In vitro studies performed by Miyake and colleagues in the year 2000 examined the effects of CLU, which they alternatively called TRPM-2. Previous findings suggest that CLU expression increased when androgen ablation occurred in the context of prostate cancer, or when estrogen ablation occurred in breast cancer [222–224]. These findings are perhaps due to CLU upregulation as a result of hormone withdrawal, as suggested by CLU's alternative name, TRPM-2. Further research by Miyake et al. showed increased, dose-dependent expression of CLU mRNA after chemotherapy treatment with paclitaxel. The authors also found that paclitaxel resistance was generated in the cell lines that overexpressed CLU *in vivo* [221].

To elucidate that the relationship was not simply correlative, but causative, CLU expression was selectively reduced using antisense oligonucleotides, resulting in no reduction in tumor growth. Concomitant treatment with paclitaxel, however, sensitized the cancer cells to the chemotherapy and enhanced tumor apoptosis validating the relationship between CLU expression and chemotherapy resistance. The mechanism by which CLU mediates chemotherapy resistance may be due to CLU's cytoprotective function, making the cells resistant to multiple apoptotic stimuli [221, 225]. Cell culture studies have demonstrated resistance to TNF as well as Fas-induced apoptosis [221, 224]. It has been proposed that CLU reduces stress-mediated protein aggregation as well [226].

Hassan et al. observed upregulation of CLU expression in ovarian cancer cells that were resistant to paclitaxel [225]. Selective knockdown of CLU expression using small interfering RNA (siRNA) sensitized OC cells to paclitaxel. Park et al. showed a similar effect in cervical cancer cell lines that initially expressed CLU at high levels and were sensitized to paclitaxel after silencing via siRNA. In patients suffering from ovarian cancer, CLU overexpression was correlated with a reduced average survival time, compared to patients with normal CLU expression [227]. Contrary to this, it was discovered that increased CLU expression was correlated to a better prognosis, specifically in advanced-stage (stage III) serous OC [228]. It appears that the role of CLU in tumorigenesis and resistance is highly complex, and that more research is needed to elucidate and understand in detail the respective mechanisms involved.

It has been shown that increased expression of CLU is associated with drug resistance in other cancers, including gastric cancer, breast cancer, and other malignancies [229]. Furthermore, chemotherapy for the treatment of lymphomas induced higher CLU levels in patients, indicating that the multidrug resistance mediated by CLU is not limited to a specific malignancy [230]. Additionally, CLU has been shown to facilitate resistance to a number chemotherapeutic agents, including dacarbazine, camptothecin, doxorubicin, 5-fluorouracil, and etoposide in cancer cells and tumors [226].

As Hassan et al. and Park et al. found, a unique way to combat multidrug resistance in gynecologic cancers may be accomplished via specific siRNAs or other gene manipulations to knockdown the CLU expression [225, 231]. Indeed, selective reduction of CLU expression increased the sensitivity to chemotherapeutic agents in several cancer cell lines. Although there was much focus placed on CLU, this was primarily because during the time of this writing, there existed the most research regarding CLU in relation to gynecologic cancers. As the data pool increases in respect to other apolipoproteins, a more complete picture will be painted, and we may see multiple other modalities to combat multidrug resistance using apolipoproteins in the near future.

## Adipocytes and lipid metabolic reprogramming

Although functionally useful for energy storage and pathologically associated with metabolic diseases, adipocytes also play a role in cancer. They may be an additive factor, enhancing the metastasis and chemotherapy resistance of various tumors in addition to promoting the invasiveness of certain gynecologic cancers, including ovarian cancer, due to adipocyte interaction with tumor cells within the abdominopelvic cavity [232]. Multiple mechanisms have been suggested, including an oxidative stress response, triggered in adipocytes that render leukemia cells resistant to anthracycline drugs [233]. A unique in vivo study was performed by Duong et al., where adipose tissue was obtained from patients and injected subcutaneously into mice, forming a lipoma. When a tumor was grafted next to the lipoma, the cancer displayed significantly increased resistance to trastuzumab versus the tumor grafted in the absence of a lipoma [234]. These examples show that adipocytes contribute a major part in cancer drug resistance.

AKT pathway activation had been demonstrated to decrease chemotherapy-induced apoptosis, using cisplatin in cisplatin-resistant OC cell lines. Conversely, AKT inhibition by the LY294002 was shown to render the sensitivity to cisplatin back to these resistant cells [235]. Advancing this concept, Yang et al. showed that adipocytes mediated chemoresistance in cancer cells through the AKT pathway, while alternate pathways may exist as identified by lipidomic analysis [236]. Although prostaglandins (PG) have been demonstrated to bestow resistance to chemotherapy in bladder cancer and pancreatic cancer models, $PGD_2$, $PGE_2$ and $PGF_2$, and Lipoxins were not found to mediate adipocyte-induced chemoresistance in the ovarian cancer model while the precursor of lipoxins and prostaglandins, arachidonic acid (AA) was subsequently implicated [237, 238]. In fact, high concentrations of AA were found to completely eliminate cisplatin-induced cytotoxicity of cancer cells, and reduced toxicity for carboplatin as well. AA was found to activate AKT, serving as one link between adipocyte medication of chemotherapy resistance as indicated by the earlier findings. While the AKT inhibitor LY294002 showed promise, it displayed toxicity problems in clinical trials. Other inhibitors are being formulated currently, and as testing progresses into multiple clinical trials, several promising AKT inhibitors may arise, allowing for reversal of cancer chemoresistance.

Previous observations revealed that adipocyte localization favored the growth of ovarian tumors, and implied adipocyte lipolysis influenced tumor growth [232]. Angiopoietin-like 4 (ANGPTL4) has previously been considered to stimulate adipocyte lipolysis, especially in response to adrenergic stimulation [239]. With respect to malignancy, ANGPTL4 levels were increased in ovarian cancer cells, located in tissues with high adipocyte content, in stark comparison to low levels in areas with a near absence of adipose tissue. Part of the proposed mechanism of ANGPTL4 is linked to its interaction with cellular adhesion molecules, mediating metastatic potential of tumors and triggering downstream effects [240, 241]. In regard to chemoresistance, ANGPTL4 was found to induce the expression of drug efflux transporters such as ABCB1, ABCC1, and ABCG2, in conjunction with increasing expression of antiapoptotic Bcl-xL [242]. To support a cause-effect relationship, silencing of ANGPTL4 by short hairpin RNA (shRNA) was shown to elevatesusceptibility level of cancer cell lines to carboplatin. When taken together, the observations from this study gave rise to the hypothesis that ANGPTL4 increased the carboplatin resistance of OC cells due to apoptosis and possibly to other lethal mechanisms induced by chemotherapeutic agents. Targeted knockdown of ANGPTL4 by shRNA may thus serve as a treatment modality to sensitize ovarian tumor cells to chemotherapy. Additionally, siRNA knockdown of ANGPTL4 may be an alternate method to combat drug resistance.

## Targeting fatty acid metabolism

Fatty acid synthase (FASN), an enzyme responsible for the production of long-chain fatty acids from carbohydrate-derived glycolytic precursors, but in cancers, its effects are multifaceted. FASN expression was correlated to tumor progression in multiple cancers, including endometrial and high-grade ovarian tumors [243, 244]. As previously discussed, dissemination of ovarian cancer is influenced by multiple factors, apparently including enhanced FASN expression [245]. In cell culture studies, FASN inhibition by cerulenin did decrease cell viability and apoptosis while sequential treatment of a cisplatin-resistant OC cell line first with cerulenin and then with cisplatin reduced the $IC_{50}$ of cisplatin, suggesting sensitization [246]. In another study, assessing cisplatin resistance in OC, FASN inhibition by orlistat led to significant delay of tumor growth [244]. These findings lend support to the possible utility of FASN inhibitors for drug-resistant tumors. Alternatively, using methods discussed earlier, knockdown of FASN via shRNA significantly decreased resistance in breast cancer cells to both Adriamycin and mitoxantrone. Although there exist multiple studies utilizing FASN knockdown via siRNA and shRNA in gynecologic cancers, the effects of chemotherapy resistance were not yet studied [247–249]. A new generation of FASN inhibitors has displayed reduced systemic toxicity in the colorectal and breast cancer models [250, 251]. These findings may apply to gynecologic cancers as well.

Fatty acid oxidation (FAO) is a pathway involving several enzymes catalyzed steps, including the rate limiting enzyme Carnitine Palmitoyltransferase 1A. Alteration of this pathway is observed in many cancers, including OC, posing it as a suitable target for therapeutic intervention [252]. The use of etomoxir and Ranolazine, which ultimately inhibit FAO, sensitized human leukemia cells to apoptosis by ABT-737 or cytosine arabinoside [253]. The effect was replicated even when the cancer cells were cultured on bone marrow-derived mesenchymal stromal cells (MSCs), which are known to enhance malignant potential. In a nasopharyngeal carcinoma model, etomoxir was found to sensitize cancer cells to radiation. It was postulated that etomoxir impaired fatty acid influx, essentially starving the cell and leaving it prone to radiation [254]. Although there is a dearth of literature directly detailing the effects of CPT1A or FAO inhibition of drug resistance, specifically in the area of gynecologic cancers, this may present an opening for future research.

## Targeting autophagy

While autophagy occurs at basal levels to ensure quality control and homeostasis in the cell, cues of cellular stress such as hypoxia, oxidative/ER stress low levels of nutrients and energy induce it to a greater extent [255]. The products of the digestive process can be recycled as building blocks in anabolic processes necessary for cell survival during stress [256]. Although it is a cell survival mechanism, extended autophagy sustained by prolonged stress can lead to caspase-dependent or -independent cell death [257]. Thus, autophagy cross-talks with apoptotic mechanisms to either activate or inhibit cell death. For example, Atg5 which is involved with the assembly of the autophagosome can mediate the release of cytochrome *c* in the cytoplasm upon cleavage by calpain. The cleaved Atg5 is able to transport to the mitochondria where it prevents the antiapoptotic activity of Bcl-xL by binding to it [258]. In contrast to this pro-apoptotic activity of autophagy agents, the forkhead box O3 (FOXO3) acts as a transcription factor for the pro-apoptotic protein Bcl-2-binding component 3 (BBC3), and FOXO3 is targeted for degradation by autophagy [259].

Reciprocally, major players of the apoptotic cascade can inhibit or activate autophagy by targeting members of the autophagy nucleation complex. Beclin1 contains a BH3 domain that facilitates its binding with antiapoptotic proteins such as Bcl-2, Bcl-xLBim, and Mcl-1. This interaction prevents the binding of lipid kinases and other components of the Beclin-1-Atg14 complex to Beclin1, and thus, inhibits the autophagosome formation [260]. As well, AMBRA1 and Beclin1 are targeted in apoptosis for caspase-mediated cleavage [261]. In the extrinsic arm of apoptosis, death-associated protein kinase (DAPK) which is recruited by death receptors has been shown to phosphorylate Beclin1. This phosphorylation of Beclin1 which occurs in its BH3 domain provokes the dissociation of Bcl-xL from Beclin1 and potentiates autophagy [262]. Overall, the cooperation between apoptosis and autophagy is an important factor in explaining the dual role of autophagy in cancer as a tumor-promoting or a tumor-suppressing mechanism, especially with regard to the modulation of drug-induced cell death [263, 264].

While the various mechanisms of action of anticancer drugs often elicits apoptosis in cancer cells, additional mutations, response mechanisms, and alternate signaling pathways emerge in cancer cells to mediate MDR and oppose this expected apoptosis outcome of anticancer agents. Several investigative efforts have reported the pro-survival and MDR-potentiating role of autophagy. Pan et al. showed that when human lung adenocarcinoma cells were treated with docetaxel, autophagy was induced via the high-mobility group box 1 (HMGB1) [265]. From the nucleus, HMGB1 translocated to the cytoplasm where it promoted the formation of the Beclin1-Class III phosphoinositide-3-kinase (PI3K-III) complex using the MEK/ERK1/2 pathway. This lipid kinase Beclin1-PI3K-III complex facilitates the nucleation and autophagosome formation, and the autophagic flux that ensues and promotes resistance to docetaxel [265]. Conversely, in human A549 cancer cells that are etoposide-resistant, the inhibition of autophagy via siRNA-mediated Beclin1 knockdown reduced Feroniellin A-induced apoptosis in these cells [266]. With mTORC1 being a central inhibitor of autophagy [267], treatment with rapamycin promoted the Feroniellin A-induced apoptosis in these cells, suggesting that activating autophagy can help overcome MDR in some cases [266].

Although autophagy mechanism in MDR is no less complex in gynecologic cancers, approaches to activate or inhibit autophagy have been employed to circumvent MDR or to directly produce autophagic cell death in gynecologic cancers [142]. In Siha cervical cancer cells, overexpression of Beclin1 and subsequent autophagy induction promoted carboplatin-induced apoptosis [268]. A similar effect from Beclin1 was reported on paclitaxel-induced cell death in cervical carcinoma Ca Ski cells [269], and increased chemosensitivity to cisplatin, 5-fluorouracil, epirubicin, paclitaxel, and cisplatin was observed in the same cell line with overexpression of Beclin1 [270]. In contrast to cervical cancer, the cytoprotective role of autophagy in OC has been investigated. Treatment with the cytotoxic agent FTY720 in several ovarian cancer cells lines induced a cytoprotective autophagy as evidenced by increased degradation of p62 and LC3. In these cell lines, inhibition of autophagy via knockdown LC3 and Beclin1 enhanced the cytotoxic effect of FTY720 [271]. Additionally, inhibiting autophagy using hydroxychloroquine and 3-methyladenine (3-MA) increased chemosensitivity to paclitaxel in MDR ovarian (SKOV3TR) cells as well as in vincristine-resistant ovarian (SKVCR) cancer cells when 3-MA and chloroquine were the autophagy inhibitors [272, 273]. These reports indicate that the inhibition of autophagy would sensitize the drug-resistant OC cells to conventional and novel anticancer agents improve treatment outcomes in OC.

As demonstrated in OC, there is evidence of the need to inhibit autophagy in endometrial cancer to overcome MDR. In human endometrial cancer JEC cells, autophagy promoted cancer cell survival, stemness and multidrug resistance [274]. When autophagy was inhibited by 3-MA or chloroquine, endometrial CSCs became more sensitive to paclitaxel. Knocking down autophagy-related proteins Atg5 and Atg7 sensitized cisplatin-resistant Ishikawa endometrial cancer cells to cisplatin [275]. Besides targeting individual proteins involved in the autophagy machinery, upstream targets have also been identified and modulating their activity has immediate effects not only on autophagy, but also on other cell survival pathways that mediate MDR. The lncRNA HOTAIR promotes autophagy and multidrug resistance via the upregulation of Beclin1, the P-gp, and MDR, and interfering with its activity in cisplatin-resistant Ishikawa cells via SiRNA increase sensitivity to cisplatin in these cells [270].

Recently, Tan et al. showed that the thioredoxin-related protein of 14 kDa (TRP14) was induced by cisplatin, and TRP14 acted upstream of autophagy through the AMPK/mTOR/p70S6K pathway in the A2780 and SKOV3 OC cells resistant to cisplatin. In endometrial cancer, the estrogen-induced gene *EIG121* was found to induce the autophagy-mediated stemness and survival of the endometrial cancer cells and CSCs [274]. The conflicting roles autophagy play in MDR may be partly explained by the influence of experimental conditions (i.e., the duration of treatment, signaling affected by treatment, types of cancer, aggressiveness, defective signaling, and MDR outlets in these cells) on autophagy and its machinery. Also, some autophagy players, such as Atg5 mentioned earlier, are involved in signaling pathways independently of autophagy and targeting them may lead to MDR modulation independently of the completion of the autophagic flux [276, 277]. Mechanistically distinguishing these nonautophagic roles from the autophagic roles of autophagy-related proteins and ensuring the completion of the autophagic flux upon treatments could clarify the importance of autophagy in MDR and cancer as well as produce additional targeting approaches.

## Role of exosomes in therapeutics

Exosomes are submicroscopic (about 30–150 nm) membrane-bound vesicles that are secreted by cells to communicate with cells in their immediate environment or at distant sites as well as to maintain intracellular homeostasis [278, 279]. They were first described in early 1980s among the extracellular vesicles secreted by normal and cancer cells [280]. Since then, extensive research has probed into their biogenesis, cargo, secretion process, and the physiological and pathophysiological events they are involved in. After the packaging of the intracellular biological cargo, exosomes are delivered in the extracellular space via the fusion of late endosomes or multivesicular bodies to the plasma membrane, and the diverse content of exosomes has been characterized to include metabolites, nucleic acids, proteins, and lipids [281, 282]. In cancer, these exosomes are released by different cells in the TME including cancer cells, CSCs, mesenchymal stem cells (MSCs), cancer-associated fibroblast (CAFs), and the tumor-associated immune cells [283, 284]. Several studies have confirmed the involvement of exosomes in facilitating MDR by extruding drugs out of cells as a cytoprotective mechanism and delivering pro-survival, antiapoptotic factors, and MDR-mediating miRNAs, lncRNAs, genes, and proteins to other cells, cancer cells and CSCs included [284–287]. As the targeting of immune checkpoint via antibodies is one of mainstay strategies in immunotherapy [288], exosomes can block the interaction of these antibodies with their target cells during their release from these cells, thereby contributing to the variable response and reduced effectiveness to these treatment modalities [285].

An example of the exosome-mediated MDR is illustrated in the study of Safaei et al. in which cisplatin-resistant human ovarian carcinoma cells(2008/C13*5.25) not only exhibited defects in their lysosomal compartment, but also increased exosome secretion compared to the cisplatin-sensitive cancer cells [289]. In addition, the exosomes secreted by these cells contained a larger amount of efflux proteins such as the MDR-associated protein 2 (MRP2), the copper-transporting ATPase (ATP7A), and ATP7B than the cisplatin-sensitive cells, suggesting that the exosomes could contain extruded drug. When exposed to the same amount of platinum, the resistant cells extruded more platinum per exosome protein than the sensitive cells [289]. The transfer of P-gp protein to chemosensitive A2780 cells by microvesicles shed by the paclitaxel-resistant A2780 produced paclitaxel and adriamycin resistance in the chemosensitive cells, indicating that this could be a nonnegligible avenue for exosomes to propagate MDR in the tumor microenvironment [290].

Besides that of MDR proteins, the transport and delivery of miRNAs to other cells via exosomes constitutes an important mechanism in the modulation of components of the tumor microenvironment [291, 292]. Exosomes can indirectly modulate P-gp expression via the delivery of tumor-promoting miRNA miR-1246 from chemo-resistant ovarian cancer cells SKOV3ip1, HeyA8, and A2780 to other cells including M2-like macrophages and other cancer cells [293]. Caveolin-1 (Cav1) being a target of miR-1246, the overexpression of Cav1 produced an inverse reaction on P-gp expression suggesting a regulatory role for Cav1 in P-gp expression or that both proteins share common signaling pathways. Treatment with miR-1246 inhibitor sensitized the chemo-resistant cells to paclitaxel [293]. Another miRNA linked to chemoresistance in ovarian cancer cells is miR-223 which targets the phosphatase and tensin homolog (PTEN)/PI3K/Akt pathway [294]. Under hypoxic conditions, exosomes derived from tumor-associated macrophages (TAMs) and carrying miR-223 rendered cisplatin resistance to OC cells. The drug resistance was impaired upon treatment of ovarian cancer cells with anti-miR-223 [294].

Thus, due to their various contributions to tumor progression and MDR, exosomes represent attractive targets for MDR inhibition, especially through their biogenesis, composition, and trafficking from donor cells to recipient cells. Yin et al. demonstrated an increased expression of annexin A3, which is a cellular protein likely involved in exosome exocytosis in the serum of platinum-sensitive OC patients compared to the platinum-responsive patients. A similar pattern was observed with the exosome of platinum-resistant and -sensitive OC cells SKOV3 and A2780 [295]. These reports suggest that annexin A3 could be targeted to inhibit exosome-mediated MDR. Lipid rafts and other proteins involved in exosome release and endocytosis such as cytoskeletal proteins (i.e., actin, dynamin) and Rab proteins have been targeted to attenuate the exosome-mediated intercellular communication [296]. Following an initial screening of selective exosome secretion inhibitors performed by Datta et al., several azole-derived compounds including neticonazole, ketoconazole, climbazole, and the farnesyl transferase inhibitor tipifarnib were shown to restrictsecretion of exosome from aggressive prostate cancer cells as evidenced by the decreased levels of exosomal markers Rab27a (the ALG-2 interacting protein X) and the neutral sphingomyelinase 2 (nSMase2) upon treatment of the cancer cells with these specific azole compounds [297]. Other azole compounds such as nitrefazole and pentylenetetrazol had the opposite effect on exosome secretion, indicating that the different azoles tested in the conditions of the study do not operate via the same mechanisms [297]. Given that ceramide, whose formation is catalyzed by nSMase, is a key player in exosome biogenesis and secretion, treatment with the nSMase inhibitor GW4869 reduced invasion and exosome secretion from the OCcells [298].

While the abovementioned inhibitors of exosome biogenesis, exocytosis, and endocytosis could possibly be employed to inhibit MDR in gynecologic cancers, the selective targeting of oncogenic and MDR-mediating exosomes is challenging as cancer cells and other cells including those that are not present in the tumor microenvironment share the same exosome-related machinery. Thus, selective delivery of these exosomes inhibitors to the tumor microenvironment is paramount to avoid adverse effects on exosomes needed for normal physiological functioning. Pertaining to the selective delivery of agents to the tumor microenvironment, exosomes themselves have been used as drug delivery vehicles to bypass MDR in cancer cells. Exosomes derived from RAW 264.7 murine macrophages successfully encapsulated paclitaxel and delivered the drug to P-gp positive MDR cells [299]. This exosome-mediated delivery of paclitaxel resulted in enhanced cytotoxicity of the drug on drug-resistant cells [299]. Semaphorin-4C (Sema4C), a receptor involved in cell-to-cell communication, was reported to promote EMT through the activation of the TGF-β1/p38 MAPK pathway in cervical cancer cells [300]. Besides reversing EMT, the downregulation of Sema4C or the upregulation of miR-25-3p, which targets Sema4C, sensitized cisplatin-resistant cervical cancer cells to cisplatin [301]. Although exosomes were not linked to miR-25-3p, these effects of the miRNA on chemoresistance and EMT suggest that miR-25-3p or a Sema4C inhibitoris an excellent candidate for exosome-based delivery to cancer cells since miRNAs are natural cargos of exosomes.

Since exosomes have an ability to interact with a diversecell population present in the TME and anticancer drugs, RNAi agents and peptides can potentially be transported by exosomes, the therapeutic opportunities unlocked by employing exosomes as delivery vehicles may outweigh those from targeting exosome biogenesis and trafficking pathways. There are still mechanisms to elucidate to fully understand the biology of exosomes, and as prementioned, exosome-targeting approaches would benefit from the identification of oncogenic factors versus normal ones involved in exosomal processes. Exosome-based delivery platforms also face challenges such as low encapsulation efficiency and the time-consuming methodology used to isolate, characterize, and manipulate exosomes to produce a specific cargo-loaded exosome [302]. Increased knowledge about the composition, biogenesis of exosomes, as well as that of cargo uptake and retention by exosomes would contribute to overcoming these challenges.

## Immunotherapy

Immunotherapy relies on the natural defensive mechanism of the immune system and seeks to boost its ability to recognize and eliminate harmful cells in the body. In the 1890s, the clinician-scientist William B. Coley administered intratumorally streptococcus bacterial products-referred to as Coley's toxins- to soft tissue and bone sarcoma patients and provided the first case report of the benefits of adding that infection component to the already-multimodal (surgery and radiation) regimen used in cancer treatment at that time [303]. Although these reports were initially met with skepticism, they were examples of cancer immunotherapy, and a progressive understanding of different mechanisms by which components of the immune system can mediate the anticancer benefits seen following stimulation of the immune system (for example, via these Coley's toxins) has greatly expanded the field [304, 305]. Now, cancer immunotherapy strategies include cancer vaccines, dendritic cell vaccines, monoclonal antibodies, adoptive T-cell transfer, cytokine-based therapy, oncolytic viruses, and the modulation of tumor-infiltrating immune cells [305–308]. Several of these immunotherapy modalities have been used to supplement chemotherapy and targeted therapy to circumvent MDR and/or sensitize chemotherapy-resistant cancer cells [309].

The regulatorsthat keep check on immune system such as programmed death-ligand 1 (PD-L1)/PD-1 and the cytotoxic T-lymphocyte-associated protein 4 (CTLA-4) by cells in the TME promotes immune suppression [310, 311]. As a result, these two checkpoints are often candidate treatment targets either alone or in combination. In preliminary reports of an ongoing phase I study (# NCT01975831), started in 2013 and expected to conclude soon) including patients with advanced OC, the combination of Durvalumab and Tremelimumab, respectively, IgG1 monoclonal antibody (mAb) against PD-L1 and IgG2 mab against CTLA-4 showed clinically relevant antitumor activity [312, 313]. A Japanese phase II clinical study (#UMIN000005714) focusing on patients with platinum-resistant OCreported that efficacy for Nivolumab, an IgG4 mab blocks the interaction between PD-1 and PD-L1 [314].

Cancer vaccines have also been developed to treat gynecologic cancers that have relapsed after chemotherapy. Similar to the prostate-specific antigen in the case of prostate cancer, CA125 is a mucin-type O-linked glycoprotein overexpressed by ovarian carcinoma cells, and its soluble form circulates in body fluids [315]. Patients with relapsed epithelial OC, peritoneal cancer, or fallopian tube cancer were vaccinated with Abagovomab or antiidiotypic antibodies that mimic CA125 in phase I clinical studies conducted in Germany [316, 317]. While the clinical efficiency was not an endpoint in these studies, the vaccinations were overall safe and elicited antibodies against CA125 as well as increased levels of CA125-specific, INFγ-positive cytotoxic CD8T cells especially in the long arm (nine injections instead of the six injections for patients in the short arm) patients [316].

A phase III clinical trial with OC patients in their first clinical remission after primary surgery and taxanes/platinum-based chemotherapy revealed that although humoral response was robustly elicited with Abagovomab, it did not produce the expected enhanced overall survival or recurrence-free survival compared to the placebo [318]. Also, there were nosignificant differences in the cellular response between the Abagovomab and placebo treatment [319]. However, it would be interesting to see if coupling Abagovomab with adoptive T-cell transfer or with cytotoxic T-lymphocyte-stimulating immunotherapy approaches could supplement the humoral response from Abagovomab and enhance effective treatment outcome.

Clinical trials evaluating the safety profile and efficacy of the combination of immunotherapy with targeted therapies for recurrent ovarian cancers have been undertaken. They include a phase II study (# NCT02659384) on the combination of the antiangiogenic agent Bevacizumab, the anti-PD-L1 Atezolizumab and acetylsalicylic acid for ovarian neoplasms treatment and a phase I study (#NCT02484404) on the combination of abovementioned Durvalumab, the poly (ADP-ribose) polymerase (PARP) inhibitors Olaparib and the antiangiogenic agent Cediranib. While results for the study NCT02659384 have yet to be disclosed, study NCT02484404 showed good tolerability and no dose-limiting toxicity in the nine patients enrolled in the study and tumor regression was seen even, suggesting clinical activity of the combination treatment within the case of platinum-resistant ovarian cancer [320]. Thus, the immunotherapy approaches by themselves or in combination with other anticancer approaches show promise in helping bypass MDR to produce tumor rejection.

Other strategies developed and evaluated in various types of cancers could benefit gynecologic cancers. The targeting of TAMs is one of them. TAMs constitute amajorpart of the tumor mass, and they have been shown to aid tumor progression in solid tumors. Due to their M2 immunosuppressive phenotype, they promote angiogenesis, provide actors for tumor progression and initiation of the premetastatic function, and mediate immunosuppression in the tumor microenvironment and chemoresistance [321]. As discussed in the exosome section, TAMs can produce exosomes containing miRNAs that can confer chemoresistance to cancer cells when these exosomes are taken up by cancer cells [294]. In colorectal cancer cells (CRCs), interleukin (IL)-6 produced by TAMs acted on cancer cells and led to the activation of the(STAT3) [322]. The tumor-suppressor miR-204-5p is a target of STAT3 which inhibit its transcription, rendering CRCs drug-resistant [322]. Thus, to eliminate the contribution of TAMs cancer progression, targeting strategies have been used on TAMs. They include: i) depleting them from the tumor microenvironment (mostly via colony stimulating factor 1 receptor (CSF-1R) inhibitors), ii) inhibit their recruitment [via the C—C chemokine receptor type 2

(CCR2) inhibitors], or iii) reversing their M2 phenotype back to an M1 immunostimulatory phenotype [via activation of stimulatory pathways of macrophages, immune checkpoint inhibitors, CD40 agonists, Signal regulatory protein 1α (SIRP1α) inhibitors] [323]. This last strategy would produce a response similar to that of using Coley's toxins described earlier and may be more promising for MDR cancers. Plexidartinib, a CSF-1R inhibitor was used with paclitaxel in a phase Ib study for advanced solid tumors including ovarian carcinoma. While the combination treatment showed a good tolerability, 45% of the patients showed a progressive disease, 34% had a stable disease, 13% showed limited response, and 3% reported a complete response [324]. The combination treatment using cholesteryl pullulan nanogels with the toll-like receptor 9 agonist CpG on murine model of CMS5a fibro-sarcoma, which does not respond to immune checkpoint inhibition, elicited antigen-presentation behavior in TAMs and slowed tumor growth [325].

Due to the selective targeting of a particular cell population that they offer, antibody-drug conjugates (ADCs) have garnered interest in recent years, and the FDA-approved ADCs such as Mylotarg (anti-CD33 ab and calicheamicin), Adcetris (anti-CD30 and monomethyl auristatin E), or Besponza (anti-CD22 and calicheamicin), respectively, for acute myeloid leukemia, systemic anaplastic large-cell lymphoma, and acute lymphocytic leukemia have shown clinical efficacy [326]. A phase III study (# NTC02631876) for which results have yet to be published assessed progression-free survival, overall response, duration of response, and survival using the ADC Mirvetuximab soravtensine (an antifolate receptor alpha ab conjugated to the tubulin inhibitor DM4) with or without chemotherapy agents (paclitaxel, topotecan) in patients with platinum-resistant primary peritoneal, epithelial OC, or primary fallopian tube cancer [327].

Although they have a short half-life, cytokines such as INFα, IL-2, IL-12, IL-15, IL-21, and TNFα have been used to stimulate different components of the immune system as an anticancer therapy strategy [328]. As cytokines help cells communicate with each other, they can affect both immune and nonimmune cells. When human colon carcinoma cells (HT 115, LoVo and SW480, LS 174 T) were treated with TNFα, IL-2, and INFγ, expression levels of P-gp decrease in some of them (HT 115, LoVo and SW480) after 48 to 72 h of cytokine exposure [329]. In addition, this decrease in P-gp expression enabled increased doxorubicin and vincristine cytotoxicity in the cells that were previously drug-resistant [329]. In an earlier study, Orengo et al. reported that in A2780 OC cells, TNFα enhanced the anticancer effect of topoisomerase I and II inhibitors; however, no enhancement was seen with cisplatin or mitomycin C [330]. Although the study conducted by Orengo et al. did not tackle chemoresistance, their studies suggest that a combination of cytokines with chemotherapy agents may produce an enhanced therapeutic response in gynecologic cancers.

Interestingly, treatment of parental cervical cancer KB-3-1 cells and its MDR counterpart KB-C1 with recombinant TNFα resulted not only in reduced P-gp expression in the MDR and parental cells, but also reduced production of endogenous TNFα, albeit this is independent of P-gp efflux activity [331]. Endogenous TNFα and P-gp were secreted from cancer cells via extracellular vesicles, and these vesicles induced proliferation of immortalized human fibroblasts and these cells may promote tumor progression [332]. Owing that it presents a double-edge sword effect as seen in this study and other reports have shown [333], treatment with TNFα as a way to circumvent MDR in anticancer treatment should be considered with caution.

Directly targeting P-gp to block efflux activity is another strategy used to combat MDR. Gatouillat et al. reported that anti-P-gp antibodies collected from mice immunized with a pegylated liposomal formulation of P-gp peptides were able to increase doxorubicin uptake in multidrug-resistant leukemia P388 cells [334]. This study presents an additional option that could possibly be applied to gynecologic cancers. However, this strategy has resulted with limited success. A recent finding showed that P-gp is important for the survival, activity of cytotoxic T-lymphocytes [335]. Thus, targeting P-gp can undermine the cytotoxic immune response from cytotoxic T-lymphocytes and contribute to the low response rate seen with MDR1 inhibitors. Perhaps, combining P-gp inhibitors with other immune-boosting agents and chemotherapy could enhance sensitivity to chemotherapy agents, activate the antigen-presenting cells for cell-mediated immunity to produce a durable anticancer response. The use of immunotherapy approaches to overcome MDR has a few challenges such as the immune-related toxicity of approaches or the possible dampening of adaptive immune response when immunotherapy is combined with chemotherapy drugs. These drugs target rapidly dividing cells, immune cells included, and thus, they may eliminate the intended contribution of immunotherapy to the treatment regimen. As well, exosomes may contribute to the removal of antibodies targeting cancer cells [286]. The use of tumor-restricted antigens for targeting and selective delivery of anticancer agents via nanoparticles may help address some of these challenges.

## Targeting epithelial-mesenchymal transition

Epithelial-mesenchymal transition (EMT) has been widely noticed for its role in chemoresistance, in addition to cancer progression and metastasis. EMT is considered to be one of the key factors contributing to the poor prognosis in several gynecologic cancers [336]. The exact mechanism linking EMT and drug resistance is yet to be elucidated. However, insights from research with CSCs have allowed a better perspective of the role of EMT in inducing drug resistance due to the recognition of EMT phenotypic cells being present in the CSCs. These CSCs are a restricted population of tumor cells that are self-renewing, multipotent, and persistently resisting and recurring after conventional chemotherapy [337]. Significantly, cells undergoing EMT have been reported to exhibit stem cell-like properties and to share important features with CSCs that are well known as major contributors to drug resistance [338, 339]. Mounting evidences suggest striking resemblance in the pathways that are activated during EMT and those that drive CSC such as Wnt, Hedgehog, and Notch pathways. Additionally, an elevated expression of drug efflux pumps and resistance to apoptosis have also been reported in EMT phenotypic cells and in CSCs [340].

The most critical factor influencing drug resistance in CSCs was observed to be the excessive drug efflux performed by the ATP binding cassette (ABC) transporters a process also observed in cells undergoing EMT [340]. Wang et al. have reported that Afatinib at nontoxic concentrations can significantly reverse ABC B1-mediated MDR in OC cells (in vitro) [341]. Because MDR involving ABC transporter protein is a major challenge to the use of several FDA-approved chemotherapeutic agents, more research efforts are urgently needed to focus on the discovery of superior MDR reversal agents to enhance the impact of molecularly resulting in more effective and safer therapeutics and better survival prospects for ovarian cancer patients.

### *Reversal of EMT using EMT-associated transcription factors*

In past few years, researchers have reported that EMT is involved in chemotherapy-induced drug resistance via EMT-associated transcription factors in several human cancers including some GCs [342–347]. Among the transcription factors identified, Zinc finger protein SNAI1 (SNAIL), Twist basic helix-loop-helix transcription factor (TWIST 1/2), and Zinc Finger *E*-box binding homeobox 1/2 (Zeb 1/2) have been found to be the major regulators of EMT in gynecologic cancers [336]. In general, TWIST1 and SNAIL are crucial EMT inducers that are highly upregulated in many different types of cancers and are responsible for increased tumor invasiveness [348]. Yoon et al. observed that tristetrapolin (TTP), a tumor suppressor that inhibits EMT in malignant cells as effective in reversing the mesenchymal phenotype in ovarian and other cancers [349]. Tristetrapolin was found to inhibit EMT by downregulating TWIST1 and SNAIL and also suppress tumor growth by downregulating cancer-associated genes. Thus, TTP can be a potent therapeutic target due to its role in preventing metastasis and inhibiting cell proliferation.

### *Reversal of EMT via kinase inhibitor*

While several approaches have been proposed to target EMT in addressing drug resistance, one of them includes reversing EMT in metastable cancer cells using noncytotoxic compounds [337]. Screening for compounds using a rat bladder carcinoma cell line resulted in compounds that can target certain kinases that can reverse the phenotype without affecting the cellular proliferation such as Activin like Kinase 5/TGF Beta Receptor 1 (ALK5/TGFBR1), Src, MAPK, and PI3K. Among these kinases, inhibition of Src using Saracatinib in OC cell line SKOV3 showed high efficiency in reversing EMT [350].

### *Reversal of EMT using miRNA therapeutics*

Some studies have indicated the presence of tumor-suppressive and tumor-enhancing roles of miRNAs in EMT. OC-associated EMT were reported to be either suppressed or elevated via a large subset of miRNAs (miR-187, miR-34a, miR-506, miR-138, miR-30c, miR-30d, miR-30e-3p, miR-370, and miR-106a) [351]. Deeper understanding of the functions of these miRNAs could help to develop new therapeutics in targeting OC. A case study in serous OC involving 459 cases identified miR-506 (by integrated genomic analysis) as a positive prognostic predictor and an EMT inhibitor, inhibiting TGF-induced EMT by targeting SNAI2 protein. Enhanced expression of miR-506 was reported to reduce characteristic cellular functions ofmalignant cells including proliferation, migration, and invasion [352]. Additionally, systemic delivery of miR-506 using1,2-DiOleoyl-sn-glycero-3-PhosphoCholine (DOPC) liposomes in orthotopic mouse models of OCdemonstrated a drasticdecrease of tumor nodules and tumor weight compared to miR-control.

Studies conducted by Jing et al. identified the importance of miR-31-3p in regulating EMT in cisplatin-resistant cervical cancer cells. In vitro studies indicated that the overexpression of miR-31-3p was inhibiting EMT in cervical cancer cells. Additionally, they observed Semaphorin 4C (SEMA4C), a transmembrane protein involved in tumor progression, to be the direct target of miR 31-3p and the increase in the expression of this miR reversed biological functions moderated by EMT including drug resistance [353]. Hence, targeting Sema4C or upregulating miR 31-3p would be a unique strategy to combat chemoresistance in cervical cancer. Studies conducted by Yeh et al., in orthotopic mouse model xenografts of OC, showed that miR-138 inhibited cancer metastasis by directly targeting SRY-related high-mobility group box 4 (SOX4) and hypoxia inducible factor-1 (HIF-1α). miR-200 showed strongcorrelation with drug resistance in OC [354]. Studies of Leskeka et al. showed a positive association between miR-200c and response to a combination of paclitaxel-carboplatin treatment in women with OC [355]. While lower expression of miR-200c indicated poor response, higher expression indicated lower relapse and progression. These studies indicate a protective role of miR-200c.Based on in vitro and in vivo studies Zhu and coworkers showed that miR-186 can be a potential candidate in reversing cisplatin resistance in EOC cells [356]. Furthermore, experiments conducted with miR-186 in EOC cells lead to reduction in the expression of TWIST1 protein expression in addition to morphological, functional, and molecular changes resembling MET and increase in apoptosis. Consequently, these changes increased the sensitivity of EOC cells to cisplatin.

Targeting miRNAs can be highly beneficial as they have the potential to target several effectors of signaling pathways involved in key cellular functions simultaneously [351]. Currently, progress of miRNA-based anticancer therapies is being madeusing targeted therapies either alone or in combination with other chemotherapeutic drugs. However, the systemic delivery of miRNAs can be a major challenge due to its susceptibility to nucleases and its possibility of triggering toxic immune response. This can be easily overcome with the use of targeted nanoparticles that can carry miRNA for supplemental therapy or inhibit miRNA using antisense oligonucleotides, and long noncoding RNA, etc. Several studies have shown the benefits of targeted nanoparticles that can efficiently deliver noncoding RNAs like siRNA, miRNA selectively to the tumor microenvironment. Studies conducted by Chen et al. have shown the potential of using reconstituted HDL, a natural and biodegradable delivery system that could be safely used for systemic delivery of miRNA, specifically in the tumor microenvironment while sparing the healthy cells [356]. Therapeutics involving the use of miRNAs can be a very effective and an efficient strategy in targeting chemoresistance in cancer. More intensive research efforts need to be directed toward investigating the use of miR-200c as therapeutic agents either alone or in combination with other targeted drugs.

## Concluding remarks and future perspectives

Ovarian cancer has been recognized as the most lethal of the gynecologic malignancies. The relatively poor patient outcomes are primarily due to recurrence of the disease, often despite surgery and intensive postsurgical treatment. The remarkable resiliency of ovarian tumors is primarily derived from their capacity to develop resistance against the anticancer agents used in maintenance therapy [357].

In the last several years, a new approach emerged, with some success, utilizing PARP inhibitors as targeted therapeutic agents to retard the progress of ovarian tumors [358]. The clinical benefits of treatment, utilizing PARP inhibitors have altered the landscape in OC therapeutics as the treatment options have broadened for many OC patients and especially for those with tumors displaying the BRCA1/2 mutation [359]. While the introduction of PARP inhibitors represents a major step forward, OC patients both with and without the BRCA1/2 mutation will eventually become resistant to PARP inhibitors. The temporary

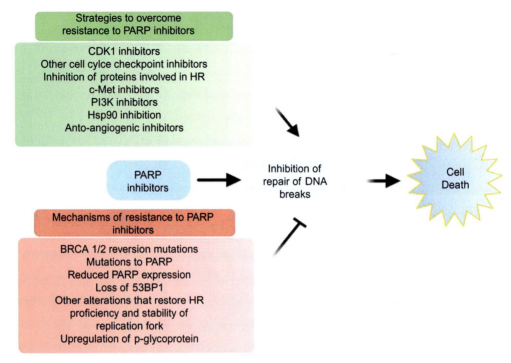

FIG. 2    Strategies to overcome resistance to PARP inhibitors: The schematic created using the *BioRender software* showing the strategies for overcoming the resistance against PARP inhibitors.

remedy perhaps lies in utilizing novel combination therapies that could enhance PARP inhibitor efficacy and overcome resistance mechanisms (see Fig. 2).

Overall, cancer therapeutics have achieved major gains in the last several years, especially in the areas of new drug development and immunotherapy. While these advances are highly significant and point to the direction of needed future research, the benefits to patients, for the most part, have been limited to the extension of survival time, often by less than 6 months. New approaches, perhaps involving the targeting of multiple, highly sensitive sites within the tumor, are needed to achieve the management and hopefully the cure for ovarian and most other cancers. Among these new approaches, targeted therapies using nanovehicles, including lipoproteins with built-in targeting potential and the opportunity to expand their focus to multiple targets, could be effective tools toward limiting or preventing drug resistance and thus pave the way for major advances in gynecologic cancers chemotherapy [360–362].

## Acknowledgments

EM was supported by the Cancer Prevention and Research Institute of Texas (CPRIT) through Grant No. RP170301. JL was supported by the National Institute of General Medical Sciences of the National Institutes of Health (NIH Award#: R25GM125587). The content is solely the responsibility of the authors and does not necessarily represent the official views of CPRIT or NIH.

# References

[1] Ferlay J, Soerjomataram I, Dikshit R, Eser S, Mathers C, Rebelo M, Parkin D, Forman D, Bray F. Cancer incidence and mortality worldwide: sources, methods and major patterns in GLOBOCAN 2012. Int J Cancer 2015;136(5):E359–86. https://doi.org/10.1002/ijc.29210.

[2] SEER Cancer Statistics Factsheets: Ovary Cancer. National Cancer Institute. Bethesda, MD; 2018. https://seer.cancer.gov/statfacts/html/ovary.html (Accessed September 2020).

[3] Siegel R, Miller K, Jemal A. Cancer statistics, 2019. CA Cancer J Clin 2019;69(1):7–34. https://doi.org/10.3322/caac.21551.

[4] SEER Cancer Statistics Factsheets: Cervix Uteri Cancer. National Cancer Institute. Bethesda, MD; 2018. https://seer.cancer.gov/statfacts/html/cervix.html (Accessed September 2020).

[5] Anon. Cervical cancer. NIH Consens Statement 1996;14(1):1–4.

[6] Torgovnik J. Cervical cancer. Retrieved September 30, 2020, from https://www.who.int/health-topics/cervical-cancer.

[7] Arbyn M, Anttila A, Jordan J, Ronco G, Schenck U, Segnan N, Wiener H, Herbert A, von Karsa L. European guidelines for quality assurance in cervical cancer screening. Second edition—summary document. Ann Oncol 2010;21(3):448–58. https://doi.org/10.1093/annonc/mdp471.

[8] Anon. Center for disease control and prevention, Genital HPV infection fact sheet. Retrieved September 29 2020 https://www.cdc.gov/std/hpv/stdfact-hpv.htm.

[9] Anon. WHO position paper. In: Human papillomavirus vaccines: Vaccins contre les papillomavirus humains: note de synthèse de l'OMS; May 2017. mai 2017 (2017) Releve epidemiologique hebdomadaire 2017; 92(19): 241–268 https://appswhoint/iris/handle/10665/255354. [Accessed 28 September 2020].

[10] Armstrong DK. 2020. Goldman-cecil medicine, 26th ed., Elsevier Inc. https://www-clinicalkey-com.proxy.unthsc.edu/#!/content/book/3-s2.0-B9780323532662001892 Gynecologic Cancers. 189, 1327–1335.e2 Retrieved September 29 2020 from.

[11] World Health Organization. Human papillomavirus (HPV) and cervical cancer. Retrieved September 29, 2020 https://www.who.int/news-room/fact-sheets/detail/human-papillomavirus-(hpv)-and-cervical-cancer; January 24, 2019.

[12] American Cancer Society. Key statistics for ovarian cancer. Retrieved September 29, 2020 https://www.cancer.org/cancer/ovarian-cancer/about/key-statistics.html; January 8, 2020.

[13] Anon. United States cancer statistics: data visualizations. Retrieved September 29, 2020, from the Centers for Disease Control and Prevention website on ovarian cancer https://gis.cdc.gov/Cancer/USCS/DataViz.html; June 2020.

[14] Anon. United States cancer statistics: data visualizations. Retrieved September 29, 2020, from the Centers for Disease Control and Prevention website on cervical cancer https://gis.cdc.gov/Cancer/USCS/DataViz.html; June 2020.

[15] American Cancer Society. Cancer Statistic center 2020 estimate for uterine corpus cancer. (n.d.)., https://cancerstatisticscenter.cancer.org/?_ga=2.1539378.2056135370.1601181636- Retrieved September 29, 2020.

[16] Hoffman B, Schorge J, Halvorson L, Hamid C, Corton M, Schaffer J, editors. Endometrial cancer. In: Williams gynecology, 4e. McGraw-Hill; 2020. https://accessmedicine-mhmedical-com.libproxy1.usc.edu/content.aspx?bookid=2658&sectionid=241073913.

[17] International Agency for Research on Cancer. World cancer report 2014. World Health Organization; 2014. Chapter 5.12. ISBN 978-92-832-0429-9.

[18] Cleland WH, Mendelson CR, Simpson ER. Aromatase activity of membrane fractions of human adipose tissue stromal cells and adipocytes. Endocrinology 1983;113(6):2155–60.

[19] National Academies of Sciences, Engineering, and Medicine. Ovarian cancers: evolving paradigms in research and care. Washington, DC: The National Academies Press; 2016. https://doi.org/10.17226/218.

[20] Bowtell DD, Böhm S, Ahmed AA, Aspuria PJ, Bast Jr RC, Beral V, Berek JS, Birrer MJ, Blagden S, Bookman MA, Brenton JD, Chiappinelli KB, Martins FC, Coukos G, Drapkin R, Edmondson R, Fotopoulou C, Galon J, Gourley C, Heong V, Huntsman DG, Iwanicki M, Karlan BY, Kaye A, Levine DA, Lu KH, IA MN, Menon U, Narod SA, Nelson BH, Nephew KP, Pharoah P, Ramos P, Romero IL, Scott CL, Sood AK, Stronach EA, Balkwill FR. Re-thinking ovarian cancer II: reducing mortality from high-grade serous ovarian cancer. Nat Rev Cancer 2015;15:668–79.

[21] Armbruster S, Coleman RL, Rauh-Hain JA. Management and treatment of recurrent epithelial ovarian cancer. Hematol Oncol Clin North Am 2018;32(6):965–82.

[22] Bitler BG, Watson ZL, Wheeler LJ, Behbakht K. PARP inhibitors: clinical utility and possibilities of overcoming resistance. Gynecol Oncol 2017;147(3):695–704.

[23] Gottesman M. Mechanisms of cancer drug resistance. Annu Rev Med 2002;53:615–27.

[24] Quintieri L, Fantin M, Vizler C. Identification of molecular determinants of tumor sensitivity and resistance to anticancer drugs. Adv Exp Med Biol 2007;593:95–104.

[25] Sikic BI. Chapter 47—Natural and acquired resistance to cancer therapies A2. In: John M, Gray JW, et al., editors. The molecular basis of cancer (fourth edition). Philadelphia: Elsevier Inc; 2015. p. 651–660.e4.

[26] Pusztai L, Siddik Z, Mills G, Bast R. Physiologic and pathologic drug resistance in ovarian carcinoma—a hypothesis based on a clonal progression model. Acta Oncol 1998;37(7–8):629–40.

[27] Yufang P, Jinge Z, Jing W, Zhong J, Zhang L, Wang Y, Yu L, Yan Z. Strategies of overcoming the physiological barriers for tumor-targeted nano-sized drug delivery systems. Curr Pharm Des 2015;21(42):6236–45.

[28] Matsumoto S, Batra S, Saito K, Yasui H, Choudhuri R, Gadisetti C, Subramanian S, Devasahayam N, Munasinghe JP, Mitchell JB, Krishna M. Antiangiogenic agent sunitinib transiently increases tumor oxygenation and suppresses cycling hypoxia. Cancer Res 2011;71(20):6350–9.

[29] Marangolo M, Bengala C, Conte P, Danova M, Pronzato P, Rosti G, Sagrada P. Dose and outcome: the hurdle of neutropenia (review). Oncol Rep 2006;16(2):233–48.

[30] Wu Q, Yang Z, Nie Y, Shi Y, Fan D. Multi-drug resistance in cancer chemotherapeutics: mechanisms and lab approaches. Cancer Lett 2014;347(2):159–66.

[31] Muriithi W, Macharia L, Heming C, Echevarria J, Nyachieo A, Filho P, Neto V. ABC transporters and the hallmarks of cancer: roles in cancer aggressiveness beyond multidrug resistance. Cancer Biol Med 2020;17(2):253–69.

[32] Sissung T, Baum C, Kirkland C, Gao R, Gardner E, Figg W. Pharmacogenetics of membrane transporters: an update on current approaches. Mol Biotechnol 2010;44:152–67.

[33] Hu Y, Li C, Li H, Li M, Shu X. Resveratrol-mediated reversal of tumor multi-drug resistance. Curr Drug Metab 2014;15(7):703–10.

[34] Liu X. Transporter-mediated drug-drug interactions and their significance. Adv Exp Med Biol 2019;1141:241–91.

[35] Locher KP. Mechanistic diversity in ATP-binding cassette (ABC) transporters. Nat Struct Mol Biol 2016;23:487–93.

[36] Robey RW, Pluchino KM, Hall MD, Fojo AT, Bates SE, Gottesman MM. Revisiting the role of ABC transporters in multidrug-resistant cancer. Nat Rev Cancer 2018;18:452–64.

[37] Zhen WH, Västermark Å, Shlykov MA, Reddy V, Sun EI, Saier MH. Evolutionary relationships of ATP-binding cassette (ABC) uptake porters. BMC Microbiol 2013;13:98.

[38] Norouzi-Barough L, Sarookhani M, Sharifi MR, Moghbelinejad S, Jangjoo S, Salehi R. Molecular mechanisms of drug resistance in ovarian cancer. Cell Physiol 2018;233(6):4546–62. https://doi.org/10.1002/jcp.26289.

[39] Chen Y, Bieber M, Teng N. Hedgehog signaling regulates drug sensitivity by targeting ABC transporters ABCB1 and ABCG2 in epithelial ovarian cancer. Mol Carcinog 2014;53:625–34.

[40] Fletcher J, Haber M, Henderson M, Norris M. ABC transporters in cancer: more than just drug efflux pumps. Nat Rev Cancer 2010;10:147–56.

[41] Sims J, Ganguly S, Bennett H, Friend J, Tepe J, Plattner R. Imatinib reverses doxorubicin resistance by affecting activation of STAT3-dependent NF-kB and HSP27/p38/AKT pathways and by inhibiting AB CB1. PLoS One 2013;8, e55509.

[42] Ween M, Armstrong M, Oehler M, Ricciardelli C. The role of ABC transporters in ovarian cancer progression and chemoresistance. Crit Rev Oncol Hematol 2015;96(2):220–56.

[43] Lancaster C, Sprowl J, Walker A, Hu S, Gibson A, Sparreboom A. Modulation of OATP1B-type transporter function alters cellular uptake and disposition of platinum chemotherapeutics. Mol Cancer Ther 2013;12:1537–44.

[44] Liu L, Liu X. Contributions of drug transporters to blood-placental barrier. Adv Exp Med Biol 2019;1141:505–48.

[45] Januchowski R, Zawierucha P, Andrzejewska M, Ruciński M, Zabel M. Microarray-based detection and expression analysis of ABC and SLC transporters in drug-resistant ovarian cancer cell lines. Biomed Pharmacother 2013;67(3):240–5.

[46] Girardi E, César-Razquin A, Lindinger S, Papakostas K, Konecka J, Hemmerich J, Kickinger S, et al. A widespread role for SLC transmembrane transporters in resistance to cytotoxic drugs. Nat Chem Biol 2020;16(4):469–78. https://doi.org/10.1038/s41589-020-0483-3.

[47] Arend R, Londoño-Joshi A, Straughn Jr J, Buchsbaum D. The Wnt/β-catenin pathway in ovarian cancer: a review. Gynecol Oncol 2013;131(3):772–9. https://doi.org/10.1016/j.ygyno.2013.09.034 [Epub 2013 Oct 11].

[48] Zhao J. Cancer stem cells and chemoresistance: the smartest survives the raid. Pharmacol Ther 2016;160:145–58. https://doi.org/10.1016/j.pharmthera.2016.02.008.

[49] Akhdar H, Legendre C, Aninat C, Morel F. Anticancer drug metabolism: chemotherapy resistance and new therapeutic approaches. In: Paxton J, editor. Topics on drug metabolism. Croatia: In Tech; 2012. p. 138–56.

[50] Housman G, Byler S, Heerboth S, Lapinska K, Longacre M, Snyder N, Sarkar S. Drug resistance in cancer: an overview. Cancer 2014;6:1769–92.

[51] Michael M, Doherty M. Tumoral drug metabolism: overview and its implications for cancer therapy. J Clin Oncol 2005;23:205–29.

[52] Saha S, Adhikary A, Bhattacharyya P, Das T, Sa G. Death by design: where curcumin sensitizes drug resistant tumors. Anticancer Res 2012;32:2567–84.

[53] Townsend DM, Tew KD. The role of Glutathione-S-transferase in anti-cancer drug resistance. Oncogene 2003;22:7369–75.

[54] Cort A, Ozben T, Saso L, De Luca C, Korkina L. Redox control of multidrug resistance and its possible modulation by antioxidants. Oxid Med Cell Longev 2016. https://doi.org/10.1155/2016/4251912, 4251912.

[55] Zanger U, Schwab M. Cytochrome P450 enzymes in drug metabolism: regulation of gene expression, enzyme activities, and impact of genetic variation. Pharmacol Ther 2013;138:103–41.

[56] Zhang T, Dai H, Liu L, Lewis D, Wei D. Classification models for predicting cytochrome p450 enzyme-substrate selectivity. Mol Inf 2012;31:53–62.

[57] Downie D, McFadyen M, Rooney P, Cruickshank M, Parkin D, Miller I, et al. Profiling cytochrome P450 expression in ovarian cancer: identification of prognostic markers. Clin Cancer Res 2005;11:7369–75.

[58] Rodriguez-Antona C, Ingelman-Sundberg M. Cytochrome P450 pharmacogenetics and cancer. Oncogene 2006;25:1679–91.

[59] McFadyen M, Cruickshank M, Miller I, McLeod H, Melvin W, Haites N, et al. Cytochrome P450 CYP1B1 overexpression in primary and metastatic ovarian cancer. Br J Cancer 2001;85:242–6.

[60] McFadyen M, Melvin W, Murray G. Cytochrome P450 enzymes: novel options for cancer therapeutics. Mol Cancer Ther 2004;3:363–71.

[61] Joseph T, Chacko P, Wesley R, Jayaprakash PG, James FV, Pillai MR. Germline genetic polymorphisms of CYP1A1, GSTM1 and GSTT1 genes in Indian cervical cancer: associations with tumor progression, age and human papillomavirus infection. Gynecol Oncol 2006;101:411–7.

[62] Singh H, Sachan R, Devi S, Pandey SN, Mittal B. Association of GSTM1, GSTT1, and GSTM3 gene polymorphisms and susceptibility to cervical cancer in a North Indian population. Am J Obstet Gynecol 2008;198. 303.e1–303.

[63] Kiran B, Karkucak M, Ozan H, Yakut T, Ozerkan K, Sag S, Ture M. GST (GSTM1, GSTT1, and GSTP1) polymorphisms in the genetic susceptibility of Turkish patients to cervical cancer. Gynecol Oncol 2010;21 (3):169–73.

[64] Yang M, Li Y, Shen X, Ruan Y, Lu Y, Jin X, Song P, Guo Y, Zhang X, Qu H, Shao Y, Quan Y. CLDN6 promotes chemoresistance through GSTP1 in human breast cancer. J Exp Clin Cancer Res 2017;36:157.

[65] Sawers L, Ferguson M, Ihrig B, Young H, Chakravarty P, Wolf C, Smith G. Glutathione S-transferase P1 (GSTP1) directly influences platinum drug chemosensitivity in ovarian tumor cell lines. Br J Cancer 2014;111:1150–8.

[66] Nagar S, Remmel R. Uridine diphosphoglucuronosyltransferase pharmacogenetics and cancer. Oncogene 2006;25(11):1659–72.

[67] Hu D, Mackenzie P, McKinnon R, Meech R. Genetic polymorphisms of human UDP-glucuronosyltransferase (UGT) genes and cancer risk. Drug Metab Rev 2016;48:47–69.

[68] Rowland A, Miners J, Mackenzie P. The UDP-glucuronosyltransferases: their role in drug metabolism and detoxification. Int J Biochem Cell Biol 2013;45(6):1121–32.

[69] Lépine J, Bernard O, Plante M, Têtu B, Pelletier G, Labrie F, Bélanger A, Guillemette C. Specificity and regioselectivity of the conjugation of estradiol, estrone, and their catecholestrogen and methoxyestrogen metabolites by human uridine diphospho-glucuronosyltransferases expressed in endometrium. J Clin Endocrinol Metab 2004;89:5222–32.

[70] Guillemette C, Lévesque É, Rouleau M. Pharmacogenomics of human uridine diphospho-glucuronosyltransferases and clinical implications. Clin Pharmacol Ther 2014;96:324–39.

[71] Kakehi M, Ikenaka Y, Nakayama SMM, Kawai YK, Watanabe KP, et al. Uridine diphosphate-glucuronosyltransferase (UGT) xenobiotic metabolizing activity and genetic evolution in pinniped species. Toxicol Sci 2015;147:360–9.

[72] McGrath M, Lepine J, Lee I-M, Villeneuve L, Buring J, Chantal Guillemette C, De Vivo I. Genetic variations in UGT1A1 and UGT2B7 and endometrial cancer risk. Pharmacogenet Genomics 2009;19(3):239–43. https://doi.org/10.1097/FPC.0b013e328323f66c.

[73] Thibaudeau J, Lepine J, Tojcic J, Duguay Y, Pelletier G, Plante M, Brisson J, Tetu B, Jacob S, Perusse L, Belanger A, Guillemette C. Characterization of common UGT1A8, UGT1A9, and UGT2B7 variants with different capacities to inactivate mutagenic 4-hydroxylated metabolites of estradiol and estrone. Cancer Res 2006;66:125–33.

[74] Hirasawa A, Zama T, Akahane T, Nomura H, Kataoka F, Saito K, et al. Polymorphisms in the UGT1A1 gene predict adverse effects of irinotecan in the treatment of gynecologic cancer in Japanese patients. J Hum Genet 2016;58:794–8.

[75] Wilson TR, Longley DB, Johnston PG. Chemoresistance in solid tumours. Ann Oncol 2006;17:315–24.

[76] Nagaraj A, Joseph P, Kovalenko O, Singh S, Armstrong A, Redline R, Resnick K, Zanotti K, Waggoner S, DiFeo A. Critical role of Wnt/β-catenin signaling in driving epithelial ovarian cancer platinum resistance. Oncotarget 2015;6(27):23720–34. https://doi.org/10.18632/oncotarget.4690.

[77] Zhao H, Zhou L, Shangguan A, Bulun S. Aromatase expression and regulation in breast and endometrial cancer. J Mol Endocrinol 2016;57(1):R19–33. https://doi.org/10.1530/JME-15-0310.

[78] Park J, Pyun W, Park H. Cancer metabolism: phenotype, signaling and therapeutic targets. Cell 2020;9(10):2308. Published online 2020 Oct 16 https://doi.org/10.3390/cells9102308.

[79] Yizhak K, Chaneton B, Gottlieb E, Ruppin E. Modeling cancer metabolism on a genome scale. Mol Cyst Biol 2015;11(6):817.

[80] Hirschey M, DeBerardinis R, Diehl A, Drew J, Frezza C, Green M, et al. Dysregulated metabolism contributes to oncogenesis. Semin Cancer Biol 2015;35:S129–50.

[81] DeBerardinis R. Is cancer a disease of abnormal cellular metabolism? New angles on an old idea. Genet Med 2009;10:767–77.

[82] Vander Heiden M, Thompson L. Understanding the Warburg effect: the metabolic requirements of cell proliferation. Science 2009;324:1029–33.

[83] Wu W, Zhao S. Metabolic changes in cancer: beyond the Warburg effect. Acta Biochim Biophys Sin 2013;45:18–26.

[84] Koppenol WH, Bounds PL, Dang CV. Otto Warburg's contributions to current concepts of cancer metabolism. Nat Rev Cancer 2011;11:325–37.

[85] Kobayashi Y, Banno K, Kunitomi H, Takahashi T, Takeda T, Nakamura K, Tsuji K, Tominaga E, Aoki D. Warburg effect in gynecologic cancers. J Obstet Gynaecol Res 2018;45:1–7.

[86] Warburg O, Wind F, Negelein E. The metabolism of tumors in the body. J Gen Physiol 1927;8(6):519–30. https://doi.org/10.1085/jgp.8.6.519.

[87] Christofk HR, Vander Heiden MG, Harris MH, et al. The M2 splice isoform of pyruvate kinase is important for cancer metabolism and tumour growth. Nature 2008;452:230–3.

[88] Le A, Cooper C, Gouw A, Dinavahi R, Maitra A, Deck M, Royer R, Vander Jagt D, Semenza G, Dang C. Inhibition of lactate dehydrogenase induces oxidative stress and inhibits tumor progression. Proc Natl Acad Sci U S A 2010;107:2037–42.

[89] Ren J, Kats L, Burgess K, Bhargava P, Signoretti S, Duffy K, Grant A, Wang X, Lorkiewicz P, Schatzman S, et al. Targeting lactate dehydrogenase—a inhibits tumorigenesis and tumor progression in mouse models of lung cancer and impacts tumor initiating cells. Cell Metab 2014;19:795–809.

[90] Negrini S, Gorgoulis VG, Halazonetis TD. Genomic instability—an evolving hallmark of cancer. Nat Rev Mol Cell Biol 2010;11:220–8.

[91] Berkers CR, Maddocks OD, Cheung EC, Mor I, Vousden KH. Metabolic regulation by p53 family members. Cell Metab 2013;18:617–33.

[92] Kruiswijk F, Labuschagne CF, Vousden KH. p53 in survival, death and metabolic health: a lifeguard with a license to kill. Nat Rev Mol Cell Biol 2015;16:393–405.

[93] Wang GL, Semenza GL. General involvement of hypoxia inducible factor 1 in transcriptional response to hypoxia. Proc Natl Acad Sci U S A 1993;90:4304–8.

[94] Denko NC. Hypoxia, HIF1 and glucose metabolism in the solid tumour. Nat Rev Cancer 2008;8:705–13.

[95] Wilde L, Roche M, Domingo-Vidal M, Tanson K, Philp N, Curry J, Martinez-Outschoorn U, et al. Metabolic coupling and the reverse Warburg effect in cancer: implications for novel biomarker and anticancer agent development. Semin Oncol 2017;44:198–203.

[96] Fu Y, Liu S, Yin S, Niu W, Xiong W, Tan M, Li G, Ming ZM. The reverse Warburg effect is likely to be an Achilles' heel of cancer that can be exploited for cancer therapy. Oncotarget 2017;8:57813–25.

[97] Kobayashi Y, Kashima H, Rahmanto YS, Banno K, Yu Y, Yusuke M, Watanabe K, Iijima M, Takeda T, Kunitomi H, et al. Drug repositioning of mevalonate pathway inhibitors as antitumor agents for ovarian cancer. Oncotarget 2017;8(42):72147–56.

[98] Kobayashi Y, Banno K, Kunitomi H, Tominaga E, Aoki D. Current state and outlook for drug repositioning anticipated in the field of ovarian cancer. J Gynecol Oncol 2019;30(1). https://doi.org/10.3802/jgo.2019.30.e10, e10 [Epub 2018 Oct 10].

[99] Gal D, MacDonald PC, Porter JC, Simpson E. Low-density lipoprotein metabolism. Int J Cancer 1981;28(3):315–9. https://doi.org/10.1002/ijc.2910280310.

[100] Yarmolinsky J, Bull C, Emma E, Vincent E, Robinson J, Walther A, Smith G, Lewis SJ, Relton C, Richard M, Martin R. Association between genetically proxied inhibition of HMG-CoA reductase and epithelial ovarian cancer. JAMA 2020;323(7):646–55. https://doi.org/10.1001/jama.2020.0150.

[101] Zeleznik O, Clish C, Kraft P, Ila-Pacheco J, Eliassen A, Tworoger S. Circulating lysophosphatidylcholines, phosphatidylcholines, ceramides, and sphingomyelins and ovarian cancer risk: a 23-year prospective study. J Natl Cancer Inst 2020;112(6):628–36. https://doi.org/10.1093/jnci/djz195.

[102] Cruz P, Mo H, McConathy W, Sabnis N, Lacko A. The role of cholesterol metabolism and cholesterol transport in carcinogenesis: a review of scientific findings, relevant to future cancer therapeutics. Front Pharmacol 2013. https://doi.org/10.3389/fphar.2013.00119.

[103] Hao Y, Daixi Li D, Yong Xu Y, Ouyang J, Wang Y, Zhang Y, Li B, Xie L, Qin G. Investigation of lipid metabolism dysregulation and the effects on immune microenvironments in pan-cancer using multiple omics data. BMC Bioinf 2019;20(Suppl. 7):195. https://doi.org/10.1186/s12859-019-2734-4.

[104] Pyragius C, Fuller M, Ricciardelli C, Oehler M. Aberrant lipid metabolism: an emerging diagnostic and therapeutic target in ovarian cancer. Int J Mol Sci 2013;14:7742–56.

[105] Baenke F, Peck B, Miess H, Schulze A. Hooked on fat: the role of lipid synthesis in cancer metabolism and tumour development. Dis Model Mech 2013;6:1353–63. aenke, Peck, Miess, & Schulze, 2013.

[106] Locasale JW. Serine, glycine and the one-carbon cycle: cancer metabolism in full circle. Nat Rev Cancer 2013;13:572–83.

[107] Phang JM, Liu W, Hancock CN, Fischer JW. Proline metabolism and cancer: emerging links to glutamine and collagen. Curr Opin Clin Nutr Metab Care 2015;18:71–7.

[108] Bhutia Y, Babu E, Ramachandran S, Ganapathy V. Amino acid transporters in cancer and their relevance to "glutamine addiction": novel targets for the design of a new class of anticancer drugs. Cancer Res 2015;75(9):1782–8. https://doi.org/10.1158/0008-5472.CAN-14-3745.

[109] Zhang L, Sui C, Yang W, Luo Q. Amino acid transporters: Emerging roles in drug delivery for tumor-targeting therapy. Asian J Pharm Sci 2020;15(2):192–206. Published online 2020 Mar 5 https://doi.org/10.1016/j.ajps.2019.12.002. PMCID: PMC7193455.

[110] Catanzaro D, Gaude OG, Giordano C, Guzzo G, Rasola A, Montopoli M. Inhibition of glucose-6-phosphate dehydrogenase sensitizes cisplatin-resistant cells to death. Oncotarget 2015;6:112.

[111] DeBerardinis R. Is cancer a disease of abnormal cellular metabolism? New angles on an old idea. Genet Med 2009;10:767–77. 30102–30114.

[112] Corbet C, Feron O. Metabolic and mind shifts: from glucose to glutamine and acetate addictions in cancer. Curr Opin Clin Nutr Metab Care 2015;18:346–53.

[113] Hudson C, Savadelis A, Nagaraj A, Joseph P, Avril S, DiFeo A, Avril N. Altered glutamine metabolism in platinum resistant ovarian cancer. Oncotarget 2016;7:41637–49.

[114] Sato M, Kawana K, Adachi K, Asaha Fujimoto A, Yoshida M, et al. Targeting glutamine metabolism and the focal adhesion kinase additively inhibits the mammalian target of the rapamycin pathway in spheroid cancer stem-like properties of ovarian clear cell carcinoma in vitro. Int J Oncol 2017;50(4):1431–8. https://doi.org/10.3892/ijo.2017.3891.

[115] Guo L, Zhou B, Liu Z, Xu Y, Lu H, Xia M. Blockage of glutaminolysis enhances the sensitivity of ovarian cancer cells to PI3K/mTOR inhibition involvement of STAT3 signaling. Tumour Biol 2016;37(8):11007–15. https://doi.org/10.1007/s13277-016-4984-3.

[116] Sinha BK. Role of oxygen and nitrogen radicals in the mechanism of anticancer drug cytotoxicity. J Cancer Sci Ther 2020;12(1):10–8 [Epub 2020 Jan 24].

[117] Kumar A, Ehrenshaft M, Tokar EJ, Mason RP, Sinha BK. Nitric oxide inhibits topoisomerase II activity and induces resistance to topoisomerase II-poisons in human tumor cells. Biochim Biophys Acta 2016;1860(7):1519–27.

[118] Cabeça T, Rafael Andrette A, de Souza LV, Morale M, Aguayo F, et al. HPV-mediated resistance to TNF and TRAIL is characterized by global alterations in apoptosis regulatory factors, dysregulation of death receptors, and induction of ROS/RNS. Int J Mol Sci 2019;20(1):198. https://doi.org/10.3390/ijms20010198.

[119] Dharmaraja A. Role of reactive oxygen species (ROS) in therapeutics and drug resistance in cancer and bacteria. J Med Chem 2017;60:3221–40.

[120] Belotte J, Fletcher N, Awonuga O, Abu-Soud M, Saed G, Diamond P, Saed M. The role of oxidative stress in the development of cisplatin resistance in epithelial ovarian cancer. Reprod Sci 2014;21:503–8.

[121] Liou G, Storz P. Reactive oxygen species in cancer. Free Radic Res 2010;44:479–96.

[122] Chen J. Reactive oxygen species and drug resistance in cancer chemotherapy. J Clin Pathol 2014;1:1017.

[123] Liu B, Huang X, Li Y, Liao W, Li M, Liu Y, He R, Feng D, Zhu R, Kurihara H. JS-K, a nitric oxide donor, induces autophagy as a complementary mechanism inhibiting ovarian cancer. BMC Cancer 2019;19:645. Published online 2019 Jul 1 https://doi.org/10.1186/s12885-019-5619-z.

[124] Chen J, Adikari M, Pallai R, Parekh R, Simpkins H. Dihydrodiol dehydrogenases regulate the generation of reactive oxygen species and the development of cisplatin resistance in human ovarian carcinoma cells. Cancer Chemother Pharmacol 2008;61(6):979–87. Published online 2007 Jul 28 https://doi.org/10.1007/s00280-007-0554-0.

[125] Wang X-J, Sun Z, Villeneuve N, Zhang S, Zhao F, Li Y, Chen W, Yi X, et al. Nrf2 enhances resistance of cancer cells to chemotherapeutic drugs, the dark side of Nrf2. Carcinogenesis 2008;29(6):1235–43. https://doi.org/10.1093/carcin/bgn095 [Epub 2008 Apr 15].

[126] Ma Q. Role of NRF2 in oxidative stress and toxicity. Annu Rev Pharmacol Toxicol 2013;53:401–26.

[127] Hassan M, Watari H, AbuAlmaaty A, Ohba Y, Sakuragi N. Apoptosis and molecular targeting therapy in cancer. Biomed Res Int 2014;2014:150845. Published online 2014 Jun 12 https://doi.org/10.1155/2014/150845.

[128] Ricci MS, Zong W-X. Chemotherapeutic approaches for targeting cell death pathways. Oncologist 2006;11:342–57.

[129] Liu G, Pei F, Yang F, Li L, Amin A, Liu S, Buchan JR, Cho W. Role of autophagy and apoptosis in non-small-cell lung cancer. Int J Mol Sci 2017;18(2):367. Published online 2017 Feb 10 https://doi.org/10.3390/ijms18020367.

[130] Smith A, Macleod K. Autophagy, cancer stem cells and drug resistance. J Pathol 2019;247(5):708–18. Published online 2019 Feb 4 https://doi.org/10.1002/path.5222.

[131] Galluzzi L, Baehrecke EH, Ballabio A, et al. Molecular definitions of autophagy and related processes. EMBO J 2017;36:1811–36.

[132] Xu Y, Kanagaratham C, Youssef M, Radzioch D. New frontiers in cancer chemotherapy-targeting cell death pathways. In: Najman S, editor. Cell biology—new insights. Intecopen Ltd, London, UK; 2016. p. 93–140.

[133] Rouschop K, van den Beucken T, Dubois L, et al. The unfolded protein response protects human tumor cells during hypoxia through regulation of the autophagy genes MAP1LC3B and ATG5. J Clin Invest 2010;120:127–41.

[134] Perera RM, Stoykova S, Nicolay BN, et al. Transcriptional control of the autophagy-lysosome system in pancreatic cancer. Nature 2015;524:361–5.

[135] Mizushima N, Yoshimori T, Ohsumi Y. The role of Atg proteins in autophagosome formation. Annu Rev Cell Dev Biol 2011;27:107–32.

[136] Kreso A, Dick JE. Evolution of the cancer stem cell model. Cell Stem Cell 2014;14:275–91.

[137] Ramos EK, Hoffmann AD, Gerson SL, et al. New opportunities and challenges to defeat cancer stem cells. Trends Cancer 2017;3:780–96.

[138] Garcia-Prat L, Martinez-Vicente M, Perdiguero E, et al. Autophagy maintains stemness by preventing senescence. Nature 2016;529:37–42.

[139] Maycotte P, Jones KL, Goodall ML, et al. Autophagy supports breast cancer stem cell maintenance by regulating IL6 secretion. Mol Cancer Res 2015;13:651–8.

[140] Bartholomeusz C, Rosen D, Wei C, et al. PEA-15 induces autophagy in human ovarian cancer cells and is associated with prolonged overall survival. Cancer Res 2008;68:9302–10.

[141] Zhou S, Zhao L, Kuang M, Zhang B, Liang Z, Yi T, Wei Y, Zhao X. Autophagy in tumorigenesis and cancer therapy: Dr. Jekyll or Mr. Hyde? Cancer Lett 2012;323:115–27.

[142] Orfanelli T, Jeong J, Doulaveris G, Holcomb K, Witkin S. Involvement of autophagy in cervical, endometrial and ovarian cancer. Int J Cancer 2014;135(3):519–28. https://doi.org/10.1002/ijc.28524. Epub 2013.

[143] Peracchio C, Alabiso O, Valente G, Isidoro C. Involvement of autophagy in ovarian cancer: a working hypothesis. J Ovarian Res 2012;5(1):22.

[144] Cotter TG. Apoptosis and cancer: the genesis of a research field. Nat Rev Cancer 2009;9(7):501–7.

[145] Kerr JF, Wyllie AH, Currie AR. Apoptosis: a basic biological phenomenon with wide-ranging implications in tissue kinetics. Br J Cancer 1972;26(4):239–57.

[146] Wong RS. Apoptosis in cancer: from pathogenesis to treatment. J Exp Clin Cancer Res 2011;30:87. https://doi.org/10.1186/1756-9966-30-87.

[147] Szegezdi E, Fitzgerald U, Samali A. Caspase-12 and ER-stress-mediated apoptosis: the story so far. Ann N Y Acad Sci 2003;1010:186–94. https://doi.org/10.1196/annals.1299.032.

[148] Danial N, Korsmeyer SJ. Cell death: critical control points. Cell 2004;116:205–19. https://doi.org/10.1016/S0092-8674(04)00046-7.

[149] Elmore S. Apoptosis: a review of programmed cell death. Toxicol Pathol 2007;35:495–516.

[150] Macpherson A, Barry S, Ricciardelli C, Oehler M. Epithelial ovarian cancer and the immune system: biology, interactions, challenges and potential advances for immunotherapy. J Clin Med 2020;9(9):2967. Published online 2020.

[151] Fraser M, Leung B, Jahani-Asl A, Yan X, Thompson W, Tsan B. Chemoresistance in human ovarian cancer: the role of apoptotic regulators. Reprod Biol Endocrinol 2003;1:66.

[152] Moldoveanu T, Follis A, Kriwacki R, Green D. Many players in BCL-2 family affairs. Trends Biochem Sci 2014;39:101–11. https://doi.org/10.1016/j.tibs.2013.12.006.

[153] Luna-Vargas M, Chipuk J. The deadly landscape of pro-apoptotic BCL-2 proteins in the outer mitochondrial membrane. FEBS J 2016;283:2676–89. https://doi.org/10.1111/febs.13624.

[154] Patch A-M, Christie E, Etemadmoghadam D, Garsed D, George J, et al. Whole-genome characterization of chemoresistant ovarian cancer. Nature 2015;521(7553):489–94.

[155] Pommier Y, Sordet O, Antony S, Hayward R, Kohn K. Apoptosis defects and chemotherapy resistance: molecular interaction maps and networks. Oncogene 2004;23:2934–49.

[156] Özes A, Miller D, Özes O, Fang F, Liu Y, Matei D, Huang T, Nephew K. NF-κB-HOTAIR axis links DNA damage response, chemoresistance and cellular senescence in ovarian cancer. Oncogene 2016;35(41):5350–61.

[157] Polo S, Jackson S. Dynamics of DNA damage response proteins at DNA breaks: a focus on protein modifications. Genes Dev 2011;25:409–33.

[158] Dietlein F, Reinhardt HC. Molecular pathways: exploiting tumor-specific molecular defects in DNA repair pathways for precision cancer therapy. Clin Cancer Res 2014;20:5882–7.

[159] Lord CJ, Tutt AN, Ashworth A. Synthetic lethality and cancer therapy: lessons learned from the development of PARP inhibitors. Annu Rev Med 2015;66:455–70.

[160] Kristeleit RS, Miller RE, and Kohn EC, 2016, Gynecologic cancers: emerging novel strategies for targeting DNA repair deficiency. Am Soc Clin Oncol Educ Book. 36. doi:https://doi.org/10.1200/EDBK_159086 Accessed 2 November 2020.

[161] Gavande N, VanderVere-Carozza P, Hinshaw H, Jalal S, Sears C, Pawelczak K, Turchi J. DNA repair targeted therapy: the past or future of cancer treatment? Pharmacol Ther 2016;160:65–83. https://doi.org/10.1016/j.pharmthera.2016.02.003.

[162] Bryant HE, Schultz N, Thomas HD, et al. Specific killing of BRCA2-deficient tumours with inhibitors of poly (ADP-ribose) polymerase. Nature 2005;434:913–7.

[163] Prakash R, Zhang Y, Feng W, et al. Homologous recombination and human health: the roles of BRCA1, BRCA2, and associated proteins. Cold Spring Harb Perspect Biol 2015;7:a016600.

[164] Patel AG, Sarkaria JN, Kaufmann SH. Nonhomologous end joining drives poly(ADP-ribose) polymerase (PARP) inhibitor lethality in homologous recombination-deficient cells. Proc Natl Acad Sci U S A 2011;108:3406–11.

[165] Damia G, Broggini M. Platinum resistance in ovarian cancer: role of DNA repair. Cancer (Basel) 2019;11 (1):119. https://doi.org/10.3390/cancers11010119.

[166] Stefanou D, Bamias A, Episkopou H, Kyrtopoulos S, Likka M, Kalampokas T, et al. Aberrant DNA damage response pathways may predict the outcome of platinum chemotherapy in ovarian cancer. PLoS One 2015;10:e0117654. https://doi.org/10.1371/journal.pone.0117654.

[167] Duensing S, Münger K. The human papillomavirus type 16 E6 and E7 oncoproteins independently induce numerical and structural chromosome instability. Cancer Res 2002;62:7075–82.

[168] Balacescu O, Balacescu L, Tudoran O, et al. Gene expression profiling reveals activation of the FA/BRCA pathway in advanced squamous cervical cancer with intrinsic resistance and therapy failure. BMC Cancer 2014;14:246.

[169] Borley J, Brown R. Epigenetic mechanisms and therapeutic targets of chemotherapy resistance in epithelial ovarian cancer. Ann Med 2015;47:359–69.

[170] Wilting R, Dannenberg J. Epigenetic mechanisms in tumorigenesis, tumor cell heterogeneity and drug resistance. Drug Resist Updat 2012;15:21–38.

[171] Bird A. DNA methylation patterns and epigenetic memory DNA methylation patterns and epigenetic memory. Genes Dev 2002;16:6–21.

[172] Balch C, Matei D, Huang T, Nephew K. Role of epigenomics in ovarian and endometrial cancers. Epigenomics 2010;2(3):419–47. https://doi.org/10.2217/epi.10.19.

[173] Yan B, Yin F, Wang QI, Zhang W, Li L. Integration and bioinformatics analysis of DNA-methylated genes associated with drug resistance in ovarian cancer. Oncol Lett 2016;12(1):157–66. https://doi.org/10.3892/ol.2016.4608.

[174] Mirzaei H, Yazdi F, Salehi R, Mirzaei H. SiRNA and epigenetic aberrations in ovarian cancer. J Cancer Res Ther 2016;12:498–508.

[175] Zeller C, Dai W, Steele N, Siddiq A, Walley A, Wilhelm-Benartzi C, et al. Candidate DNA methylation drivers of acquired cisplatin resistance in ovarian cancer identified by methylome and expression profiling. Oncogene 2012;31:4567–76.

[176] Papakonstantinou E, Androutsopoulos G, Logotheti S, Adonakis G, Maroulis I, Tzelepi V. DNA methylation in epithelial ovarian cancer: current data and future perspectives. Curr Mol Pharmacol 2020. https://doi.org/10.2174/1874467213666200810141858.

[177] Stefansson O, Villanueva A, Vidal A, Martí L, Esteller M. BRCA1 epigenetic inactivation predicts sensitivity to platinum-based chemotherapy in breast and ovarian cancer. Epigenetics 2012;7:1225–9.

[178] Seeber LM, van Diest PJ. Epigenetics in ovarian cancer. Methods Mol Biol 2012;863:253–69. https://doi.org/10.1007/978-1-61779-612-8_15.

[179] Sawan C, Herceg Z. Histone modifications and cancer. Adv Genet 2010;70:57–85.

[180] Kim J, Lee S, Miller K. Preserving genome integrity and function: the DNA damage response and histone modifications. Crit Rev Biochem Mol Biol 2019;54(3):208–41. https://doi.org/10.1080/10409238.2019.1620676.

[181] Chervona Y, Costa M. Histone modifications and cancer: biomarkers of prognosis? Am J Cancer Res 2012;2:589–97.

[182] Gyparaki M, Papavassiliou A. Epigenetic pathways offer targets for ovarian cancer treatment. Clin Breast Cancer 2017;18(3):189–91. https://doi.org/10.1016/j.clbc.2017.09.009.

[183] Ali M, Cacan E, Liu Y, Pierce Y, Creasman J, Pierce Y, et al. Transcriptional suppression, DNA methylation, and histone deacetylation of the regulator of g-protein signaling 10 (RGS10) gene in ovarian cancer cells. PLoS One 2013;8, e60185.

[184] Marsh D, Shah J, Cole AJ. Histones and their modifications in ovarian cancer drivers of disease and therapeutic targets. Front Oncol 2014;4:144.

[185] Yang C, Zhang J, Ma Y, Chunfu W, Cui W, Wang L. Histone methyltransferase and drug resistance in cancers. J Exp Clin Cancer Res 2020;39, 173.

[186] Donzelli S, Mori F, Biagioni F, Bellissimo T, Pulito C, Muti P, Strano S, Blandino G. MicroRNAs: short noncoding players in cancer chemoresistance. Mol Cell Ther 2014;2:16. https://doi.org/10.1186/2052-8426-2-16 [eCollection 2014].

[187] He L, Hannon G. MicroRNAs: small RNAs with a big role in gene regulation. Nat Rev Genet 2004;5(7):522–31.

[188] Stopa N, Krebs J, Shechter D. The PRMT5 arginine methyltransferase: many roles in development, cancer and beyond. Cell Mol Life Sci 2015;72(11):2041–59.

[189] Gaur A, Jewell DA, Liang Y, et al. Characterization of microRNA expression levels and their biological correlates in human cancer cell lines. Cancer Res 2007;67(6):2456–68.

[190] Blower P, Chung J, Verducci J, et al. microRNAs modulate the chemo sensitivity of tumor cells. Mol Cancer Ther 2008;7(1):1–9.

[191] Calin GA, Croce CM. microRNA signatures in human cancers. Nature 2006;6(11):857–66.

[192] Zhang L, Huang J, Yang N, Greshock J, Megraw MS, Giannakakis A, Liang S, Naylor TL, Barchetti A, Ward MR, Yao G, Medina A, O'Brien-Jenkins A, Katsaros D, Hatzigeorgiou A, Gimotty PA, Weber BL, Coukos G. microRNAs exhibit high frequency genomic alterations in human cancer. Proc Natl Acad Sci U S A 2006;103:9136–41.

[193] Srivastava S, Zubair A, Miree O, Singh S, Rocconi R, Scalici J, Singh A. MicroRNAs in gynecological cancers: small molecules with big implications. Cancer Lett 2017;407:123–38.

[194] Ponting CP, Oliver PL, Reik W. Evolution and functions of long noncoding RNAs. Cell 2009;136:629–41.

[195] Hosseini E, Meryet-Figuiere M, Sabzalipoor H, Kashani H, Nikzad H, Asemi Z. Dysregulated expression of long noncoding RNAs in gynecologic cancers. Mol Cancer 2017;16(1):107. https://doi.org/10.1186/s12943-017-0671-2.

[196] Schmitt AM, Chang HY. Long noncoding RNAs in cancer pathways. Cancer Cell 2016;29:452–63.

[197] Deng H, Zhang J, Shi J, Guo Z, He C, Ding L, Tang JH, Hou Y. Role of long non-coding RNA in tumor drug resistance. Tumour Biol 2016;37(9):11623–31. https://doi.org/10.1007/s13277-016-5125-8.

[198] Li Z, Zhao L, Wang Q. Overexpression of long non-coding RNA HOTTIP increases chemoresistance of osteosarcoma cell by activating the Wnt/β-catenin pathway. Am J Transl Res 2016;8:2385–93.

[199] Prensner JR, Chinnaiyan AM. The emergence of lncRNAs in cancer biology. Cancer Discov 2011;1(5):391–407. https://doi.org/10.1158/2159-8290.CD-11-0209.

[200] Gupta R, Shah N, Wang K, Kim J, Horlings H, Wong D, Tsai M, Hung T, Argani P, Rinn J, Wang Y, Brzoska P, Kong B, Li R, West RB, van de Vijver MJ, Sukumar S, Chang HY. Long non-coding RNA HOTAIR reprograms chromatin state to promote cancer metastasis. Nature 2010;464(7291):1071–6.

[201] J-J Q, Lin Y-Y, L-C Y, J-X D, W-W F, H-Y J, Zhang Y, Li Q, K-Q H. Overexpression of long non-coding RNA HOTAIR predicts poor patient prognosis and promotes tumor metastasis in epithelial ovarian cancer. Gynecol Oncol 2014;134(1):121–8. https://doi.org/10.1016/j.ygyno.2014.03.556.

[202] Huang L, Liao L-M, Liu A-W, Wu J-B, Cheng X-L, Lin J-X, Zheng M. Overexpression of long noncoding RNA HOTAIR predicts a poor prognosis in patients with cervical cancer. Arch Gynecol Obstet 2014;290(4):717–23. https://doi.org/10.1007/s00404-014-3236-2.

[203] He X, Bao W, Li X, Chen Z, Che Q, Wang H, Wan X-P. The long non-coding RNA HOTAIR is upregulated in endometrial carcinoma and correlates with poor prognosis. Int J Mol Med 2014;33(2):325–32.

[204] Niinuma T, Suzuki H, Nojima M, Nosho K, Yamamoto H, Takamaru H, Yamamoto E, Maruyama R, Nobuoka T, et al. Upregulation of miR-196a and HOTAIR drive malignant character in gastrointestinal stromal tumors. Cancer Res 2012;72(5):1126–36. https://doi.org/10.1158/0008-5472.CAN-11-1803.

[205] Sheng X, Li J, Yang L, Zhiyi Chen Z, Zhao Q, Tan L, Zhou Y, Li J. Promoter hypermethylation influences the suppressive role of maternally expressed 3, a long non-coding RNA, in the development of epithelial ovarian cancer. Oncol Rep 2014;32(1):277–85.

[206] Yoshimoto R, Mayeda A, Yoshida M, Nakagawa S. MALAT1 long non-coding RNA in cancer. Biochim Biophys Acta (BBA) 2016;1859(1):192–9. https://doi.org/10.1016/j.bbagrm.2015.09.012.

[207] Vicente-Dueñas C, Romero-Camarero I, Cobaleda C, Sánchez-García I. Function of oncogenes in cancer development: a changing paradigm. EMBO J 2013;32:1502–13.

[208] Makashov A, Malov S, Kozlov A. Oncogenes, tumor suppressor and differentiation genes represent the oldest human gene classes and evolve concurrently. Sci Rep 2019;9:16410.

[209] Yu D. The role of oncogenes in drug resistance. Cytotechnology 1998;27:283–92.

[210] Spandidos D, Dokianakis D, Kallergi G, Aggelakis E. Molecular basis of gynecological cancer. Ann N Y Acad Sci 2000;900:56–64. https://doi.org/10.1111/j.1749-6632.2000.tb06216.x.

[211] Liu X, Gao Y, Lu Y, Zhang J, Li L. Oncogenes associated with drug resistance in ovarian cancer. J Cancer Res Clin Oncol 2014;141:381–95.

[212] Patch A-M, Christie E, Etemadmoghadam D, Garsed D, George J, Fereday S, et al. Whole-genome characterization of chemoresistant ovarian cancer. Nature 2015;521(7553):489–94.

[213] Ushijima K. Treatment for recurrent ovarian cancer-at first relapse. J Oncol 2010;2010:497429.

[214] Agarwal R, Kaye SB. Ovarian cancer: strategies for overcoming resistance to chemotherapy. Nat Rev Cancer 2003;3(7):502–16.

[215] Celestino AT, Levy D, Maria Ruiz JL, Bydlowski SP. ABCB1, ABCC1, and LRP gene expressions are altered by LDL, HDL, and serum deprivation in a human doxorubicin-resistant uterine sarcoma cell line. Biochem Biophys Res Commun 2015;457(4):664–8.

[216] Kamau SW, Krämer SD, Günthert M, Wunderli-Allenspach H. Effect of the modulation of the membrane lipid composition on the localization and function of P-glycoprotein in MDR1-MDCK cells. In Vitro Cell Dev Biol Anim 2005;41(7):207–16.

[217] Siti ZS, Seoparjoo A, Shahrul H. Lipoproteins modulate growth and P-glycoprotein expression in drug-resistant HER2-overexpressed breast cancer cells. Heliyon 2019;5(4), e01573.

[218] Devaraj S, Semaan JR, Jialal I. Biochemistry, apolipoprotein B. StatPearls [Internet]. Treasure Island (FL): StatPearls Publishing; 2020. https://www.ncbi.nlm.nih.gov/books/NBK538139. [Accessed 1 October 2020].

[219] Watari H, Ohta Y, Hassan MK, Xiong Y, Tanaka S, Sakuragi N. Clusterin expression predicts survival of invasive cervical cancer patients treated with radical hysterectomy and systematic lymphadenectomy. Gynecol Oncol 2008;108(3):527–32.

[220] Ren L, Yi J, Li W, Zheng X, Liu J, Wang J, Du G. Apolipoproteins and cancer. Cancer Med 2019;8(16):7032–43.

[221] Miyake H, Hara S, Zellweger T, Kamidono S, Gleave ME, Hara I. Acquisition of resistance to Fas-mediated apoptosis by overexpression of clusterin in human renal-cell carcinoma cells. Mol Urol 2001;5(3):105–11.

[222] Kyprianou N, English HF, Davidson NE, Isaacs JT. Programmed cell death during regression of the MCF-7 human breast cancer following estrogen ablation. Cancer Res 1991;51(1):162–6.

[223] Montpetit ML, Lawless KR, Tenniswood M. Androgen-repressed messages in the rat ventral prostate. Prostate 1986;8(1):25–36.

[224] Sensibar JA, Griswold MD, Sylvester SR, Buttyan R, Bardin CW, Cheng CY, Dudek S, Lee C. Prostatic ductal system in rats: regional variation in localization of an androgen-repressed gene product, sulfated glycoprotein-2. Endocrinology 1991;128(4):2091–102.

[225] Hassan MK, Watari H, Han Y, Mitamura T, Hosaka M, Wang L, Tanaka S, Sakuragi N. Clusterin is a potential molecular predictor for ovarian cancer patient's survival: targeting clusterin improves response to paclitaxel. J Exp Clin Cancer Res 2011;30(1):113.

[226] Zhong B, Sallman DA, Gilvary DL, Pernazza D, Sahakian E, Fritz D, Cheng JQ, Trougakos I, Wei S, Djeu JY. Induction of clusterin by AKT—role in Zhong et al. Cytoprotection against docetaxel in prostate tumor cells. Mol Cancer Ther 2010;9(6):1831–41.

[227] Yang GF, Li XM, Xie D. Overexpression of clusterin in ovarian cancer is correlated with impaired survival. Int J Gynecol Cancer 2009;19(8):1342–6.

[228] Partheen K, Levan K, Osterberg L, Claesson I, Fallenius G, Sundfeldt K, Horvath G. Four potential biomarkers as prognostic factors in stage III serous ovarian adenocarcinomas. Int J Cancer 2008;123(9):2130–7.

[229] Kang HC, Kim IJ, Park JH, Shin Y, Ku JL, Jung MS, Yoo BC, Kim HK, Park JG. Identification of genes with differential expression in acquired drug-resistant gastric cancer cells using high-density oligonucleotide microarrays. Clin Cancer Res 2004;10(1 Pt. 1):272–84.

[230] Chow KU, Nowak D, Kim SZ, Schneider B, Komor M, Boehrer S, Mitrou PS, Hoelzer D, Weidmann E, Hofmann WK. In vivo drug-response in patients with leukemic non-Hodgkin's lymphomas is associated with in vitro chemosensitivity and gene expression profiling. Pharmacol Res 2006;53(1):49–61.

[231] Park DC, Yeo SG, Shin EY, Mok SC, Kim DH. Clusterin confers paclitaxel resistance in cervical cancer. Gynecol Oncol 2006;103(3):996–1000. https://doi.org/10.1016/j.ygyno.2006.06.037.

[232] Nieman KM, Kenny HA, Penicka CV, Ladanyi A, Buell-Gutbrod R, Zillhardt MR, Romero IL, Carey MS, Mills GB, Hotamisligil GS, Yamada SD, Peter ME, Gwin K, Lengyel E. Adipocytes promote ovarian cancer metastasis and provide energy for rapid tumor growth. Nat Med 2011;17(11):1498–503.

[233] Sheng X, Tucci J, Parmentier JH, Ji L, Behan JW, Heisterkamp N, Mittelman SD. Adipocytes cause leukemia cell resistance to daunorubicin via oxidative stress response. Oncotarget 2016;7(45):73147–59.

[234] Duong MN, Cleret A, Matera EL, Chettab K, Mathé D, Valsesia-Wittmann S, Clémenceau B, Dumontet C. Adipose cells promote resistance of breast cancer cells to trastuzumab-mediated antibody-dependent cellular cytotoxicity. Breast Cancer Res 2015;17(1):57.

[235] Dai Y, Jin S, Li X, Wang D. The involvement of Bcl-2 family proteins in AKT-regulated cell survival in cisplatin resistant epithelial ovarian cancer. Oncotarget 2017;8(1):1354–68.

[236] Yang J, Zaman MM, Vlasakov I, Roy R, Huang L, Martin CR, Freedman SD, Serhan CN, Moses MA. Adipocytes promote ovarian cancer chemoresistance. Sci Rep 2019;9(1):13316.

[237] Kurtova AV, Xiao J, Mo Q, Pazhanisamy S, Krasnow R, Lerner SP, Chen F, Roh TT, Lay E, Ho PL, Chan KS. Blocking PGE2-induced tumour repopulation abrogates bladder cancer chemoresistance. Nature 2015;517 (7533):209–13.

[238] Angst E, Reber HA, Hines OJ, Eibl G. Mononuclear cell-derived interleukin-1 beta confers chemoresistance in pancreatic cancer cells by upregulation of cyclooxygenase-2. Surgery 2008;144(1):57–65.

[239] Koliwad SK, Gray NE, Wang JC. Angiopoietin-like 4 (Angptl4): a glucocorticoid-dependent gatekeeper of fatty acid flux during fasting. Adipocyte 2012;1(3):182–7.

[240] Huang RL, Teo Z, Chong HC, Zhu P, Tan MJ, Tan CK, Lam CR, Sng MK, Leong DT, Tan SM, Kersten S, Ding JL, Li HY, Tan NS. ANGPTL4 modulates vascular junction integrity by integrin signaling and disruption of intercellular VE-cadherin and claudin-5 clusters. Blood 2011;118(14):3990–4002.

[241] Santos AR, Corredor RG, Obeso BA, Trakhtenberg EF, Wang Y, Ponmattam J, Dvoriantchikova G, Ivanov D, Shestopalov VI, Goldberg JL, Fini ME, Bajenaru ML. β1 integrin-focal adhesion kinase (FAK) signaling modulates retinal ganglion cell (RGC) survival. PLoS One 2012;7(10), e48332.

[242] Zhou S, Wang R, Xiao H. Adipocytes induce the resistance of ovarian cancer to carboplatin through ANGPTL4. Oncol Rep 2020;44(3):927–38.

[243] Kuo CY, Ann DK. When fats commit crimes: fatty acid metabolism, cancer stemness and therapeutic resistance. Cancer Commun (Lond) 2018;38(1):47.

[244] Papaevangelou E, Almeida GS, Box C, de Souza NM, Chung YL. The effect of FASN inhibition on the growth and metabolism of a cisplatin-resistant ovarian carcinoma model. Int J Cancer 2018;143(4):992–1002.

[245] Jiang L, Wang H, Li J, Fang X, Pan H, Yuan X, Zhang P. Up-regulated FASN expression promotes transcoelomic metastasis of ovarian cancer cell through epithelial-mesenchymal transition. Int J Mol Sci 2014;15(7):11539–54.

[246] Bauerschlag DO, Maass N, Leonhardt P, Verburg FA, Pecks U, Zeppernick F, Morgenroth A, Mottaghy FM, Tolba R, Meinhold-Heerlein I, Bräutigam K. Fatty acid synthase overexpression: target for therapy and reversal of chemoresistance in ovarian cancer. J Transl Med 2015;13:146.

[247] Jiang L, Fang X, Wang H, Li D, Wang X. Ovarian cancer-intrinsic fatty acid synthase prevents anti-tumor immunity by disrupting tumor-infiltrating dendritic cells. Front Immunol 2018;9:2927.

[248] Ueda SM, Yap KL, Davidson B, Tian Y, Murthy V, Wang TL, Visvanathan K, Kuhajda FP, Bristow RE, Zhang H, Shih IM. Expression of fatty acid synthase depends on NAC1 and is associated with recurrent ovarian serous carcinomas. J Oncol 2010;2010:285191.

[249] Wagner R, Stübiger G, Veigel D, Wuczkowski M, Lanzerstorfer P, Weghuber J, Karteris E, Nowikovsky K, Wilfinger-Lutz N, Singer CF, Colomer R, Benhamú B, López-Rodríguez ML, Valent P, Grunt TW. Multi-level suppression of receptor-PI3K-mTORC1 by fatty acid synthase inhibitors is crucial for their efficacy against ovarian cancer cells. Oncotarget 2017;8(7):11600–13.

[250] Menendez JA, Lupu R. Fatty acid synthase (FASN) as a therapeutic target in breast cancer. Expert Opin Ther Targets 2017;21(11):1001–16.

[251] Zaytseva YY, Rychahou PG, Le AT, Scott TL, Flight RM, Kim JT, Harris J, Liu J, Wang C, Morris AJ, Sivakumaran TA, Fan T, Moseley H, Gao T, Lee EY, Weiss HL, Heuer TS, Kemble G, Evers M. Preclinical evaluation of novel fatty acid synthase inhibitors in primary colorectal cancer cells and a patient-derived xenograft model of colorectal cancer. Oncotarget 2018;9(37):24787–800.

[252] Zhao G, Cardenas H, Matei D. Ovarian cancer-why lipids matter. Cancers (Basel) 2019;11(12):1870.

[253] Samudio I, Harmancey R, Fiegl M, Kantarjian H, Konopleva M, Korchin B, Kaluarachchi K, Bornmann W, Duvvuri S, Taegtmeyer H, Andreeff M. Pharmacologic inhibition of fatty acid oxidation sensitizes human leukemia cells to apoptosis induction. J Clin Invest 2010;120(1):142–56.

[254] Tan Z, Xiao L, Tang M, Bai F, Li J, Li L, Shi F, Li N, Li Y, Du Q, Lu J, Weng X, Yi W, Zhang H, Fan J, Zhou J, Gao Q, Onuchic JN, Bode AM, Luo X, Cao Y. Targeting CPT1A-mediated fatty acid oxidation sensitizes nasopharyngeal carcinoma to radiation therapy. Theranostics 2018;8(9):2329–47.

[255] Eskelinen E. Autophagy: supporting cellular and organismal homeostasis by self-eating. Int J Biochem Cell Biol 2019;111:1–10. https://doi.org/10.1016/j.biocel.2019.03.010.

[256] Hewitt G, Korolchuk VI. Repair, reuse, recycle: the expanding role of autophagy in genome maintenance. Trends Cell Biol 2017;27(5):340–51.

[257] Ryter S, Mizumura K, Choi A. The impact of autophagy on cell death modalities. Int J Cell Biol 2014;2014:502676.

[258] Yousefi S, Perozzo R, Schmid I, Ziemiecki A, Schaffner T, Scapozza L, Brunner T, Simon HU. Calpain-mediated cleavage of Atg5 switches autophagy to apoptosis. Nat Cell Biol 2006;8(10):1124–32.

[259] Fitzwalter BE, Towers CG, Sullivan KD, Andrysik Z, Hoh M, Ludwig M, O'Prey J, Ryan KM, Espinosa JM, Morgan MJ, Thorburn A. Autophagy inhibition mediates apoptosis sensitization in cancer therapy by relieving FOXO3a turnover. Dev Cell 2018;44(5):555–65 [e3].

[260] Mukhopadhyay S, Panda PK, Sinha N, Das DN, Bhutia SK. Autophagy and apoptosis: where do they meet? Apoptosis 2014;19(4):555–66.

[261] Marino G, Niso-Santano M, Baehrecke EH, Kroemer G. Self-consumption: the interplay of autophagy and apoptosis. Nat Rev Mol Cell Biol 2014;15(2):81–94.

[262] Zalckvar E, Berissi H, Mizrachy L, Idelchuk Y, Koren I, Eisenstein M, Sabanay H, Pinkas-Kramarski R, Kimchi A. DAP-kinase-mediated phosphorylation on the BH3 domain of beclin 1 promotes dissociation of beclin 1 from Bcl-XL and induction of autophagy. EMBO Rep 2009;10(3):285–92.

[263] Towers CG, Wodetzki D, Thorburn A. Autophagy and cancer: modulation of cell death pathways and cancer cell adaptations. J Cell Biol 2020;219(1):e201909033.

[264] Kwan BLY, Wai VWK. Autophagy in multidrug-resistant cancers. In: Gorbunov NV, Schneider M, editors. Autophagy in current trends in cellular physiology and pathology. Rijeka, Croatia: In Tech; 2016. p. 435–54. https://doi.org/10.5772/61911.

[265] Pan B, Chen D, Huang J, Wang R, Feng B, Song H, Chen L. HMGB1-mediated autophagy promotes docetaxel resistance in human lung adenocarcinoma. Mol Cancer 2014;13:165.

[266] Kaewpiboon C, Surapinit S, Malilas W, Moon J, Phuwapraisirisan P, Tip-Pyang S, Johnston RN, Koh SS, Assavalapsakul W, Chung YH. Feroniellin A-induced autophagy causes apoptosis in multidrug-resistant human A549 lung cancer cells. Int J Oncol 2014;44(4):1233–42.

[267] Dossou AS, Basu A. The emerging roles of mTORC1 in macromanaging autophagy. Cancers (Basel) 2019; 11(10):1422–39.

[268] Sun Y, Zhang J, Peng ZL. Beclin1 induces autophagy and its potential contributions to sensitizes SiHa cells to carboplatin therapy. Int J Gynecol Cancer 2009;19(4):772–6.

[269] Sun Y, Jin L, Liu J, Lin S, Yang Y, Sui Y, Shi H. Effect of autophagy on paclitaxel-induced CaSki cell death. Zhong Nan Da Xue Xue Bao Yi Xue Ban 2010;35(6):557–65.

[270] Sun Y, Liu JH, Jin L, Lin SM, Yang Y, Sui YX, Shi H. Over-expression of the Beclin1 gene upregulates chemosensitivity to anti-cancer drugs by enhancing therapy-induced apoptosis in cervix squamous carcinoma CaSki cells. Cancer Lett 2010;294(2):204–10.

[271] Zhang N, Qi Y, Wadham C, Wang L, Warren A, Di W, Xia P. FTY720 induces necrotic cell death and autophagy in ovarian cancer cells: a protective role of autophagy. Autophagy 2010;6(8):1157–67.

[272] Wang J, Garbutt C, Ma H, Gao P, Hornicek FJ, Kan Q, Shi H, Duan Z. Expression and role of autophagy-associated p62 (SQSTM1) in multidrug resistant ovarian cancer. Gynecol Oncol 2018;150(1):143–50.

[273] Liang B, Liu X, Liu Y, Kong D, Liu X, Zhong R, Ma S. Inhibition of autophagy sensitizes MDR-phenotype ovarian cancer SKVCR cells to chemotherapy. Biomed Pharmacother 2016;82:98–105.

[274] Ran X, Zhou P, Zhang K. Autophagy plays an important role in stemness mediation and the novel dual function of EIG121 in both autophagy and stemness regulation of endometrial carcinoma JEC cells. Int J Oncol 2017;51 (2):644–56.

[275] Fukuda T, Oda K, Wada-Hiraike O, Sone K, Inaba K, Ikeda Y, Miyasaka A, Kashiyama T, Tanikawa M, Arimoto T, Kuramoto H, Yano T, Kawana K, Osuga Y, Fujii T. The anti-malarial chloroquine suppresses proliferation and overcomes cisplatin resistance of endometrial cancer cells via autophagy inhibition. Gynecol Oncol 2015;137(3):538–45.

[276] Cadwell K, Debnath J. Beyond self-eating: the control of nonautophagic functions and signaling pathways by autophagy-related proteins. J Cell Biol 2018;217(3):813–22.

[277] Subramani S, Malhotra V. Non-autophagic roles of autophagy-related proteins. EMBO Rep 2013;14(2):1431–51.

[278] Takahashi A, Okada R, Nagao K, Kawamata Y, Hanyu A, Yoshimoto S, Takasugi M, Watanabe S, Kanemaki MT, Obuse C, Hara E. Exosomes maintain cellular homeostasis by excreting harmful DNA from cells. Nat Commun 2017;8:15287.

[279] Thery C, Zitvogel L, Amigorena S. Exosomes: composition, biogenesis and function. Nat Rev Immunol 2002;2 (8):569–79.

[280] Trams EG, Lauter CJ, Salem Jr N, Heine U. Exfoliation of membrane ecto-enzymes in the form of micro-vesicles. Biochim Biophys Acta 1981;645(1):63–70.

[281] Doyle LM, Wang MZ. Overview of extracellular vesicles, their origin, composition, purpose, and methods for exosome isolation and analysis. Cell 2019;8(7):727.

[282] Gurunathan S, Kang MH, Jeyaraj M, Qasim M, Kim JH. Review of the isolation, characterization, biological function, and multifarious therapeutic approaches of exosomes. Cell 2019;8(4):307.

[283] Mashouri L, Yousefi H, Aref AR, Ahadi AM, Molaei F, Alahari SK. Exosomes: composition, biogenesis, and mechanisms in cancer metastasis and drug resistance. Mol Cancer 2019;18(1):75.

[284] Dai J, Su Y, Zhong S, Cong L, Liu B, Yang J, Tao Y, He Z, Chen C, Jiang Y. Exosomes: key players in cancer and potential therapeutic strategy. Signal Transduct Target Ther 2020;5(1):145.

[285] Panno ML, Giordano F. Effects of psoralens as anti-tumoral agents in breast cancer cells. World J Clin Oncol 2014;5(3):348–58.

[286] Eguchi T, Taha EA, Calderwood SK, Ono KA. Novel model of cancer drug resistance: oncosomal release of cytotoxic and antibody-based drugs. Biology (Basel) 2020;9(3):47.

[287] Steinbichler TB, Dudas J, Skvortsov S, Ganswindt U, Riechelmann H, Skvortsova II. Therapy resistance mediated by exosomes. Mol Cancer 2019;18(1):58.

[288] Pan C, Liu H, Robins E, Song W, Liu D, Li Z, Zheng L. Next-generation immuno-oncology agents: current momentum shifts in cancer immunotherapy. J Hematol Oncol 2020;13(1):29.

[289] Safaei R, Larson BJ, Cheng TC, Gibson MA, Otani S, Naerdemann W, Howell SB. Abnormal lysosomal trafficking and enhanced exosomal export of cisplatin in drug-resistant human ovarian carcinoma cells. Mol Cancer Ther 2005;4(10):1595–604.

[290] Zhang FF, Zhu YF, Zhao QN, Yang DT, Dong YP, Jiang L, Xing WX, Li XY, Xing H, Shi M, Chen Y, Bruce IC, Jin J, Ma X. Microvesicles mediate transfer of P-glycoprotein to paclitaxel-sensitive A2780 human ovarian cancer cells, conferring paclitaxel-resistance. Eur J Pharmacol 2014;738:83–90.

[291] Yousafzai NA, Wang H, Wang Z, Zhu Y, Zhu L, Jin H, Wang X. Exosome mediated multidrug resistance in cancer. Am J Cancer Res 2018;8(11):2210–26.

[292] Guo QR, Wang H, Yan YD, Liu Y, Su CY, Chen HB, Yan YY, Adhikari R, Wu Q, Zhang JY. The role of Exosomal microRNA in Cancer drug resistance. Front Oncol 2020;10:472.

[293] Kanlikilicer P, Bayraktar R, Denizli M, Rashed MH, Ivan C, Aslan B, Mitra R, Karagoz K, Bayraktar E, Zhang X, Rodriguez-Aguayo C, El-Arabey AA, Kahraman N, Baydogan S, Ozkayar O, Gatza ML, Ozpolat B, Calin GA, Sood AK, Lopez-Berestein G. Exosomal miRNA confers chemo resistance via targeting Cav1/p-gp/M2-type macrophage axis in ovarian cancer. EBioMedicine 2018;38:100–12.

[294] Zhu X, Shen H, Yin X, Yang M, Wei H, Chen Q, Feng F, Liu Y, Xu W, Li Y. Macrophages derived exosomes deliver miR-223 to epithelial ovarian cancer cells to elicit a chemoresistant phenotype. J Exp Clin Cancer Res 2019;38(1):81.

[295] Yin J, Yan X, Yao X, Zhang Y, Shan Y, Mao N, Yang Y, Pan L. Secretion of annexin A3 from ovarian cancer cells and its association with platinum resistance in ovarian cancer patients. J Cell Mol Med 2012;16(2):337–48.

[296] Bastos N, Ruivo CF, da Silva S, Melo SA. Exosomes in cancer: use them or target them? Semin Cell Dev Biol 2018;78:13–21.

[297] Datta A, Kim H, McGee L, Johnson AE, Talwar S, Marugan J, Southall N, Hu X, Lal M, Mondal D, Ferrer M, Abdel-Mageed AB. High-throughput screening identified selective inhibitors of exosome biogenesis and secretion: a drug repurposing strategy for advanced cancer. Sci Rep 2018;8(1):8161.

[298] Nakamura K, Sawada K, Kinose Y, Yoshimura A, Toda A, Nakatsuka E, Hashimoto K, Mabuchi S, Morishige KI, Kurachi H, Lengyel E, Kimura T. Exosomes promote ovarian cancer cell invasion through transfer of CD44 to peritoneal mesothelial cells. Mol Cancer Res 2017;15(1):78–92.

[299] Kim MS, Haney MJ, Zhao Y, Mahajan V, Deygen I, Klyachko NL, Inskoe E, Piroyan A, Sokolsky M, Okolie O, Hingtgen SD, Kabanov AV, Batrakova EV. Development of exosome-encapsulated paclitaxel to overcome MDR in cancer cells. Nanomedicine 2016;12(3):655–64.

[300] Yang L, Yu Y, Xiong Z, Chen H, Tan B, Hu H. Downregulation of SEMA4C inhibit epithelial-mesenchymal transition (EMT) and the invasion and metastasis of cervical cancer cells via inhibiting transforming growth factor-beta 1 (TGF-beta1)-induced hela cells p38 mitogen-activated protein kinase (MAPK) activation. Med Sci Monit 2020;26, e918123.

[301] Song J, Li Y. miR-25-3p reverses epithelial-mesenchymal transition via targeting Sema4C in cisplatin-resistance cervical cancer cells. Cancer Sci 2017;108(1):23–31.

[302] Li X, Corbett AL, Taatizadeh E, Tasnim N, Little JP, Garnis C, Daugaard M, Guns E, Hoorfar M, Li ITS. Challenges and opportunities in exosome research-perspectives from biology, engineering, and cancer therapy. APL Bioeng 2019;3(1), 011503.

[303] Coley WB. The diagnosis and treatment of bone sarcoma. Glasgow Med J 1936;126(3):128–34.

[304] McCarthy EF. The toxins of William B. Coley and the treatment of bone and soft-tissue sarcomas. Iowa Orthop J 2006;26:154–8.

[305] Zhang Y, Zhang Z. The history and advances in cancer immunotherapy: understanding the characteristics of tumor-infiltrating immune cells and their therapeutic implications. Cell Mol Immunol 2020;17(8):807–21.

[306] Chandra A, Pius C, Nabeel M, Nair M, Vishwanatha JK, Ahmad S, Basha R. Ovarian cancer: current status and strategies for improving therapeutic outcomes. Cancer Med 2019;8(16):7018–31.

[307] Jiang T, Zhou C. The past, present and future of immunotherapy against tumor. Transl Lung Cancer Res 2015;4(3):253–64.

[308] Waldman AD, Fritz JM, Lenardo MJ. A guide to cancer immunotherapy: from T cell basic science to clinical practice. Nat Rev Immunol 2020;20(11):651–68.

[309] Curiel TJ. Immunotherapy: a useful strategy to help combat multidrug resistance. Drug Resist Updat 2012;15 (1–2):106–13.

[310] Robert C. A decade of immune-checkpoint inhibitors in cancer therapy. Nat Commun 2020;11(1):3801.

[311] Wei SC, Duffy CR, Allison JP. Fundamental mechanisms of immune checkpoint blockade therapy. Cancer Discov 2018;8(9):1069–86.

[312] Callahan MK, Ott PA, Odunsi K, Bertolini SV, Pan LS, Venhaus RR, Karakunnel JJ, Hodi FS, Wolchok JD. A phase 1 study to evaluate the safety and tolerability of MEDI4736, an anti–PD-L1 antibody, in combination with tremelimumab in patients with advanced solid tumors. J Clin Oncol 2014;32(15_suppl):TPS3120.

[313] Callahan MK, Odunsi K, Sznol M, Nemunaitis JJ, Ott PA, Dillon PM, Park AJ, Schwarzenberger P, Ricciardi T, Macri MJ, Ryan A, Venhaus RR, Wolchok JD. Phase 1 study to evaluate the safety and tolerability of MEDI4736 (durvalumab, DUR) + tremelimumab (TRE) in patients with advanced solid tumors. J Clin Oncol 2017;35 (15_suppl):3069.

[314] Hamanishi J, Mandai M, Ikeda T, Minami M, Kawaguchi A, Matsumura N, Abiko K, Baba T, Yamaguchi K, Ueda A, Kanai M, Mori Y, Matsumoto S, Murayama T, Chikuma S, Morita S, Yokode M, Shimizu A, Honjo T, Konishi I. Efficacy and safety of anti-PD-1 antibody (Nivolumab: BMS-936558, ONO-4538) in patients with platinum-resistant ovarian cancer. J Clin Oncol 2014;32(15_suppl):5511.

[315] Scholler N, Urban N. CA125 in ovarian cancer. Biomark Med 2007;1(4):513–23.

[316] Pfisterer J, du Bois A, Sehouli J, Loibl S, Reinartz S, Reuss A, Canzler U, Belau A, Jackisch C, Kimmig R, Wollschlaeger K, Heilmann V, Hilpert F. The anti-idiotypic antibody abagovomab in patients with recurrent ovarian cancer. A phase I trial of the AGO-OVAR. Ann Oncol 2006;17(10):1568–77.

[317] Reinartz S, Kohler S, Schlebusch H, Krista K, Giffels P, Renke K, Huober J, Mobus V, Kreienberg R, DuBois A, Sabbatini P, Wagner U. Vaccination of patients with advanced ovarian carcinoma with the anti-idiotype ACA125: immunological response and survival (phase Ib/II). Clin Cancer Res 2004;10(5):1580–7.

[318] Sabbatini P, Harter P, Scambia G, Sehouli J, Meier W, Wimberger P, Baumann KH, Kurzeder C, Schmalfeldt B, Cibula D, Bidzinski M, Casado A, Martoni A, Colombo N, Holloway RW, Selvaggi L, Li A, del Campo J, Cwiertka K, Pinter T, Vermorken JB, Pujade-Lauraine E, Scartoni S, Bertolotti M, Simonelli C, Capriati A, Maggi CA, Berek JS, Pfisterer J. Abagovomab as maintenance therapy in patients with epithelial ovarian cancer: a phase III trial of the AGO OVAR, COGI, GINECO, and GEICO—the MIMOSA study. J Clin Oncol 2013;31 (12):1554–61.

[319] Buzzonetti A, Fossati M, Catzola V, Scambia G, Fattorossi A, Battaglia A. Immunological response induced by abagovomab as a maintenance therapy in patients with epithelial ovarian cancer: relationship with survival-a substudy of the MIMOSA trial. Cancer Immunol Immunother 2014;63(10):1037–45.

[320] Zimmer AS, Nichols E, Cimino-Mathews A, Peer C, Cao L, Lee MJ, Kohn EC, Annunziata CM, Lipkowitz S, Trepel JB, Sharma R, Mikkilineni L, Gatti-Mays M, Figg WD, Houston ND, Lee JM. A phase I study of the PD-L1 inhibitor, durvalumab, in combination with a PARP inhibitor, olaparib, and a VEGFR1-3 inhibitor, cediranib, in recurrent women's cancers with biomarker analyses. J Immunother Cancer 2019;7(1):197.

[321] Nowak M, Klink M. The role of tumor-associated macrophages in the progression and chemoresistance of ovarian cancer. Cell 2020;9(5):1299.

[322] Yin Y, Yao S, Hu Y, Feng Y, Li M, Bian Z, Zhang J, Qin Y, Qi X, Zhou L, Fei B, Zou J, Hua D, Huang Z. The immune-microenvironment confers chemoresistance of colorectal cancer through macrophage-derived IL6. Clin Cancer Res 2017;23(23):7375–87.

[323] Anfray C, Ummarino A, Andon FT, Allavena P. Current strategies to target tumor-associated-macrophages to improve anti-tumor immune responses. Cell 2019;9(1):46.

[324] Wesolowski R, Sharma N, Reebel L, Rodal MB, Peck A, West BL, Marimuthu A, Severson P, Karlin DA, Dowlati A, Le MH, Coussens LM, Rugo HS. Phase Ib study of the combination of pexidartinib (PLX3397), a CSF-1R inhibitor, and paclitaxel in patients with advanced solid tumors. Ther Adv Med Oncol 2019;11, 1758835919854238.

[325] Muraoka D, Seo N, Hayashi T, Tahara Y, Fujii K, Tawara I, Miyahara Y, Okamori K, Yagita H, Imoto S, Yamaguchi R, Komura M, Miyano S, Goto M, Sawada SI, Asai A, Ikeda H, Akiyoshi K, Harada N, Shiku H. Antigen delivery targeted to tumor-associated macrophages overcomes tumor immune resistance. J Clin Invest 2019;129(3):1278–94.

[326] Khongorzul P, Ling CJ, Khan FU, Ihsan AU, Zhang J. Antibody-drug conjugates: a comprehensive review. Mol Cancer Res 2020;18(1):3–19.

[327] Moore KN, Vergote I, Oaknin A, Colombo N, Banerjee S, Oza A, Pautier P, Malek K, Birrer MJ. FORWARD I: a phase III study of mirvetuximab soravtansine versus chemotherapy in platinum-resistant ovarian cancer. Future Oncol 2018;14(17):1669–78.

[328] Berraondo P, Sanmamed MF, Ochoa MC, Etxeberria I, Aznar MA, Perez-Gracia JL, Rodriguez-Ruiz ME, Ponz-Sarvise M, Castanon E, Melero I. Cytokines in clinical cancer immunotherapy. Br J Cancer 2019;120(1):6–15.

[329] Walther W, Stein U. Influence of cytokines on mdr1 expression in human colon carcinoma cell lines: increased cytotoxicity of MDR relevant drugs. J Cancer Res Clin Oncol 1994;120(8):471–8.

[330] Orengo G, Noviello E, Cimoli G, Pagnan G, Parodi S, Venturini M, Conte P, Schenone F, Conzi G, Russo P. Potentiation of topoisomerase I and II inhibitors cell killing by tumor necrosis factor: relationship to DNA strand breakage formation. Jpn J Cancer Res 1992;83(11):1132–6.

[331] Berguetti T, Quintaes LSP, Hancio T, Robaina MC, Cruz ALS, Maia RC, de Souza PS. TNF-alpha modulates P-glycoprotein expression and contributes to cellular proliferation via extracellular vesicles. Cell 2019;8(5):500.

[332] Liu T, Han C, Wang S, Fang P, Ma Z, Xu L, Yin R. Cancer-associated fibroblasts: an emerging target of anti-cancer immunotherapy. J Hematol Oncol 2019;12(1):86.

[333] Montfort A, Colacios C, Levade T, Andrieu-Abadie N, Meyer N, Segui B. The TNF paradox in cancer progression and immunotherapy. Front Immunol 2019;10:1818.

[334] Gatouillat G, Odot J, Balasse E, Nicolau C, Tosi PF, Hickman DT, Lopez-Deber MP, Madoulet C. Immunization with liposome-anchored pegylated peptides modulates doxorubicin sensitivity in P-glycoprotein-expressing P388 cells. Cancer Lett 2007;257(2):165–71.

[335] Chen ML, Sun A, Cao W, Eliason A, Mendez KM, Getzler AJ, Tsuda S, Diao H, Mukori C, Bruno NE, Kim SY, Pipkin ME, Koralov SB, Sundrud MS. Physiological expression and function of the MDR1 transporter in cytotoxic T lymphocytes. J Exp Med 2020;217(5), e20191388.

[336] Campo L, Zhang C, Breuer EK. EMT-inducing molecular factors in gynecological cancers. Biomed Res Int 2015;2015:420891.

[337] Voon DC, Huang RY, Jackson RA, Thiery JP. The EMT spectrum and therapeutic opportunities. Mol Oncol 2017;11(7):878–91.

[338] Mani SA, Guo W, Liao MJ, Eaton EN, Ayyanan A, Zhou AY, Brooks M, Reinhard F, Zhang CC, Shipitsin M, Campbell LL, Polyak K, Brisken C, Yang J, Weinberg RA. The epithelial-mesenchymal transition generates cells with properties of stem cells. Cell 2008;133(4):704–15.

[339] Singh A, Settleman J. EMT, cancer stem cells and drug resistance: an emerging axis of evil in the war on cancer. Oncogene 2010;29(34):4741–51.

[340] Du B, Shim JS. Targeting epithelial–mesenchymal transition (EMT) to overcome drug resistance in cancer. Molecules (Basel, Switzerland) 2016;21(7):965.

[341] Wang SQ, Liu ST, Zhao BX, Yang FH, Wang YT, Liang QY, Sun YB, Liu Y, Song ZH, Cai Y, Li GF. Afatinib reverses multidrug resistance in ovarian cancer via dually inhibiting ATP binding cassette subfamily B member 1. Oncotarget 2015;6(28):26142–60.

[342] Shang Y, Cai X, Fan D. Roles of epithelial-mesenchymal transition in cancer drug resistance. Curr Cancer Drug Targets 2013;13:915–29.

[343] Nuti SV, Mor G, Li P, Yin G. TWIST and ovarian cancer stem cells: implications for chemoresistance and metastasis. Oncotarget 2014;5:7260–71.

[344] Du F, Wu X, Liu Y, Wang T, Qi X, Mao Y, Jiang L, Zhu Y, Chen Y, Zhu R, Han X, Jin J, Ma X, Hua D. Acquisition of paclitaxel resistance via PI3K dependent epithelial mesenchymal transition in A2780 human ovarian cancer cells. Oncol Rep 2013;30(3):1113–8.

[345] Wu Q, Wang R, Yang Q, Hou X, Chen S, Hou Y, Chen C, Yang Y, Miele L, Sarkar FH, Chen Y, Wang Z. Chemoresistance to gemcitabine in hepatoma cells induces epithelial-mesenchymal transition and involves activation of PDGF-D pathway. Oncotarget 2013;4(11):1999–2009.

[346] Hollier BG, Tinnirello AA, Werden SJ, Evans KW, Taube JH, Sarkar TR, Sphyris N, Shariati M, Kumar SV, Battula VL, Herschkowitz JI, Guerra R, Chang JT, Miura N, Rosen JM, Mani SA. FOXC2 expression links epithelial-mesenchymal transition and stem cell properties in breast cancer. Cancer Res 2013;73(6):1981–92.

[347] Liu X, Wang D, Liu H, Feng Y, Zhu T, Zhang L, Zhu B, Zhang Y. Knockdown of astrocyte elevated gene-1 (AEG-1) in cervical cancer cells decreases their invasiveness, epithelial to mesenchymal transition, and chemoresistance. Cell Cycle 2014;13(11):1702–7.

[348] Zheng H, Kang Y. Multilayer control of the EMT master regulators. Oncogene 2014;33:1755–63.

[349] Yoon NA, Jo HG, Lee UH, Park JH, Yoon JE, Ryu J, Kang SS, Min YJ, Ju SA, Seo EH, Huh IY, Lee BJ, Park JW, Cho WJ. Tristetraprolin suppresses the EMT through the down-regulation of Twist1 and Snail1 in cancer cells. Oncotarget 2016;7(8):8931–43.

[350] Huang RY, Wong MK, Tan TZ, Kuay KT, Ng AH, Chung VY, Chu YS, Matsumura N, Lai HC, Lee YF, Sim WJ, Chai C, Pietschmann E, Mori S, Low JJ, Choolani M, Thiery JP. An EMT spectrum defines an anoikis-resistant and spheroidogenic intermediate mesenchymal state that is sensitive to e-cadherin restoration by a src-kinase inhibitor, saracatinib (AZD0530). Cell Death Dis 2013;4(11), e915.

[351] Koutsaki M, Spandidos DA, Zaravinos A. Epithelial-mesenchymal transition-associated miRNAs in ovarian carcinoma, with highlight on the miR-200 family: prognostic value and prospective role in ovarian cancer therapeutics. Cancer Lett 2014;351(2):173–81.

[352] Yang D, Sun Y, Hu L, Zheng H, Ji P, Pecot CV, Zhao Y, Reynolds S, Cheng H, Rupaimoole R, Cogdell D, Nykter M, Broaddus R, Rodriguez-Aguayo C, Lopez-Berestein G, Liu J, Shmulevich I, Sood AK, Chen K, Zhang W. Integrated analyses identify a master microRNA regulatory network for the mesenchymal subtype in serous ovarian cancer. Cancer Cell 2013;23(2):186–99.

[353] Jing L, Wang B, Yourong F, Wang T, Wang S, Wu M. Sema4C mediates EMT inducing chemotherapeutic resistance of miR-31–3p in cervical cancer cells. Sci Rep 2019;9:17727. https://doi.org/10.1038/s41598-019-54177-z.

[354] Yeh YM, Chuang CM, Chao KC, Wang LH. MicroRNA-138 suppresses ovarian cancer cell invasion and metastasis by targeting SOX4 and HIF-1α. Int J Cancer 2013;133(4):867–78.

[355] Leskelä S, Leandro-García LJ, Mendiola M, Barriuso J, Inglada-Pérez L, Muñoz I, Martínez-Delgado B, Redondo A, de Santiago J, Robledo M, Hardisson D, Rodríguez-Antona C. The miR-200 family controls beta-tubulin III expression and is associated with paclitaxel-based treatment response and progression-free survival in ovarian cancer patients. Endocr Relat Cancer 2010;18(1):85–95.

[356] Chen X, Mangala LS, Mooberry L, Bayraktar E, Dasari SK, Ma S, Ivan C, Court KA, Rodriguez-Aguayo C, Bayraktar R, Raut S, Sabnis N, Kong X, Yang X, Lopez-Berestein G, Lacko AG, Sood AK. Identifying and targeting angiogenesis-related microRNAs in ovarian cancer. Oncogene 2019;38(33):6095–108.

[357] Christie EL, Bowtell DDL. Acquired chemotherapy resistance in ovarian cancer. Ann Oncol 2017;28(Suppl_8): viii13–5.

[358] Walsh C. Targeted therapy for ovarian cancer: the rapidly evolving landscape of PARP inhibitor use. Minerva Ginecol 2018;70(2):150–70. https://doi.org/10.23736/S0026-4784.17.04152-1.

[359] Klotz DM, Wimberger P. Overcoming PARP inhibitor resistance in ovarian cancer: what are the most promising strategies? Arch Gynecol Obstet 2020;302(5):1087–102. https://doi.org/10.1007/s00404-020-05677-1.

[360] Johnson R, Sabnis N, Sun X, Ahluwalia R, Lacko AG. SR-B1-targeted nano-delivery of anti-cancer agents: a promising new approach to treat triple-negative breast cancer. Breast Cancer 2017;9:383–92. https://doi.org/10.2147/BCTT.S131038 [Dove Med Press].

[361] Mooberry LK, Sabnis NA, Panchoo M, Nagarajan B, Lacko AG. Targeting the SR-B1 receptor as a gateway for cancer therapy and imaging. Front Pharmacol 2016;7:466.

[362] Lacko A, Sabnis N, Nagarajan B, McConathy WJ. HDL as a drug and nucleic acid delivery vehicle. Front Pharmacol 2015;6:247. https://doi.org/10.3389/fphar.2015.00247.

# Current treatment modalities in major gynecologic cancers: Emphasis on response rates

*Maya Nair[a], Lorna A. Brudie[b], Vikas Venkata Mudgapalli[c], V. Gayathri[d], Anjali Chandra[e], Sarfraz Ahmad[f], and Riyaz Basha[g]*

[a]Environmental Health & Safety Department, University of North Texas Health Science Center, Fort Worth, TX, United States [b]Orlando Health Cancer Institute, Gynecologic Oncology Center, Orlando, FL, United States [c]New York Institute of Technology, College of Osteopathic Medicine, Jonesboro, AR, United States [d]Bioengineering Department, SRM Institute of Science & Technology, Chennai, Tamil Nadu, India [e]Des Moines University College of Podiatric Medicine and Surgery, Des Moines, IA, United States [f]Gynecologic Oncology, AdventHealth Cancer Institute, Florida State University, University of Central Florida, Orlando, FL, United States [g]Department of Pediatrics and Women's Health, Texas College of Osteopathic Medicine, University of North Texas Health Science Center, Fort Worth, TX, United States

## Abstract

Gynecologic cancers, regardless of type, are responsible for causing tremendous life stress that has affected many women and their families. The diagnosis followed by after-effects of treatment drastically change the lives of cancer patients and their partners. Although many cases have gone into remission with current pharmaceutical and therapeutic measures, metastasis and relapse continue to impact the response rates in these patients, which also affects quality-of-life (QoL) among survivors. For this reason, research has aimed at looking at how to improve therapeutic outcomes and minimize serious adverse events during and after treatments. In this chapter, we focused on three major gynecologic cancers: cervical, ovarian, and endometrial. The survival rates for all three of these gynecologic malignancies and the role of various components that affect the survival rate and QoL are discussed. Platinum-based chemotherapeutic agents and taxanes are the mainstay treatment in most gynecologic cancers, however, in time cause toxicity and severe adverse effects that may be long-lasting. Overall, while administration of both paclitaxel and carboplatin is the current standard first-line

R. Basha, S. Ahmad (eds.)
*Overcoming Drug Resistance in Gynecologic Cancers*
ISBN 978-0-12-824299-5
https://doi.org/10.1016/B978-0-12-824299-5.00020-4

chemotherapeutic treatment for epithelial ovarian cancer (platinum-sensitive type), there is an ongoing investigation regarding different modes of treatment to be utilized upfront. Immunotherapy is currently being explored as an innovative therapeutic option for gynecologic cancers. Many ongoing studies are currently in the process of investigating novel agents or immunotherapies and combined treatment with traditional chemotherapy. We also look forward to more widespread vaccination programs worldwide for the prevention of HPV to decrease the risk of HPV-related cancer and disease.

# Abbreviations

| | |
|---|---|
| AUC | area under the curve |
| BMI | body mass index |
| BRCA | breast cancer |
| BSO | bilateral salpingo-oophorectomy |
| CC | cervical cancer |
| CCSC | cervical cancer stem cells |
| CD133 | prominin-1 |
| CD49f | integrin α6 |
| CIN | cervical intraepithelial neoplasia |
| CSCs | cancer stem cells |
| CT | chemotherapy |
| CVF | cervicovaginal fluids |
| DC | disease control |
| DCRT | conformal radiation therapy |
| DFS | disease-free survival |
| DNA | deoxyribo nucleic acid |
| EBRT | external beam radiotherapy |
| EC | endometrial cancer |
| EGFR | epidermal growth factor receptor |
| EMT | epithelial-mesenchymal transition |
| EOC | epithelial ovarian cancer |
| FDA | Food and Drug administration |
| FIGO | International Federation of Gynecology and Obstetrics |
| GOG | Gynecologic Oncology Group |
| HER-2/neu | human epidermal growth factor receptor 2 |
| HER2 | human epidermal growth factor receptor 2 |
| HPV | human papilloma virus |
| IMRT | intensity-modulated radiotherapy |
| ISGP | International Society of Gynecologic Pathologists |
| KIf4 | kinesin-4 kinesin family |
| MAPK | mitogen-activated protein kinase |
| MI | myometrial invasion |
| MIS | minimally invasive surgery |
| MMR | mismatch repair |
| MSI | microsatellite instability |
| mTOR | mammalian target of rapamycin |
| NANOG | nanog home box |
| NCCN | National Comprehensive Cancer Network |
| OC | ovarian cancer |
| OCT4 | octamer-binding protein |
| ORR | objective response rate |
| OS | overall survival |
| PARP | poly-ADP ribose polymerase |
| PD | death protein |
| PDL | death protein ligand |

| PFI | platinum-free interval |
| PI3K-AKT | phosphatidylinositol 3-kinase |
| PLD | pegylated liposomal doxorubicin |
| POLE | DNA polymerase-epsilon |
| PORTEC | postoperative radiation therapy in endometrial carcinoma |
| QoL | quality-of-life |
| RR | response rate |
| RTOG | Radiation Therapy Oncology Group |
| SCJ | squamo-columnar junction |
| SD | stable disease |
| SLN | sentinel lymph node |
| SNAI1 | snail family transcriptional repressor 1 |
| SNAI2 | snail family transcriptional repressor 2 |
| SOX2 | sex-determining region Y-box2 |
| TWIST 1 | twist family BHLH transcription factor 1 |
| USC | uterine serous carcinoma |
| VBT | vaginal brachytherapy |
| VEGF | vascular endothelial growth factor |
| WPRT | whole-pelvic radiation therapy |

## Conflict of interest

No potential conflicts of interest were disclosed.

## Introduction

Gynecologic cancers have been responsible for causing widespread illnesses that have affected the lives of many women. Although many cases have gone into remission with current pharmaceutical and therapeutic measures, metastasis and relapse are common and impact the response rates in these patients which in turn affects their quality of life (QoL). For this reason, research has aimed at looking at how to improve therapeutic outcomes and minimize serious complications during and after the treatments. In this chapter, we have focused on three major gynecologic cancers: cervical, ovarian, and endometrial.

## Cervical cancer

Cervical cancer (CC) is among the top five leading causes of cancer-related deaths in women worldwide. Over 0.5 million new cases are detected, and more than 300,000 deaths are reported annually [1]. High-risk strains of the Human Papilloma Virus (HPV) are the major cause of disease. The disease is preventable, and currently, there has been reduced incidence in developed countries [2]. This reduction is primarily due to screening practices, and further decline is expected with the usage of the HPV vaccine. In underdeveloped countries, the incidence and death is relatively high due to poor screening, lack of resources, and poor vaccination rate.

In general, cervical cancer is among those cancers that are fairly resistant to chemotherapy [2]. In 2020, there were 13,800 cases of uterine cervical cancer identified and 4290 deaths from uterine cervical cancer reported [3]. Despite the advances in screening and HPV vaccination, CC continues to be the foremost reason for cancer mortality globally due to the unavailability of resources. Most (99.7%) of aggressive cervical cancer is associated with persistent HPV infection. Prophylactic HPV vaccination offers a strong opportunity to prevent cervical cancer but currently is not a form of treatment for the disease. Treatment options depend on the stage at time of diagnosis. Surgery, chemotherapy, or combination thereof is the first-line treatment approach; and preservation of fertility is one of the key factors taken into account before making treatment recommendations. For locally advanced CC, less toxic treatment options like intensity-modulated radiotherapy (IMRT) are a viable option given less toxicity associated with the treatment.

## Current treatment options and response rates

Bevacizumab has been added to standard chemotherapy for the treatment of advanced or metastatic disease. Based on data from Gynecologic Oncology Group (GOG)-240, overall survival (OS) improved from 13 to 17 months in patients treated with bevacizumab [4, 5]. Even with this additional agent, patients with advanced-stage disease have only 55% of overall survival for 4-years [5]. We are still in need of novel treatment plans that address the challenges associated with advanced, metastatic, and difficult-to-treat CC [5].

Staging of cancer is an important step to assess the severity and progression of the disease and to determine an effective treatment plan [6]. The staging of gynecologic cancers follows the recommendations and guidelines from the FIGO (International Federation of Gynecology and Obstetrics). Initially, staging of CC has been performed clinically. Physical exam by the physician followed by biopsies, imaging, and other diagnostic tests like chest X-ray cystoscopy and proctoscopy was used for the determination of clinical stages. Surgical samples are useful for determining the pathological stage, but previously, the clinical stage did not change and remained the same. Cervical cancer staging, like all gynecologic malignancies, is noted as stage I through IV [2]. Stage IV correlating with advanced disease with distant metastasis (such as lungs and bone) and can be challenging to treat.

FIGO presented a review of the staging system for cervical cancer in 2018 explaining the reasons for the changes and summarizing suitable diagnostic methods and treatment options for particular stages of disease according to current guidelines [7]. Statistical data showing poor survival rate for 5-year disease-free survival (DFS) and OS rates align well with the FIGO staging scheme published in 2018 [7, 8]. One of the major changes of the 2018 staging guidelines compared to previous guidelines is now the addition of Positron emission tomography (PET)/computed tomography (CT) and/or pathologic data which incorporates lymph node involvement. With this improvement, there would be an expected improvement in predicting more realistic survival results compared to the prior guidelines that did not take this into account.

Once the stage has been determined, treatment planning can ensue. Early-stage cervical cancer is usually treated surgically, depending on the depth of stromal invasion and tumor size. Surgery can be tailored depending on desired fertility. After surgery, adjuvant therapy may be recommended depending on certain risk factors present in pathology. Radiation +/− Cisplatin may be indicated. Combination chemoradiation is used for locally advanced CC while triple agent systemic therapy is employed for advanced/metastatic disease. Cisplatin used in combination with Paclitaxel has been shown to have a high objective response rate

(ORR), but there is no improvement in terms of overall survival. The combination of Topotecan and Cisplatin was the first to show a significant overall success, compared to Cisplatin alone [2]. Staging guides the treatment of choice, with outcomes worse in those who have pelvic or para-aortic node involvement, as would be expected [9, 10]. Clinical trials were conducted to assess the efficacy and safety of combination therapy bevacizumab, carboplatin, and paclitaxel (BCP) in CC patients with advanced-stage or recurrent disease. Thirty-four patients participated in the study, t ORR was 88% with prolonged overall survival [11–14]. Bevacizumab, an antiangiogenic agent that targets vascular endothelial growth factor 2 (VEGF-2), added to standard chemotherapy in cervical cancer showed significant improvement on survival. Novel compounds with ensuring results and newer treatment strategies to overcome the challenges in this field are continuing [15].

It is important to recognize that even though Cisplatin is used as the primary agent for the treatment of cervical cancer, its use is not without concern. Patients may develop severe toxicity related to treatment as well as platinum resistance, wherein the drug is unable to accumulate within the cell, due to drug efflux. Furthermore, there may be increases in repairing damaged DNA and inactivation of apoptosis, allowing the cancer cell lines to evade the drug and persist [16].

Concerning recurrent disease, studies have demonstrated that usage of Pembrolizumab as a treatment for previously treated cervical cancers has been successful in patients who express programmed death-ligand 1 (PD-L1) on tissue samples. It works as a monoclonal antibody programmed cell death protein-1 (PD-1)/PD-L1 inhibitor. The inhibition of PD-1 activity has been shown to decrease tumor growth, in syngeneic mouse models [17]. Based on the statistical analysis conducted [8], patients had a partial response to treatment with this immunotherapy agent [18], patients had a complete response to the drug, and the ORR was 12.2% [2]. These values are all in line with or higher in success compared to other standard treatments provided in similar settings. Furthermore, the response duration was also increased with the usage of Pembrolizumab, and reduction in tumor size was seen in more than half the patient base [2].

## Precision medicine

In more recent times, the use of molecular profiling has also become important in the treatment of CC, especially those with recurrent disease. This may be helpful in guiding treatment options for cytotoxic chemotherapy as well as novel or targeted agents.

Several mutations are detected that are indicative of the malignant phenotype. Driver mutations are responsible for tumor propagation and progression. This information is crucial in developing targeted therapy. Next-generation sequencing (NGS) of tumors provides valuable information of such mutations that help to develop personalized targeted therapy. Tayshetye et al. [19] analyzed the genomic profile of 157 patients and found that 82% of patients had a mutation that had an FDA-approved targeted agent available [19]. Scaranti et al. [20] conducted a study in 37 CC patients and the results substantiated the association between *KRAS* (Ki-ras2 Kirsten rat sarcoma viral oncogene homolog) mutation with worse CC prognosis. The high rate of potentially actionable mutation indicates the increasing scope for targeted treatment trials in CC [20].

## Stem cell treatment

Cervical cancer stem cells (CCSC) present disconcerted growth properties that lead to cancer initiation and metastasis. Cancer stem cell prototypes support heterogeneity of cancer, which results in therapeutic recalcitrancy. CCSC is associated with chemotherapeutic drug resistance such as resistance to cisplatin and resistance to radiation. Mendoza-Almanza et al. [21] explained the schemes that HPV-16 and HPV-18 used to change normal epithelial cells into cancer cells. The authors also explained the role of CCSC biomarkers and their cellular pathways.

An important factor for CC development is the epithelial-mesenchymal transition (EMT). The role of EMT in the generation of invasive cells and metastasis is very important. The proteins (regulators) of this pathway include SNAI1, SNAI2, and TWIST1. Chemoresistance and metastasis are initiated by the EMT changes like deletion of E-cadherin expression and increase in the expression of N-cadherin and Vimentin [22–25]. Current therapies are not directed towards abolishing CCSCs, wherein their presence leads to poor prognosis in CC. This situation prompted researchers to develop combination therapies that could target CCSCs and cancer cells. Targeted therapy would likely be developed with the information/insights gained from CCSC specific biomarkers.

The microbiota of the vagina may play an important role in cervical cancer and intraepithelial neoplasia. Studies have reported an association between abnormal vaginal microbiota and late-stage CC [26]. Lactobacillus is the common species found in association with HPV. However, *Shuttleworthia satelles*, *Sneathia* spp., and *Megasphaera elsdenii* were frequently found in cervical intraepithelial neoplasia (CIN) cases [27, 28]. *Fusobacterium* spp. is another microorganism abundantly associated with CC cases [27]. The changes in the vaginal/cervical microbiota through the progression of CIN I to CC have also been investigated by researchers [21]. Studies by Norenhag et al. [29] suggested a correlation between the change in microbiota with the progression of HPV-related cervical cancer. This information is important in identifying potential biomarkers for CC [27].

Several studies have been conducted to gain knowledge in the area of the molecular pathogenesis of CC. Most of these studies focused on the involvement of HPV in CC. Lately, some studies shine light on the involvement of cancer stem cells in CC development and its complexity. The HPV oncoproteins E6 and E7 infect the stem cells and modify them to CSCs. Proteins like SOX2, NANOG, OCT4, Klf4, and Nestin maintain the stemness of CSCs. Additionally, CD44, CD133, and CD49f could be excellent targets to direct therapeutic efforts and block Hedgehog, PI3K/Akt/mTOR, Wnt, or Notch signaling pathways in CCSC [21].

The cervical Squamo-Columnar Junction (SCJ) alteration region microanatomy is distinctive compared to the HPV-infected anogenital sections including the vagina, vulva, and anus. This distinctive undifferentiated cervical SCJ is coupled with a higher risk for cancer formation [30].

## Targeted therapy

Budhwani et al. [31] in their review discussed genes that are involved in cervical cancer prediction. A total of 18 genes were identified to be dysregulated in CC. The authors discussed about 27 new genes that play significant roles in the central Focal Adhesion pathway. Overexpression of these genes leads to poor survival outcomes, and some of these genes are associated with the HPV infection pathway also [31]. Targeting these genes involved in the carcinogenic pathways can lead to novel treatment plans with better outcomes for CC patients. A combination of these targeted therapies with immunotherapy that provide a strong immune response specific to HPV may result in tremendous improvement for CC treatment [32].

Inhibitors of angiogenesis block new blood vessel growth around the tumor by inhibiting vascular endothelial growth factor (VEGF). As noted earlier, bevacizumab (Avastin) is a monoclonal antibody that is synthesized to target VEGF and block angiogenesis. Bevacizumab can be used to treat an aggressive form of cervical cancer. Bevacizumab is part of combination therapy in several clinical trials currently active globally.

## Immunotherapy

Immunotherapy is a novel promising strategy for cancer patients with advanced-stage and recurrent disease. For highly aggressive metastatic cancers, combination of immunotherapy with radiation therapy increases the tumor response rate through the abscopal effect. This strategy of priming the immune system also helps enhance the control of cancer at the site of irradiation. This explains the importance of understanding the tumor microenvironment and genes associated with the immune system [4].

Immune checkpoint inhibitors have demonstrated antitumor activity in many tumors. The PD-1 is a T-cell co-inhibitory receptor that works under immunoregulatory capacity in normal conditions. Specifically, PD-1 works to assist in the ability of a tumor to evade immunosurveillance. The PD-L1 expression has been highly correlated alongside PD-1 in HPV-associated cervical epithelial malignancies [2]. Pembrolizumab is a highly selective immunotherapeutic agent that is a fully humanized monoclonal antibody. Its mechanism of action works by inhibiting the prevention of the interaction between PD-1 and its ligands PD-L1 and PD-L2.2. Pembrolizumab monotherapy has shown effectiveness in previously treated PD-L1 positive advanced-stage cervical cancer [2] by antitumor activity and possessing a manageable safety profile. This immunotherapy in conjunction with Bevacizumab is being studied to determine its efficacy in patients with recurrent, persistent, or metastatic cervical cancers [2]. Other studies are looking at its combined effects with concurrent chemoradiation and concurrent versus sequential chemoradiotherapy in patients with locally advanced-stage cervical cancer [2]. A summary of the global clinical trials for CC is provided in Table 1.

## Ovarian cancer

Ovarian cancer (OC) is the foremost cause of death due to gynecologic malignancies in the United States, impacting over 22,000 women per year [33]. The majority of ovarian cancer cases are epithelial (95%), with the rest derived from other cell types: germ-cell and sex-cord stromal [34]. High-grade serous carcinoma is the main subtype of epithelial ovarian cancer (EOC) [34–36]. Ovarian cancer response to treatment can be further classified into platinum-sensitive and platinum-resistant disease. Platinum sensitivity in OC patients is

TABLE 1   A summary of the global clinical trials for cervical, ovarian and endometrial (uterine) cancers.

| Disease | Therapy | Total number of studies | Number of studies completed | Reference/source |
|---|---|---|---|---|
| Cervical cancer | Observational | 666 | 276 | https://www.clinicaltrials.gov/ct2/results?term=observational&cond=Cervical+Cancer&Search=Apply&recrs=e&age_v=&gndr=&type=&rslt= |
| | Targeted therapy | 309 | 149 | |
| | Immunotherapy | 133 | 26 | |
| | Combination therapy | 508 | 172 | |
| | Cellular therapy | 993 | 111 | |
| Ovarian cancer | Observational | 535 | 209 | https://www.clinicaltrials.gov/ct2/results?term=observational&cond=Ovarian+Cancer&Search=Apply&recrs=e&age_v=&gndr=&type=&rslt= |
| | Targeted therapy | 320 | 158 | |
| | Immunotherapy | 153 | 35 | |
| | Combination therapy | 601 | 65 | |
| | Cellular therapy | 840 | 369 | |
| Endometrial cancer | Observational | 197 | 88 | https://www.clinicaltrials.gov/ct2/results?recrs=&cond=Endometrial+Cancer&term=&cntry=&state=&city=&dist= |
| | Targeted therapy | 96 | 40 | |
| | Immunotherapy | 28 | 3 | |
| | Combination therapy | 140 | 140 | |
| | Cellular therapy | 344 | 344 | |

stratified based on the platinum-free interval (PFI), which is the interval between the platinum-based treatment and the relapse of the disease. For patients with a PFI greater than or equal to 6-months, they are classified as having "platinum-sensitive" ovarian cancer. For patients with a PFI of less than 6-months, they are classified as having "platinum-resistant" ovarian cancer.

## Risk factors

There have been studies supporting the associated risk factors of ovarian cancer. The reported possible risk factors of ovarian cancer are increasing age, early menarche or late menopause, family history, BReast CAncer (BRCA) gene variants, Lynch syndrome, endometriosis, and asbestos exposure [37–48].

## Protective factors

Along with possible factors increasing the risk of incidence of ovarian cancer in patients, there have been well-studied protective factors that can help reduce the risk of ovarian cancer in one's lifetime. The factors strongly associated with a decreased risk include bilateral salpingo-oophorectomy (BSO), oral contraceptive use, tubal ligation, hysterectomy, breastfeeding, and multiparity [39, 49–58].

## Statistics

In 2020, there were estimated to be 21,750 incidences of ovarian cancer in the United States, and the estimated deaths from ovarian cancer were reported to be 13,940 [3, 59]. EOC along with peritoneal and fallopian tube cancers is considered to be the deadliest gynecologic cancer. Five-year survival is usually <50% for OC patients. The majority of new cases are found at an advanced stage due to ineffective screening measures. The recurrence rate is still >80% despite all the advances in therapy over the years. Even though data from ovarian cancer patients with BRCA mutation treated with poly-ADP ribose polymerase (PARP) inhibitors are encouraging, prognosis remains poor overall. Immunotherapy has become promising in the treatment of ovarian cancer [34, 35].

## Current treatment options and response rates

The gold standard of treatment for ovarian cancer is hysterectomy, bilateral salpingo-oophorectomy, omentectomy, pelvic and para-aortic lymphadenectomy (if enlarged nodes or stage I disease), and intra-peritoneal biopsies [33]. Depending on the stage and pathologic findings, chemotherapy may be administered. Initial findings at presentation would guide recommendations to deliver chemotherapy before cytoreductive surgery (neoadjuvant) or after (adjuvant) [33]. Many factors must be considered while treating ovarian cancer, such as stage, volume of disease, resectability of the tumor in addition to patients' age, risk factors, and comorbidities. The mainstay of treatment includes six cycles of carboplatin and paclitaxel [33]. This regimen has shown a 70%–80% response rate for ovarian cancer. However, most patients treated with platinum-based chemotherapy may develop the recurrent disease with reduced disease-free periods and lower efficacy [33].

For platinum-resistant EOC, there are several active treatment options, but no ideal treatment option is available. First-line therapy for platinum-resistant EOC being treated for the first time is weekly paclitaxel. In patients with progressive platinum-resistant EOC (despite paclitaxel or those who are not eligible for paclitaxel), pegylated liposomal doxorubicin (PLD) is administered every 4 weeks [60, 61]. For few patients with the recurrent platinum-resistant disease, single-agent chemotherapy plus bevacizumab can be given [60]. For a majority of patients with EOC, the approach to management currently indicates surgical staging and cytoreduction followed by adjuvant chemotherapy [62]. Adjuvant platinum- or taxane-based chemotherapy can follow surgery, or neoadjuvant chemotherapy can precede surgery for EOC in patients with a Mullerian malignancy [62].

In a recent study [1], a safe, well-tolerated targeted anticancer gene therapy, VB-111 (Ofranergene), was tested in a phase I/II study in patients with platinum-resistant ovarian cancer. VB-111 functions via a dual mechanism, which involves targeting tumor vasculature and antitumor immune effects. The results of this phase I/II study featured favorable response rates (58% CA-125 response rates). Furthermore, VB-111 in combination with paclitaxel has a better overall survival effect at therapeutic dose compared to subtherapeutic dose [1].

## Resistance to treatment

Despite the widespread use of carboplatin and paclitaxel as a first-line chemotherapeutic regimen, recurrence of ovarian cancer occurs in 25% of patients with early-stage ovarian cancer and in greater than 80% of patients with advanced-stage disease [63]. The wide array of tumor cells can lead to alterations in signaling pathways, activation of oncogenes, inactivation of tumor suppressors, and more molecular abnormalities [64]. Studies into the resistance of ovarian cancer to various chemotherapeutic regimens have given rise to many questions. First, there are two types of chemotherapy resistance: (i) intrinsic, and (ii) acquired [64]. The intrinsic chemotherapy resistance is induced by cancer cells having inherent alterations in their functions [65]. For instance, this could include modifications in drug uptake, drug efflux, apoptosis, and more [65]. The acquired chemotherapy resistance can be caused by genetic or epigenetic modifications during treatment. These modifications can help cancer cells survive the course of chemotherapeutic treatment [65].

Since there are various mechanisms in intrinsic and acquired chemotherapeutic resistance, more research needs to be done into what biomarkers indicate which types of resistance occur in advanced-stage ovarian cancer. In particular, these questions have accelerated investigations into biomarkers of chemotherapy resistance in advanced-stage ovarian cancer. Studies in the area of transcriptomics, genomics, and proteomics have indicated much progress in our understanding of the molecular basis of chemotherapy responses [65]. Proteomics has a promising future in the investigation of biomarkers to help monitor chemotherapeutic response in advanced-stage ovarian cancer patients [65].

In a recent study [65], some observations were made regarding the mechanisms of chemoresistance in ovarian cancer. The results from their study indicated that the chemoresistance was mainly intrinsic. This would indicate that the resistance mechanisms that determine the cellular phenotypes already are in place in tumor cells before the treatment [65]. Additionally, after chemoresistance, proteins, chaperones, transcription regulators, and cytoskeletal proteins were found to be increased in the tissues of ovarian cancer patients [65]. Post-translational modifications can also play a role in chemo-response of carboplatin and paclitaxel in ovarian cancer. These alterations have an impact on protein folding, cellular integrity, apoptosis, and metabolism in tumor cells [65], which results in a varied response of tumor cells to chemotherapy.

## Ovarian cancer stem cell and chemoresistance

Several studies have already provided information on the role of stem cells in tumor development and prognosis [66]. Peritoneal ascites accumulate during the OC progression. It provides pro-inflammatory cancerous background that coordinates cellular and molecular variations leading to aggressive disease progression. The anoikis resistance is a key feature of these stem cells [67]. Tumor relapse occurs due to the resistance developed in these CSCs towards the chemotherapeutic agents [68]. Ovarian cancer stem cells cultured to form a multilayer spheroid produced new tumors in mice [69]. Studies provide information on high glycolytic functions as specific metabolic property of these CSCs [70–73]. Such unique functions can lead to drug resistance. Liao et al. showed that in their rodent tumor model ovarian CSCs with higher glycolysis in comparison with parental cells, exhibit chemo-resistance [74]. Some of the major drug resistance mechanisms associated with CSCs are summarized in Table 2. These studies shed light on the fact that a combination of tumor resection and CSCs targeted therapy would be more effective for OC management [95]. Currently, there are more than 500 clinical trials that are recruiting participants for various combination therapies, and a summary of global clinical trials for OC is provided in Table 1.

# Endometrial cancer

Endometrial cancer (EC) is the most common gynecologic cancer and the sixth leading cancer in women worldwide. Endometrial cancer develops from the lining of the uterine corpus (endometrium). Uterine sarcomas develop from the muscular layer (myometrium). Carcinosarcomas also exist, which is a combination of these two cancers with an aggressive nature

TABLE 2   Major drug resistance mechanisms associated with cancer stem cells.

| Number | Chemo-resistance process associated with cancer stem cells | Reference(s) |
| --- | --- | --- |
| 1 | ABC transporters, Aldehyde dehydrogenase (ALDH), DNA repair, and signaling pathways | [48, 72, 73, 75] |
| 2 | Hoechst 3342—ABC transporters. P-glycoprotein (MDR1) and breast cancer resistance protein (ABCG2) | [63, 64] |
| 3 | ABCB1 and ABCG2for doxorubicin | [65] |
| 4 | P-glycoprotein (MDR1) for Paclitaxel | [62, 65] |
| 5 | Higher expression level of ABCG2 and ABCB1 in OC and BC | [66, 67] |
| 6 | Higher expression levels of ABCA1, ABCB5, and ABCC3/MRP3 | [68] |
| 7 | ABCA1, ABCB1/MDR1/P-GP, and ABCG2/BCRP expression in ovarian CSCs | [67, 69–71] |
| 8 | Role of ALDH in EMT process of OC | [76] |
| 9 | B-cell lymphoma-2 (BCL-2) protein family | [77–80] |
| 10 | Bcl-xl over expression in chemoresistance OC | [81–83] |
| 11 | WNT/β-catenin and NOTCH | [84–88] |
| 12 | Correlation between WNT pathway and Cisplatin resistance | [89] |
| 13 | Role of NOTCH signaling pathway in tumor progression | [90–94] |

TABLE 3    Current treatment options for gynecologic cancers.

Surgery: Removal of cancerous cells by surgical procedures

Radiation therapy: Cancer treatment in which high doses of radiation is used to irradiate cancer cells and shrink tumors

Chemotherapy: Cancer treatment that uses pharmaceuticals to kill cancer cells

Immunotherapy: It is a type of cancer treatment that helps your immune system fight cancer

Targeted therapy: It is a type of cancer treatment that targets the changes in cancer cells that help them grow, divide, and spread

Hormone therapy: It is a treatment that slows or stops the growth of breast and prostate cancers that use hormones to grow

Stem cell transplants: These are procedures that restore blood-forming stem cells in cancer patients who have had theirs destroyed by very high doses of chemotherapy or radiation therapy

Precision medicine: This helps doctors/providers to select treatments that are most likely to help patients based on a genetic/biomarker profiles and understanding of their disease

and need for treatment regardless of stage [96]. A summary of current treatment for gynecological cancers are given in Table 3.

## Statistics

The highest incidence rates of EC are observed in women from 45 to 74 years [97, 98]. The incidence of EC is increasing annually by an estimated 1%–2%. There is an increasing number of deaths observed for EC with the declining mortality rate for ovarian cancer cases [99–101]. The projected estimate for EC in the year 2021 is 66,570 new cases and 12,940 deaths [102, 103]. Among the gynecologic cancers, uterine cancer or EC is the most common cancer. In the United States, around 3.1% of the female population develop EC in their lifetime.

The highest prevalence rates of EC are observed in Western countries and the lowest prevalence rates are reported in Sub-Saharan Africa, South Central Asia, and the Middle East [104]. EC also stands fourth most common cancer among women in the United Kingdom with over 9000 cases every year observed, and an increase in incidence rates by almost 20% in the past decade is reported [105]. Patients with EC generally have a 5-year relative survival rate of 95% if they are diagnosed at an early stage [97, 106]. A large number of patients with EC are postmenopausal women of ~60 years [107]. However, up to 14% of cases are in premenopausal women with increased body mass index (BMI) [108, 109]. High BMI is increasing in the US population over the last few decades, hence more EC cases prevalence is expected [110, 111]. Five-year OS is determined by stage proves a ≥95% 5-year OS for patients who have tumor localized to the uterus. The 5-year survival rates decrease with advanced-stage disease. Approximately 69% of these patients have localized metastasis and 17% patients have distant metastasis [112]. Majority of the EC are infrequent, and some of the EC cases are formed from mutations; and mismatch repair (MMR) is one of the known mutations that leads to EC [113].

## Treatment options

The EC classification is based on histological kinds (Types I and II) [114]. Type I EC are normally low-grade, estrogen-mediated cancers that are correlated with excess endogenous estrogen production and obesity. Type I EC constitutes almost 70%–80% of the new cases. Type I EC is characterized by microsatellite instability (MSI), which results from higher rates of mutations or changes of *K-ras* and *PTEN* expression levels and defects in the MMR genes [115–119]. Estrogen-independent Type II cancers are marked as high grade. Conversely, Type II EC generally occurs in older, leaner female and are aggressive [114, 120]. Type II EC tumors are known to have *p53* gene mutations. They are also characterized by high expression levels of human epidermal growth factor receptor 2 (HER-2/neu) and show aneuploidy [98, 121–126]. Early diagnosis is associated with improved results with the treatment, in particular for early-stage cancer. In most cases, endometrial cancer would be cured surgically with hysterectomy and BSO with or without lymphadenectomy [98, 121–126]. Several studies are being conducted to improve the efficiency of treatment strategies in EC, particularly regarding novel minimally invasive surgical (MIS) approaches with sentinel lymph node (SLN) mapping. There is mounting evidence that SLN mapping as described in the National Comprehensive Cancer Network (NCCN) surgical algorithm can replace comprehensive lymphadenectomy in patients with apparently uterine-confined EC, and greatly reduces the morbidity associated with lymphadenectomy. Prospective studies are desirable to determine the effect of SLN biopsy on disease recurrence, survival, cost, and QoL.

Adjuvant radiotherapy or chemotherapy, or combination thereof, is offered to women with high-risk diseases to decrease the risk of relapse [123, 124]. The significance of adjuvant chemotherapy has become contentious in EC. External beam pelvic radiotherapy, vaginal brachytherapy (VBT), chemotherapy, and combined chemotherapy with radiotherapy are the adjuvant treatment strategies for EC, which have been studied in several randomized trials [127–130]. Adjuvant therapy is presently based on the presence of clinicopathological risk factors.

Adjuvant chemotherapy is offered alone or in combination with VBT in patients with stage IA myometrial invasion (MI) or IB uterine serous carcinoma (USC) [127–130]. Adjuvant VBT is suggested to minimize the side effect for patients with high-risk EC cases. It provides maximum local control with minimum impact on the QoL for such patients. Pelvic radiotherapy, after surgery, significantly reduces the loco-regional (vaginal and/or pelvic) relapse were reported in many clinical trials. But in many patients treated with pelvic radiotherapy showed a greater risk of toxicity mainly in the gastrointestinal tract [127–130]. In both (PORTEC)-1 and GOG-99 trials, a higher risk of recurrence without additional therapy was found in 30% of the patients [128–131]. This high- or intermediate-risk group had considerable benefit from pelvic radiotherapy in terms of pelvic control. The 5-year risk of pelvic recurrence was reduced from 23% to 5% for this testing group. Patients who received brachytherapy had improved QoL, due to the decreased toxicity-related side effects from the treatment [128–131]. Many studies showed that the standard of care for patients with high-intermediate risk endometrial cancer is adjuvant VBT as this provides maximal vaginal control without significant morbidity [132].

Overall survival is considerably improved with the introduction of chemotherapy [133]. The major issue with chemotherapy has been the high toxicity and high relapse rate. A total of 660 patients with high-risk histology participated in this international PORTEC-3 trial. In this study, an overall survival benefit of 5% was observed with adjuvant chemotherapy. This study compared pelvic radiotherapy alone with pelvic radiotherapy with two concurrent cycles of cisplatin followed by four cycles of adjuvant carboplatin and paclitaxel at 3-week intervals, based on a previous phase II Radiation Therapy Oncology Group (RTOG) trial [134, 135].

For cases with extensive involvement of lymph nodes and deeper myometrial invasion, sequential radiotherapy is suggested. A "Sandwich" regimen of chemotherapy followed by radiation therapy then more chemotherapy may also be utilized for treatment. Management strategy for stage IA USC patients without MI who had gone through comprehensive surgery continues to be controversial. A combination of vaginal brachytherapy and monitoring is the best option. Patients with stage III disease would benefit from EBRT in combination with chemotherapy. Sequential adjuvant chemotherapy followed by RT or as a "sandwich" regimen might be offered to stage I–III uterine carcinosarcoma patients.

For EC patients with stage I, II, and III disease, locoregional control can be improved using adjuvant whole-pelvic radiation therapy (WPRT). IMRT might have better patient outcomes compared to the standard 3D conformal radiation therapy (3DCRT). This option needs to be further validated from the data of the RTOG-1203 trail [136, 137].

The first randomized trial (RTOG 1203) compared IMRT with 3D conformal techniques in relation to acute patient-reported toxicity. The results are indicative of considerably less gastrointestinal and urinary morbidity for patients in the IMRT arm. The greatest toxicities were observed in both the GOG-258 and PORTEC-3 trials that were related to chemotherapy [136]. The current scenario reinforces using chemoradiation for maximizing recurrence-free survival and improve overall survival as well. More data collection and analyses are required to evaluate better treatment options in combination with chemotherapy.

In addition to the challenges associated with the treatment regimens, another key factor in understanding EC is the lack of guidelines to process specimens collected from the surgical treatment of patients suffering from the disease. The International Society of Gynecologic Pathologists (ISGP) has developed a set of recommendations for EC specimen collection and processing [138]. The purpose of these suggestions is for standardizing the handling of endometrial cancer specimens. This standardization process is crucial for proper reporting of the pathological stage that can help for proper understanding of this disease and its management.

## Resistance to treatment

Advanced-stage endometrial cancer is very antagonistic. The low survival rate of advanced-stage endometrial cancer is mainly due to the resistance developed towards chemotherapy. Various pathways are identified to explain the mechanisms of acquired chemoresistance. Some of these mechanisms involve efflux pumps, signaling pathways like PI3K/AKT, MAPK, EGFR, mTOR, repair mechanisms, and tumor suppressors like P53 and Par-4 [139]. Novel strategies of combination therapies and targeted therapy are possible solutions to these challenges evolving from chemoresistance.

## Immunotherapy and targeted therapy

Abnormally increased levels of circulating estrogen are due to obesity, late menopause, nulliparity, early menarche, and exposure to tamoxifen [139]. About 2%–5% of EC cases are due to familial mutations highlighting specific germline in the MMR genes [140]. Understanding the involvement of genetic components in EC is gaining more and more attention among cancer researchers and providers. Potential targeted treatment regimen has been evolving in recent years from these studies. The information gained on various mutations, polymorphisms, and the signaling pathways altered and affected by tumorigenesis, changes in the microenvironment, and its role in tumor growth shed light on potential targeted therapies. Tucci et al. summarizes currently available immunotherapeutic strategies for EC and report the results of the clinical trials and ongoing studies on targeted therapy [141]. The review was carried out collecting preclinical and clinical findings globally. There has been no significant improvement in survival for EC patients in the last decade. Death rates for patients with uterine cancer have increased. Hence, it is crucial to identify significant predictive risk factors that can help to develop novel therapeutics for better outcomes of EC patients with aggressive/advanced-stage disease. The detection of genetic modifications, involving somatic mutations and microsatellite instability are important steps towards novel targeted therapy. Furthermore, evaluation of changes in intracellular signaling pathways that leads to tumor development provides newer choices for immunotherapy and targeted therapy.

Recent genomic studies have provided genomic changes present in EC and shed light on the pathogenesis of the disease. A major challenge is in the development of effective guidelines in classification, staging, and diagnostic process leading to better clinical management [107]. Adding molecular testing with the clinical diagnostic process is highly recommended since The Cancer Genome Atlas (TCGA) data was published in 2013 [142]. Currently, there is no consensus about what test needs to be done. Some potential targets reported by Yen et al. in their review are POLE, and HER2, due to the potential efficacy of immune checkpoint inhibitors [107]. Genomics-based testing will lead to personalized therapy and prognostic tools for early detection methods. An all-inclusive understanding of the molecular modifications accountable for the initiation, relapse, and resistance patterns of EC will help in the improvement of patient outcomes.

## Role of metabolites and biomarkers

Njoku et al. are exploring the viability of metabolites as biomarkers for EC [143] and provide the current stage of EC metabolomic biomarker studies and challenges involved in the biomarker breakthrough and verification. All major classes of biomolecules like lipid, hormones, and amino acid metabolites are potential biomarkers for EC detection and prediction. The authors claimed that the results are encouraging for translational studies. The sampling process is rather minimally invasive and can easily be incorporated into the clinical screening process.

There is mounting interest in the development of EC biomarkers with minimally invasive methodologies [144]. Cervicovaginal fluids (CVF) are one of such metabolites which can be used as a potential source for metabolites specific to EC [145]. There are several reports of using CVF for detecting inflammatory and cancerous situations of the lower genital tract, involving cervical cancer and bacterial vaginosis [146, 147]. Cheng et al., compared the CVF metabolomic profile of 21 EC cases and 33 non-EC controls using nuclear magnetic resonance spectroscopy. Their studies showed overexpression of choline, formate, fumarate, phosphocholine, and malate in EC specimens [145]. Significant downregulation of phenylalanine, isoleucine, aspartate, pyruvate, and asparagine is also reported. Diagnostic performance for phosphocholine with an area under the curve (AUC) of 0.82 [95% confidence interval (CI) 0.69–0.93], is reported to be the best of all the metabolites listed. Phosphocholine, malate, and asparagine had the potential to not only predict EC but also other gynecologic malignancies [145]. Phosphocholine, a surrogate for cell proliferation, has been linked to high-grade EC [148]. More studies on bigger sample sizes need to be done to further validate these findings. A summary of global clinical trials for EC is provided in Table 1.

## Conclusions and future perspectives

Gynecologic cancer patients have to face unique challenges in life. The diagnosis followed by the after-effect of treatment drastically change the lives of cancer patients and their partners. Women face negative impacts in terms of sexual, psychological, and social functioning [149]. Treatment-related modifications such as the occurrence of body image disorders, poor quality of life and depression and anxiety disorders are common among patients. In addition to this, the relationship with their life partner is negatively influenced which leads to adverse effects on the social activity. Cancer is a family experience—it affects partners of sick women in terms of psychological and sexual functioning [149].

Prophylactic HPV vaccination represents a strong opportunity for disease prevention for cervical cancer, and strategies should be developed to increase the vaccination rates in underdeveloped regions. Healthcare workers in partnership with community members can develop educational materials to help the underrepresented population that can benefit from HPV vaccination.

Conventional/standard therapeutic options are still largely used in the treatment of gynecologic cancers, however, recent advancements in the technology facilitated the utilization of the tools such as NGS and other sequencing methods (e.g., Sanger sequencing and pyrosequencing), genomic analyses in treating certain patients. Some of the gynecologic cancers are classified as Rare Gynecologic Tumors (RGT) have a poor prognosis [150]. It is imperative that understanding the diversity of tumors is necessary to address the gap in prognostic response. Huang et al. reported the changes adopted in the treatment options for 152 patients utilizing the molecular profiling of patient tumors and matching with the available therapeutic agents including the investigational agents in clinical trials targeting the distinct signatures [151]. In this study, the majority (76.3%) of patients were identified with some functional mutations. Interestingly, only about 1/3rd patients received a change in the treatment plan (95% patients with FDA-approved drugs; 5% patients enrolled in clinical trials). This further emphasizes that even though the functional mutation is identified through advanced technology, the availability of drugs to target such mutations is still needed. High throughput screening and -omics technologies are being used in drug development and discovery, hence developing effective therapeutic agents for appropriate targets are crucial for improving the outcomes of patients with gynecologic malignancies.

Immunotherapy is currently being explored as an innovative therapeutic option for gynecologic cancers, such as ovarian cancer. Research is actively ongoing with immunotherapy to help patients with gynecologic cancers [75]. There are a few published and ongoing studies exploring the utility of cancer vaccines and immune checkpoint inhibitor therapy for gynecologic cancers [75]. Studies have been published investigating the use of nivolumab in platinum-resistant epithelial ovarian cancer response rate (RR) of 15%, disease control (DC) rate of 45%, the safety and activity of avelumab in platinum-resistant ovarian cancer [ORR 9.7%, stable disease (SD) rate 44.4%], and the safety and activity of single-agent pembrolizumab (ORR 11.5%, SD rate 34.6%) [75]. In patients with concurrent BRCA mutations and ovarian cancer, combination therapy with immune checkpoint inhibitors and PARP inhibitors is thought to increase the efficacy of PARP inhibitors [75].

Staging of cancer is an important step to assess the severity and progress of the disease and to determine the effective treatment plan [6]. The recommendations and guidelines from the FIGO is followed for the staging of gynecologic cancers. As shown in Fig. 1, the statistical indicators of the survival rates for these three major gynecologic cancers are very low. The low survival rate is the main challenge for patients diagnosed with the aggressive stage of cancer, even though they are treated with aggressive therapies. Exploring alterations in the cancer microenvironment and studying the molecular level changes involved in gynecologic cancers may provide chances for: (i) improving upon the current histologic classification system, (ii) increase diagnostic testing modalities, and (iii) personalize treatments using targeted therapies. The NIH initiative "All of US" [152] has a goal to improve healthcare through research. It is building a diverse database that can inform thousands of studies on a variety of health conditions. Information from such a database can provide data that can lead to personalized treatment. An all-inclusive understanding of the molecular modifications accountable for the initiation, relapse, and resistance patterns of EC will help in the improvement of patient outcomes and QoL.

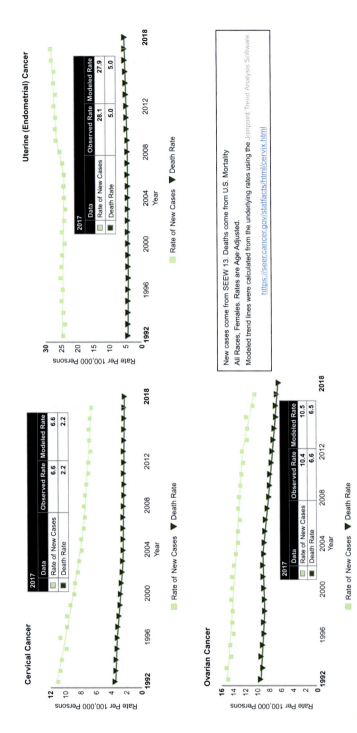

FIG. 1 Gynecologic cancers statistics 1982–2017.

# Acknowledgments

RB is supported by grants from the National Institute on Minority Health and Health Disparities (#1S21MD012472-01; 2U54 MD006882-06) and National Cancer Institute (#P20CA233355-01).

# References

[1] Cohen PA, Jhingran A, Oaknin A, Denny L. Cervical cancer. Lancet 2019;393(10167):169–82. 30638582.

[2] Tewari KS, Monk BJ. Evidence-based treatment paradigms for management of invasive cervical carcinoma. J Clin Oncol 2019;37(27):2472–89. 31403858.

[3] Siegel RL, Miller KD, Jemal A. Cancer statistics, 2020. CA Cancer J Clin 2020;70(1):7–30. https://doi.org/10.3322/caac.21590. Epub 2020 Jan 8 31912902.

[4] Samuels S, Balint B, von der Leyen H, Hupé P, de Koning L, Kamoun C, Luscap-Rondof W, Wittkop U, Bagrintseva K, Popovic M, Kereszt A, Berns E, Kenter GG, Jordanova ES, Kamal M, Scholl S. Precision medicine in cancer: challenges and recommendations from an EU-funded cervical cancer biobanking study. Br J Cancer 2016;115(12):1575–83. 27875525.

[5] Jackson MW, Rusthoven CG, Fisher CM, Schefter TE. Clinical potential of bevacizumab in the treatment of metastatic and locally advanced cervical cancer: current evidence. Onco Targets Ther 2014;7:751–9. 24876784.

[6] Anon, https://www.cancer.org/cancer/cervical-cancer/detection-diagnosis-staging/staged.html.

[7] Zeng J, Qu P, Hu Y, Sun P, Qi J, Zhao G, Gao Y. Clinicopathological risk factors in the light of the revised 2018 International Federation of Gynecology and Obstetrics staging system for early cervical cancer with staging IB: a single center retrospective study. Medicine (Baltimore) 2020;99(16). https://doi.org/10.1097/MD.0000000000019714, e19714. 32311956. PMC7440235.

[8] Horn LC, Brambs CE, Opitz S, Ulrich UA, Höhn AK. FIGO-Klassifikation für das Zervixkarzinom 2019—was ist neu? [the 2019 FIGO classification for cervical carcinoma-what's new?]. Pathologe 2019;40(6):629–35. 31612260.

[9] Delgado G, Bundy B, Zaino R, Sevin BU, Creasman WT, Major F. Prospective surgical-pathological study of disease-free interval in patients with stage IB squamous cell carcinoma of the cervix: a Gynecologic Oncology Group Study. Gynecol Oncol 1990;38(3):352–7. https://doi.org/10.1016/0090-8258(90)90072-s. 2227547.

[10] Grigsby PW, Lu JD, Mutch DG, Kim RY, Eifel PJ. Twice-daily fractionation of external irradiation with brachytherapy and chemotherapy in carcinoma of the cervix with positive para-aortic lymph nodes: phase II study of the radiation therapy oncology group 92-10. Int J Radiat Oncol Biol Phys 1998;41(4):817–22. https://doi.org/10.1016/s0360-3016(98)00132-1. 9652843.

[11] Zhang L, Zheng C-y, Cao J-h, Luo S-l. Efficacy of paclitaxel, carboplatin, and bevacizumab for cervical cancer, a protocol for systematic review and meta-analysis. Medicine 2020;99(24). https://doi.org/10.1097/MD.0000000000020558, e20558.

[12] Suzuki K, Nagao S, Shibutani T, Yamamoto K, Jimi T, Yano H, Kitai M, Shiozaki T, Matsuoka K, Yamaguchi S. Phase II trial of paclitaxel, carboplatin, and bevacizumab for advanced or recurrent cervical cancer. Gynecol Oncol 2019;154(3):554–7. https://doi.org/10.1016/j.ygyno.2019.05.018. Epub 2019 Jul 5 31285082.

[13] Rosen VM, Guerra I, McCormack M, Nogueira-Rodrigues A, Sasse A, Munk VC, Shang A. Systematic review and network meta-analysis of bevacizumab plus first-line topotecan-paclitaxel or cisplatin-paclitaxel versus non-bevacizumab-containing therapies in persistent, recurrent, or metastatic cervical cancer. Int J Gynecol Cancer 2017;27(6):1237–46. https://doi.org/10.1097/IGC.0000000000001000. 28448304. PMC5499964.

[14] Kumar L, Gupta S. Integrating chemotherapy in the management of cervical cancer: a critical appraisal. Oncology 2016;91(Suppl 1):8–17. https://doi.org/10.1159/000447576. Epub 2016 Jul 28 27464068.

[15] Marquina G, Manzano A, Casado A. Targeted agents in cervical cancer: beyond bevacizumab. Curr Oncol Rep 2018;20(5):40. https://doi.org/10.1007/s11912-018-0680-3. 29611060.

[16] Zhu H, Luo H, Zhang W, Shen Z, Hu X, Zhu X. Molecular mechanisms of cisplatin resistance in cervical cancer. Drug Des Devel Ther 2016;10:1885–95. https://doi.org/10.2147/DDDT.S106412. 27354763. PMC4907638.

[17] Anon, https://ascopost.com/issues/september-25-2018/pembrolizumab-for-advanced-cervical-cancer/.

[18] Chemoradiotherapy for Cervical Cancer Meta-analysis Collaboration (CCCMAC). Reducing uncertainties about the effects of chemoradiotherapy for cervical cancer: individual patient data meta-analysis. Cochrane Database Syst Rev 2010;2010(1). https://doi.org/10.1002/14651858.CD008285, CD008285. 20091664. PMC7105912.

[19] Tayshetye P, Miller K, Monga D, Brem C, Silverman JF, Finley GG. Molecular profiling of advanced malignancies: a community oncology network experience and review of literature. Front Med 2020. https://doi.org/10.3389/fmed.2020.00314.

[20] Scaranti M, Caldwell R, Miralles S, Shinde R, Pal A, Ang JE, Biondo A, Guo C, Cojocaru E, Gennatas S, Lockie F, Bertan C, Baker C, Carreira S, Banerjee S, Kaye S, de Bono JS, Banerji U, Minchom A, Lopez J. 1055P—clinical impact of molecular profiling of cervical cancer (CC) patients (pts) in a dedicated phase I (P1) unit. Ann Oncol 2019;30(5):v430.

[21] Mendoza-Almanza G, Ortíz-Sánchez E, Rocha-Zavaleta L, Rivas-Santiago C, Esparza-Ibarra E, Olmos J. Cervical cancer stem cells and other leading factors associated with cervical cancer development. Oncol Lett 2019;18(4):3423–32. https://doi.org/10.3892/ol.2019.10718. Epub 2019 Aug 6 31516560. PMC6733009.

[22] López J, Ruíz G, Organista-Nava J, Gariglio P, García-Carrancá A. Human papillomavirus infections and cancer stem cells of tumors from the uterine cervix. Open Virol J 2012;6:232–40. https://doi.org/10.2174/1874357901206010232. Epub 2012 Dec 28 23341858. PMC3547319.

[23] Lin J, Liu X, Ding D. Evidence for epithelial-mesenchymal transition in cancer stem-like cells derived from carcinoma cell lines of the cervix uteri. Int J Clin Exp Pathol 2015;8(1):847–55. 25755785. PMC4348812.

[24] Yang MH, Imrali A, Heeschen C. Circulating cancer stem cells: the importance to select. Chin J Cancer Res 2015;27(5):437–49. https://doi.org/10.3978/j.issn.1000-9604.2015.04.08. 26543330. PMC4626824.

[25] Batlle E, Clevers H. Cancer stem cells revisited. Nat Med 2017;23(10):1124–34. https://doi.org/10.1038/nm.4409. 28985214.

[26] Audirac-Chalifour A, Torres-Poveda K, Bahena-Román M, Téllez-Sosa J, Martínez-Barnetche J, Cortina-Ceballos B, López-Estrada G, Delgado-Romero K, Burguete-García AI, Cantú D, García-Carrancá A, Madrid-Marina V. Cervical microbiome and cytokine profile at various stages of cervical cancer: a pilot study. PLoS One 2016;11(4). https://doi.org/10.1371/journal.pone.0153274, e0153274. 27115350. PMC4846060.

[27] Chase D, Goulder A, Zenhausern F, Monk B, Herbst-Kralovetz M. The vaginal and gastrointestinal microbiomes in gynecologic cancers: a review of applications in etiology, symptoms and treatment. Gynecol Oncol 2015;138(1):190–200. https://doi.org/10.1016/j.ygyno.2015.04.036. Epub 2015 May 5 25957158.

[28] Linhares IM, Summers PR, Larsen B, Giraldo PC, Witkin SS. Contemporary perspectives on vaginal pH and lactobacilli. Am J Obstet Gynecol 2011;204(2):120. e1-5 https://doi.org/10.1016/j.ajog.2010.07.010. Epub 2010 Sep 15 20832044.

[29] Norenhag J, Du J, Olovsson M, Verstraelen H, Engstrand L, Brusselaers N. The vaginal microbiota, human papillomavirus and cervical dysplasia: a systematic review and network meta-analysis. BJOG 2020;127(2):171–80. https://doi.org/10.1111/1471-0528.15854. Epub 2019 Jul 17 31237400.

[30] Doorbar J, Griffin H. Refining our understanding of cervical neoplasia and its cellular origins. Papillomavirus Res 2019;7:176–9. https://doi.org/10.1016/j.pvr.2019.04.005. Epub 2019 Apr 8 30974183. PMC6477515.

[31] Budhwani M, Lukowski SW, Porceddu SV, Frazer IH, Chandra J. Dysregulation of stemness pathways in HPV mediated cervical malignant transformation identifies potential oncotherapy targets. Front Cell Infect Microbiol 2020. https://doi.org/10.3389/fcimb.2020.00307.

[32] Frazer IH, Chandra J. Immunotherapy for HPV associated cancer. Papillomavirus Res 2019;8. https://doi.org/10.1016/j.pvr.2019.100176, 100176. Epub 2019 Jul 13 31310819. PMC6639647.

[33] Arend RC, Beer HM, Cohen YC, Berlin S, Birrer MJ, Campos SM, Rachmilewitz Minei T, Harats D, Wall JA, Foxall ME, Penson RT. Ofranergene obadenovec (VB-111) in platinum-resistant ovarian cancer; favorable response rates in a phase I/II study are associated with an immunotherapeutic effect. Gynecol Oncol 2020;157 (3):578–84. https://doi.org/10.1016/j.ygyno.2020.02.034. Epub 2020 Apr 5 32265057.

[34] Pokhriyal R, Hariprasad R, Kumar L, Hariprasad G. Chemotherapy resistance in advanced ovarian cancer patients. Biomark Cancer 2019;11, 1179299X19860815. Published 2019 Jul 5 https://doi.org/10.1177/1179299X19860815.

[35] Lacey JV, Sherman ME. Ovarian neoplasia. In: Robboy SL, Mutter GL, Prat J, et al., editors. Robboy's pathology of the female reproductive tract. 2nd ed. Oxford: Churchill Livingstone Elsevier; 2009. p. 60.

[36] Ben-Baruch G, Sivan E, Moran O, Rizel S, Menczer J, Seidman DS. Primary peritoneal serous papillary carcinoma: a study of 25 cases and comparison with stage III-IV ovarian papillary serous carcinoma. Gynecol Oncol 1996;60(3):393–6. https://doi.org/10.1006/gyno.1996.0060. 8774644.

[37] Berek JS, Crum C, Friedlander M. Cancer of the ovary, fallopian tube, and peritoneum. Int J Gynaecol Obstet 2012;119(Suppl 2):S118–29. https://doi.org/10.1016/S0020-7292(12)60025-3. 22999503.

[38] Gates MA, Rosner BA, Hecht JL, Tworoger SS. Risk factors for epithelial ovarian cancer by histologic subtype. Am J Epidemiol 2010;171(1):45–53. https://doi.org/10.1093/aje/kwp314. Epub 2009 Nov 12 19910378. PMC2796984.

[39] Tsilidis KK, Allen NE, Key TJ, et al. Oral contraceptive use and reproductive factors and risk of ovarian cancer in the European prospective investigation into cancer and nutrition. Br J Cancer 2011;105(9):1436–42. https://doi.org/10.1038/bjc.2011.371. Epub 2011 Sep 13 21915124. PMC3241548.

[40] Kerlikowske K, Brown JS, Grady DG. Should women with familial ovarian cancer undergo prophylactic oophorectomy? Obstet Gynecol 1992;80:700.

[41] Jhamb N, Lambrou NC. Epidemiology and clinical presentation of ovarian cancer. In: Yang SC, editor. Early diagnosis and treatment of cancer series: ovarian cancer. Philadelphia: Saunders Elsevier; 2010.

[42] Boyd J. Specific keynote: hereditary ovarian cancer: what we know. Gynecol Oncol 2003;88(1 Pt 2):S8–10. discussion S11–3 https://doi.org/10.1006/gyno.2002.6674, 12586076.

[43] Li AJ, Karlan BY. Genetic factors in ovarian carcinoma. Curr Oncol Rep 2001;3(1):27–32. https://doi.org/10.1007/s11912-001-0039-y. 11123866.

[44] Risch HA, McLaughlin JR, Cole DE, et al. Prevalence and penetrance of germline BRCA1 and BRCA2 mutations in a population series of 649 women with ovarian cancer. Am J Hum Genet 2001;68:700.

[45] Pal T, Permuth-Wey J, Betts JA, Krischer JP, Fiorica J, Arango H, LaPolla J, Hoffman M, Martino MA, Wakeley K, Wilbanks G, Nicosia S, Cantor A, Sutphen R. BRCA1 and BRCA2 mutations account for a large proportion of ovarian carcinoma cases. Cancer 2005;104(12):2807–16. https://doi.org/10.1002/cncr.21536. 16284991.

[46] Rubin SC, Blackwood MA, Bandera C, Behbakht K, Benjamin I, Rebbeck TR, Boyd J. BRCA1, BRCA2, and hereditary nonpolyposis colorectal cancer gene mutations in an unselected ovarian cancer population: relationship to family history and implications for genetic testing. Am J Obstet Gynecol 1998;178(4):670–7. https://doi.org/10.1016/s0002-9378(98)70476-4. 9579428.

[47] Camargo MC, Stayner LT, Straif K, Reina M, Al-Alem U, Demers PA, Landrigan PJ. Occupational exposure to asbestos and ovarian cancer: a meta-analysis. Environ Health Perspect 2011;119(9):1211–7. https://doi.org/10.1289/ehp.1003283. Epub 2011 Jun 3 21642044. PMC3230399.

[48] Grandi G, Toss A, Cortesi L, Botticelli L, Volpe A, Cagnacci A. The association between endometriomas and ovarian cancer: preventive effect of inhibiting ovulation and menstruation during reproductive life. Biomed Res Int 2015;2015. https://doi.org/10.1155/2015/751571, 751571. Epub 2015 Aug 30 26413541. PMC4568052.

[49] Pearce CL, Templeman C, Rossing MA, Lee A, Near AM, Webb PM, Nagle CM, Doherty JA, Cushing-Haugen KL, Wicklund KG, Chang-Claude J, Hein R, Lurie G, Wilkens LR, Carney ME, Goodman MT, Moysich K, Kjaer SK, Hogdall E, Jensen A, Goode EL, Fridley BL, Larson MC, Schildkraut JM, Palmieri RT, Cramer DW, Terry KL, Vitonis AF, Titus LJ, Ziogas A, Brewster W, Anton-Culver H, Gentry-Maharaj A, Ramus SJ, Anderson AR, Brueggmann D, Fasching PA, Gayther SA, Huntsman DG, Menon U, Ness RB, Pike MC, Risch H, Wu AH, Berchuck A, Ovarian Cancer Association Consortium. Association between endometriosis and risk of histological subtypes of ovarian cancer: a pooled analysis of case-control studies. Lancet Oncol 2012;13(4):385–94. https://doi.org/10.1016/S1470-2045(11)70404-1. Epub 2012 Feb 22 22361336. PMC3664011.

[50] Berek JS, Chalas E, Edelson M, Moore DH, Burke WM, Cliby WA, Berchuck A, Society of Gynecologic Oncologists Clinical Practice Committee. Prophylactic and risk-reducing bilateral salpingo-oophorectomy: recommendations based on risk of ovarian cancer. Obstet Gynecol 2010;116(3):733–43. https://doi.org/10.1097/AOG.0b013e3181ec5fc1. 20733460.

[51] Whittemore AS, Harris R, Itnyre J. Characteristics relating to ovarian cancer risk: collaborative analysis of 12 US case-control studies. II. Invasive epithelial ovarian cancers in white women. Collaborative Ovarian Cancer Group. Am J Epidemiol 1992;136(10):1184–203. https://doi.org/10.1093/oxfordjournals.aje.a116427. 1476141.

[52] Havrilesky LJ, Moorman PG, Lowery WJ, Gierisch JM, Coeytaux RR, Urrutia RP, Dinan M, McBroom AJ, Hasselblad V, Sanders GD, Myers ER. Oral contraceptive pills as primary prevention for ovarian cancer: a systematic review and meta-analysis. Obstet Gynecol 2013;122(1):139–47. https://doi.org/10.1097/AOG.0b013e318291c235. 23743450.

[53] Iodice S, Barile M, Rotmensz N, Feroce I, Bonanni B, Radice P, Bernard L, Maisonneuve P, Gandini S. Oral contraceptive use and breast or ovarian cancer risk in BRCA1/2 carriers: a meta-analysis. Eur J Cancer 2010;46 (12):2275–84. https://doi.org/10.1016/j.ejca.2010.04.018. Epub 2010 May 27 20537530.

[54] Moorman PG, Havrilesky LJ, Gierisch JM, Coeytaux RR, Lowery WJ, Peragallo Urrutia R, Dinan M, McBroom AJ, Hasselblad V, Sanders GD, Myers ER. Oral contraceptives and risk of ovarian cancer and breast cancer

among high-risk women: a systematic review and meta-analysis. J Clin Oncol 2013;31(33):4188–98. https://doi.org/10.1200/JCO.2013.48.9021. Epub 2013 Oct 21 24145348.

[55] Li DP, Du C, Zhang ZM, Li GX, Yu ZF, Wang X, Li PF, Cheng C, Liu YP, Zhao YS. Breastfeeding and ovarian cancer risk: a systematic review and meta-analysis of 40 epidemiological studies. Asian Pac J Cancer Prev 2014;15(12):4829–37. https://doi.org/10.7314/apjcp.2014.15.12.4829. 24998548.

[56] Whiteman DC, Murphy MF, Cook LS, Cramer DW, Hartge P, Marchbanks PA, Nasca PC, Ness RB, Purdie DM, Risch HA. Multiple births and risk of epithelial ovarian cancer. J Natl Cancer Inst 2000;92(14):1172–7. https://doi.org/10.1093/jnci/92.14.1172. 10904091.

[57] Lambe M, Wuu J, Rossing MA, Hsieh CC. Twinning and maternal risk of ovarian cancer. Lancet 1999;353 (9168):1941. https://doi.org/10.1016/S0140-6736(99)02000-0. 10371582.

[58] Olsen J, Storm H. Pregnancy experience in women who later developed oestrogen-related cancers (Denmark). Cancer Causes Control 1998;9(6):653–7. https://doi.org/10.1023/a:1008831802805. 10189052.

[59] Pike MC, Pearce CL, Peters R, Cozen W, Wan P, Wu AH. Hormonal factors and the risk of invasive ovarian cancer: a population-based case-control study. Fertil Steril 2004;82(1):186–95. https://doi.org/10.1016/j.fertnstert.2004.03.013. Erratum in: Fertil Steril. 2004;82(2):516 15237010.

[60] Pakish JB, Jazaeri AA. Immunotherapy in gynecologic cancers: are we there yet? Curr Treat Options Oncol 2017;18(10):59. https://doi.org/10.1007/s11864-017-0504-y. 28840453. PMC5936621.

[61] Pujade-Lauraine E, Hilpert F, Weber B, Reuss A, Poveda A, Kristensen G, Sorio R, Vergote I, Witteveen P, Bamias A, Pereira D, Wimberger P, Oaknin A, Mirza MR, Follana P, Bollag D, Ray-Coquard I. Bevacizumab combined with chemotherapy for platinum-resistant recurrent ovarian cancer: the AURELIA open-label randomized phase III trial. J Clin Oncol 2014;32(13):1302–8. https://doi.org/10.1200/JCO.2013.51.4489. Epub 2014 Mar 17. Erratum in: J Clin Oncol. 2014 Dec 10;32(35):4025 24637997.

[62] Parmar MK, Ledermann JA, Colombo N, du Bois A, Delaloye JF, Kristensen GB, Wheeler S, Swart AM, Qian W, Torri V, Floriani I, Jayson G, Lamont A, Tropé C; ICON and AGO Collaborators. Paclitaxel plus platinum-based chemotherapy versus conventional platinum-based chemotherapy in women with relapsed ovarian cancer: the ICON4/AGO-OVAR-2.2 trial. Lancet 2003;361(9375):2099–106. https://doi.org/10.1016/s0140-6736(03)13718-x. 12826431.

[63] O'Toole O'Leary SJ. Ovarian cancer chemoresistance. In: Schwab M, editor. Encyclopedia of cancer. Berlin: Springer; 2011.

[64] Gottesman MM. Mechanisms of cancer drug resistance. Annu Rev Med 2002;53:615–27. https://doi.org/10.1146/annurev.med.53.082901.103929. 11818492.

[65] Rubin SC, Randall TC, Armstrong KA, Chi DS, Hoskins WJ. Ten-year follow-up of ovarian cancer patients after second-look laparotomy with negative findings. Obstet Gynecol 1999;93(1):21–4. https://doi.org/10.1016/s0029-7844(98)00334-2. 9916949.

[66] Keyvani V, Farshchian M, Esmaeili SA, Yari H, Moghbeli M, Nezhad S. Ovarian cancer stem cells and targeted therapy. J Ovarian Res 2019;12:120–31.

[67] Dontu G, et al. In vitro propagation and transcriptional profiling of human mammary stem/progenitor cells. Genes Dev 2003;17(10):1253–70.

[68] Valent P, Bonnet D, De Maria R, Lapidot T, Copland M, Melo JV, Chomienne C, Ishikawa F, Schuringa JJ, Stassi G, Huntly B, Herrmann H, Soulier J, Roesch A, Schuurhuis GJ, Wöhrer S, Arock M, Zuber J, Cerny-Reiterer S, Johnsen HE, Andreeff M, Eaves C. Cancer stem cell definitions and terminology: the devil is in the details. Nat Rev Cancer 2012;12(11):767–75. https://doi.org/10.1038/nrc3368. Epub 2012 Oct 11 23051844.

[69] Bapat SA, Mali AM, Koppikar CB, Kurrey NK. Stem and progenitor-like cells contribute to the aggressive behavior of human epithelial ovarian cancer. Cancer Res 2005;65(8):3025–9. https://doi.org/10.1158/0008-5472.CAN-04-3931. 15833827.

[70] Deshmukh A, Deshpande K, Arfuso F, Newsholme P, Dharmarajan A. Cancer stem cell metabolism: a potential target for cancer therapy. Mol Cancer 2016;15(1):69. https://doi.org/10.1186/s12943-016-0555-x. 27825361. PMC5101698.

[71] Liu PP, Liao J, Tang ZJ, Wu WJ, Yang J, Zeng ZL, Hu Y, Wang P, Ju HQ, Xu RH, Huang P. Metabolic regulation of cancer cell side population by glucose through activation of the Akt pathway. Cell Death Differ 2014;21(1):124–35. https://doi.org/10.1038/cdd.2013.131. Epub 2013 Oct 4 24096870. PMC3857620.

[72] Palorini R, Votta G, Balestrieri C, Monestiroli A, Olivieri S, Vento R, Chiaradonna F. Energy metabolism characterization of a novel cancer stem cell-like line 3AB-OS. J Cell Biochem 2014;115(2):368–79. https://doi.org/10.1002/jcb.24671. 24030970.

[73] Zhou Y, Zhou Y, Shingu T, Feng L, Chen Z, Ogasawara M, Keating MJ, Kondo S, Huang P. Metabolic alterations in highly tumorigenic glioblastoma cells: preference for hypoxia and high dependency on glycolysis. J Biol Chem 2011;286(37):32843–53. https://doi.org/10.1074/jbc.M111.260935. Epub 2011 Jul 27 21795717. PMC3173179.

[74] Liao J, Qian F, Tchabo N, Mhawech-Fauceglia P, Beck A, Qian Z, Wang X, Huss WJ, Lele SB, Morrison CD, Odunsi K, Ovarian cancer spheroid cells with stem cell-like properties contribute to tumor generation, metastasis and chemotherapy resistance through hypoxia-resistant metabolism. 2014. https://doi.org/10.1371/journal.pone.0084941.

[75] Dizon DS, Weitzen S, Rojan A, Schwartz J, Miller J, Disilvestro P, Gordinier ME, Moore R, Tejada-Berges T, Pires L, Legare R, Granai CO. Two for good measure: six versus eight cycles of carboplatin and paclitaxel as adjuvant treatment for epithelial ovarian cancer. Gynecol Oncol 2006;100(2):417–21. https://doi.org/10.1016/j.ygyno.2005.10.031. Epub 2005 Dec 5 16336992.

[76] Shapiro AB, Corder AB, Ling V. P-glycoprotein-mediated Hoechst 33342, transport out of the lipid bilayer. Eur J Biochem 1997;250(1):115–21.

[77] Litman T, Brangi M, Hudson E, Fetsch P, Abati A, Ross DD, Miyake K, Resau JH, Bates SE. The multidrug-resistant phenotype associated with overexpression of the new ABC half-transporter, MXR (ABCG2). J Cell Sci 2000;113(Pt 11):2011–21. 10806112.

[78] Chow EK, Fan LL, Chen X, Bishop JM. Oncogene-specific formation of chemoresistant murine hepatic cancer stem cells. Hepatology 2012;56(4):1331–41. https://doi.org/10.1002/hep.25776. Epub 2012 Aug 27 22505225. PMC3418440.

[79] Chuthapisith S, Eremin J, El-Sheemey M, Eremin O. Breast cancer chemoresistance: emerging importance of cancer stem cells. Surg Oncol 2010;19(1):27–32. https://doi.org/10.1016/j.suronc.2009.01.004. Epub 2009 Feb 28 19251410.

[80] Eyre R, Harvey I, Stemke-Hale K, Lennard TW, Tyson-Capper A, Meeson AP. Reversing paclitaxel resistance in ovarian cancer cells via inhibition of the ABCB1 expressing side population. Tumour Biol 2014;35(10):9879–92. https://doi.org/10.1007/s13277-014-2277-2. Epub 2014 Jul 4. PMID.

[81] Hu L, McArthur C, Jaffe R. Ovarian cancer stem-like side-population cells are tumourigenic and chemoresistant. Br J Cancer 2010;102(8):1276.

[82] Kim DK, Seo EJ, Choi EJ, Lee SI, Kwon YW, Jang IH, Kim SC, Kim KH, Suh DS, Seong-Jang K, Lee SC, Kim JH. Crucial role of HMGA1 in the self-renewal and drug resistance of ovarian cancer stem cells. Exp Mol Med 2016;48(8), e255.

[83] Ranganathan P, Weaver KL, Capobianco AJ. Notch signaling in solid tumors: a little bit of everything but not all the time. Nat Rev Cancer 2011;11(5):338–51. https://doi.org/10.1038/nrc3035. Epub 2011 Apr 14 21508972.

[84] Moghbeli M, Rad A, Farshchian M, Taghehchian N, Gholamin M, Abbaszadegan MR. Correlation between Meis1 and Msi1 in esophageal squamous cell carcinoma. J Gastrointest Cancer 2016;47(3):273–7. https://doi.org/10.1007/s12029-016-9824-6. 27142513.

[85] Moghbeli M, Sadrizadeh A, Forghanifard MM, Mozaffari HM, Golmakani E, Abbaszadegan MR. Role of Msi1 and PYGO2 in esophageal squamous cell carcinoma depth of invasion. J Cell Commun Signal 2016;10(1):49–53. https://doi.org/10.1007/s12079-015-0314-6. Epub 2015 Dec 7 26643817. PMC4850136.

[86] Abbaszadegan MR, Moghbeli M. Role of MAML1 and MEIS1 in esophageal squamous cell carcinoma depth of invasion. Pathol Oncol Res 2018;24(2):245–50.

[87] Meng RD, Shelton CC, Li YM, Qin LX, Notterman D, Paty PB, Schwartz GK. Gamma-Secretase inhibitors abrogate oxaliplatin-induced activation of the Notch-1 signaling pathway in colon cancer cells resulting in enhanced chemosensitivity. Cancer Res 2009;69(2):573–82. https://doi.org/10.1158/0008-5472.CAN-08-2088. 19147571. PMC3242515.

[88] Li Y, Chen T, Zhu J, Zhang H, Jiang H, Sun H. High ALDH activity defines ovarian cancer stem-like cells with enhanced invasiveness and EMT progress which are responsible for tumor invasion. Biochem Biophys Res Commun 2018;495(1):1081–8. https://doi.org/10.1016/j.bbrc.2017.11.117. Epub 2017 Nov 21 29170132.

[89] Opferman JT, Kothari A. Anti-apoptotic BCL-2 family members in development. Cell Death Differ 2018;25(1):37–45. https://doi.org/10.1038/cdd.2017.170. Epub 2017 Nov 3 29099482. PMC5729530.

[90] Kim R, Emi M, Tanabe K. Role of mitochondria as the gardens of cell death. Cancer Chemother Pharmacol 2006;57(5):545–53.

[91] Pegoraro L, Palumbo A, Erikson J, Falda M, Giovanazzo B, Emanuel BS, Rovera G, Nowell PC, Croce CM. A 14;18 and an 8;14 chromosome translocation in a cell line derived from an acute B-cell leukemia. Proc Natl Acad Sci U S A 1984;81(22):7166–70. https://doi.org/10.1073/pnas.81.22.7166. 6334305. PMC392098.

[92] Graninger WB, Seto M, Boutain B, Goldman P, Korsmeyer SJ. Expression of Bcl-2 and Bcl-2-Ig fusion transcripts in normal and neoplastic cells. J Clin Invest 1987;80(5):1512–5. https://doi.org/10.1172/JCI113235. 3500184. PMC442413.

[93] Madjd Z, Mehrjerdi AZ, Sharifi AM, Molanaei S, Shahzadi SZ, Asadi-Lari M. CD44 + cancer cells express higher levels of the anti-apoptotic protein Bcl-2 in breast tumours. Cancer Immun 2009;9:4. 19385591. PMC2935767.

[94] Williams J, Lucas PC, Griffith KA, Choi M, Fogoros S, Hu YY, Liu JR. Expression of Bcl-xL in ovarian carcinoma is associated with chemoresistance and recurrent disease. Gynecol Oncol 2005;96(2):287–95. https://doi.org/10.1016/j.ygyno.2004.10.026. 15661210.

[95] McAuliffe SM, Morgan SL, Wyant GA, Tran LT, Muto KW, Chen YS, Chin KT, Partridge JC, Poole BB, Cheng KH, Daggett Jr J, Cullen K, Kantoff E, Hasselbatt K, Berkowitz J, Muto MG, Berkowitz RS, Aster JC, Matulonis UA, Dinulescu DM. Targeting Notch, a key pathway for ovarian cancer stem cells, sensitizes tumors to platinum therapy. Proc Natl Acad Sci U S A 2012;109(43):E2939–48. https://doi.org/10.1073/pnas.1206400109. Epub 2012 Sep 27 23019585. PMC3491453.

[96] Colombo N, Preti E, Landoni F, Carinelli S, Colombo A, Marini C, Sessa C, ESMO Guidelines Working Group. Endometrial cancer: ESMO clinical practice guidelines for diagnosis, treatment and follow-up. Ann Oncol 2011;22(Suppl 6):vi35–9. https://doi.org/10.1093/annonc/mdr374. 21908501.

[97] National Cancer Institute, n.d.Cancer stat facts: uterine cancer, https://seer.cancer.gov/Statfacts/html/corp. html, [Accessed 22 October 2019].

[98] Colombo N, Creutzberg C, Amant F, Bosse T, González-Martín A, Ledermann J, Marth C, Nout R, Querleu D, Mirza MR, Sessa C. ESMO-ESGO-ESTRO consensus conference on endometrial cancer: diagnosis, treatment and follow-up. Int J Gynecol Cancer 2016;26(1):2–30. https://doi.org/10.1097/IGC.0000000000000609. 26645990. PMC4679344.

[99] Khazaei Z, Goodarzi E, Sohrabivafa M, Naemi H, Mansori K. Association between the incidence and mortality rates for corpus uteri cancer and human development index (HDI): a global ecological study. Obstet Gynecol Sci 2020;63(2):141–9. https://doi.org/10.5468/ogs.2020.63.2.141. Epub 2020 Feb 20 32206653. PMC7073362.

[100] American Cancer Society. Cancer Facts and Figures, https://www.cancer.org/content/dam/cancer-org/research/cancer-facts-and-statistics/annual-cancer-facts-and-figures/2016/cancer-facts-and-figures-2016. pdf; 2016. [Accessed 11 July 2017].

[101] National Cancer Institute Surveillance, n.d. Epidemiology, and end results program. Cancer stat facts: endometrial cancer https://seer.cancer.gov/statfacts/html/corp.html. [Accessed 11 July 2017].

[102] Charo LM, Plaxe SC. Recent advances in endometrial cancer: a review of key clinical trials from 2015 to 2019. F1000Res 2019;8. https://doi.org/10.12688/f1000research.17408.1. F1000 Faculty Rev-849 31231511. PMC6567288.

[103] Siegel RL, Miller KD, Jemal A. Cancer statistics, 2019. CA Cancer J Clin 2019;69(1):7–34. https://doi.org/10.3322/caac.21551. Epub 2019 Jan 8 30620402.

[104] Lortet-Tieulent J, Ferlay J, Bray F, Jemal A. International patterns and trends in endometrial cancer incidence, 1978-2013. J Natl Cancer Inst 2018;110(4):354–61. https://doi.org/10.1093/jnci/djx214. 29045681.

[105] CRUK. Cancer Research United Kingdom: uterine cancer incidence statistics. Available online at www.cancerresearchuk.org/health-professional/cancer-statistics/statistics-by-cancer-type/uterine-cancer/incidence; 2020. February 2020.

[106] Cancer Research UK, n.d. Uterine cancer statistics, https://www.cancerresearchuk.org/health-professional/cancer-statistics/statistics-by-cancer-type/uterine-cancer, [Accessed 22 October 2019].

[107] Yen TT, Wang TL, Fader AN, Shih IM, Gaillard S. Molecular classification and emerging targeted therapy in endometrial cancer. Int J Gynecol Pathol 2020;39(1):26–35. https://doi.org/10.1097/PGP.0000000000000585. 30741844. PMC6685771.

[108] Garg K, Soslow RA. Endometrial carcinoma in women aged 40 years and younger. Arch Pathol Lab Med 2014;138(3):335–42. https://doi.org/10.5858/arpa.2012-0654-RA. 24576029.

[109] Wise MR, Gill P, Lensen S, Thompson JM, Farquhar CM. Body mass index trumps age in decision for endometrial biopsy: cohort study of symptomatic premenopausal women. Am J Obstet Gynecol 2016;215(5):598. e1–8. https://doi.org/10.1016/j.ajog.2016.06.006. Epub 2016 Jun 8 27287687.

[110] Soliman PT, Oh JC, Schmeler KM, Sun CC, Slomovitz BM, Gershenson DM, Burke TW, Lu KH. Risk factors for young premenopausal women with endometrial cancer. Obstet Gynecol 2005;105(3):575–80. https://doi.org/10.1097/01.AOG.0000154151.14516.f7. 15738027.

[111] Ogden CL, Carroll MD, Fryar CD, Flegal KM. Prevalence of obesity among adults and youth: United States, 2011-2014. NCHS Data Brief 2015;219:1–8. 26633046.

[112] American Cancer Society. Cancer facts & figures 2017; 2017. Atlanta.

[113] Sonoda K. Molecular biology of gynecological cancer. Oncol Lett 2016;11(1):16–22. https://doi.org/10.3892/ol.2015.3862. Epub 2015 Nov 5 26834851. PMC4727069.

[114] Bokhman JV. Two pathogenetic types of endometrial carcinoma. Gynecol Oncol 1983;15(1):10–7. https://doi.org/10.1016/0090-8258(83)90111-7. 6822361.

[115] Maxwell GL, Risinger JI, Gumbs C, Shaw H, Bentley RC, Barrett JC, Berchuck A, Futreal PA. Mutation of the PTEN tumor suppressor gene in endometrial hyperplasias. Cancer Res 1998;58:2500–3.

[116] Basil JB, Goodfellow PJ, Rader JS, Mutch DG, Herzog TJ. Clinical significance of microsatellite instability in endometrial carcinoma. Cancer 2000;89:1758–64.

[117] Mutter GL, Lin MC, Fitzgerald JT, Kum JB, Baak JP, Lees JA, et al. Altered PTEN expression as a diagnostic marker for the earliest endometrial precancers. J Natl Cancer Inst 2000;92:924–30.

[118] Bilbao C, Rodriguez G, Ramirez R, Falcon O, Leon L, Chirino R, et al. The relationship between microsatellite instability and PTEN gene mutations in endometrial cancer. Int J Cancer 2006;119:563–70.

[119] Hecht JL, Mutter GL. Molecular and pathologic aspects of endometrial carcinogenesis. J Clin Oncol 2006;24:4783–91.

[120] Suarez AA, Felix AS, Cohn DE. Bokhman redux: endometrial cancer "types" in the 21st century. Gynecol Oncol 2017;144(2):243–9. https://doi.org/10.1016/j.ygyno.2016.12.010. Epub 2016 Dec 16 27993480.

[121] Sundar S, Balega J, Crosbie E, Drake A, Edmondson R, Fotopoulou C, Gallos I, Ganesan R, Gupta J, Johnson N, Kitson S, Mackintosh M, Martin-Hirsch P, Miles T, Rafii S, Reed N, Rolland P, Singh K, Sivalingam V, Walther A. BGCS uterine cancer guidelines: recommendations for practice. Eur J Obstet Gynecol Reprod Biol 2017;213:71–97. https://doi.org/10.1016/j.ejogrb.2017.04.015. Epub 2017 Apr 13 28437632.

[122] Funston G, O'Flynn H, Ryan NAJ, Hamilton W, Crosbie EJ. Recognizing gynecological cancer in primary care: risk factors, red flags, and referrals. Adv Ther 2018;35(4):577–89. https://doi.org/10.1007/s12325-018-0683-3. Epub 2018 Mar 7. Erratum in: Adv Ther. 2018 Apr 4 29516408. PMC5910472.

[123] Wortman BG, Creutzberg CL, Putter H, Jürgenliemk-Schulz IM, Jobsen JJ, LCHW L, van der Steen-Banasik EM, JWM M, Slot A, MCS K, van Triest B, Nijman HW, Stelloo E, Bosse T, de Boer SM, van Putten WLJ, Smit VTHBM, Nout RA, PORTEC Study Group. Ten-year results of the PORTEC-2 trial for high-intermediate risk endometrial carcinoma: improving patient selection for adjuvant therapy. Br J Cancer 2018;119(9):1067–74. https://doi.org/10.1038/s41416-018-0310-8. Epub 2018 Oct 25 30356126. PMC6219495.

[124] de Boer SM, Powell ME, Mileshkin L, Katsaros D, Bessette P, Haie-Meder C, Ottevanger PB, Ledermann JA, Khaw P, Colombo A, Fyles A, Baron MH, Jürgenliemk-Schulz IM, Kitchener HC, Nijman HW, Wilson G, Brooks S, Carinelli S, Provencher D, Hanzen C, LCHW L, VTHBM S, Singh N, Do V, D'Amico R, Nout RA, Feeney A, Verhoeven-Adema KW, Putter H, Creutzberg CL, PORTEC study group. Adjuvant chemoradiotherapy versus radiotherapy alone for women with high-risk endometrial cancer (PORTEC-3): final results of an international, open-label, multicentre, randomised, phase 3 trial. Lancet Oncol 2018;19(3):295–309. https://doi.org/10.1016/S1470-2045(18)30079-2. Epub 2018 Feb 12. Erratum in: Lancet Oncol. 2018 Apr;19(4):e184 29449189. PMC5840256.

[125] NICE Gynaecological cancers, recognition and referral. n.d.Available online at:https://cks.nice.org.uk/topics/gynaecological-cancers-recognition-referral.

[126] Hamilton W, Hajioff S, Graham J, Schmidt-Hansen M. Suspected cancer (part 2–adults): reference tables from updated NICE guidance. BMJ 2015;350. https://doi.org/10.1136/bmj.h3044, h3044. Erratum in: BMJ. 2015;351:h3935 26104465.

[127] Keys HM, Roberts JA, Brunetto VL, Zaino RJ, Spirtos NM, Bloss JD, Pearlman A, Maiman MA, Bell JG. Gynecologic oncology group. A phase III trial of surgery with or without adjunctive external pelvic radiation therapy in intermediate risk endometrial adenocarcinoma: a Gynecologic Oncology Group Study. Gynecol Oncol 2004;92(3):744–51. https://doi.org/10.1016/j.ygyno.2003.11.048. Erratum in: Gynecol Oncol 2004 Jul;94(1):241–2 14984936.

[128] Creutzberg CL, van Putten WL, Koper PC, Lybeert ML, Jobsen JJ, Wárlám-Rodenhuis CC, De Winter KA, Lutgens LC, van den Bergh AC, van de Steen-Banasik E, Beerman H, van Lent M. Surgery and postoperative radiotherapy versus surgery alone for patients with stage-1 endometrial carcinoma: multicentre randomised

trial. PORTEC study group. Post operative radiation therapy in endometrial carcinoma. Lancet 2000;355 (9213):1404–11. https://doi.org/10.1016/s0140-6736(00)02139-5. 10791524.

[129] ASTEC/EN.5 Study Group, Blake P, Swart AM, Orton J, Kitchener H, Whelan T, Lkka H, Eisenhauer E, Bacon M, Tu D, Parmar MK, Amos C, Murray C, Qian W. Adjuvant external beam radiotherapy in the treatment of endometrial cancer (MRC ASTEC and NCIC CTG EN.5 randomised trials): pooled trial results, systematic review, and meta-analysis. Lancet 2009;373(9658):137–46. https://doi.org/10.1016/S0140-6736(08)61767-5. Epub 2008 Dec 16 19070891. PMC2646125.

[130] Sorbe B, Horvath G, Andersson H, Boman K, Lundgren C, Pettersson B. External pelvic and vaginal irradiation versus vaginal irradiation alone as postoperative therapy in medium-risk endometrial carcinoma—a prospective randomized study. Int J Radiat Oncol Biol Phys 2012;82(3):1249–55. https://doi.org/10.1016/j.ijrobp.2011.04.014. Epub 2011 Jun 14 21676554.

[131] Scholten AN, van Putten WL, Beerman H, Smit VT, Koper PC, Lybeert ML, Jobsen JJ, Wárlám-Rodenhuis CC, De Winter KA, Lutgens LC, van Lent M, Creutzberg CL, PORTEC Study Group. Postoperative radiotherapy for stage 1 endometrial carcinoma: long-term outcome of the randomized PORTEC trial with central pathology review. Int J Radiat Oncol Biol Phys 2005;63(3):834–8. https://doi.org/10.1016/j.ijrobp.2005.03.007. Epub 2005 May 31 15927414.

[132] Plaut EA. National Health Insurance includes mental health. Am J Public Health 1978;68(5):443–4. https://doi.org/10.2105/ajph.68.5.443. 645990. PMC1653882.

[133] Randall ME, Filiaci VL, Muss H, Spirtos NM, Mannel RS, Fowler J, Thigpen JT, Benda JA, Gynecologic Oncology Group Study. Randomized phase III trial of whole-abdominal irradiation versus doxorubicin and cisplatin chemotherapy in advanced endometrial carcinoma: a Gynecologic Oncology Group Study. J Clin Oncol 2006;24 (1):36–44. https://doi.org/10.1200/JCO.2004.00.7617. Epub 2005 Dec 5 16330675.

[134] Greven K, Winter K, Underhill K, Fontenesci J, Cooper J, Burke T. Final analysis of RTOG 9708: adjuvant postoperative irradiation combined with cisplatin/paclitaxel chemotherapy following surgery for patients with high-risk endometrial cancer. Gynecol Oncol 2006;103(1):155–9. https://doi.org/10.1016/j.ygyno.2006.02.007. Epub 2006 Mar 20 16545437.

[135] de Boer SM, Powell ME, Mileshkin L, Katsaros D, Bessette P, Haie-Meder C, Ottevanger PB, Ledermann JA, Khaw P, D'Amico R, Fyles A, Baron MH, Jürgenliemk-Schulz IM, Kitchener HC, Nijman HW, Wilson G, Brooks S, Gribaudo S, Provencher D, Hanzen C, Kruitwagen RF, VTHBM S, Singh N, Do V, Lissoni A, Nout RA, Feeney A, Verhoeven-Adema KW, Putter H, Creutzberg CL, PORTEC Study Group. Adjuvant chemoradiotherapy versus radiotherapy alone in women with high-risk endometrial cancer (PORTEC-3): patterns of recurrence and post-hoc survival analysis of a randomised phase 3 trial. Lancet Oncol 2019;20(9):1273–85. https://doi.org/10.1016/S1470-2045(19)30395-X. Epub 2019 Jul 22. Erratum in: Lancet Oncol. 2019 Sep;20(9):e468 31345626. PMC6722042.

[136] de Boer SM, Powell ME, Mileshkin L, Katsaros D, Bessette P, Haie-Meder C, Ottevanger PB, Ledermann JA, Khaw P, Colombo A, Fyles A, Baron MH, Kitchener HC, Nijman HW, Kruitwagen RF, Nout RA, Verhoeven-Adema KW, Smit VT, Putter H, Creutzberg CL, PORTEC study group. Toxicity and quality of life after adjuvant chemoradiotherapy versus radiotherapy alone for women with high-risk endometrial cancer (PORTEC-3): an open-label, multicentre, randomised, phase 3 trial. Lancet Oncol 2016;17(8):1114–26. https://doi.org/10.1016/S1470-2045(16)30120-6. Epub 2016 Jul 7 27397040.

[137] Ta MH, Schernberg A, Giraud P, Monnier L, Darai É, Bendifallah S, Schlienger M, Touboul E, Orthuon A, Challand T, Huguet F, Rivin Del Campo E. Comparison of 3D conformal radiation therapy and intensity-modulated radiation therapy in patients with endometrial cancer: efficacy, safety and prognostic analysis. Acta Oncol 2019;58(8):1127–34. https://doi.org/10.1080/0284186X.2019.1599136. Epub 2019 Apr 24 31017032.

[138] Malpica A, Euscher ED, Hecht JL, Ali-Fehmi R, Quick CM, Singh N, Horn LC, Alvarado-Cabrero I, Matias-Guiu X, Hirschowitz L, Duggan M, Ordi J, Parkash V, Mikami Y, Ruhul Quddus M, Zaino R, Staebler A, Zaloudek C, McCluggage WG, Oliva E. Endometrial carcinoma, grossing and processing issues: recommendations of the International Society of Gynecologic Pathologists. Int J Gynecol Pathol 2019;38(Suppl 1(Iss 1 Suppl 1)): S9–S24. https://doi.org/10.1097/PGP.0000000000000552. 30550481. PMC6296844.

[139] Brasseur K, Gévry N, Asselin E. Chemoresistance and targeted therapies in ovarian and endometrial cancers. Oncotarget 2017;8(3):4008–42. https://doi.org/10.18632/oncotarget.14021. 28008141. PMC5354810.

[140] Lynch HT, Snyder CL, Shaw TG, Heinen CD, Hitchins MP. Milestones of Lynch syndrome: 1895-2015. Nat Rev Cancer 2015;15(3):181–94. https://doi.org/10.1038/nrc3878. Epub 2015 Feb 12 25673086.

[141] Di Tucci C, Capone C, Galati G, Iacobelli V, Schiavi MC, Di Donato V, Muzii L, Panici PB. Immunotherapy in endometrial cancer: new scenarios on the horizon. J Gynecol Oncol 2019;30(3). https://doi.org/10.3802/jgo.2019.30.e46, e46. 30887763. PMC6424849.

[142] Cancer Genome Atlas Research Network, Kandoth C, Schultz N, Cherniack AD, Akbani R, Liu Y, Shen H, Robertson AG, Pashtan I, Shen R, Benz CC, Yau C, Laird PW, Ding L, Zhang W, Mills GB, Kucherlapati R, Mardis ER, Levine DA. Integrated genomic characterization of endometrial carcinoma. Nature 2013;497(7447):67–73. https://doi.org/10.1038/nature12113. Erratum in: Nature. 2013 Aug 8;500(7461):242 23636398. PMC3704730.

[143] Njoku K, Sutton CJ, Whetton AD, Crosbie EJ. Metabolomic biomarkers for detection, prognosis and identifying recurrence in endometrial cancer. Metabolites 2020;10(8):314. https://doi.org/10.3390/metabo10080314. 32751940. PMC7463916.

[144] Costas L, Frias-Gomez J, Guardiola M, Benavente Y, Pineda M, Pavón MÁ, Martínez JM, Climent M, Barahona M, Canet J, Paytubi S, Salinas M, Palomero L, Bianchi I, Reventós J, Capellà G, Diaz M, Vidal A, Piulats JM, Aytés Á, Ponce J, Brunet J, Bosch FX, Matias-Guiu X, Alemany L, de Sanjosé S, Screenwide Team. New perspectives on screening and early detection of endometrial cancer. Int J Cancer 2019;145(12):3194–206. https://doi.org/10.1002/ijc.32514. Epub 2019 Jun 28 31199503.

[145] Altadill T, Dowdy TM, Gill K, Reques A, Menon SS, Moiola CP, Lopez-Gil C, Coll E, Matias-Guiu X, Cabrera S, Garcia A, Reventos J, Byers SW, Gil-Moreno A, Cheema AK, Colas E. Metabolomic and lipidomic profiling identifies the role of the rna editing pathway in endometrial carcinogenesis. Sci Rep 2017;7(1):8803. https://doi.org/10.1038/s41598-017-09169-2. 28821813. PMC5562852.

[146] Njoku K, Crosbie EJ. Does the vaginal microbiome drive cervical carcinogenesis? BJOG Int J Obstet Gynaecol 2020;127:181. https://doi.org/10.1111/1471-0528.15867.

[147] Zegels G, Van Raemdonck GAA, Tjalma WAA, Van Ostade XWM. Use of cervicovaginal fluid for the identification of biomarkers for pathologies of the female genital tract. Proteome Sci 2010;8:63. https://doi.org/10.1186/1477-5956-8-63.

[148] Ytre-Hauge S, Husby JA, Magnussen IJ, Werner HMJ, Salvesen O, Bjørge L, Trovik J, Stefansson IM, Salvesen HB, Haldorsen IS. Preoperative tumor size at MRI predicts deep myometrial invasion, lymph node metastases, and patient outcome in endometrial carcinomas. Int J Gynecol Cancer 2015;25:459–66. https://doi.org/10.1097/IGC.0000000000000367.

[149] Iżycki D, Woźniak K, Iżycka N. Consequences of gynecological cancer in patients and their partners from the sexual and psychological perspective. Prz Menopauzalny 2016;15(2):112–6. https://doi.org/10.5114/pm.2016.61194. Epub 2016 Jul 22 27582686. PMC4993986.

[150] Di Fiore R, Suleiman S, Ellul B, O'Toole SA, Savona-Ventura C, Felix A, Napolioni V, Conlon NT, Kahramanoglu I, Azzopardi MJ, Dalmas M, Calleja N, Brincat MR, Muscat-Baron Y, Sabol M, Dimitrievska V, Yordanov A, Vasileva-Slaveva M, von Brockdorff K, Micallef RA, Kubelac P, Achimaş-Cadariu P, Vlad C, Tzortzatou O, Poka R, Giordano A, Felice A, Reed N, Herrington CS, Faraggi D, Calleja-Agius J. GYNOCARE update: modern strategies to improve diagnosis and treatment of rare gynecologic tumors—current challenges and future directions. Cancers (Basel) 2021;13(3):493. https://doi.org/10.3390/cancers13030493. 33514073. PMC7865420.

[151] Huang M, Hunter T, Slomovitz B, Schlumbrecht M. Impact of molecular testing in clinical practice in gynecologic cancers. Cancer Med 2019;8(5):2013–9. https://doi.org/10.1002/cam4.2064. Epub 2019 Mar 7 30848097. PMC6536929.

[152] https://allofus.nih.gov/about/all-us-research-program-overview n.d. [Accessed 30 March 2021].

# Further reading

Kagabu M, Nagasawa T, Sato C, Fukagawa Y, Kawamura H, Tomabechi H, Takemoto S, Shoji T, Baba T. Immunotherapy for uterine cervical cancer using checkpoint inhibitors: future directions. Int J Mol Sci 2020;21(7):2335–47. 32230938.

Abdullah LN, Chow EK. Mechanisms of chemoresistance in cancer stem cells. Clin Transl Med 2013;2(1):3–12. https://doi.org/10.1186/2001-1326-2-3. 23369605. PMC3565873.

Kuroda T, Hirohashi Y, Torigoe T, Yasuda K, Takahashi A, Asanuma H, Morita R, Mariya T, Asano T, Mizuuchi M, Saito T, Sato N. ALDH1-high ovarian cancer stem-like cells can be isolated from serous and clear cell adenocarcinoma cells, and ALDH1 high expression is associated with poor prognosis. PLoS One 2013;8(6). https://doi.org/10.1371/journal.pone.0065158, e65158. 23762304. PMC3675199.

Dalton WS, Crowley JJ, Salmon SS, Grogan TM, Laufman LR, Weiss GR, Bonnet JD. A phase III randomized study of oral verapamil as a chemosensitizer to reverse drug resistance in patients with refractory myeloma. A Southwest Oncology Group study. Cancer 1995;75(3):815–20. https://doi.org/10.1002/1097-0142(19950201)75:3<815::aid-cncr2820750311>3.0.co;2-r. 7828131.

Sládek NE. Human aldehyde dehydrogenases: potential pathological,pharmacological, and toxicological impact. J Biochem Mol Toxicol 2003;17(1):7–23.

Scharenberg CW, Harkey MA, Torok-Storb B. The ABCG2 transporter is an efficient Hoechst 33342 efflux pump and is preferentially expressed by immature human hematopoietic progenitors. Blood 2002;99(2):507–12.

Uhlén M, Fagerberg L, Hallström BM, Lindskog C, Oksvold P, Mardinoglu A, Sivertsson Å, Kampf C, Sjöstedt E, Asplund A, Olsson I, Edlund K, Lundberg E, Navani S, Szigyarto CA, Odeberg J, Djureinovic D, Takanen JO, Hober S, Alm T, Edqvist PH, Berling H, Tegel H, Mulder J, Rockberg J, Nilsson P, Schwenk JM, Hamsten M, von Feilitzen K, Forsberg M, Persson L, Johansson F, Zwahlen M, von Heijne G, Nielsen J, Proteomics PF, Tissue-based map of the human proteome. Science 2015;347(6220). https://doi.org/10.1126/science.1260419, 1260419. 25613900.

Chou JL, Huang RL, Shay J, Chen LY, Lin SJ, Yan PS, Chao WT, Lai YH, Lai YL, Chao TK, Lee CI, Tai CK, Wu SF, Nephew KP, Huang TH, Lai HC, Chan MW. Hypermethylation of the TGF-β target, ABCA1 is associated with poor prognosis in ovarian cancer patients. Clin Epigenetics 2015;7(1):1. https://doi.org/10.1186/s13148-014-0036-2. 25628764. PMC4307187.

Wong M, Tan N, Zha J, Peale FV, Yue P, Fairbrother WJ, Belmont LD. Navitoclax (ABT-263) reduces Bcl-x(L)-mediated chemoresistance in ovarian cancer models. Mol Cancer Ther 2012;11(4):1026–35. https://doi.org/10.1158/1535-7163.MCT-11-0693. Epub 2012 Feb 1 22302098.

Witham J, Valenti MR, De-Haven-Brandon AK, Vidot S, Eccles SA, Kaye SB, Richardson A. The Bcl-2/Bcl-XL family inhibitor ABT-737 sensitizes ovarian cancer cells to carboplatin. Clin Cancer Res 2007;13(23):7191–8. https://doi.org/10.1158/1078-0432.CCR-07-0362. 18056200.

Reya T, Duncan AW, Ailles L, Domen J, Scherer DC, Willert K, Hintz L, Nusse R, Weissman IL. A role for Wnt signalling in self-renewal of haematopoietic stem cells. Nature 2003;423(6938):409–14. https://doi.org/10.1038/nature01593. Epub 2003 Apr 27 12717450.

Zhao C, Blum J, Chen A, Kwon HY, Jung SH, Cook JM, Lagoo A, Reya T. Loss of beta-catenin impairs the renewal of normal and CML stem cells in vivo. Cancer Cell 2007;12(6):528–41. https://doi.org/10.1016/j.ccr.2007.11.003. 18068630. PMC2262869.

Bisson I, Prowse DM. WNT signaling regulates self-renewal and differentiation of prostate cancer cells with stem cell characteristics. Cell Res 2009;19(6):683–97. https://doi.org/10.1038/cr.2009.43. 19365403.

Capodanno Y, Buishand FO, Pang LY, Kirpensteijn J, Mol JA, Argyle DJ. Notch pathway inhibition targets chemoresistant insulinoma cancer stem cells. Endocr Relat Cancer 2018;25(2):131–44. https://doi.org/10.1530/ERC-17-0415. Epub 2017 Nov 24 29175872.

Abbaszadegan MR, Riahi A, Forghanifard MM, Moghbeli M. WNT and NOTCH signaling pathways as activators for epidermal growth factor receptor in esophageal squamous cell carcinoma. Cell Mol Biol Lett 2018;23:42. https://doi.org/10.1186/s11658-018-0109-x. 30202417. PMC6122622.

Yang W, Yan HX, Chen L, Liu Q, He YQ, Yu LX, Zhang SH, Huang DD, Tang L, Kong XN, Chen C, Liu SQ, Wu MC, Wang HY. Wnt/beta-catenin signaling contributes to activation of normal and tumorigenic liver progenitor cells. Cancer Res 2008;68(11):4287–95. https://doi.org/10.1158/0008-5472.CAN-07-6691. 18519688.

# Drug resistance in gynecologic cancers: Emphasis on noncoding RNAs and drug efflux mechanisms

*Rama Rao Malla[a], Kiranmayi Patnala[b], Deepak Kakara Gift Kumar[a], and Rakshimitha Marni[a]*

[a]Cancer Biology Lab, Department of Biochemistry and Bioinformatics, Institute of Science, Gandhi Institute of Technology and Management (GITAM, Deemed University), Visakhapatnam, Andhra Pradesh, India [b]Department of Biotechnology, Institute of Science, Gandhi Institute of Technology and Management (GITAM, Deemed University), Visakhapatnam, Andhra Pradesh, India

## Abstract

Gynecologic cancers are the cancers that originate in different reproductive organs of a woman. They are common between the age of 30 and 75 years. The major aggressive gynecologic cancers include cervical, ovarian, and endometrial cancers. Every type of gynecologic cancers are having distinctive signs, symptoms, and risk factors as well as various strategies for avoiding the disease conditions. All women are at threat for these cancers, and the danger increases with age and lifestyle activities. Chemotherapy is the first-line treatment option for major gynecologic cancers. However, the most common risk factor for major gynecologic cancers is the development of drug resistance to first-line therapeutic management/strategies. Noncoding RNAs are nonprotein coding transcripts which are differentially expressed among the malignant and benign tumors of gynecologic cancers as well as normal tissues. Deregulated expression of noncoding RNAs promotes or suppresses cancer initiation or progression as well as drug resistance. In addition, drug efflux mechanisms and their signaling mechanisms are also responsible for drug resistance in gynecologic cancers. This chapter attempts to describe the findings and drug resistance mechanisms with special reference to noncoding RNAs and drug efflux mechanisms in three aggressive gynecologic cancers of women.

## Abbreviations

| | |
|---|---|
| **AKT** | protein kinase B |
| **BRCA** | breast cancer gene |
| **BRWD1** | bromodomain and WD repeat domain containing 1 |
| **BTG3** | BTG antiproliferation factor 3 |

R. Basha, S. Ahmad (eds.)
*Overcoming Drug Resistance in Gynecologic Cancers*
ISBN 978-0-12-824299-5
https://doi.org/10.1016/B978-0-12-824299-5.00018-6

| | |
|---|---|
| CC | cervical cancer |
| CIS | cisplatin |
| CSCs | cancer stem cells |
| DUOXA1 | double oxidase maturation factor 1 |
| EGFR | epidermal growth factor receptor |
| EMC | endometrial cancer |
| EMT | epithelial to mesenchymal transition |
| EREs | estrogen response elements |
| FBXL5 | F-box and leucine rich repeat protein |
| FOXO3 | forkhead box O3 |
| GSH | glutathione |
| GSK | glycogen synthase kinase |
| HK2 | hexokinase 2 |
| HPV | human papilloma virus |
| iASPP | inhibitor of apoptosis-stimulating protein of p53 |
| IGFR | insulin-like growth factor 1 |
| JAK/STAT | janus kinase/signal transducers and activators of transcription |
| lncRNAs | long noncoding RNAs |
| MALAT1 | metastasis associated lung adenocarcinoma transcript 1 |
| MAPK | mitogen-activated protein kinase |
| MDR1 | multidrug resistance protein 1 |
| MLH1 | mutL alpha |
| MMPs | matrix metallopeptidases |
| MMR | mismatch repair |
| MSI | microsatellite instability |
| OC | ovarian cancer |
| PAP | papanicolaou test |
| PARP-1 | poly (ADP) ribose polymerase 1 |
| Pgp | p-glycoprotein |
| PI3K | phosphoinositide 3-kinase |
| PTEN | phosphatase and tensin homolog |
| PTPRJ | protein tyrosine phosphatase receptor J |
| PTX | paclitaxel |
| ROS | reactive oxygen species |
| SLC3A2 | solute carrier family 3 member 2 |
| TCF | transcription factor |
| UTR | untranslated region |
| VEGF | vascular endothelial growth factor |
| XIAP | X-linked inhibitor of apoptosis protein |
| ZEB1 | zinc finger E-box-binding homeobox 1 |

## Conflict of interest

No potential conflicts of interest were disclosed.

## Introduction

Gynecologic cancers (GyCs) are the cancers that originate in different reproductive organs of a woman. In developing countries, patients with GyCs are adversely affected in terms of prognosis and clinical outcomes due to potential differences in the pathology and lack of awareness of screening facilities. They are extremely common in age among 30 and 75 years with a mean age of 45 years. The major aggressive gynecologic cancers include cervical, ovarian, and endometrial cancers. Every type of gynecologic cancer is having distinctive signs, symptoms, and risk factors as well as various management strategies for avoiding the disease conditions. Squamous cell carcinoma is by far the most common histologic type in cervical cancer, while adenocarcinomas are perhaps the most common histologic type in ovarian and endometrial cancers [1]. All women are at potential threat for these cancers, and this danger increases with age and lifestyle activities. However, the most widespread risk factor for all the major aggressive cancers is the development of drug resistance to first-line therapeutic management strategies.

Noncoding RNAs are nonprotein coding transcripts including long noncoding RNA (lncRNAs) with more than 200 nucleotides and microRNAs (miRNAs) with less than 200 nucleotides. The expression patterns of noncoding RNAs are different among the malignant and benign tumors as well as normal tissues. Deregulated expression of noncoding RNAs promotes or suppresses cancer initiation or progression as well as drug resistance [2]. Recently, aberrant expression of lncRNAs, as well as miRNAs, were reported in ovarian, cervical, and endometrial cancers, implying a significant role in their treatment. In addition, drug efflux mechanisms and their signaling mechanisms are also responsible for drug resistance in these gynecologic cancers. This chapter attempts to describe the finding and drug resistance mechanisms with special reference to noncoding RNAs and drug mechanisms in three aggressive gynecologic cancers of women.

## Cervical cancer

Cancer arising from the cervix is known as cervical cancer (CC). The cervix joins the vagina to the upper region of the uterus. Most of the women above the age of 30 years are prone to CC. Long-term infection due to the human papillomavirus (HPV) transmitted sexually leads to CC. CC ranks third among the leading gynecologic cancers in the developed countries [3–5]. The increasing incidences of CC is due to unusual sexual practices, intercourse at a very early age, multiple partners, less use of condoms, and also due to immune suppression about HIV infection [6]. Women infected with HIV are more prone to CC than non-HIV individuals.

CC is diagnosed by the Papanicolaou test (PAP) and human papillomavirus test (HPV). The PAP test helps in early diagnosis and also helps in detecting the early cancerous and cancerous cell lesions for better treatment. HPV test could also help in detecting infections caused by HPV that eventually leads to cancer. Along with these two tests, different molecular assays such as DNA and RNA-based tests are available to detect CC [7]. As per the guidelines laid down by the American Cancer Society (ACS), screening of CC should be carried out in women individually at an early age of 21 years. Women within the age group range of 21–29 are to be screened once in 3 years with PAP test. The co-testing strategy (PAP test and HPV test) is to be carried out in women aging between 30 and 59 years every 5 years [8]. Various drugs, viz., cisplatin, methotrexate, irinotecan, mitolactol, vincristine, melphalan, and cyclophosphamide, etc., are used to treat CC. Resistance to the anticancer drugs in CC is one of the major reasons for treatment failure and drug resistance in CC. Drug resistance can be due to several mutations in drug targets, increased drug detoxification systems, resistance towards apoptosis, and aberrant activation of various signaling pathways.

The therapeutic management of CC includes surgery, chemotherapy, and/or radiation therapy [9]. Even though there are many advances in early screening and diagnosis of CC, drug resistance, invasion, and metastasis leads to treatment failure and CC-related deaths [10]. Cervical cancer stem cells are one of the main causes for resistance to drugs (chemo- and radio-resistance) in CC. Various other pathways that are responsible for drug resistance in CC include Wnt signaling pathway [11, 12]; Sonic Hedgehog signaling pathway [13]; Notch signaling pathway [14–17]; EGFR signaling pathway [18]; JAK-STAT signaling pathway [19]; NF-kβ signaling pathway [20]; RAS signaling [21]; and few other miRNAs induce drug resistance in CC. The precise mechanism by which CC cells acquire chemoresistance is rather unclear.

Wnt signaling pathway is very complex and has a prominent role in the differentiation of cells, proliferation, and cell migration [22, 23]. DNA methylation and mutations in DNA in Wnt signaling pathway induce molecular changes that result in abnormal Wnt signaling pathway activation in CC cells [24–26]. The abnormal Wnt signaling pathway is due to the mutations in ligands, inhibitors, membrane receptors, and intracellular mediators. Peer-reviewed published studies have revealed that Wnt signaling genes are responsible for the pathogenesis of CC. Various components of Wnt signaling and Wnt ligands are characterized in CC. Cell lines and tissue samples were obtained from CC showed the downregulation of Wnt7A [27]. Also, Wnt10B mRNA [28], Wnt5A [29], Wnt11 [30], and Wnt14 [31] were found to be highly expressed in CC cell line, Hela S3. Pérez-Plasencia et al. considered the expression of mRNAs in a large number in HPV16 cells and results indicated a sharp rise in Wnt4 and Wnt8A when compared to the normal cells [32]. Simultaneously, the same group also stated a drastic increase in the Fzd2 and Fzd10 receptors expression and reduced expression of Fzd7 in cancer cells [33]. The work conducted by Rampias et al. and Bello et al. showed localization of β-catenin in the nucleus and cytoplasm of HPV-positive cells, whereas in the HPV-negative cells, β-catenin was accumulated in the plasma membrane and leads to tumor progression in CC [34–36]. The active and inactive forms of GSK3β, pGSK3β-Try$^{216}$, and pGSK3β-Ser$^9$, respectively, were studied by Rath et al. In CC cells, the active form of GSK3β was highly expressed in normal tissues, whereas in cancerous and precancerous lesions, it was less expressed. Microarray studies indicated the deregulated expression of Wnt signaling pathway genes, TCF7 and TCF3. Notably, TCF7 was highly upregulated and TCF3 was downregulated in CC cells and CC tumors [27]. The use of conventional chemotherapeutic drugs in advanced stages of CC leads to drug resistance due to the activation of various signaling pathways [37].

Two pathways, viz., death receptor pathway and mitochondrial pathway are involved in cell apoptosis. E5, E6, and E7 proteins decreased apoptosis in cancer cells by altering several signaling pathways. And HPV mobilizes various oncoproteins to alter the signaling pathways to decrease apoptosis in the host cells, and thus, leading to drug resistance [38, 39]. Cancer stem cells (CSCs) associated with CC are responsible for chemo-radio-resistance leading to metastasis, relapse of tumors, and treatment failure. CSCs in CC are minor masses of the cells found in the tumor which can proliferate rapidly and are inherently resistant to therapy [40]. Any form of therapy (chemo- or radiotherapy) can show its effects on well-differentiated cancerous cells, resulting in tumor shrinkage but CSCs remain a challenge. Studies have shown that active transcription factor AP-1 is a key regulator in the expression of HPV oncogenes E6 and E7. AP-1 activity is established by the members of Jun or Fos family proteins [41–43]. The CSCs isolated from the HPV16 CC cell line showed higher expression of P-1 in CSCs when compared to non-CSCs on the exposure to UV radiation. The proteins from the AP-1 family, viz., c-Fos, c-Jun, JunD, and JunB were highly expressed in CSCs when related to the non-CSCs contributing to radio-resistance [44].

# Drug resistance mechanisms in cervical cancer

Cisplatin (CIS) is the common chemotherapeutic drug used to treat CC. However, most of the women foster resistance to cisplatin over the due course of treatment. Long noncoding RNAs (lncRNAs) are the sequences of RNA having an approximate length of 200 nucleotides. These lncRNAs show a significant part in the emergence of resistance to drugs in CC cells by interacting with key apoptosis regulators and induce drug resistance. MALAT1 (a lncRNA) is extensively studied in CC cells as it is responsible for the activation of multiple signaling pathways [45–47]. MALAT1 promotes CIS resistance by resists apoptosis in cancerous cells by upregulating Bcl-2 and Bcl-xL, and by upregulating caspase-3 and caspase-8 [48]. Several studies have shown that PI3K/AKT pathway induces drug resistance by affecting many drug-resistant and antiapoptosis proteins [49]. Peer-reviewed studies suggested that downregulation of MALAT1 in CC cells sensitized CC cells to cisplatin and vice-versa by regulating BRWD1 and apoptosis. The major PI3K/AKT signaling pathway-related proteins, p-PI3K and p-AKT were upregulated due to the overexpression of MALAT1. Tumor suppressor lncRNA GAS5 promotes CIS resistance in CC cells by targeting pAkt via regulating miRNA-21 [50]. These studies suggest that lncRNAs promote cisplatin resistance in CC cells.

The microRNAs are small noncoding RNAs, which promote drug resistance in CC cells. For example, miRNA-7-5p promotes CIS resistance in CC cells by inhibiting the DNA repair mechanisms via modulating PARP-1 expression as well as enhancing the autophagy via reducing Bcl-2 expression [51]. The EMT inducer (iASPP) promotes CIS resistance in CC cells by increasing the expression of miRNA-21a via reducing the expression of FBXL5 and BTG3. This study suggests that miRNA-21a mediates iASPP-dependent CIS resistance in CC cells [52]. AJUBA is a LIM domain protein functioning in promoting cisplatin resistance in CC by upregulating the YAP and TAZ, the downstream targets [53]. Cancer cells develop CIS resistance by increasing the DNA repairability. CIS and PARP-1 inhibitor PJ34 amplified cisplatin sensitivity in CC by means of reducing cell propagation, increasing cell cycle arrest, and apoptosis as well as reducing invasion and metastasis. The combination also hindered β-catenin signaling as well as its targets c-Myc, cyclin D1 and MMPs suggesting a probable link amid drug resistance and damage repair mechanisms as well as WNT signaling in CC [54].

Paclitaxel (PTX) is a commonly used chemotherapeutic agent for CC treatment. Microarray analysis showed that downregulation of miRNA-125 promotes PTX resistance in CC cells by reducing the expression of STAT3 [55]. Autophagy enhances PTX resistance by promoting Warburg effect via the HIF-1α signaling in CC cells [56]. PVT1 promotes PTX resistance by enhancing the epigenetic silencing of miRNA-195 as well as modulating the EMT in CC cells [57]. Protein tyrosine phosphatase receptor J (PTPRJ) negatively regulates 5-fluorouracil resistance via JAK1/STAT3 pathway as well as its targets such as cyclin D1, BAX, VEGF, and MMP2 in CC cells [58]. These studies reveal that CC cells develop resistance to chemotherapeutic drugs through different mechanisms.

# Ovarian cancer

Ovarian cancer (OC) involves ovaries, fallopian tubes, and peritoneal layers/cavity; but not involving the mucosal epithelium [59]. As being a mostly asymptomatic cancer, early detection is rare and can be life-threatening due to late diagnosis [60, 61]. Concerning the ACS statistics, ovarian cancer is the fifth prominent cause for cancer-causing deaths than any other female gynecologic cancers [62]. The risk factors for developing ovarian cancer in a woman's lifetime include family history, breast cancer gene mutations, and/or infertility that occurs between the early menarche and late menopause [63]. Cytoreduction followed by the administration of platinum-based chemotherapeutic drugs is given as the standard therapy for patients with OC [64]. Early detection of ovarian cancer can provide a highly responsive treatment strategy [65]. Chemotherapy drugs such as alkylating agents, natural products, and platinum compounds are the foremost concerning medications for the treatment of patients with ovarian malignancy [66]. Resistance to the above class of chemotherapeutic drugs may be acquired due to the relative dose intensity, induction of the drug efflux receptor proteins such as p-glycoproteins, increase in phase 2 metabolic enzyme levels such as GSH, and increase in the DNA repair process also confers to the development of resistance to drugs [67]. Resistance development to chemotherapy within 2 years after the initial administration has remained an important concern for patients with OC [68].

## Drug resistance mechanisms in ovarian cancer

Resistance to cisplatin was the main hitch for the prediction of ovarian cancer. Nrf2 (nuclear factor, erythroid 2 like 2), a regulator of antioxidant proteins during oxidative stress and protect cells from oxidative damage. Recent studies reported that Nrf2 mediates CIS resistance in OCs by enhancing the expression of CD99 [69]. However, hexokinase 2 (HK2), which regulates the phosphorylation of glucose in the critical step in glycolysis. Zhang et al. [70] reported that HK2 functionally associates with CIS resistance by enhancing the phosphorylation of ERK1/2 as well as autophagy in OCs [70]. Cui et al. [71] demonstrated that zinc finger E-box-binding homeobox 1 (ZEB1), which is a vital promoter of various cancers, is also responsible for CIS resistance in OC cells. The authors claimed that ZEB1 confers CIS resistance in OCs by reducing the expression of SLC3A2 expression [71]. In contrast, FOXO3 reverses the CIS resistance in OCs by promoting the arrest of cell cycle and apoptosis through the enhanced transcription of pro-apoptotic proteins FasL and Bim [72]. Meng et al. [73] asserted that double oxidase maturation factor 1(DUOXA1) is involved in the CIS resistance in OC cells via ROS production. The increased production of ROS increases the CIS resistance by enhancing the activity of the ATR-Chk1 pathway [73].

Advanced studies reported that in OCs CIS resistance is modulated by miRNAs. Zhang et al. [74] showed that miRNA-1294 mediates the underlying mechanism of CIS resistance in OCs. The authors demonstrated that miRNA-1294 promotes CIS resistance by regulating IGFR expression as well as enhancing EMT, invasion, and migration of tumor cells [74]. In contrast, miRNA-199a-3p reverses the CIS resistance in OC cells by inducing arrest at the G1 phase in the cell cycle and reduced expression of integrin $\beta 8$ via binding at its 3-UTR [75]. Similarly, miRNA-139 reverses the CIS resistance in OC cells by decreasing the expression of c-Jun and Bcl2 via inhibiting the interaction with ATF2 [76]. These studies suggest that miRNAs differentially regulate the CIS resistance in OC cells via divergent mechanisms.

# Endometrial cancer

The malignancy of the uterus that originates in the lining of the epithelium of the uterus is called endometrial cancer (EMC). Even though the diagnosis of EMC is decent, and most of the affected cases are identified at the progressive stage with a tumor and followed by metastases [77]. The significant problem for these patients is, even if they are cured with chemo- and/or hormonal therapies, the existence proportion is too low. Overall, EMC is the common reproductive cancer and positions as the fourth common neoplasm and the eighth most important cause of death in females. Most of the uterine (endometrial) cancers are related to this and come under adenocarcinoma, type 1. This is characterized by unopposed estrogen exposure and the presence of estrogen leads to the commencement of the premalignant stage of the disease. The type 2 uterine malignancies are not as frequent and of nonendometrioid histology.

The rampant increase of EMC in the western world has lead biologists and oncologists to recognize active preventive measures and molecular markers. The foremost treatment options for advanced-stage and recurrent EMC are chemo- and/or hormone therapies. In recent years, there have been important developments in treating patients with EMC. Despite these developments in the management of EMC, the total existence of patients was not enhanced as a substantial number of affected cases have tumor refractory to these treatments. Henceforth, resolving the sensitivity/resistance is becoming all the time more significant for the individualization of EMC therapy.

# Drug resistance mechanisms in endometrial cancer

Chemo-resistance is responsible for the failure of treatment and causing a substantial mortality rate in patients with advanced stage of cancer [78]. This obstacle can be overcome by different modifications like, the upsurge of efflux transporters to discard drugs and a decline in the cell division that limits the effect of drug compounds that are targeted to the arrest of mitosis. In the following section, we shall consider the peer-reviewed literature of identified mechanisms of chemoresistance specific to EMC. The overexpression of the MDR1 gene is linked with resistance against paclitaxel. This attainment is clearly described by an increase in the level of the P-glycoprotein (P-gp) efflux pump, thereby excluding the occurrence of chemotherapeutic drugs in EMC [79]. The increase in the levels of copper pumps is related to resistance by the platinum compounds. Concerning endometrial cancer, copper-transporter ATP7B is reported to be overexpressed and related to CIS resistance [80].

From a long time, the DNA repair mechanisms are accompanying with chemoresistance in EMC. The standard norm of a mismatch repair (MMR) is to identify a mismatched or an unmatched DNA base, repair it, and correctly unite the DNA [81]. During the therapy with platinum compounds, the mismatch repair system is becoming incapable to comprehend the repairs of mismatched DNA, thereby causing apoptosis. The MLH1 gene is silenced by the hypermethylation of the promoter, which in turn, initiates the microsatellite instability (MSI) in EMC. Works done by Kanaya et al. in 2003, put forward that hypermethylation of MLH1 promoter is recurrent in normal endometrium neighboring to the cancer cells, thereby backing up the idea that hypermethylation of MMR genes is one of the starting steps which activates several genetic happenings in EMC [82]. The double-strand DNA breaks are repaired by the BRCA1 and BRCA2 genes involving the repair mechanism known as homologous recombination. Remarkably, gene mutations of BRCA1 and BRCA2 are also accompanying by an augmented risk of EMC. It was recurrently observed in women who were treated with tamoxifen [83].

Various signaling pathways play important parts in the mechanisms of drug resistance in endometrial cancers. The PI3K/AKT signaling pathway is one of the key cascades, that is either regularly mutated or hyperactivated at diverse levels in EMCs. It was also shown that only AKT1 and AKT2 isoforms are accountable for the gaining of resistance alongside paclitaxel and cisplatin, whereas all isoforms of AKT rise resistance to doxorubicin in EMCs [84]. PTEN is a tumor suppressor lipid phosphatase performing undesirably on the PI3K path through its capability to dephosphorylate PIP3 to PIP2, thereby monitoring the action of PI3K downstream targets. It is important to note that PTEN undergoes a high percentage of modifications in EMC. Downregulation of PTEN clues to an intensification of resistance against the platinum compounds [85, 86]. XIAP, an apoptotic inhibitor, is involved in the PI3K/AKT pathway to guard cells by performing as a promoter of AKT activity through collaboration with PTEN and regulating negatively at the protein level. XIAP is also intricate in the resistance against cisplatin and doxorubicin [87].

One more type of oncogenes we can take into consideration in these cancers are the EGFR (HER-1) and ErbB2 (HER-2), which are receptors tyrosine kinases present on the cell surface. Both the genes are overexpressed in advanced stages of endometrial cancers. These two receptors can trigger several signal-related pathways, with both the PI3K and MAPK that provide cell growth to tumors and add to the drug resistance in reproductive organs' cancers. The overexpression of EGFR and ErbB2 alongside the initiation of PI3K and MAPK pathways enhances the resistance to paclitaxel and cisplatin in EMC [88]. Estrogen goes and binds to receptor, wherein it dimerizes and helps in moving from cytoplasm to the nucleus. In the nucleus, it binds to EREs and acts as a transcription factor. Estrogen also binds to a G-coupled protein receptor (GPR), the GPR30 receptor individually without the intervention of estrogen receptors. The GRP78 is activated in the cells affected by EMC due to the activation of estrogen, thus avoiding apoptosis and leading to resistance for cisplatin and paclitaxel [89].

Changes in p53 are also related to platinum compounds' resistance in EMCs. In general, p53 is a well-studied tumor suppressor in endometrial cancer and its prominence in various cancer types is undisputable. An improved understanding of all the proposed mechanisms may permit well-organized therapies that can be given to the affected people. Long noncoding RNA, HOTAIR would confer CIS resistance in EMCs by regulating autophagy through changing the expression of MDR, Beclin 1, and P-gp genes [90] (Table 1).

TABLE 1   Mechanisms involved in resistance to drugs for endometrial cancers.

| Mechanism of action | Protein affected | Regulation | Drug(s) involved | References |
| --- | --- | --- | --- | --- |
| Efflux transporters | P-glycoprotein | Up | Taxanes, Doxorubicin | [79] |
| Efflux transporters | ATP7A/ATP7B | Up | Platinum | [80] |
| DNA repair | MMR | Down | Platinum | [82, 83] |
| Signaling pathways | AKT | Up | Doxorubicin, Taxanes, Platinum | |
| Signaling pathways | PTEN | Down | Platinum | [85, 86] |
| Signaling pathways | XIAP | Up | Platinum, Taxanes, Doxorubicin | [87] |
| Signaling pathways | ErbB2 | Up | Platinum, Taxanes | [88] |
| Signaling pathways | GRP78 | Up | Platinum, Taxanes | [89] |
| Signaling pathways | BCL-2 | Up | Platinum | [91] |
| Long noncoding | HOTAI | Up | Cisplatin | [90] |

# Conclusions

The cancers of various reproductive organs of a woman are known as gynecologic cancers. They mainly include cervical, ovarian, and endometrial cancers with distinctive signs, symptoms, and risk factors. Available chemotherapy options with cisplatin, paclitaxel, and 5-fluorocil are the first-line treatments for these gynecologic cancers. However, drug resistance development is the most common risk factor for these gynecologic cancers. Noncoding RNAs are nonprotein-coding transcripts, which regulate drug resistance through different mechanisms/targets. Apart from that, drug efflux mechanisms and their signaling mechanisms also promote drug resistance in these gynecologic cancers. Overall, the understanding of recent findings and drug resistance mechanisms with special reference to noncoding RNAs and drug efflux mechanisms in these aggressive gynecologic cancers of women are useful strategies for improved design of therapeutic options.

## Acknowledgments

The authors are grateful to the Institute of Science, Gandhi Institute of Technology and Management, Visakhapatnam, Andhra Pradesh, India, for research support.

## Funding

The authors thank the UGC-DAE-CSR-KC/CRS/19/RB-04/1047, Kolkata, for providing financial assistance.

# References

[1] Dhakal HP, Pradhan M. Histological pattern of gynecological cancers. JNMA J Nepal Med Assoc 2009;48:301–5.

[2] Zhao M, Qiu Y, Yang B, Sun L, Hei K, Du X, Li Y. Long non-coding RNAs involved in gynecological cancer. Int J Gynecol Cancer 2014;24:1140–5.

[3] Tsikouras P, Zervoudis S, Manav B, Tomara E, Iatrakis G, Romanidis C, Bothou A, Galazios G. Cervical cancer: screening, diagnosis and staging. J BUON 2016;21:320–5.

[4] Moshkovich O, Lebrun-Harris L, Makaroff L, Chidambaran P, Chung M, Sripipatana A, Lin SC. Challenges and opportunities to improve cervical cancer screening rates in US health centers through patient-centered medical home transformation. Adv Prev Med 2015;2015:182073.

[5] Maguire R, Kotronoulas G, Simpson M, Paterson C. A systematic review of the supportive care needs of women living with and beyond cervical cancer. Gynecol Oncol 2015;136:478–90.

[6] Gustafsson L, Pontén J, Bergström R, Adami HO. International incidence rates of invasive cervical cancer before cytological screening. Int J Cancer 1997;71:159–65.

[7] Lie AK, Kristensen G. Human papillomavirus E6/E7 mRNA testing as a predictive marker for cervical carcinoma. Expert Rev Mol Diagn 2008;8:405–15.

[8] Saslow D, Solomon D, Lawson HW, Killackey M, Kulasingam SL, Cain J, Garcia FA, Moriarty AT, Waxman AG, Wilbur DC, Wentzensen N, Downs Jr LS, Spitzer M, Moscicki AB, Franco EL, Stoler MH, Schiffman M, Castle PE, Myers ER. American Cancer Society, American Society for Colposcopy and Cervical Pathology, and American Society for Clinical Pathology screening guidelines for the prevention and early detection of cervical cancer. Am J Clin Pathol 2012;137:516–42.

[9] Barra F, Lorusso D, Maggiore ULR, Ditto A, Bogani G, Raspagliesi F, Ferrero S. Investigational drugs for the treatment of cervical cancer. Expert Opin Investig Drugs 2017;26:389–402.

[10] Wang L, Dai G, Yang J, Wu W, Zhang W. Cervical cancer cell growth, drug resistance, and epithelial-mesenchymal transition are suppressed by γ-secretase inhibitor RO4929097. Med Sci Monit 2018;24:4046.

[11] Xu H, Sun Y, Zeng L, Li Y, Hu S, He S, Chen H, Zou Q, Luo B. Inhibition of cytosolic phospholipase A2 alpha increases chemosensitivity in cervical carcinoma through suppressing β-catenin signaling. Cancer Biol Ther 2019;20:912–21.

[12] Zhang Y, Liu B, Zhao Q, Hou T, Huang X. Nuclear localizaiton of β-catenin is associated with poor survival and chemo-/radioresistance in human cervical squamous cell cancer. Int J Clin Exp Pathol 2014;7:3908–17.

[13] Sharma A, De R, Javed S, Srinivasan R, Pal A, Bhattacharyya S. Sonic hedgehog pathway activation regulates cervical cancer stem cell characteristics during epithelial to mesenchymal transition. J Cell Physiol 2019;23415726–41.

[14] Wang Z, Li Y, Kong D, Sarkar FH. The role of Notch signaling pathway in epithelial-mesenchymal transition (EMT) during development and tumor aggressiveness. Curr Drug Targets 2010;11:745–51.

[15] Li LC, Peng Y, Liu YM, Wang LL, Wu XL. Gastric cancer cell growth and epithelial-mesenchymal transition are inhibited by γ-secretase inhibitor DAPT. Oncol Lett 2014;7:2160–4.

[16] Yang J, Guo W, Wang L, Yu L, Mei H, Fang S, Chen A, Liu Y, Xia K, Liu G. Notch signaling is important for epithelial-mesenchymal transition induced by low concentrations of doxorubicin in osteosarcoma cell lines. Oncol Lett 2017;13:2260–8.

[17] Wang L, Dai G, Yang J, Wu W, Zhang W. Cervical cancer cell growth, drug resistance, and epithelial-mesenchymal transition are suppressed by y-secretase inhibitor RO4929097. Med Sci Monit 2018;24:4046–53.

[18] Hugo de Almeida V, Guimarães IDS, Almendra LR, Rondon AMR, Tilli TM, de Melo AC, Sternberg C, Monteiro RQ. Positive crosstalk between EGFR and the TF-PAR2 pathway mediates resistance to cisplatin and poor survival in cervical cancer. Oncotarget 2018;9:30594–609.

[19] Rawlings JS, Rosler KM, Harrison DA. The JAK/STAT signaling pathway. J Cell Sci 2004;117:1281–3.

[20] Maruyama W, Shirakawa K, Matsui H, Matsumoto T, Yamazaki H, Sarca AD, Kazuma Y, Kobayashi M, Shindo K, Takaori-Kondo A. Classical NF-κB pathway is responsible for APOBEC3B expression in cancer cells. Biochem Biophys Res Commun 2016;478:1466–71.

[21] Gurung AB, Bhattacharjee A. Significance of ras signaling in cancer and strategies for its control. Oncol Hematol Rev 2015;11:147–152.

[22] Niehrs C. The complex world of WNT receptor signalling. Nat Rev Mol Cell Biol 2012;13:767–79.

[23] Guo Y, Xiao L, Sun L, Liu F. Wnt/beta-catenin signaling: a promising new target for fibrosis diseases. Physiol Res 2012;61:337–46.

[24] Rodríguez-Sastre MA, González-Maya L, Delgado R, Lizano M, Tsubaki G, Mohar A, García-Carrancá A. Abnormal distribution of E-cadherin and beta-catenin in different histologic types of cancer of the uterine cervix. Gynecol Oncol 2005;97:330–6.

[25] Clevers H, Nusse R. Wnt/β-catenin signaling and disease. Cell 2012;149:1192–205.

[26] Yang M, Wang M, Li X, Xie Y, Xia X, Tian J, Zhang K, Tang A. Wnt signaling in cervical cancer? J Cancer 2018;9:1277–86.

[27] Ramos-Solano M, Meza-Canales ID, Torres-Reyes LA, Alvarez-Zavala M, Alvarado-Ruíz L, Rincon-Orozco B, Garcia-Chagollan M, Ochoa-Hernández AB, Ortiz-Lazareno PC, Rösl F, Gariglio P, Jave-Suárez LF, Aguilar-Lemarroy A. Expression of WNT genes in cervical cancer-derived cells: implication of WNT7A in cell proliferation and migration. Exp Cell Res 2015;335:39–50.

[28] Kirikoshi H, Katoh M. Expression and regulation of WNT10B in human cancer: up-regulation of WNT10B in MCF-7 cells by beta-estradiol and down-regulation of WNT10B in NT2 cells by retinoic acid. Int J Mol Med 2002;10:507–11.

[29] Lin L, Liu Y, Zhao W, Sun B, Chen Q. Wnt5A expression is associated with the tumor metastasis and clinical survival in cervical cancer. Int J Clin Exp Pathol 2014;7:6072–8.

[30] Wei H, Wang N, Zhang Y, Wang S, Pang X, Zhang J, Luo Q, Su Y, Zhang S. Clinical significance of Wnt-11 and squamous cell carcinoma antigen expression in cervical cancer. Med Oncol 2014;31:933.

[31] Kirikoshi H, Sekihara H, Katoh M. Expression of WNT14 and WNT14B mRNAs in human cancer, up-regulation of WNT14 by IFNgamma and up-regulation of WNT14B by beta-estradiol. Int J Oncol 2001;19:1221–5.

[32] Pérez-Plasencia C, Vázquez-Ortiz G, López-Romero R, Piña-Sanchez P, Moreno J, Salcedo M. Genome wide expression analysis in HPV16 cervical cancer: identification of altered metabolic pathways. Infect Agent Cancer 2007;2:16.

[33] Perez-Plasencia C, Duenas-Gonzalez A, Alatorre-Tavera B. Second hit in cervical carcinogenesis process: involvement of wnt/beta catenin pathway. Int Arch Med 2008;1:10.

[34] Rampias T, Boutati E, Pectasides E, Sasaki C, Kountourakis P, Weinberger P, Psyrri A. Activation of Wnt signaling pathway by human papillomavirus E6 and E7 oncogenes in HPV16-positive oropharyngeal squamous carcinoma cells. Mol Cancer Res 2010;8:433–43.

[35] Bello JO, Nieva LO, Paredes AC, Gonzalez AM, Zavaleta LR, Lizano M. Regulation of the Wnt/β-catenin signaling pathway by human papillomavirus E6 and E7 oncoproteins. Viruses 2015;7:4734–55.

[36] Shinohara A, Yokoyama Y, Wan X, Takahashi Y, Mori Y, Takami T, Shimokawa K, Tamaya T. Cytoplasmic/nuclear expression without mutation of exon 3 of the beta-catenin gene is frequent in the development of the neoplasm of the uterine cervix. Gynecol Oncol 2001;82:450–5.

[37] Chen J. Signaling pathways in HPV-associated cancers and therapeutic implications. Rev Med Virol 2015;25 (Suppl 1):24–53.

[38] Xie X, Piao L, Bullock BN, Smith A, Su T, Zhang M, Teknos TN, Arora PS, Pan Q. Targeting HPV16 E6-p300 interaction reactivates p53 and inhibits the tumorigenicity of HPV-positive head and neck squamous cell carcinoma. Oncogene 2014;33:1037–46.

[39] Butz K, Denk C, Ullmann A, Scheffner M, Hoppe-Seyler F. Induction of apoptosis in human papillomavirus-positive cancer cells by peptide aptamers targeting the viral E6 oncoprotein. Proc Natl Acad Sci U S A 2000;97:6693–7.

[40] Pajonk F, Vlashi E, McBride WH. Radiation resistance of cancer stem cells: the 4 R's of radiobiology revisited. Stem Cells 2010;28:639–48.

[41] Prusty BK, Das BC. Constitutive activation of transcription factor AP-1 in cervical cancer and suppression of human papillomavirus (HPV) transcription and AP-1 activity in HeLa cells by curcumin. Int J Cancer 2005;113:951–60.

[42] Butz K, Hoppe-Seyler F. Transcriptional control of human papillomavirus (HPV) oncogene expression: composition of the HPV type 18 upstream regulatory region. J Virol 1993;67:6476–86.

[43] Angel P, Karin M. The role of Jun, Fos and the AP-1 complex in cell-proliferation and transformation. Biochim Biophys Acta 1991;1072:129–57.

[44] Tyagi A, Vishnoi K, Kaur H, Srivastava Y, Roy BG, Das BC, Bharti AC. Cervical cancer stem cells manifest radioresistance: association with upregulated AP-1 activity. Sci Rep 2017;7:4781.

[45] Wang N, Hou MS, Zhan Y, Shen XB, Xue HY. MALAT1 promotes cisplatin resistance in cervical cancer by activating the PI3K/AKT pathway. Eur Rev Med Pharmacol Sci 2018;22:7653–9.

[46] Wu XS, Wang XA, Wu WG, Hu YP, Li ML, Ding Q, Weng H, Shu YJ, Liu TY, Jiang L, Cao Y, Bao RF, Mu JS, Tan ZJ, Tao F, Liu YB. MALAT1 promotes the proliferation and metastasis of gallbladder cancer cells by activating the ERK/MAPK pathway. Cancer Biol Ther 2014;15:806–14.

[47] Guo F, Li Y, Liu Y, Wang J, Li Y, Li G. Inhibition of metastasis-associated lung adenocarcinoma transcript 1 in CaSki human cervical cancer cells suppresses cell proliferation and invasion. Acta Biochim Biophys Sin Shanghai 2010;42:224–9.

[48] Kremer KN, Peterson KL, Schneider PA, Meng XW, Dai H, Hess AD, Smith BD, Rodriguez-Ramirez C, Karp JE, Kaufmann SH, Hedin KE. CXCR4 chemokine receptor signaling induces apoptosis in acute myeloid leukemia cells via regulation of the Bcl-2 family members Bcl-XL, Noxa, and Bak. J Biol Chem 2013;288:22899–914.

[49] Fu X, Feng J, Zeng D, Ding Y, Yu C, Yang B. PAK4 confers cisplatin resistance in gastric cancer cells via PI3K/Akt- and MEK/ERK-dependent pathways. Biosci Rep 2014;34:59–67.

[50] Wen Q, Liu Y, Lyu H, Xu X, Wu Q, Liu N, Yin Q, Li J, Sheng X. Long noncoding RNA GAS5, which acts as a tumor suppressor via microRNA 21, regulates cisplatin resistance expression in cervical cancer. Int J Gynecol Cancer 2017;27:1096–108.

[51] Yang F, Guo L, Cao Y, Li S, Li J, Liu M. MicroRNA-7-5p promotes cisplatin resistance of cervical cancer cells and modulation of cellular energy homeostasis by regulating the expression of the PARP-1 and BCL2 genes. Med Sci Monit 2018;24:6506–16.

[52] Xiong Y, Sun F, Dong P, Watari H, Yue J, Yu MF, Lan CY, Wang Y, Ma ZB. iASPP induces EMT and cisplatin resistance in human cervical cancer through miR-20a-FBXL5/BTG3 signaling. J Exp Clin Cancer Res 2017;36:48.

[53] Bi L, Ma F, Tian R, Zhou Y, Lan W, Song Q, Cheng X. AJUBA increases the cisplatin resistance through hippo pathway in cervical cancer. Gene 2018;644:148–54.

[54] Mann M, Kumar S, Sharma A, Chauhan SS, Bhatla N, Kumar S, Bakhshi S, Gupta R, Kumar L. PARP-1 inhibitor modulate β-catenin signaling to enhance cisplatin sensitivity in cancer cervix. Oncotarget 2019;10:4262–75.

[55] Fan Z, Cui H, Yu H, Ji Q, Kang L, Han B, Wang J, Dong Q, Li Y, Yan Z, Yan X, Zhang X, Lin Z, Hu Y, Jiao S. MiR-125a promotes paclitaxel sensitivity in cervical cancer through altering STAT3 expression. Oncogenesis 2016;5, e197.

[56] Peng X, Gong F, Chen Y, Jiang Y, Liu J, Yu M, Zhang S, Wang M, Xiao G, Liao H. Autophagy promotes paclitaxel resistance of cervical cancer cells: involvement of Warburg effect activated hypoxia-induced factor 1-α-mediated signaling. Cell Death Dis 2014;5, e1367.

[57] Shen CJ, Cheng YM, Wang CL. LncRNA PVT1 epigenetically silences miR-195 and modulates EMT and chemoresistance in cervical cancer cells. J Drug Target 2017;25:637–44.

[58] Yan CM, Zhao YL, Cai HY, Miao GY, Ma W. Blockage of PTPRJ promotes cell growth and resistance to 5-FU through activation of JAK1/STAT3 in the cervical carcinoma cell line C33A. Oncol Rep 2015;33:1737–44.

[59] PDQ Adult Treatment Editorial Board. Ovarian epithelial, fallopian tube, and primary peritoneal cancer treatment (PDQ®). Patient Version, Bethesda, MD: PDQ Cancer Information Summaries, National Cancer Institute (US); 2002.

[60] Momenimovahed Z, Tiznobaik A, Taheri S, Salehiniya H. Ovarian cancer in the world: epidemiology and risk factors. Int J Womens Health 2019;11:287–99.

[61] Halkia E, Spiliotis J, Sugarbaker P. Diagnosis and management of peritoneal metastases from ovarian cancer. Gastroenterol Res Pract 2012;2012:541842.

[62] Torre LA, Trabert B, DeSantis CE, Miller KD, Samimi G, Runowicz CD, Gaudet MM, Jemal A, Siegel RL. Ovarian cancer statistics, 2018. CA Cancer J Clin 2018;68:284–96.

[63] Daniilidis A, Karagiannis V. Epithelial ovarian cancer. Risk factors, screening and the role of prophylactic oophorectomy. Hippokratia 2007;11:63–6.

[64] Petrillo M, Anchora LP, Scambia G, Fagotti A. Cytoreductive surgery plus platinum-based hyperthermic intraperitoneal chemotherapy in epithelial ovarian cancer: a promising integrated approach to improve locoregional control. Oncologist 2016;21:532–4.

[65] Chandra A, Pius C, Nabeel M, Nair M, Vishwanatha JK, Ahmad S, Basha R. Ovarian cancer: current status and strategies for improving therapeutic outcomes. Cancer Med 2019;8:7018–31.

[66] McGuire III WP, Markman M. Primary ovarian cancer chemotherapy: current standards of care. Br J Cancer 2003;89(Suppl 3):S3–8.

[67] Johnson SW, Ozols RF, Hamilton TC. Mechanisms of drug resistance in ovarian cancer. Cancer 1993;71:644–9.

[68] Cristea M, Han E, Salmon L, Morgan RJ. Practical considerations in ovarian cancer chemotherapy. Ther Adv Med Oncol 2010;2:175–87.

[69] Wu J, Zhang L, Li H, Wu S, Liu Z. Nrf2 induced cisplatin resistance in ovarian cancer by promoting CD99 expression. Biochem Biophys Res Commun 2019;518:698–705.

[70] Zhang XY, Zhang M, Cong Q, Zhang MX, Zhang MY, Lu YY, Xu CJ. Hexokinase 2 confers resistance to cisplatin in ovarian cancer cells by enhancing cisplatin-induced autophagy. Int J Biochem Cell Biol 2018;95:9–16.

[71] Cui Y, Qin L, Tian D, Wang T, Fan L, Zhang P, Wang Z. ZEB1 promotes chemoresistance to cisplatin in ovarian cancer cells by suppressing SLC3A2. Chemotherapy 2018;63:262–71.

[72] Lu M, Chen X, Xiao J, Xiang J, Yang L, Chen D. FOXO3a reverses the cisplatin resistance in ovarian cancer. Arch Med Res 2018;49:84–8.

[73] Meng Y, Chen CW, Yung MMH, Sun W, Sun J, Li Z, Li J, Li Z, Zhou W, Liu SS, Cheung ANY, Ngan HYS, Braisted JC, Kai Y, Peng W, Tzatsos A, Li Y, Dai Z, Zheng W, Chan DW, Zhu W. DUOXA1-mediated ROS production promotes cisplatin resistance by activating ATR-Chk1 pathway in ovarian cancer. Cancer Lett 2018;428:104–16.

[74] Zhang Y, Huang S, Guo Y, Li L. MiR-1294 confers cisplatin resistance in ovarian cancer cells by targeting IGF1R. Biomed Pharmacother 2018;106:1357–63.

[75] Cui Y, Wu F, Tian D, Wang T, Lu T, Huang X, Zhang P, Qin L. miR-199a-3p enhances cisplatin sensitivity of ovarian cancer cells by targeting ITGB8. Oncol Rep 2018;39:1649–57.

[76] Jiang Y, Jiang J, Jia H, Qiao Z, Zhang J. Recovery of miR-139-5p in ovarian cancer reverses cisplatin resistance by targeting C-Jun. Cell Physiol Biochem 2018;51:129–41.

[77] Plataniotis G, Castiglione M. Endometrial cancer: ESMO clinical practice guidelines for diagnosis, treatment and follow-up. Ann Oncol 2010;21(Suppl 5):v41–5.

[78] Longley DB, Johnston PG. Molecular mechanisms of drug resistance. J Pathol 2005;205:275–92.

[79] Schneider J, Efferth T, Centeno MM, Mattern J, Rodríguez-Escudero FJ, Volm M. High rate of expression of multidrug resistance-associated P-glycoprotein in human endometrial carcinoma and normal endometrial tissue. Eur J Cancer 1993;29a:554–8.

[80] Samimi G, Katano K, Holzer AK, Safaei R, Howell SB. Modulation of the cellular pharmacology of cisplatin and its analogs by the copper exporters ATP7A and ATP7B. Mol Pharmacol 2004;66:25–32.

[81] Kunkel TA, Erie DA. DNA mismatch repair. Annu Rev Biochem 2005;74:681–710.

[82] Kanaya T, Kyo S, Maida Y, Yatabe N, Tanaka M, Nakamura M, Inoue M. Frequent hypermethylation of MLH1 promoter in normal endometrium of patients with endometrial cancers. Oncogene 2003;22:2352–60.

[83] Segev Y, Iqbal J, Lubinski J, Gronwald J, Lynch HT, Moller P, Ghadirian P, Rosen B, Tung N, Kim-Sing C, Foulkes WD, Neuhausen SL, Senter L, Singer CF, Karlan B, Ping S, Narod SA. The incidence of endometrial cancer in women with BRCA1 and BRCA2 mutations: an international prospective cohort study. Gynecol Oncol 2013;130:127–31.

[84] Girouard J, Lafleur MJ, Parent S, Leblanc V, Asselin E. Involvement of Akt isoforms in chemoresistance of endometrial carcinoma cells. Gynecol Oncol 2013;128:335–43.

[85] Oda K, Stokoe D, Taketani Y, McCormick F. High frequency of coexistent mutations of PIK3CA and PTEN genes in endometrial carcinoma. Cancer Res 2005;65:10669–73.

[86] Tashiro H, Blazes MS, Wu R, Cho KR, Bose S, Wang SI, Li J, Parsons R, Ellenson LH. Mutations in PTEN are frequent in endometrial carcinoma but rare in other common gynecological malignancies. Cancer Res 1997;57:3935–40.

[87] Turner S, Gagnon V, Parent S, Asselin E. X-linked inhibitor of apoptosis protein and Akt: two anti-apoptotic proteins involved in chemoresistance of endometrial cancer cells. Cancer Res 2007;67:3589.

[88] Mori N, Kyo S, Nakamura M, Hashimoto M, Maida Y, Mizumoto Y, Takakura M, Ohno S, Kiyono T, Inoue M. Expression of HER-2 affects patient survival and paclitaxel sensitivity in endometrial cancer. Br J Cancer 2010;103:889–98.

[89] Luvsandagva B, Nakamura K, Kitahara Y, Aoki H, Murata T, Ikeda S, Minegishi T. GRP78 induced by estrogen plays a role in the chemosensitivity of endometrial cancer. Gynecol Oncol 2012;126:132–9.

[90] Sun MY, Zhu JY, Zhang CY, Zhang M, Song YN, Rahman K, Zhang LJ, Zhang H. Autophagy regulated by lncRNA HOTAIR contributes to the cisplatin-induced resistance in endometrial cancer cells. Biotechnol Lett 2017;39:1477–84.

[91] Rouette A, Parent S, Girouard J, Leblanc V, Asselin E. Cisplatin increases B-cell-lymphoma-2 expression via activation of protein kinase C and Akt2 in endometrial cancer cells. Inter J Cancer 2012;130:1755–67.

CHAPTER

# 7

# Genetic polymorphisms in gynecologic cancers

*Ketevani Kankava, Eka Kvaratskhelia, and Elene Abzianidze*

Department of Molecular and Medical Genetics, Tbilisi State Medical University, Tbilisi, Georgia

## Abstract

The role of genetic predisposition in cancer pathogenesis is well known. Despite the fact that in many cases exact mechanisms cannot be understood, overall observational and experimental data confirm the role of hereditary factors in cancer development and progression. Gynecologic cancers are not an exception. Increased risk for these tumors in the families of affected individuals lead to intensive research aggravated by huge data from GWAS. It seems like an insight into genetic pathways not only helps in understanding the disease development, but also designs more precise treatment options with better results and less side effects. Currently, there are no many clinical decisions in gynecologic cancer management that need to be based on genetic predisposition analysis, although increasing interest and availability of new technologies will probably soon make incorporation of hereditary factors in management possible. As expected, the genetic variants causing markedly elevated risk are rather rare, but those, associated with mild or moderate risk are numerous, quite common and especially important when analyzed in clusters. There are peculiarities of large-scale population studies, which determine a need of time for development and translation of significant scientific knowledge, but at this point, increased interest has resulted in rapid accumulation of reproducible data. The main approach of medicine is treating people, not the diseases. Heredity is a part of each patient and given the current progress in science and medicine, the era when understanding the genetic background will become a part of every physician's life approaching.

## Abbreviations

| | |
|---|---|
| **DNA** | deoxyribonucleic acid |
| **GWAS** | Genome-wide association studies |
| **HPV** | Human papilloma virus |
| **lncRNA** | long noncoding RNA |
| **MHC** | major histocompatibility complex |
| **MMR** | mismatch repair |
| **RNA** | ribonucleic acid |
| **PCOS** | polycystic ovary syndrome |
| **SNP** | single nucleotide polymorphism |

R. Basha, S. Ahmad (eds.)
*Overcoming Drug Resistance in Gynecologic Cancers*
ISBN 978-0-12-824299-5
https://doi.org/10.1016/B978-0-12-824299-5.00009-5

## Conflict of interest

No potential conflicts of interest were disclosed.

# Introduction

Identification of genetic contribution in carcinogenesis was an important step in understanding the diversity in tumor progression, response to treatment, or prognosis. A small percentage of gynecologic cancers have known genetic contributors [1]. This number is expected to be far more if research succeeds to identify all possible polygenic risk determinants in cancers that are believed to be sporadic [2]. Genome-wide association studies (GWAS) have brought some light into the complicated field of genomic background of gynecologic cancers and at the same time raised topics for further investigation. Most significant genes and loci currently known for associations with gynecologic cancers are shown in Fig. 1.

Interpretation of huge data from population studies remains a challenge. Especially when number of confounding effects need to be investigated independently and in combinations. Large-scale studies are mostly limited to populations with European ancestry [6] and common tumor types, with an exception of cervical cancer studies, where research of Asian population predominates [7–12]. Some recent studies investigating shared and unique risk loci in different populations do also appear [13–15]. Current understanding of carcinogenesis clearly delineates pathogenetic backgrounds of tumors arising at the same site but having different histology. Whether or not the roots of these molecular mechanisms could reach genetic polymorphisms still needs to be discovered.

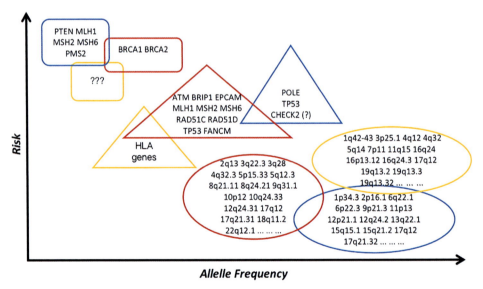

FIG. 1    Schematic diagram of genes and loci associated with gynecologic cancer risk. (Blue shapes show risk loci for endometrial carcinoma, red shapes—ovarian carcinoma, yellow shapes—cervical carcinoma. Rectangle contains high-risk low-frequency genes, triangles—moderate-risk moderate-frequency genes, circles—low-risk high-frequency loci) [3–5].

Same is true for the attempts to explain a big difference in response to treatment or prognosis within seemingly homogenous patient groups. Current knowledge allows mostly hypothesizing genetic variants to be causative in this scenario.

Another issue that GWAS has to deal with is interpretation of the effect of an identified polymorphism. For most of the loci, nearby genes are supposed to be affected, but in fact, chromatin organization can ensure interplay of neighboring regions [3]. So all genomic data will need to be proved on gene expression level.

This chapter summarizes genetic polymorphisms reported in association with gynecologic tumors and their effects on pathways in carcinogenesis.

## Types of gynecologic cancers

### Endometrial cancer

Endometrial cancer can develop as a part of genetic syndromes (such as Lynch syndrome or Cowden syndrome). The mutations, associated with these conditions are high-risk pathogenic variants of mismatch repair genes (MMR) ((MLH1, MSH2, MSH6, and PMS2) or PTEN, respectively. Microsatellite instability, determined by MMR mutations or epigenetic silencing, has been described in almost one-third of endometrial cancers [16]. It has been proposed that MMR silencing could even be predisposed genetically [17], although this hypothesis is still not confirmed in larger cohorts [16]. MLH1 variants, other than the mutations known to cause Lynch syndrome, have also been described in endometrial carcinomas [18]. In addition, significance of other tumor-suppressor genes in endometrial cancerogenesis has been studied. CHECK2, STK11, and other Cowden syndrome-associated genes are among them. These conditions are responsible for only small percentage of endometrial cancers [19] and cannot fully explain the significant familial association of these tumors [20, 21].

Current classification of endometrial tumors has incorporated a number of molecular changes and POLE mutation is one among them. POLE encodes DNA polymerase and mutations are thought to empair proofreading during the replication process. POLE pathogenic germline variants have been found to increase endometrial cancer risk and single cases of ovarian cancer associated with POLE variants have also been described [19].

Studies on CHECK2—another cell cycle regulator gene proposed to be associated with endometrial cancer show rather controversial data and not as convincing as for other tumor types [19].

GWAS analyses have identified a number of common genetic variants associated with endometrial cancer. Importantly, there is still a significant gap in assessing different histological subtypes of endometrial tumors. Most of the studies report cumulative data for all histological types. Although endometrioid and nonendometrioid endometrial carcinomas are believed to have distinct molecular genetic backgrounds, relatively small number of nonendometrioid cancer cases are available not allowing to receive genome-wide significant data for those groups alone. Continuous amendments to histologic classification of endometrial tumors complicate comparison among the studies from different points of time. The latest classification considers mucinous endometrial carcinoma as a pattern of endometrioid cancer. In addition, incorporation of common molecular defects becomes increasingly important. So the association of genetic variants to certain cancer classes might require revision [22].

One of the first genes to be associated with endometrial cancer was HNF1B. At least three SNPs in regions in linkage disequilibrium with the first four exons of this gene on chromosome 17q12 were found to carry genome-wide significant risk for endometrioid-type endometrial carcinoma [23]. HNF1B is a transcription factor known to a serve as a transcriptional activator in isoforms A and B, while isoform C is associated with transcriptional repression [24]. The risk alleles are linked to increased promoter activity or gene overexpression. Oncogenic potential of HNF1B has been related to different tumors mainly of urogenital origin. At the same time, rs11651052 variant of the same locus was found to increase risk of type-2 diabetes mellitus and be protective against endometrial carcinoma [25], probably underlining the opposite effects on immunity in these distinct settings. The significance of these SNPs in nonendometrioid tumors and individuals of non-European ancestry is somewhat controversial due to smaller number of cases available for testing [3, 26–28]. There have even been attempts to correlate polymorphisms with survival, revealing rs4430796 genotype to be an independent prognostic factor for endometrial carcinoma treated with radio-chemotherapy [29]. It is still unclear, whether this has anything to do with chemosensitivity.

There are other loci on chromosome 17 linked to endometrial cancer risk. At least 5 candidate genes have been identified in the region 17q21.32 [30]. CBX1, HOXB2, HOXB8, MIR1203, and SNX11 are among them. CBX1 is a member of CBX family known to affect gene regulation and tumorigenesis [31]. SNX11 is a member of sortin nexins. Their function is largely unclear, though they are thought to participate in endosomal membrane trafficking and lysosomal degradation of plasma membrane ion channels [32, 33].

17q12 is another risk locus associated with candidate genes EVI2A, NF1, and SUZ12. NF1 is a well-known tumor-suppressor gene [34] described as a target for somatic mutations for rare endometrial and ovarian cancers [35–37]. EVI2A represents a MEK/ERK pathway activator and due to proximity to NF1 gene has been described to be affected in NF1 microdeletion cases. SUZ12 protein participates in epigenetic regulation of gene expression by histone methylation, chromatin remodeling, and transcriptional silencing [38]. The corresponding gene is a crucial in endometrial stromal sarcoma pathogenesis. Translocations involving SUZ12 are considered diagnostic for certain types of the stromal tumors in uterus [39].

GWAS-significant polymorphisms were found in 8q24.21 locus well known for containing the MYC gene (Fig. 2). In addition, several microRNA genes are located in this complex region including MIR1204, MIR1205, MIR1206, and MIR1208 [3, 40, 41]. MYC is a transcription factor, the role of which in cancerogenesis is indisputable, though the ways of its interaction with chromatin are still being under investigation [42]. Given the fact that GWAS variants do not always regulate the gene closest to them and chromatin structure can ensure regulation across longer fragments, the exact mechanism of how the risk variants affect the MYC activity still needs to be investigated [3]. MYC suppression, especially epigenetic interaction, is considered to have a potential therapeutic implication [43] and could be discussed as a promising strategy in long-term perspective. Expression of microRNAs of the region 8q24.21 has also been linked to gastric, breast, and ovarian cancer [44–46]. A study on pancreatic cancer proposes MIR1207 overexpression as a determinant of chemosensitivity [47].

Variants in 15q21.2 region have been reproducibly associated with cancer risk [40, 41, 48]. The gene CYP19A1 located in this region encodes for aromatase, an enzyme that converts androstenedione to estrone and testosterone to estradiol (Fig. 3). Variants with A alleles of rs749292 and rs727479 in CYP19A1 are associated with increased estrogen synthesis and thus increased endometrial stimulation. These variants have been found to correlate with

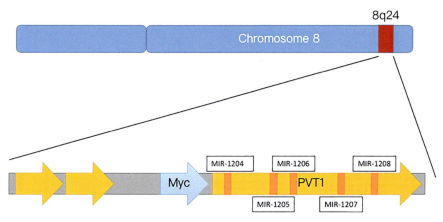

FIG. 2 Schematic diagram of 8q24.21 locus with Myc gene and microRNA genes. *Adapted from Boloix A, Masanas M, Jiménez C, et al., Long non-coding RNA PVT1 as a prognostic and therapeutic target in pediatric cancer. Front Oncol. 2019;9:1173.*

FIG. 3 CYP11A1, CYP17, and CYP19 involvement in estrogen metabolism.

endometrial cancer risk in a dose-dependent manner [49]. Interestingly the association of the rs727479 polymorphism with endometrial cancer risk is higher in older patients and patients with high BMI [48]. Several other polymorphisms in the same locus have been reported in smaller studies, such as 3 bp insertion/deletion and intron 4 [TTTA]n repeat polymorphism [50]. Aromatase inhibitors have long been used for breast cancer treatment, but not that well developed for endometrial cancer management. Studies conducted on breast cancer patients identified that certain CYP19A1 polymorphisms (most significantly rs727479) are associated with poor prognosis to aromatase inhibitor therapy [51, 52]. This could be taken into account when selecting the treatment approach.

CYP11A1 and CYP17A1 are two other enzymes involved in estrogen metabolism, polymorphisms of which have been found to increase risk of endometrial cancer. This has not been supported by genome-wide significant data [50] and has been questioned by other studies [53].

As mentioned, most risk-related loci for endometrial cancer are shared with other tumors. 2p16.1 is one of them. This locus has GWAS association with both endometrioid and nonendometrioid cancers [30]. *BCL11A* gene is considered a candidate gene in this region, first identified to be important for lymphocyte development and having tumor-suppressor effect [54]. On the other hand, its effect in lung cancer is controversial showing oncogenic activity [55] as well as an opposite effect causing genome instability in case of reduced expression [56]. The product of BCL11A was found to interact with a number of known driver proteins: TP53, ESR1, PTEN, RB1, AKT1, KRAS, PIK3R2, and POLE are among them. Endo-metrial cancer studies support the concept of its tumor-suppressor capabilities as the variants associated with increased cancer risk have been correlated with decreased expression of BCL11A [30].

The expression of these genes, targeted by BCL11A, can also be altered by nearby SNPs. ESR1 must have most prominent functional association with gynecologic cancer, as ESR1 en-codes for estrogen receptor. But in fact, most SNPs described in this region (6q25) remain con-troversial and are not related to overexpression of this gene [57].

TP53 is the major tumor-suppressor gene involved in cell cycle, apoptosis, and regulatory processes. Several meta-analyses confirmed the role of SNPs not only in this, but also func-tionally related genes MDM2 and P14ARF in endometrial cancer development [58–60]. P14ARF interacts with MDM2 and among other; this interaction results in inhibition of p53. Coexistence of MDM2 and some P14ARF polymorphisms has been found to be protec-tive. In fact, MDM2 polymorphisms have been studied in other gynecologic tumors (ovarian and cervical cancer) with some controversial results [61], questioned by some large-scale studies [59]. SNPT390G polymorphism has been found to have opposite effects on ovarian cancer risk in Caucasian and Chinese populations [62, 63].

Another well-known region for polymorphisms associated with a wide spectrum of dis-eases is 9p21.3. Several variants in a number of candidate genes in this locus are found to in-crease risk for cardiovascular disease [64, 65] and different types of tumors [66]. The most important cancer-associated genes in this locus are CDKN2A, CDKN2B, CDKN2BAS, and MTAP. CDKN2A and CDKN2B are cyclin-dependent kinase inhibitors encoding for p14, p15, and p16 proteins with crucial role in cell cycle regulation and senescence (Fig. 4). CDKN2BAS is a noncoding RNA gene from this region, which regulates CDKN2A and CDKN2B expression. One more gene from the same locus, MTAP (methylthioadenosine phosphorylase), is involved in polyamine metabolism and has earlier been reported as a tar-get gene for melanoma risk variants [67]. Interestingly, preliminary data show that tumors with MTAP loss could benefit from metabolism-based therapies [68] or be more chemosensitive [69]. Variants in 9p21.3 locus related to increased endometrial cancer suscep-tibility are thought to affect CDKN2A, CDKN2B, and CDKN2BAS expression.

Tumor-suppressor gene SH2B3 in locus 12p24.12 is also a candidate risk gene for endome-trial cancer. A missense variant SNP rs3184504 in this region was confirmed to be genome-wide significant for both endometrial and colorectal carcinomas [70]. There are three addi-tional known SNPs of the same region. The SH2B adaptor protein 3 (SH2B3) gene encodes a negative regulator of cytokine signaling. Its germline and somatic mutations are seen in acute lymphoblastic leukemia. Reduced expression of SH2B3 leads to increased Jak2 and STAT3 phosphorylation and increased cell proliferation [71]. rs3184504 polymorphism has been described in autoimmune conditions, but notably with different risk allele just like in case of HNF1B [70].

Another interesting polymorphic region, proposed to be shared by endometrial and colon cancers, is TERT-CLPTM1L locus on 5p15. TERT encodes a telomerase subunit. Whether or not the identified SNPs reach genome-wide significance level is controversial, but the ten-dency has been detected in a number of studies [70, 72]. TERT polymorphisms are associated with longer telomeres and have also been reported in breast, prostate, and colon cancer [70, 73–75]. Interestingly, the risk allele for colorectal carcinoma—(A) of the rs2736100 SNP

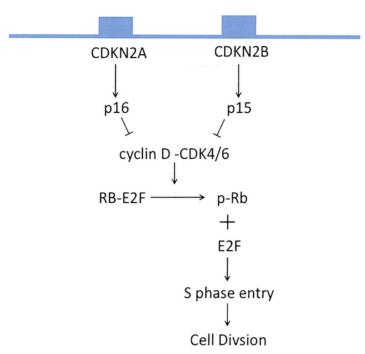

FIG. 4   Effect of loss of expression of CDKN2A and CDKN2B. *Adapted from Almontashiri NAM, et al. The 9p21.3 risk locus for coronary artery disease: A 10-year search for its mechanism. J Taibah Univ Med Sci. 2017;12(3):199–204.*

turned out to be protective against endometrial cancer. And that is not the only exclusion, another polymorphism near gene TSHZ1 shows alleles with opposite effects on these tumors [70].

Intergenic SNPs at 13q22.1 have shown genome-wide significant association with endometrial cancer [40]. This locus is suggested to regulate the activity of Kruppel-like Factors KLF5 expression, which is located upstream from this region together with KLF12. KLF5 is a transcription factor, which plays a role in uterine development and cancerogenesis [76]. Elevated KLF5 levels are strongly correlated with activating KRAS mutations. KLF5 itself is upregulated by long noncoding RNA UCA1 facilitating cancer development in endometrium [77]. Overexpression of KLF5 in cancer cells have even been proposed as a determinant of poor prognosis [78].

Additional potential target loci for endometrial cancer risk are shown in the Table 1.

The idea of endometrioid carcinoma both in endometrium and ovary being linked with endometriosis has long been discussed but is still somewhat controversial. A cross-disease meta-analysis by Painter et al. [79] confirmed these diseases to share risk variants. Most important SNP reported in this study lies in intron 2 of PTPRD gene encoding a thyrosine phosphatase on chromosome 9p23. Downregulation of the gene has been seen in gliomas [80], gastric cancers, and a number of other tumors frequently coexisting with CDKN2A mutations [81] due to proximity or other unexplained factors for pressured codeletion. CDKN2A locus alone has also been described to be linked with endometriosis [3]. Somatic mutations of

TABLE 1   Additional loci and target genes, variants of which are thought to contribute to endometrial cancer development [3, 41].

| Risk locus | Candidate target gene | Transcript function |
| --- | --- | --- |
| 1p34.3 | CDCA8 | Cell division regulation |
| 6q22.3 | SOX4 | Protein–protein and protein–DNA interactions |
| | CASC15 | lncRNA regulating proliferation and migration |
| 6q22.31 | HEY2 | Helix–loop–helix transcriptional repressor in the Notch pathway |
| | NCOA7 | Modulates the activity of the estrogen receptor via direct binding |
| 11p13 | CCDC73 | Unclear |
| | EIF3M | Translation initiation |
| | RCN1 | Endoplasmic reticulum protein |
| | TCP11L1 | Protein binding, signal transduction |
| | WT1-AS | Antisense lncRNA regulating WT1 expression |
| 12p12.1 | SSPN | Connects cytoskeleton with extracellular matrix |
| 12q24.21 | SNORA27 | Small nucleolar RNA |
| 15q15.1 | BMF | Sensing intracellular damage, apoptosis |
| | GPR176 | G-protein-coupled receptor |
| | SRP14-AS1 | Antisense lncRNA regulating intracellular processes |

PTPRD are readily identifiable in endometrial cancers [82]. Loss of PTPRD function affects carcinogenesis through STAT3 activation sharing downstream steps of pathogenesis with SH2B3. Other SNPs shared by endometriosis and endometrial cancer have been linked to the genes SKAP1, KITLG, DUSP6, TLR1,TLR6, WNT4, and ZBTB40 [79]. DUSP6 regulates ERK signaling, which could serve as a target in endometrial carcinoma treatment [83] and therefore might be a promising factor in personalized treatment approaches.

## Ovarian cancer

Epithelial tumors of ovary exist in many different histological types, most commonly serous (high and low grade), mucinous, endometrioid, and clear cell variants. The fact that most types might present as cystadenomas, borderline tumors, or carcinomas raises discussion about the genetic alterations shared by these lesions with different clinical course. Most important question that still needs to be answered based on molecular genetic data is whether the borderline tumors are precursors of carcinomas arising from the same main genetic hit.

Genetic risk for ovarian cancer is determined by well-known high-risk genes in more than 10% of cases [84]. But even without these known mutations, relatives of women with ovarian cancer have increased risk; this supports the expectation of far more risk variants still need to be identified [85]. The estimated contribution of more common susceptibility variants in less substantial elevation in ovarian cancer risk is steadily increasing and is currently estimated to be around 3.1% [86].

High-risk genetic variants determining ovarian cancer are related to *BRCA1* and *BRCA2* genes. Pathogenic variants in these tumor-suppressor genes are numerous and they variably affect the carcinogenic potential of the cells [4]. Epigenetic changes in these genes involved in repair of DNA double-strand breaks have also long been recognized as carcinogenic and targeted by hypomethylating agents. There is a weak evidence that BRCA1 and BRCA2 risk alleles determine prognosis to some extent [87].

Other germline mutations associated with ovarian cancer risk are RAD51C, RAD51D, BRIP1 (FANCJ), ATM, and FANCM [87, 88]. RAD51 family genes are thought to be involved in DNA repair through multiple mechanisms, some of which consider interaction with BRCA genes. ATM is important for homologous recombination and cell cycle regulation [5]. FANCM is a Fanconi anemia core complex protein, participating in DNA damage detection. Recent studies suggest that polygenic risk score calculation could be beneficial for risk prediction and prevention strategies [85]. And in terms of this polygenic risk BRIP1, another Fanconi anemia gene is considered to be important along with BRCA1 and BRCA2 as the lifetime risk for ovarian cancer in mutation carriers was estimated to be 5.8% [87, 89].BRIP1 encodes a helicase, which also acts through interaction with BRCA1. Interestingly, all of these mutations have mostly been associated with high-grade serous carcinoma risk [87, 89].

Unlike the case of endometrial cancers, ovarian tumor pools available for GWAS analyses are more diverse, thus allowing identification of more variants with increased risk for certain subtypes [15, 73, 86, 87, 90–96] (Table 2). But in fact, the data are still stronger for the most common type—serous. Some relatively older studies do not provide stratification of high- and low-grade serous cancer risks, although the genetic background of these two tumors is believed to be distinct.

Similarities between epidemiology and histopathologic features of endometrial and ovarian tumors triggered interest toward common molecular genetic features. Mutations in sporadic carcinomas showed many overlaps, but the frequencies were not similar. A significant number of risk loci are shared between ovarian and endometrial cancers, some of them being confirmed by large-scale meta-analysis [91] (Table 3).

GWAS data have shown that SNPs in 2q13, 8q24.21, and 12q24.31 are associated with serous carcinoma risk in patients with germline BRCA1/2 mutations. These variants affect genes with crucial effects in carcinogenesis, like BCL211, MYC, PVT1, and HNF1A mostly by epigenetic interactions [90].

10q24.33 was found to be an important risk locus for borderline serous ovarian tumor and low-grade serous carcinoma. Functional studies revealed a gene OBFC1 to be affected by this polymorphism. OBFC1 is a component of the telomere binding complex and a subunit of an alpha accessory factor that stimulates the activity of DNA polymerase-alpha-primase, which initiates DNA polymerization [90, 97]. It has been shown to be linked to telomere length.

A region with SNPs linked to ovarian cancer is 9p22.2. The exact functional characterization of this region is rather complicated, as it includes several genes and some of the SNPs are located outside the reading frame or intronic regions. Though the most interesting gene proposed to be affected by this polymorphism is BNC2, it encodes a protein believed to be

**TABLE 2** Loci of SNPs, shown to correlate with tumor type [15, 73, 86, 87, 90–96].

| | | | Histologic type | | | | | | |
|---|---|---|---|---|---|---|---|---|---|
| Locus | Serous | Serous Low Grade | Serous High Grade | Mucinous | Endometrioid | Clear cell | Serous Borderline | Mucinous Borderline | Reference |
| 1p36 | ▦ | | | | | Marginal | | | [86] |
| 1p34.3 | | | | | | ▦ | | | [86] |
| 2p16.1 | | | | | | | | | [91] |
| 2q13 | | | ▦ | | | | | | [90] |
| 2q14.1 | ▦ | | | | | | | | [94] |
| 2q31 | | ▦ | | ▦ | Marginal | | | | [93, 94] |
| 3q22.3 | | | | | | | | ▦ | [90] |
| 3q25.31 | | | | | | | | | [93, 94] |
| 3q28 | | | ▦ | | | | ▦ | | [90] |
| 4q26 | ▦ | | | | | | | | [86] |
| 4q32.3 | | | | | | | ▦ | | [90] |
| 5p15.15 | | ▦ | ▦ | | | | | | [87] |
| 5p15.33 | ▦ | | ▦ | | | | ▦ | | [73, 91] |
| 5q12.3 | | | | | ▦ | | | | [90] |
| 6p22.1 | ▦ | | | Marginal | | | | | [86] |
| 7p12.1 | ▦ | | | | ▦ | | | | [15, 92] |
| 7q22.1 | | | ▦ | | | ▦ | ▦ | | [91] |
| 8q21.11 | | ▦ | ▦ | Marginal | | | | | [90] |
| 8q21.13 | | | | | ▦ | | | | [94] |
| 8q24.21 | ▦ | | ▦ | Marginal | Marginal | | | | [90, 93] |
| 9p12 | ▦ | | | | | | | | [91] |
| 9p22.2 | | | | | | | | | [94] |
| 9q31.1 | | | ▦ | ▦ | | | | ▦ | [90] |

## Histologic type

| Locus | Serous | Serous Low Grade | Serous High Grade | Mucinous | Endometrioid | Clear cell | Serous Borderline | Mucinous Borderline | Reference |
|---|---|---|---|---|---|---|---|---|---|
| 9q34.2 | ■ | | ■ | | Marginal | | | | [86, 91] |
| 10p12.31 | | | | | | | | | [91] |
| 10q24.33 | | ■ | | | | ■ | ■ | | [90] |
| 12q22 | | | ■ | | | | ■ | | [15, 92] |
| 12q24.31 | | | | | | | | | [90] |
| 13q14.2 | | | | ■ | | | | | [15, 92] |
| 14q24.1 | | | | | | | | | [15, 92] |
| 17q11.2 | ■ | | | | Marginal | | | | [86] |
| 17q12 | | | | | | ■ | | | [91, 94] |
| 17q21.31 | | | ■ | | Marginal | | | | [91, 95] |
| 17q21.32 | | ■ | ■ | | | ■ | ■ | | [91, 94] |
| 18q11.2 | | | | | | | | | [90] |
| 19p13.11 | ■ | | | ■ | | | | | [94, 96] |
| 19p13.2 | ■ | | | | | | | | [94] |
| 22q12.1 | ■ | ■ | | | | | | | [90] |
| 22q12.2 | | ■ | | | | | | | [15, 92] |

TABLE 3    Regions and target cells with loci, significantly associated with the risk of both endometrial and ovarian tumors [91].

| Region | Candidate target gene/s |
| --- | --- |
| 2p16.1 | BCL11A |
| 5p15.33 | TERT |
| 7q22.1 | CYP3A43 |
| 7p22.2 | COX19, ENSG00000229043, GPER1, ZFAND2A |
| 8q24.21 | MYC |
| 9p12 | Not known |
| 9q34.2 | ABO, CACFD1 |
| 10p12.31 | CASC10, MIR1915, MLLT10, SKIDA1 |
| 11q13.3 | CCND1, MYEOV |
| 17q12 | HNF1B |
| 17q21.32 | CBX1, HOXB2, HOXB8, MIR1203, SNX11 |

important transcription regulator [98]. This gene is expressed in reproductive tissues, but data about its importance in cancer is rather limited [99].

17q21 locus already discussed to show contain SNPs associated with endometrial cancer contain different variants, related with ovarian cancer risk (not reaching genome-wide significant level) [93]. These are SNPs intronic to SKAP1 gene, a cell cycle regulator and inhibitor of RAS and RAF activation, also known to be a candidate for endometriosis and endometrial cancer risk. Other SNPs in the nearby region 17q21.31 have more than 15 candidate genes, function of which they might affect [95].

Another region, showing relatively weaker association with ovarian cancer is located in 3q25 region and is intronic to TiPARP [93]. The product of this gene is involved in repair of single-strand DNA breaks. PARP inhibition is an approved treatment for BRCA1-mutated ovarian carcinoma [100, 101]. So it might be beneficial to investigate these polymorphisms in relation to treatment effects.

A relatively recent meta-analysis found association of 8q21, 10p12, and 17q12 with ovarian cancer risk. A number of putative candidate genes in this loci do not yet allow exact characterization of the effects [102].

It is not surprising that many genetic pathways linked to ovarian cancer are related to BRCA1. MERIT40 is a gene that interacts with BRCA1, probably retaining it at the cites of double-strand breaks. And polymorphism in the locus 19p13.11 with this gene showed genome-wide significant association with ovarian cancer risk [96].

Polymorphisms associated with WNT4, RSPO1, SYNPO2, GPX6, ABO, ATAD5, COL15A1/TGFBR1, NAMPTL, and SRGAP1 genes have been reported [86, 103], and some differences were identified between groups with and without BRCA1/2 mutations [86]. So BRCA1/2 carrier status can determine different set of SNPs to increase the ovarian cancer risk. Most of these genes are known for their effects in tumorigenesis and seeing ABO; blood

TABLE 4   SNPs affecting overall survival (OS) and relapse-free survival (RFS) [109].

| SNP | Chromosome | Nearest gene | Significant effect on OS/RFS |
|-----|-----------|--------------|------------------------------|
| rs6674079 | 1q22 | RP11–284F21.8 | OS |
| rs7950311 | 11p15.4 | HBG2 | OS |
| rs4910232 | 11p15.3 | RP11-179A10.1 | PFS |
| rs2549714 | 16q23 | RP11–314O13.1 | OS |
| rs3795247 | 19p12 | ZNF100 | PFS |

group determining gene is not a surprise as it has been confirmed to be related to other tumors as well [104]. Data on SNPs in 4q32.2 locus are reported in some studies to be exclusive for BRCA1 mutation carriers [105].

Some of the genes related to estrogen effects have already been mentioned previously. Expression of another estrogen receptor gene, ESR2, has been correlated with a number of tumors including ovarian cancer. Polymorphisms in the locus 14q32.2 affecting this gene are related to not only increased risk, but also prognosis of ovarian carcinomas [106].

Noncoding RNA studies showed a number of putative noncoding RNA genes affected by polymorphisms implicated in ovarian cancer [13]. The antisense intragenic SNPs of HOX noncoding RNA (HOTAIR) have also been proven to determine ovarian and cervical cancer risk [107]. They are thought to affect the function of the crucial genes, involved in oncogenesis, most of which are mentioned previously.

An interesting study from 2019 by Harris et al. [108] looked into the genetically predicted risks for polycystic ovary syndrome (PCOS) and ovarian cancer. Although it somewhat contradicts other observational data (stating that PCOS increases the risk of carcinoma) [109], the results show inverse relationship between genetically determined PCOS and endometrioid ovarian carcinoma.

The idea that outcome variability in ovarian cancer patients can have roots in genetic predispositions raised a need in GWAS oriented for identification variants influencing prognosis and determining treatment sensitivity. Clinical outcome seems to be influenced by at least 46 SNPs [110], which lie in regions related to genes with wide variety of functions, including cell cycle regulation and DNA repair. A study by Johnatty et al. [111] explored 5 significant SNPs related to ovarian cancer outcome. Three of these SNPs are related to HOTAIR, already mentioned as a gynecologic cancer risk locus. They determine progression-free survival independent from chemotherapy regime used (Table 4). The same paper interestingly looked into the pathways associated with progression-free survival and overall survival of ovarian cancer (Table 5), proposing polygenic panels for prognostication [111].

Another group of SNPs in *ABCC2* gene was found to affect ovarian cancer sensitivity to paclitaxel and carboplatin treatment. This gene functions in bile excretion and is thought to affect drug metabolism [112].

A lot of promising targets and treatment-related perspectives can be picked up in this complex field of huge number of variants with complicated nomenclature and functional associations. But even though the methodology becomes more and more standardized, there is still room for improvement of reproducibility and reliability of the data from such complex assays.

TABLE 5   Gene sets and pathways related to overall survival (OS) and relapse-free survival (RFS) of patients with ovarian carcinoma [109].

| Pathway | Number of genes | Most important genes |
| --- | --- | --- |
| *Pathways associated with progression-free survival* | | |
| YAP1 and WWTR1 (TAZ) stimulated gene expression (REACTOME database) | 23 | CTGF,TBL1X,NCOA6,TEAD3,MED1,PPARA, TEAD1,NCOA3, KAT2B |
| G0 and early G1 (REACTOME database) | 23 | RBL2,CDC25A,MYBL2,LIN9,HDAC1,CCNA1, LIN52 |
| Amine derived hormones (REACTOME database) | 15 | CGA,TPO,SLC5A5,TH |
| Formation of incision complex in GG NER (REACTOME database) | 21 | ERCC2,RAD23B,GTF2H1,GTF2H2,RPA1,ERCC1, DDB2,XPA,DDB1 |
| G protein activation (REACTOME database) | 27 | GNB2,GNAT1,GNAI2,GNAI1,POMC,GNB3, GNG4,GNGT2,GNAO1,GNG8,GNG3 |
| Lysosome vesicle biogenesis (REACTOME database) | 22 | CLTA,AP1B1,AP1S1,DNAJC6,AP1G1,GNS,M6PR, VAMP8,BLOC1S1 |
| Inhibition of insulin secretion by adrenaline noradrenaline (REACTOME database) | 25 | GNB2,GNAI2,CACNB2,GNAI1,ADRA2A,GNB3, GNG4,GNGT2,GNAO1,GNG8,GNG3 |
| Cyclin A B1 Associated events during G2 M transition (REACTOME database) | 15 | CDC25A,PLK1,CCNA1,WEE1,CDC25B,PKMYT1, XPO1 |
| *Pathways associated with overall survival* | | |
| HDL mediated lipid transport (REACTOME database) | 15 | BMP1,CETP,APOA1,APOC3,ABCG1 |
| Xenobiotics (REACTOME database) | 15 | CYP2A13,CYP2B6,CYP2F1 |
| Lipoprotein metabolism (REACTOME database) | 28 | BMP1,CETP,APOA1,APOC3,APOA5,ABCG1 |
| Insulin synthesis and processing (REACTOME database) | 20 | SNAP25,INS,EXOC5,ERO1L,PCSK1,EXOC4,PCSK2 |
| MTA3 pathway (BIOCARTA database) | 19 | TUBA1A,TUBA1C,HDAC1,MBD3,ALDOA,CDH1, MTA1,SNAI2,TUBA3C |
| Acetylcholine binding and downstream events (REACTOME database) | 15 | CHRNG,CHRND |
| Synthesis of bile acids and bile salts via 7 alpha hydroxycholesterol (REACTOME database) | 15 | SLC27A5, HSD17B4, AKR1D1, SLC27A2, CYP27A1, ACOX2,HSD3B7, ABCB11 |
| KEGG maturity onset diabetes of the young (KEGG database) | 23 | ONECUT1, INS, HNF1A, BHLHA15, NR5A2, FOXA3 |
| Immunoregulatory interactions between a lymphoid and a nonlymphoid cell (REACTOME database) | 56 | CD96, CD8A, CD8B, IFITM1, KIR3DL2, CRTAM, ICAM,KIR3DL1, FCGR3A, LILRB2, CD19, LILRB5, LILRB3,CD200R1, RAET1E, FCGR2B, SELL,ULBP2, ULBP1,KIR2DL4, B2M, CDH1, CD81 |

## Cervical cancer

Data on genetic risk factors of cervical cancer are the least studied among common gynecologic tumors. HPV infection explains majority of cases, screening, and vaccination succeeded to reduce the frequency of cervical cancer cases and remove this entity from the most common tumor category in developed world. But not all cases of cervical cancer could be attributed to HPV and in addition not all infections have identical pattern of progression, so there is still lack of knowledge about genetic contributors. Histologic diversity plays an important role in understanding diverse molecular backgrounds. Both squamous cell carcinoma and adenocarcinoma of cervix may be HPV-associated or HPV-independent and both are known to develop from precursors. In case of adenocarcinomas, HPV-independent cases show wide morphologic variability (gastric, clear cell, mesonephric, and other subtypes). HPV-independent squamous tumors often present with KRAS, ARID1A, and PTEN mutations, while adenocarcinomas may in addition show STK11 (may also be part of Peutz–Jeghers syndrome), TP53, ERBB2, and MDM2 mutations {20}. Somatic copy number alterations have also been reported analyzing cervical cancer databases [113]. Most of the studies do not report the histologic types of the tumors included; moreover, the classification has not been constant through the past years so results are cumulative.

As HPV is still a major contributor in cervical cancer pathogenesis, genetic predisposition is mainly explained by altered expression of genes involved in inflammatory and immune mechanisms. The most powerful genetic association of cervical cancer risk was found with HLA haplotypes. A large number of SNPs in HLA-DPB1, HLA-DQA1, HLA-DOB, HLA-DQA2, and HLA-DRA in MHC region on 6p21.3 (Fig. 5) have all been found important [114–121]. It has been proposed that this association is attributable to alteration in inflammatory responses [122].

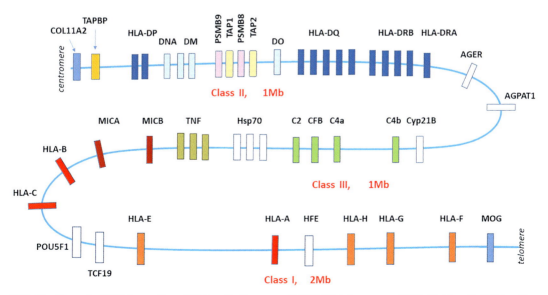

FIG. 5    Genes in MHC locus. From Wikimedia Commons [MHC HLA locus.png. (2020, October 9). Wikimedia Commons, the free media repository. *Retrieved 11:33, February 13, 2021 from https://commons.wikimedia.org/w/index. php?title=File:MHC_HLA_locus.png&oldid=485020961].*

IRF3 is a type I interferon gene located on 19q13.3 locus with exonic variants associated with cervical cancer risk on one hand with and HPV persistence on the other hand [123]. Polymorphisms in another immune-response–related locus 4q32 with toll-like receptor gene TLR2 are found to determine progression to CIN3 or cancer [123]. TLR4 and TLR9 polymorphisms do also have a role [124]. Components of innate immune system such as IL-1b affect cervical cancer risk and it has been confirmed by analyzing IL-1b levels in blood [104], polymorphisms of functional regulators, or variants of IL-1B gene itself [125, 126]. Other candidates of immune system components affecting cervical cancer risk are IL-12 [8, 127], IL-10 [128–130], IFNG [9, 131], TNFA [126, 131–136], MICA [115, 137], and PRF1 [138], although data for most of specific SNPs in these loci are somewhat controversial. Many studies in this range are restricted by rather specific small populations and still need to be reproduced in larger and more representative groups. The location of TNFA in MHC region further complicates the SNPs with individual significance [135]. In addition to TNFA, VEGF receptor gene FTL1 polymorphisms have been found significant in a small Chinese population [139].

Other loci related to HPV persistence and progression are located in exons of DNA repair genes FANCA, EXO1, CYBA, XRCC1, HRAS, ERCC1, ERCC4, and XPC (16q24.3, 1q42–43, 16q24, 19q13.2, 11q15, 19q13.32, 16p13.12, and 3p25.1, respectively) [123, 138, 140].

Another large group of susceptibility genes participate in cell cycle and apoptosis-related processes. These are genes, encoding for FAS, FASL, p53, MDM2, cyclin-dependent kinases, and Caspases. The data on these loci are even more conflicting [141–143].

A study restricted to east Asian population identified two significant loci associated with cervical cancer risk: 5q14 and 7p11 [144]. ARRDC3 is considered a target gene in 5q14 locus. It regulates G protein-mediated signaling by sorting receptors to endosomes. Experiment showed that AARDC3 knockdown decreases susceptibility to HPV infection. This gene has also been investigated in breast cancer studies [144].

Substantial number of studies report metabolism-associated gene variants to affect cervical cancer risk. Cytochrome P450 gene CYP1A1 is one of them. The correlation seems not to be straightforward and uniform among populations, still meta-analysis confirms the significance of SNPs in this region [11, 145, 146]. Another enzyme-coding gene, MTHFR (Methylentetrahydrofolate Reductase), shows conflicting data on the effects of both C677T and A1298C polymorphisms on cancer risk. MTHFR is a major enzyme in folate metabolism, confirmed to participate in pathogenesis of a number of diseases through affecting numerous cellular processes including epigenetic regulation [147, 148]. Thymidylate Synthase gene has also recently been related to cervical cancer risk [149].

A number of noncoding RNA gene variants have been shown to be related to cervical cancer risk [150]. Long noncoding RNAs HOX, HOTAIR, discussed for ovarian cancer is among them [151, 152]. Some studies even suggest that HOTAIR polymorphisms determine prognosis—recurrence or death [153]. Some noncoding RNA expression has been proposed as a treatment sensitivity determining factor, so polymorphisms in these loci could affect management decisions [154]. Some evidence demonstrates the role of microRNA variants on cervical cancer risk [10, 119].

If we summarize all variants proposed to explain cervical cancer risk, it will end up in a list of more than 70 different loci with far more SNPs in them [155]. Interestingly, the GWAS data show rather restricted results with most SNPs in MHC region and two additional significant loci identified in Chinese population: 4q12 and 17q12 [156]. A candidate gene for 4q12 is

EXOC1. This result has been questioned by another study on a population with more diverse ancestries [137]. 17q21 is a complex region with a number of endometrial carcinoma susceptibility genes mentioned previously. The SNP associated with cervical cancer affects GSDMB expression [156].

Association of cervical risk with other gynecologic conditions is rather low (Fig. 6). Understandably, it can be explained by different set of genes, affected by genetic variability determining HPV-related carcinogenesis [157]. A novel approach searches for variants predisposing to high-risk HPV infection, which appears promising for determination of risk groups and effective screening [158].

Despite the fact that at the moment with current data, the evidence on genetic polymorphisms is not strong enough to influence the management, there are some data already available. Four SNPs have been reported to affect response to neoadjuvant platinum-based chemotherapy. The most significant effect was seen with a variant in 4q34.3 locus with no target gene identified, so functional impact of this SNP is not clear [159].

Some genetic variants might be promising for development of novel therapeutic approaches. One of them is MICA. Its expression on epithelial cells triggers immune response, while shedding is associated with more susceptibility to cancer. This shedding process is mediated by ADAM10 enzyme [155]. Inhibition of ADAM10 is an effective therapeutic approach [160], which doesnot work in case of MICA variants, which encode for protein that sheds from cell surface independent from ADAM10 activity [161].

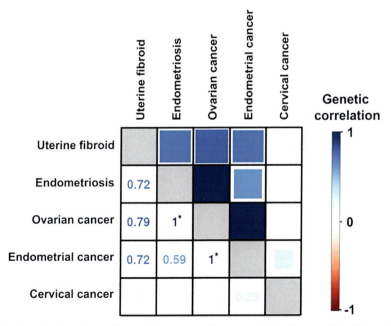

FIG. 6  Cross-trait evaluation of genetic correlation among five gynecologic diseases. Correlation is expressed by the color and size of square on the right upper triangle, while represented in digits on the left lower triangle. Asterisks indicate that the real output value exceeded one but was set to one for display purpose. *From Masuda T, Low SK, Akiyama M, et al. GWAS of five gynecologic diseases and cross-trait analysis in Japanese. Eur J Hum Genet. 2020;28(1):95–107.*

# Conclusion

The genetic background of gynecologic cancer risk is rather complicated, although given the data already existing and number of studies performed each year, it seems like the time when polymorphisms will guide screening and treatment planning is not far away.

# References

[1] Holman LL, Lu KH. Genetic risk and gynecologic cancers. Hematol Oncol Clin North Am 2012;26(1):13–29. https://doi.org/10.1016/j.hoc.2011.11.003.

[2] Lu Y, Ek WE, Whiteman D, et al. Most common 'sporadic' cancers have a significant germline genetic component. Hum Mol Genet 2014;23(22):6112–8. https://doi.org/10.1093/hmg/ddu312.

[3] O'Mara TA, Glubb DM, Kho PF, Thompson DJ, Spurdle AB. Genome-wide association studies of endometrial Cancer: latest developments and future directions. Cancer Epidemiol Biomarkers Prev 2019;28(7):1095–102. https://doi.org/10.1158/1055-9965.EPI-18-1031. Epub 2019 Apr 30 31040137.

[4] PDQ Cancer Genetics Editorial Board. Genetics of breast and gynecologic cancers (PDQ): Health professional version. In: PDQ cancer information summaries [internet]. Bethesda (MD): National Cancer Institute (US); 2020 Nov 4. 2002-.聽Available from: https://www.ncbi.nlm.nih.gov/books/NBK65767/.

[5] Lu H-M, Li S, Black MH, et al. Association of breast and ovarian cancers with predisposition genes identified by large-scale sequencing. JAMA Oncol 2019;5(1):51–7.

[6] Chen MM, Crous-Bou M, Setiawan VW, et al. Exome-wide association study of endometrial cancer in a multiethnic population. PLoS One 2014;9(5):e97045 [Published 2014 May 8] https://doi.org/10.1371/journal.pone.0097045.

[7] Qian N, Chen X, Han S, et al. Circulating IL-1beta levels, polymorphisms of IL-1B, and risk of cervical cancer in Chinese women. J Cancer Res Clin Oncol 2010;136(5):709–16.

[8] Chen X, Han S, Wang S, et al. Interactions of IL-12A and IL-12B polymorphisms on the risk of cervical cancer in Chinese women. Clin Cancer Res 2009;15(1):400–5.

[9] Gangwar R, Pandey S, Mittal RD. Association of interferon-gamma +874A polymorphism with the risk of developing cervical cancer in north-Indian population: IFN-γ in cervical cancer. BJOG 2009;116(12):1671–7.

[10] Zhou X, Chen X, Hu L, et al. Polymorphisms involved in the miR-218-LAMB3 pathway and susceptibility of cervical cancer, a case-control study in Chinese women. Gynecol Oncol 2010;117(2):287–90.

[11] Sengupta D, Guha U, Mitra S, Ghosh S, Bhattacharjee S, Sengupta M. Meta-analysis of polymorphic variants conferring genetic risk to cervical cancer in Indian women supports CYP1A1 as an important associated locus. Asian Pac J Cancer Prev 2018;19(8):2071–81.

[12] Miura K, Mishima H, Kinoshita A, et al. Genome-wide association study of HPV-associated cervical cancer in Japanese women: GWAS of cervical Cancer. J Med Virol 2014;86(7):1153–8.

[13] Manichaikul A, Peres LC, Wang X-Q, et al. Identification of novel epithelial ovarian cancer loci in women of African ancestry. Int J Cancer 2020;146(11):2987–98.

[14] Lawrenson K, Song F, Hazelett DJ, et al. Genome-wide association studies identify susceptibility loci for epithelial ovarian cancer in east Asian women. Gynecol Oncol 2019;153(2):343–55.

[15] Yodsurang V, Tang Y, Takahashi Y, et al. Genome-wide association study (GWAS) of ovarian cancer in Japanese predicted regulatory variants in 22q13.1. PLoS One 2018;13(12):e0209096.

[16] Meyer LA, Westin SN, Lu KH, Milam MR. Genetic polymorphisms and endometrial cancer risk. Expert Rev Anticancer Ther 2008;8(7):1159–67. https://doi.org/10.1586/14737140.8.7.1159.

[17] Chen H, Taylor NP, Sotamaa KM, et al. Evidence for heritable predisposition to epigenetic silencing of MLH1. Int J Cancer 2007;120(8):1684–8.

[18] Russell H, Kedzierska K, Buchanan DD, et al. The MLH1 polymorphism rs1800734 and risk of endometrial cancer with microsatellite instability. Clin Epigenetics 2020;12(1):102.

[19] Spurdle AB, Bowman MA, Shamsani J, Kirk J. Endometrial cancer gene panels: clinical diagnostic vs research germline DNA testing. Mod Pathol 2017;30(8):1048–68. https://doi.org/10.1038/modpathol.2017.20 [Epub 2017 Apr 28] 28452373.

[20] Mucci LA, Hjelmborg JB, Harris JR, et al. Familial risk and heritability of Cancer among twins in Nordic countries [published correction appears in JAMA. 2016 Feb 23;315(8):822]. JAMA 2016;315(1):68–76. https://doi.org/10.1001/jama.2015.17703.

[21] Haidopoulos D, Simou M, Akrivos N, et al. Risk factors in women 40 years of age and younger with endometrial carcinoma. Acta Obstet Gynecol Scand 2010;89(10):1326–30. https://doi.org/10.3109/00016349.2010.515666. 20846065.

[22] WHO Classification of Tumours Editorial Board. Female genital tumours [Internet]. Lyon (France): International Agency for Research on Cancer. 5th ed. vol. 4. WHO classification of tumours series; 2020. Available from: https://tumourclassification.iarc.who.int/chapters/34.

[23] Spurdle AB, Thompson DJ, Ahmed S, et al. Genome-wide association study identifies a common variant associated with risk of endometrial cancer. Nat Genet 2011;43(5):451–4. https://doi.org/10.1038/ng.812.

[24] Bach I, Yaniv M. More potent transcriptional activators or a transdominant inhibitor of the HNF1 homeoprotein family are generated by alternative RNA processing [published correction appears in EMBO J 1994 Jan 15; 13(2):492]. EMBO J 1993;12(11):4229–42.

[25] Nead KT, Sharp SJ, Thompson DJ, et al. Evidence of a causal association between insulinemia and endometrial cancer: a mendelian randomization analysis. J Natl Cancer Inst 2015;107(9):djv178 [Published 2015 Jul 1] https://doi.org/10.1093/jnci/djv178.

[26] Setiawan VW, Haessler J, Schumacher F, et al. HNF1B and endometrial cancer risk: results from the PAGE study. PLoS One 2012;7(1):e30390. https://doi.org/10.1371/journal.pone.0030390.

[27] Long J, Zheng W, Xiang Y-B, et al. Genome-wide association study identifies a possible susceptibility locus for endometrial cancer. Cancer Epidemiol Biomarkers Prev 2012;21(6):980–7.

[28] Fadare O, Liang SX. Diagnostic utility of hepatocyte nuclear factor 1-beta immunoreactivity in endometrial carcinomas: lack of specificity for endometrial clear cell carcinoma. Appl Immunohistochem Mol Morphol 2012; 20(6):580–7. https://doi.org/10.1097/PAI.0b013e31824973d1. 22495362.

[29] Mandato VD, Farnetti E, Torricelli F, et al. HNF1B polymorphism influences the prognosis of endometrial cancer patients: a cohort study. BMC Cancer 2015;15:229 [Published 2015 Apr 7] https://doi.org/10.1186/s12885-015-1246-5.

[30] O'Mara TA, Spurdle AB, Glubb DM. Endometrial cancer association consortium. analysis of promoter-associated chromatin interactions reveals biologically relevant candidate target genes at endometrial cancer risk loci. Cancers (Basel) 2019;11(10):1440 [Published 2019 Sep 26] https://doi.org/10.3390/cancers11101440.

[31] Xu Y, Pan S, Song Y, Pan C, Chen C, Zhu X. The prognostic value of the chromobox family in human ovarian cancer. J Cancer 2020;11(17):5198–209 [Published 2020 Jul 6] https://doi.org/10.7150/jca.44475.

[32] Li C, Ma W, Yin S, et al. Sorting nexin 11 regulates lysosomal degradation of plasma membrane TRPV3: SNX11 regulates lysosomal degradation of TRPV3. Traffic 2016;17(5):500–14.

[33] Liu T, Li J, Liu Y, et al. SNX11 identified as an essential host factor for SFTS virus infection by CRISPR knockout screening. Virol Sin 2019;34(5):508–20. https://doi.org/10.1007/s12250-019-00141-0.

[34] Cichowski K, Jacks T. NF1 tumor suppressor gene function: narrowing the GAP. Cell 2001;104(4):593–604. https://doi.org/10.1016/s0092-8674(01)00245-8. 11239415.

[35] Wang Y, Yu M, Yang JX, et al. Genomic comparison of endometrioid endometrial carcinoma and its precancerous lesions in chinese patients by high-depth next generation sequencing. Front Oncol 2019;9:123 [Published 2019 Mar 4] https://doi.org/10.3389/fonc.2019.00123.

[36] Er TK, Su YF, Wu CC, et al. Targeted next-generation sequencing for molecular diagnosis of endometriosis-associated ovarian cancer. J Mol Med 2016;94(7):835–47.

[37] Minkiewicz I, Wilbrandt-Szczepańska E, Jendrzejewski J, Sworczak K, Korwat A, Śledziński M. Co-occurrence of adrenocortical carcinoma and gastrointestinal stromal tumor in a patient with neurofibromatosis type 1 and a history of endometrial cancer. Acta Endocrinol (Buchar) 2020;16(3):353–8.

[38] Hrzenjak A. JAZF1/SUZ12 gene fusion in endometrial stromal sarcomas. Orphanet J Rare Dis 2016;11:15 [Published 2016 Feb 16] https://doi.org/10.1186/s13023-016-0400-8.

[39] Conklin CM, Longacre TA. Endometrial stromal tumors: the new WHO classification. Adv Anat Pathol 2014; 21(6):383–93. https://doi.org/10.1097/PAP.0000000000000046. 25299308.

[40] Cheng TH, Thompson DJ, O'Mara TA, et al. Five endometrial cancer risk loci identified through genome-wide association analysis. Nat Genet 2016;48(6):667–74.

[41] O'Mara TA, Glubb DM, Amant F, et al. Identification of nine new susceptibility loci for endometrial cancer. Nat Commun 2018;9(1). https://doi.org/10.1038/s41467-018-05427-7.

[42] Dang CV. MYC on the path to cancer. Cell 2012;149(1):22–35. https://doi.org/10.1016/j.cell.2012.03.003.

[43] Allen-Petersen BL, Sears RC. Mission possible: advances in MYC therapeutic targeting in Cancer. BioDrugs 2019;33(5):539–53. https://doi.org/10.1007/s40259-019-00370-5.

[44] Anauate AC, Leal MF, Wisnieski F, Santos LC, Gigek CO, Chen ES, Calcagno DQ, Assumpção PP, Demachki S, Arasaki CH, Artigiani R, Rasmussen LT, Payão SM, Burbano RR, MAC S. Analysis of 8q24.21 miRNA cluster expression and copy number variation in gastric cancer. Future Med Chem 2019;11(9):947–58. https://doi.org/10.4155/fmc-2018-0477 [Epub 2019 May 29] 31141411.

[45] Peña-Chilet M, Martínez MT, Pérez-Fidalgo JA, et al. MicroRNA profile in very young women with breast cancer. BMC Cancer 2014;14:529 [Published 2014 Jul 21] https://doi.org/10.1186/1471-2407-14-529.

[46] Wu G, Liu A, Zhu J, et al. MiR-1207 overexpression promotes cancer stem cell-like traits in ovarian cancer by activating the Wnt/β-catenin signaling pathway. Oncotarget 2015;6(30):28882–94. https://doi.org/10.18632/oncotarget.4921.

[47] You L, Wang H, Yang G, et al. Gemcitabine exhibits a suppressive effect on pancreatic cancer cell growth by regulating processing of PVT1 to miR1207. Mol Oncol 2018;12(12):2147–64. https://doi.org/10.1002/1878-0261.12393.

[48] Thompson DJ, O'Mara TA, Glubb DM, et al. CYP19A1 fine-mapping and Mendelian randomization: estradiol is causal for endometrial cancer. Endocr Relat Cancer 2016;23(2):77–91.

[49] Setiawan VW, Doherty JA, Shu X-O, et al. Two estrogen-related variants in CYP19A1 and endometrial cancer risk: a pooled analysis in the epidemiology of endometrial cancer consortium. Cancer Epidemiol Biomarkers Prev 2009;18(1):242–7.

[50] Olson SH, Orlow I, Bayuga S, et al. Variants in hormone biosynthesis genes and risk of endometrial cancer. Cancer Causes Control 2008;19(9):955–63. https://doi.org/10.1007/s10552-008-9160-7.

[51] Glubb DM, O'Mara TA, Shamsani J, Spurdle AB. The association of CYP19A1 variation with circulating estradiol and aromatase inhibitor outcome: can CYP19A1 variants be used to predict treatment efficacy? Front Pharmacol 2017;8:218 [Published 2017 Apr 25] https://doi.org/10.3389/fphar.2017.00218.

[52] Liu L, Bai YX, Zhou JH, et al. A polymorphism at the 3'-UTR region of the aromatase gene is associated with the efficacy of the aromatase inhibitor, anastrozole, in metastatic breast carcinoma. Int J Mol Sci 2013;14(9):18973–88 [Published 2013 Sep 13] https://doi.org/10.3390/ijms140918973.

[53] Gaudet MM, Lacey Jr JV, Lissowska J, et al. Genetic variation in CYP17 and endometrial cancer risk. Hum Genet 2008;123:155–62.

[54] Liu P, Keller JR, Ortiz M, et al. Bcl11a is essential for normal lymphoid development. Nat Immunol 2003;4 (6):525–32.

[55] Lazarus KA, Hadi F, Zambon E, et al. BCL11A interacts with SOX2 to control the expression of epigenetic regulators in lung squamous carcinoma. Nat Commun 2018;9(1):3327.

[56] Huang HT, Chen SM, Pan LB, Yao J, Ma HT. Loss of function of SWI/SNF chromatin remodeling genes leads to genome instability of human lung cancer. Oncol Rep 2015;33(1):283–91. https://doi.org/10.3892/or.2014.3584 [Epub 2014 Nov 3] 25370573.

[57] O'Mara TA, Glubb DM, Painter JN, et al. Comprehensive genetic assessment of the ESR1 locus identifies a risk region for endometrial cancer. Endocr Relat Cancer 2015;22(5):851–61. https://doi.org/10.1530/ERC-15-0319.

[58] Wujcicka W, Zając A, Stachowiak G. Impact of MDM2, TP53 and P14ARF polymorphisms on endometrial cancer risk and onset. In Vivo 2019;33(3):917–24. https://doi.org/10.21873/invivo.11559.

[59] Bafligil C, Thompson DJ, Lophatananon A, et al. Association between genetic polymorphisms and endometrial cancer risk: a systematic review. J Med Genet 2020;57(9):591–600. https://doi.org/10.1136/jmedgenet-2019-106529.

[60] Zając A, Stachowiak G, Pertyński T, Romanowicz H, Wilczyński J, Smolarz B. Association between MDM2 SNP309 polymorphism and endometrial cancer risk in polish women. Pol J Pathol 2012;4:278–83. https://doi.org/10.5114/pjp.2012.32776.

[61] Zhang J, Zhang Y, Zhang Z. Association of rs2279744 and rs117039649 promoter polymorphism with the risk of gynecological cancer: a meta-analysis of case-control studies. Medicine (Baltimore) 2018;97(2):e9554. https://doi.org/10.1097/MD.0000000000009554.

[62] Kang S, Wang D-J, Li W-S, et al. Association of p73 and MDM2 polymorphisms with the risk of epithelial ovarian cancer in Chinese women. Int J Gynecol Cancer 2009;19(4):572–7.

[63] Knappskog S, Bjørnslett M, Myklebust LM, et al. The MDM2 promoter SNP285C/309G haplotype diminishes Sp1 transcription factor binding and reduces risk for breast and ovarian cancer in Caucasians. Cancer Cell 2011;19(2):273–82.

[64] Malakar AK, Choudhury D, Halder B, Paul P, Uddin A, Chakraborty S. A review on coronary artery disease, its risk factors, and therapeutics: MALAKAR et al. J Cell Physiol 2019;234(10):16812–23.

[65] Gallo JE, Ochoa JE, Warren HR, et al. Hypertension and the roles of the 9p21.3 risk locus: Classic findings and new association data. Int J Cardiol Hypertens 2020;7(100050):100050.

[66] Gu F, Pfeiffer RM, Bhattacharjee S, et al. Common genetic variants in the 9p21 region and their associations with multiple tumours. Br J Cancer 2013;108(6):1378–86. https://doi.org/10.1038/bjc.2013.7.

[67] Bishop DT, Demenais F, Iles MM, et al. Genome-wide association study identifies three loci associated with melanoma risk. Nat Genet 2009;41(8):920–5. https://doi.org/10.1038/ng.411.

[68] Chaturvedi S, Hoffman RM, Bertino JR. Exploiting methionine restriction for cancer treatment. Biochem Pharmacol 2018;154:170–3. https://doi.org/10.1016/j.bcp.2018.05.003 [Epub 2018 May 4] 29733806.

[69] Fedoriw A, Rajapurkar SR, O'Brien S, et al. Anti-tumor activity of the type I PRMT inhibitor, GSK3368715, synergizes with PRMT5 inhibition through MTAP loss. Cancer Cell 2019;36(1):100–14 [e25].

[70] Cheng TH, Thompson D, Painter J, et al. Meta-analysis of genome-wide association studies identifies common susceptibility polymorphisms for colorectal and endometrial cancer near SH2B3 and TSHZ1. Sci Rep 2015;5:17369 [Published 2015 Dec 1] https://doi.org/10.1038/srep17369.

[71] Perez-Garcia A, Ambesi-Impiombato A, Hadler M, et al. Genetic loss of SH2B3 in acute lymphoblastic leukemia. Blood 2013;122(14):2425–32. https://doi.org/10.1182/blood-2013-05-500850.

[72] Carvajal-Carmona LG, O'Mara TA, Painter JN, et al. Candidate locus analysis of the TERT-CLPTM1L cancer risk region on chromosome 5p15 identifies multiple independent variants associated with endometrial cancer risk. Hum Genet 2015;134(2):231–45. https://doi.org/10.1007/s00439-014-1515-4.

[73] Bojesen SE, Pooley KA, Johnatty SE, et al. Multiple independent variants at the TERT locus are associated with telomere length and risks of breast and ovarian cancer. Nat Genet 2013;45(4):371–384e3842. https://doi.org/10.1038/ng.2566.

[74] Kote-Jarai Z, Saunders EJ, Leongamornlert DA, et al. Fine-mapping identifies multiple prostate cancer risk loci at 5p15, one of which associates with TERT expression. Hum Mol Genet 2013;22(20):4239.

[75] Prescott J, McGrath M, Lee IM, Buring JE, De Vivo I. Telomere length and genetic analyses in population-based studies of endometrial cancer risk. Cancer 2010;116(18):4275–82. https://doi.org/10.1002/cncr.25328.

[76] Simmen RCM, Pabona JMP, Velarde MC, Simmons C, Rahal O, Simmen FA. The emerging role of Krüppel-like factors in endocrine-responsive cancers of female reproductive tissues. J Endocrinol 2010;204(3):223–31.

[77] Liu T, Wang X, Zhai J, Wang Q, Zhang B. Long Noncoding RNA UCA1 facilitates endometrial cancer development by regulating KLF5 and RXFP1 gene expressions. Cancer Biother Radiopharm 2020. https://doi.org/10.1089/cbr.2019.3278 [Epub ahead of print] 32412793.

[78] Viola L, Londero AP, Bertozzi S, et al. Prognostic role of Krüppel-like factors 5, 9, and 11 in endometrial endometrioid cancer. Pathol Oncol Res 2020;26(4):2265–72.

[79] Painter JN, et al. Genetic overlap between endometriosis and endometrial cancer: evidence from cross-disease genetic correlation and GWAS meta-analyses. Cancer Med 2018;7(5):1978–87. https://doi.org/10.1002/cam4.1445.

[80] Ortiz B, Fabius AW, Wu WH, et al. Loss of the tyrosine phosphatase PTPRD leads to aberrant STAT3 activation and promotes gliomagenesis. Proc Natl Acad Sci U S A 2014;111(22):8149–54. https://doi.org/10.1073/pnas.1401952111.

[81] Ortiz B, White JR, Wu WH, Chan TA. Deletion of Ptprd and Cdkn2a cooperate to accelerate tumorigenesis. Oncotarget 2014;5(16):6976–82. https://doi.org/10.18632/oncotarget.2106.

[82] Catalogue of Somatic Mutations in Cancer (COSMIC) database, http://cancer.sanger.ac.uk/cosmic. [Accessed 2 February 2021].

[83] Westin SN, Broaddus RR. Personalized therapy in endometrial cancer: challenges and opportunities. Cancer Biol Ther 2012;13(1):1–13. https://doi.org/10.4161/cbt.13.1.18438.

[84] Prat J, Ribé A, Gallardo A. Hereditary ovarian cancer. Hum Pathol 2005;36(8):861–70.

[85] Jervis S, Song H, Lee A, et al. Ovarian cancer familial relative risks by tumour subtypes and by known ovarian cancer genetic susceptibility variants. J Med Genet 2014;51(2):108–13.

[86] Kuchenbaecker KB, Ramus SJ, Tyrer J, et al. Identification of six new susceptibility loci for invasive epithelial ovarian cancer. Nat Genet 2015;47(2):164–71. https://doi.org/10.1038/ng.3185.

[87] Jones MR, Kamara D, Karlan BY, Pharoah PDP, Gayther SA. Genetic epidemiology of ovarian cancer and prospects for polygenic risk prediction. Gynecol Oncol 2017;147(3):705–13.

[88] Loveday C, Turnbull C, Ramsay E, et al. Germline mutations in RAD51D confer susceptibility to ovarian cancer. Nat Genet 2011;43(9):879–82.

[89] Norquist BM, Harrell MI, Brady MF, et al. Inherited mutations in women with ovarian carcinoma. JAMA Oncol 2016;2(4):482–90. https://doi.org/10.1001/jamaoncol.2015.5495.

[90] Phelan CM, Kuchenbaecker KB, Tyrer JP, et al. Identification of 12 new susceptibility loci for different histotypes of epithelial ovarian cancer. Nat Genet 2017;49(5):680–91. https://doi.org/10.1038/ng.3826.

[91] Glubb DM, Thompson DJ, Aben KKH, et al. Cross-cancer genome-wide association study of endometrial cancer and epithelial ovarian cancer identifies genetic risk regions associated with risk of both cancers. Cancer Epidemiol Biomarkers Prev 2021;30(1):217–28.

[92] Earp MA, Kelemen LE, Magliocco AM, et al. Genome-wide association study of subtype-specific epithelial ovarian cancer risk alleles using pooled DNA. Hum Genet 2014;133(5):481–97.

[93] Goode EL, Chenevix-Trench G, Song H, et al. A genome-wide association study identifies susceptibility loci for ovarian cancer at 2q31 and 8q24. Nat Genet 2010;42(10):874–9. https://doi.org/10.1038/ng.668.

[94] Kar SP, Berchuck A, Gayther SA, et al. Common genetic variation and susceptibility to ovarian cancer: current insights and future directions. Cancer Epidemiol Biomarkers Prev 2018;27(4):395–404.

[95] Permuth-Wey J, Lawrenson K, Shen HC, et al. Identification and molecular characterization of a new ovarian cancer susceptibility locus at 17q21.31. Nat Commun 2013;4:1627. https://doi.org/10.1038/ncomms2613.

[96] Bolton KL, Tyrer J, Song H, et al. Corrigendum: common variants at 19p13 are associated with susceptibility to ovarian cancer. Nat Genet 2016;48(1):101.

[97] Codd V, Nelson CP, Albrecht E, et al. Identification of seven loci affecting mean telomere length and their association with disease. Nat Genet 2013;45(4):422–427e4272. https://doi.org/10.1038/ng.2528.

[98] Song H, Ramus SJ, Tyrer J, et al. A genome-wide association study identifies a new ovarian cancer susceptibility locus on 9p22.2. Nat Genet 2009;41(9):996–1000. https://doi.org/10.1038/ng.424.

[99] O'Driscoll L, McMorrow J, Doolan P, et al. Investigation of the molecular profile of basal cell carcinoma using whole genome microarrays. Mol Cancer 2006;5:74 [Published 2006 Dec 15] https://doi.org/10.1186/1476-4598-5-74.

[100] O'Cearbhaill RE. Using PARP inhibitors in advanced ovarian Cancer. Oncology (Williston Park) 2018;32 (7):339–43.

[101] Veneris JT, Matulonis UA, Liu JF, Konstantinopoulos PA. Choosing wisely: selecting PARP inhibitor combinations to promote anti-tumor immune responses beyond BRCA mutations. Gynecol Oncol 2020;156(2):488–97. https://doi.org/10.1016/j.ygyno.2019.09.021. Epub 2019 Oct 17 31630846.

[102] Pharoah PDP, Tsai Y-Y, Ramus SJ, et al. GWAS meta-analysis and replication identifies three new susceptibility loci for ovarian cancer. Nat Genet 2013;45(4):362–70 [370e1-2].

[103] Chen K, Ma H, Li L, et al. Genome-wide association study identifies new susceptibility loci for epithelial ovarian cancer in Han Chinese women. Nat Commun 2014;5(1):4682.

[104] Duan Y-F, Zhu F, Li X-D, et al. Association between ABO gene polymorphism (rs505922) and cancer risk: a meta-analysis. Tumour Biol 2015;36(7):5081–7.

[105] Couch FJ, Wang X, McGuffog L, et al. Genome-wide association study in BRCA1 mutation carriers identifies novel loci associated with breast and ovarian cancer risk. PLoS Genet 2013;9(3):e1003212. https://doi.org/10.1371/journal.pgen.1003212.

[106] Feng Y, Peng Z, Liu W, et al. Evaluation of the epidemiological and prognosis significance of ESR2 rs3020450 polymorphism in ovarian cancer. Gene 2019;710:316–23.

[107] Li J, Liu R, Tang S, et al. The effect of long noncoding RNAs HOX transcript antisense intergenic RNA single-nucleotide polymorphisms on breast cancer, cervical cancer, and ovarian cancer susceptibility: a meta-analysis: LI et al. J Cell Biochem 2019;120(5):7056–67.

[108] Pinto R, Assis J, Nogueira A, Pereira C, Pereira D, Medeiros R. Rethinking ovarian cancer genomics: where genome-wide association studies stand? Pharmacogenomics 2017;18(17):1611–25.

[109] Johnatty SE, Tyrer JP, Kar S, et al. Genome-wide analysis identifies novel loci associated with ovarian cancer outcomes: findings from the ovarian cancer association consortium. Clin Cancer Res 2015;21(23):5264–76.

[110] Gao B, Lu Y, Nieuweboer AJM, et al. Genome-wide association study of paclitaxel and carboplatin disposition in women with epithelial ovarian cancer. Sci Rep 2018;8(1):1508.

[111] Harris HR, Terry KL. Polycystic ovary syndrome and risk of endometrial, ovarian, and breast cancer: a systematic review. Fertil Res Pract 2016;2:14 [Published 2016 Dec 5] https://doi.org/10.1186/s40738-016-0029-2.

[112] Harris HR, Cushing-Haugen KL, Webb PM, et al. Association between genetically predicted polycystic ovary syndrome and ovarian cancer: a Mendelian randomization study. Int J Epidemiol 2019;48(3):822–30. https://doi.org/10.1093/ije/dyz113.

[113] Luo H, Xu X, Yang J, et al. Genome-wide somatic copy number alteration analysis and database construction for cervical cancer. Mol Genet Genomics 2020;295(3):765–73.

[114] vansson E, Juko-Pecirep I, Erlich H, et al. Pathway-based analysis of genetic susceptibility to cervical cancer in situ: HLA-DPB1 affects risk in Swedish women. Genes Immun 2011;12:605–14. https://doi.org/10.1038/gene.2011.40.

[115] Chen D, Juko-Pecirep I, Hammer J, et al. Genome-wide association study of susceptibility loci for cervical cancer. J Natl Cancer Inst 2013;105(9):624–33.

[116] Kuguyo O, Tsikai N, Thomford NE, et al. Genetic susceptibility for cervical cancer in African populations: what are the host genetic drivers? OMICS 2018;22(7):468–83.

[117] Cheng L, Guo Y, Zhan S, Xia P. Association between HLA-DP gene polymorphisms and cervical cancer risk: a meta-analysis. Biomed Res Int 2018;2018:7301595 [Published 2018 Jun 13] https://doi.org/10.1155/2018/7301595.

[118] Yang YC, Chang TY, Chen TC, Lin WS, Lin CL, Lee YJ. Replication of results from a cervical cancer genome-wide association study in Taiwanese women. Sci Rep 2018;8(1):15319 [Published 2018 Oct 17] https://doi.org/10.1038/s41598-018-33430-x.

[119] Chen H, Wang T, Huang S, Zeng P. New novel non-MHC genes were identified for cervical cancer with an integrative analysis approach of transcriptome-wide association study. J Cancer 2021;12(3):840–8 [Published 2021 Jan 1] https://doi.org/10.7150/jca.47918.

[120] Leo PJ, Madeleine MM, Wang S, et al. Defining the genetic susceptibility to cervical neoplasia-A genome-wide association study [published correction appears in PLoS Genet. 2018 Mar 1;14 (3):e1007257]. PLoS Genet 2017;13(8):e1006866 [Published 2017 Aug 14] https://doi.org/10.1371/journal.pgen.1006866.

[121] Ramachandran D, Schürmann P, Mao Q, et al. Association of genomic variants at the human leukocyte antigen locus with cervical cancer risk, HPV status and gene expression levels. Int J Cancer 2020;147(9):2458–68.

[122] Chen D, Enroth S, Ivansson E, Gyllensten U. Pathway analysis of cervical cancer genome-wide association study highlights the MHC region and pathways involved in response to infection. Hum Mol Genet 2014;23(22):6047–60.

[123] Wang SS, Bratti MC, Rodríguez AC, et al. Common variants in immune and DNA repair genes and risk for human papillomavirus persistence and progression to cervical cancer. J Infect Dis 2009;199(1):20–30. https://doi.org/10.1086/595563.

[124] Pandey NO, Chauhan AV, Raithatha NS, et al. Association of TLR4 and TLR9 polymorphisms and haplotypes with cervical cancer susceptibility [published correction appears in Sci Rep. 2019 Dec 4;9(1):18658]. Sci Rep 2019;9(1):9729 [Published 2019 Jul 5] https://doi.org/10.1038/s41598-019-46077-z.

[125] Chen X, Jiang J, Shen H, Hu Z. Genetic susceptibility of cervical cancer. J Biomed Res 2011;25(3):155–64. https://doi.org/10.1016/S1674-8301(11)60020-1.

[126] Wang S, Sun H, Jia Y, et al. Association of 42 SNPs with genetic risk for cervical cancer: an extensive meta-analysis. BMC Med Genet 2015;16:25 [Published 2015 Apr 15] https://doi.org/10.1186/s12881-015-0168-z.

[127] Tamandani DMK, Shekari M, Suri V. Interleukin-12 gene polymorphism and cervical cancer risk. Am J Clin Oncol 2009;32(5):524–8.

[128] Turner DM, Williams DM, Sankaran D, Lazarus M, Sinnott PJ, Hutchinson IV. An investigation of polymorphism in the interleukin-10 gene promoter. Eur J Immunogenet 1997;24(1):1–8.

[129] Zoodsma M, Nolte IM, Schipper M, et al. Interleukin-10 and Fas polymorphisms and susceptibility for (pre)neoplastic cervical disease. Int J Gynecol Cancer 2005;15(s3):282–90.

[130] Wang K, Jiao Z, Chen H, et al. The association between rs1800872 polymorphism in interleukin-10 and risk of cervical cancer: a meta-analysis. Medicine (Baltimore) 2021;100(3):e23892.

[131] Ivansson EL, Juko-Pecirep I, Gyllensten UB. Interaction of immunological genes on chromosome 2q33 and IFNG in susceptibility to cervical cancer. Gynecol Oncol 2010;116:544–8.

[132] Wang Y, Yang J, Huang J, Tian Z. Tumor necrosis factor-alpha polymorphisms and cervical cancer: evidence from a meta-analysis. Gynecol Obstet Invest 2020;85(2):153–8.

[133] Singh H, Jain M, Sachan R, Mittal B. Association of TNFA (−308G>A) and IL-10 (-819C>T) promoter polymorphisms with risk of cervical cancer. Int J Gynecol Cancer 2009;19:1190–4.

[134] Calhoun ES, McGovern RM, Janney CA, et al. Host genetic polymorphism analysis in cervical cancer. Clin Chem 2002;48(8):1218–24.

[135] Johanneson B, Chen D, Enroth S, Cui T, Gyllensten U. Systematic validation of hypothesis-driven candidate genes for cervical cancer in a genome-wide association study. Carcinogenesis 2014;35(9):2084–8.

[136] Duvlis S, Dabeski D, Cvetkovski A, Mladenovska K, Plaseska-Karanfilska D. Association of TNF-a (rs361525 and rs1800629) with susceptibility to cervical intraepithelial lesion and cervical carcinoma in women from republic of North Macedonia. Int J Immunogenet 2020;47(6):522–8.

[137] McKay J, Tenet V, Franceschi S, et al. Immuno-related polymorphisms and cervical cancer risk: The IARC multicentric case-control study [published correction appears in PLoS One. 2017 Jul 7;12 (7):e0181285]. PLoS One 2017;12(5):e0177775 [Published 2017 May 15] https://doi.org/10.1371/journal.pone.0177775.

[138] Joo J, Omae Y, Hitomi Y, et al. The association of integration patterns of human papilloma virus and single nucleotide polymorphisms on immune- or DNA repair-related genes in cervical cancer patients. Sci Rep 2019;9(1):13132 [Published 2019 Sep 11] https://doi.org/10.1038/s41598-019-49523-0.

[139] Han L, Husaiyin S, Ma C, Wang L, Niyazi M. TNFAIP8L1 and FLT1 polymorphisms alter the susceptibility to cervical cancer amongst uyghur females in China. Biosci Rep 2019;39(7). https://doi.org/10.1042/BSR20191155. BSR20191155. [Published 2019 Jul 18].

[140] Das S, Naher L, Aka TD, et al. The ECCR1 rs11615, ERCC4rs2276466, XPC rs2228000 and XPC rs2228001 polymorphisms increase the cervical cancer risk and aggressiveness in the Bangladeshi population. Heliyon 2021; 7(1):e05919 [Published 2021 Jan 9] https://doi.org/10.1016/j.heliyon.2021.e05919.

[141] Tan SC, Ankathil R. Genetic susceptibility to cervical cancer: role of common polymorphisms in apoptosis-related genes. Tumour Biol 2015;36(9):6633–44.

[142] Martínez-Nava GA, Fernández-Niño JA, Madrid-Marina V, Torres-Poveda K. Cervical cancer genetic susceptibility: a systematic review and meta-analyses of recent evidence. PLoS One 2016;11(7):e0157344 [Published 2016 Jul 14] https://doi.org/10.1371/journal.pone.0157344.

[143] Hashemi M, Aftabi S, Moazeni-Roodi A, Sarani H, Wiechec E, Ghavami S. Association of CASP8 polymorphisms and cancer susceptibility: a meta-analysis. Eur J Pharmacol 2020;881(173201):173201.

[144] Takeuchi F, Kukimoto I, Li Z, et al. Genome-wide association study of cervical cancer suggests a role for ARRDC3 gene in human papillomavirus infection. Hum Mol Genet 2019;28(2):341–8.

[145] Ding B, Sun W, Han S, Cai Y, Ren M, Shen Y. Cytochrome P450 1A1 gene polymorphisms and cervical cancer risk: a systematic review and meta-analysis. Medicine (Baltimore) 2018;97(13):e0210.

[146] Wongpratate M, Ishida W, Phuthong S, Natphopsuk S, Ishida T. Genetic polymorphisms of the human cytochrome P450 1A1 (CYP1A1) and cervical cancer susceptibility among Northeast Thai women. Asian Pac J Cancer Prev 2020;21(1):243–8 [Published 2020 Jan 1] 10.31557/APJCP.2020.21.1.243.

[147] Gong JM, Shen Y, Shan WW, He YX. The association between MTHFR polymorphism and cervical cancer. Sci Rep 2018;8(1):7244 [Published 2018 May 8] https://doi.org/10.1038/s41598-018-25726-9.

[148] Sohrabi A, Bassam-Tolami F, Imani M. The impact of MTHFR 1298 a>C and 677 C>T gene polymorphisms as susceptibility risk factors in cervical intraepithelial neoplasia related to HPV and sexually transmitted infections. J Obstet Gynaecol India 2020;70(6):503–9. https://doi.org/10.1007/s13224-020-01363-z.

[149] Silva NNT, Santos ACS, Nogueira VM, Carneiro CM, Lima AA. 3'UTR polymorphism of Thymidylate Synthase gene increased the risk of persistence of pre-neoplastic cervical lesions. BMC Cancer 2020;20(1):323 [Published 2020 Apr 15] https://doi.org/10.1186/s12885-020-06811-7.

[150] Weng SL, Ng SC, Lee YC, et al. The relationships of genetic polymorphisms of the long noncoding RNA growth arrest-specific transcript 5 with uterine cervical cancer. Int J Med Sci 2020;17(9):1187–95.

[151] Hsu W, Liu L, Chen X, Zhang Y, Zhu W. LncRNA CASC11 promotes the cervical cancer progression by activating Wnt/beta-catenin signaling pathway. Biol Res 2019;52(1):33 [Published 2019 Jun 29] https://doi.org/10.1186/s40659-019-0240-9.

[152] Zhang J, Liu X, You LH, Zhou RZ. Significant association between long non-coding RNA HOTAIR polymorphisms and cancer susceptibility: a meta-analysis. Onco Targets Ther 2016;9:3335–43 [Published 2016 Jun 1] https://doi.org/10.2147/OTT.S107190.

[153] Weng SL, Wu WJ, Hsiao YH, Yang SF, Hsu CF, Wang PH. Significant association of long non-coding RNAs HOTAIR genetic polymorphisms with cancer recurrence and patient survival in patients with uterine cervical cancer. Int J Med Sci 2018;15(12):1312–9 [Published 2018 Aug 6] https://doi.org/10.7150/ijms.27505.

[154] Zou S-H, Du X, Lin H, Wang P-C, Li M. Paclitaxel inhibits the progression of cervical cancer by inhibiting autophagy via lncRNARP11-381N20.2. Eur Rev Med Pharmacol Sci 2018;22(10):3010–7.

[155] Chen D, Gyllensten U. Lessons and implications from association studies and post-GWAS analyses of cervical cancer. Trends Genet 2015;31(1):41–54.

[156] Shi Y, Li L, Hu Z, et al. A genome-wide association study identifies two new cervical cancer susceptibility loci at 4q12 and 17q12. Nat Genet 2013;45(8):918–22.

[157] Masuda T, Low SK, Akiyama M, et al. GWAS of five gynecologic diseases and cross-trait analysis in Japanese. Eur J Hum Genet 2020;28(1):95–107. https://doi.org/10.1038/s41431-019-0495-1.

[158] Adebamowo SN, Adeyemo AA, Rotimi CN, et al. Genome-wide association study of prevalent and persistent cervical high-risk human papillomavirus (HPV) infection. BMC Med Genet 2020;21(1):231 [Published 2020 Nov 23] https://doi.org/10.1186/s12881-020-01156-1.

[159] Li X, Huang K, Zhang Q, et al. Genome-wide association study identifies four SNPs associated with response to platinum-based neoadjuvant chemotherapy for cervical cancer. Sci Rep 2017;7:41103 [Published 2017 Jan 25] https://doi.org/10.1038/srep41103.

[160] Salih HR, Rammensee H-G, Steinle A. Cutting edge: down-regulation of MICA on human tumors by proteolytic shedding. J Immunol 2002;169(8):4098–102.

[161] Ashiru O, Boutet P, Fernández-Messina L, et al. Natural killer cell cytotoxicity is suppressed by exposure to the human NKG2D ligand MICA*008 that is shed by tumor cells in exosomes. Cancer Res 2010;70(2):481–9. https://doi.org/10.1158/0008-5472.CAN-09-1688.

# Overcoming drug resistance in cervical cancer: Chemosensitizing agents and targeted therapies

*Anum Jalil[a], James Wert[a], Aimen Farooq[a], and Sarfraz Ahmad[b]*

[a]Department of Internal Medicine, AdventHealth, Orlando, FL, United States [b]Gynecologic Oncology, AdventHealth Cancer Institute, Florida State University, University of Central Florida, Orlando, FL, United States

## Abstract

There are an estimated 13,800 new cases of cervical cancer that will be diagnosed in the United States in 2020 alone with an estimated number of deaths reaching up to 4290. Surgery, radiation therapy, and chemotherapy are well-established treatment options for cervical cancer. Cisplatin has long been a first-line chemotherapeutic agent, first used in the early 1970s for the treatment of advanced cervical cancer. As a result of long ubiquitous use of cisplatin in the past decades, drug resistance has presented with a new challenge in the treatment of cervical cancer. To overcome drug resistance and potentially develop novel therapies, understanding the drug resistance mechanisms is required. New therapies designed to circumvent resistance among these different mechanistic pathways include *Salmonella typhimurium* A1-R, erlotinib, and cyclin-dependent kinase inhibitors. This chapter is a discussion from the most up-to-date data on overcoming drug resistance in cervical cancer and their related mechanisms. The understanding of these pathways and the ability to create novel therapies to target them will optimistically allow for the overcoming of drug resistance.

## Abbreviations

| | |
|---|---|
| **5-FU** | 5-fluorouracil |
| **ATM** | ataxia telangiectasia mutated serine/threonine kinase |
| **CCRT** | concurrent chemoradiotherapy |
| **CDK** | cyclin-dependent kinases |
| **CHK2** | checkpoint kinase 2 |
| **CIN** | cervical intraepithelial neoplasia |
| **CSCs** | cancer stem cells |
| **CTR1** | copper transporter-1 |
| **CYP** | cytochrome P |
| **DDR** | DNA damage repair |

R. Basha, S. Ahmad (eds.)
*Overcoming Drug Resistance in Gynecologic Cancers*
ISBN 978-0-12-824299-5
https://doi.org/10.1016/B978-0-12-824299-5.00010-1

| EGFR-TKI | epidermal growth factor receptor tyrosine kinase inhibitor |
|---|---|
| HPV | human papilloma virus |
| IL1 | interleukin 1 |
| LEEP | loop electrosurgical excision procedure |
| MAPK | mitogen-activated protein kinase |
| MDC1 | mediator of DNA damage checkpoint 1 |
| MDR | multi-drug resistance |
| mTOR | mammalian target of rapamycin |
| NF-κB | nuclear factor kappa-B |
| P-gp | P-glycoprotein |
| PI3K | phosphatidylinositol-3-kinase |
| ROS | reactive oxygen species |
| RT | radiation therapy |
| TNFα | tumor necrosis factor α |
| VEGF | vascular endothelial growth factor |

## Conflict of interest

No potential conflicts of interest were disclosed.

## Introduction/Background

Cancer of the uterine cervix is one of the preventable human malignancies, but still remains a major cause of death among women worldwide, resulting in more than 300,000 deaths annually. It is the fourth most common type of malignancy in women and mostly affects women living in low- or middle-income countries likely with secondary to limited access to health care including screening, vaccination, and treatment opportunities [1–3]. Persistent infection with certain types of human papillomavirus (HPV) is the cause for almost all cervical cancers. HPV infection, although common in healthy women, only rarely causes cervical cancer. Other risk factors that may contribute to the persistence of HPV infection as well as development of cervical cancer are suppressed immune system, a high number of childbirths, multiple sexual partners, long-term use of contraceptive pills, and cigarette smoking.

Preinvasive lesions in the cervix usually have no symptoms, but once the abnormal cells develop into cancer and invade local tissue, the most common symptom is abnormal vaginal bleeding. Most precancers of the cervix develop gradually, and cervical cancer is usually preventable if a woman gets regular screening. Screening prevents cervical cancer by early detection and treatment of precancerous lesions, detects invasive cancer earlier in the course, and provides a better opportunity for successful treatment. The cervical cancer death rate in the year 2015 (2.3 per 100,000 population) was less than half than that in 1975 (5.6 per 100,000 population) due to declines in incidence and the early detection of cancer through screening [4].

The Pap smear test is a simple procedure in which a sample of cells is collected from the cervical area and examined under a microscope. The HPV test detects HPV infection associated with the cancer and can predict the risk of cervical cancer many years in advance. The HPV tests also identify women at risk for adenocarcinoma, a type of cervical cancer, often missed by the Pap test, that accounts for 28% of the total cases of cervical cancer. The American Cancer Society, the American Society for Colposcopy and Cervical Pathology, and the American Society for Clinical Pathology recommend screening for women ages 21 to 65 years and the incorporation of the HPV test with the Pap test for ages 30 to 65 years. Vaccines that protect against the types of HPV that cause 90% of cervical cancers are available, but unfortunately, the rate of immunization continues to be low in the United States; in 2016, 50% of girls ages 13 to 17 years—and only 36% at age 13 years—were up to date with the HPV vaccination series [4]. The HPV vaccines neither protect against all types of HPV, nor do they protect against already established infection, which is the reason why vaccinated women should still undergo screening for cervical cancer.

Cervical lesions that are precancerous can be treated with a loop electrosurgical excision procedure (LEEP), cryotherapy, laser ablation, or conization. Invasive cancers are treated with surgery or radiation therapy usually combined with chemotherapy. Advanced-stage disease is often treated by chemotherapy alone. However, for metastatic, persistent, or recurrent cervical cancer, targeted therapy is added to the standard chemotherapy to improve overall survival outcomes.

In this chapter, we sought to succinctly discuss most up-to-date data pertaining to overcoming drug resistance in cervical cancer and their related mechanisms. The understanding of these pathways/mechanisms and the ability to create novel therapies to target them will optimistically allow for the overcoming of drug resistance in cervical cancer.

## Pathophysiology of cervical cancer

Human papillomavirus (HPV) is a small deoxyribonucleic acid (DNA) virus consisting of approximately 7900 base pairs. It plays the central role in the development of cervical neoplasia and is detected in upto 99.7% of cervical cancer cases [5]. More than 40 genital mucosal HPV types have been identified, among which approximately 15 are known to have oncogenic potential. Types 16 and 18 are the most common types of HPV, isolated in over 70% of all cervical cancers, among which upto 50% are HPV type 16 [6]. However, it is important to note that not all infections with these high-risk HPV types (16 or 18) progress to cancer. Diverse oncogenic potential can be seen even within a single oncogenic type of HPV [7]. Major steps involved in cervical cancer development [8] are infection of the epithelium at the *Cervical Transformation* zone, which is the junction between the glandular epithelium of the endocervical canal and the squamous epithelium of ectocervix by the oncogenic variety of HPV. Persistence of the infection results in the progression of a certain group of epithelial cells to precancer, which then develops into cancer and invades through the basement membrane.

Although HPV infection is very common in the genital tract, progression to cancer is seen in a comparatively small proportion of women infected with the virus. It is estimated that upto 75%–80% of the adults who are sexually active will be infected with HPV before the age of 50 years [9, 10]. However, most infections do not persist, and the virus alone without other risk factors is not a sufficient cause of cervical neoplasia. If the infection persists, it takes an average of 15 years from acquiring the initial infection to progression to high-grade cervical intraepithelial neoplasia (CIN) and, eventually, invasive cancer [11].HPV infection in the genital tract can also cause a variety of other diseases including genital warts and cancers of the vulva, vagina, anus, and penis [12, 13].

## Cervical cancer treatment options

Treatment of cervical cancer depends on several factors, most importantly the stage of cancer and the individual patient parameters. Early stage cervical cancer is usually managed with surgery including fertility sparing surgery [14]. Radiotherapy (RT) is an option. The pros and cons of surgery versus RT for early stage cervical cancer have been studied, with results showing equal effectiveness. Decision is made by keeping in mind the patient factors, age, incidence of posttreatment sexual dysfunction, complications, and morbidity associated with each of the options [15]. For the locally advanced-stage disease or metastatic cervical cancer, the gold-standard treatment is concurrent chemoradiotherapy (CCRT). The standard treatment usually includes cisplatin-based chemotherapy. Many gynecologic cancers are being treated with platinum-based compounds like paclitaxel, cisplatin, and carboplatin in combination with other therapeutic agents [16, 17]. Cisplatin can be used as a singleagent or in combination with other therapeutic agents, e.g., fluorouracil. Cisplatin is usually given once a week to improve medical compliance and to reduce potential side effects and toxicity [18]. In addition, one trial has been done with female patients with locally advanced-stage or recurrent cervical cancer to evaluate the concurrent weekly paclitaxel plus RT regimen. It showed 8 of 13 patients with complete response [19]. Paclitaxel stabilizes microtubules and is also a front-line agent in the treatment of cervical cancer. Radiotherapy options include external-beam radiotherapy or brachytherapy and are given in combination with chemotherapy, as noted earlier [20].

Recurrent or metastatic cervical cancer disease carries a relatively poor prognosis. Now, we have treatment regimens combining chemotherapy with bevacizumab, showing promising results and improved overall survival for these patients. Bevacizumab is a monoclonal antibody against vascular endothelial growth factor (VEGF) [21, 22].

## Resistance to chemotherapeutic agents and ways to overcome resistance

Chemotherapy is an integral component for the management of cervical cancer, most importantly for the unfortunate patients with recurrent, advanced-stage, or metastatic disease. Cancer cells can develop resistance to chemotherapeutic agents especially when tumor recurs, thus decreasing the efficacy of the medication [23]. Resistance to chemotherapeutic agents is one of the main obstacles that oncologists are facing. With the advancements in biomolecular sciences, there has been a remarkable increase in the number of chemotherapeutic agents, leading to alarming levels of resistance in cancer cells. More exposure to medications makes tumor cells even more prone to developing resistance and leads to widespread metastasis.

Cisplatin is a widely used chemotherapy agent that plays an important role in the treatment of many kinds of malignancies including cervical cancer. It is a cytotoxic agent that interacts with DNA molecule to form crosslinked compounds named as "adducts." Cisplatin enters cells via several transporters including copper transporter-1 (CTR1). Cisplatin-induced DNA damage leads to activation of several downstream molecular pathways like p53 and MAPK, ultimately resulting in apoptosis. Formation of reactive oxygen species (ROS) is also involved in this process. On the other hand, Akt is a member of a family of kinases and plays an important role in cell survival as depicted in Fig. 1. Inhibitors of Akt can potentially lead to cellular sensitization to cisplatin-induced DNA damage [24]. There are several different mechanisms that are involved in the development of resistance to cisplatin (Fig. 1). Increased repair of DNA is a major mechanism of drug resistance. Decreased uptake of drug into the tumor cells and increased inactivation of drug are other major mechanisms.

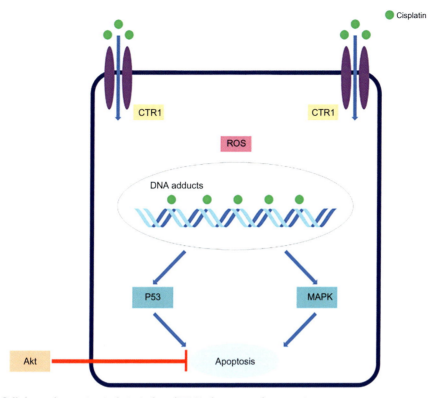

FIG. 1    Cellular pathways in cisplatin-induced DNA damage and apoptosis.

These resistance mechanisms lead to failure of induction of apoptosis in cancer cells, resulting in unchecked tumor replication and spread of malignancy [25, 26]. In addition, abnormal DNA methylation also plays a part in the development of drug resistance as it inactivates various genes that play a crucial role in the cytotoxic activity of drug [27].

The intracellular regulatory pathways that are involved in the unchecked cellular proliferation and survival are valuable targets for novel chemosensitizing agents. Our knowledge of these signaling cascades can be utilized to help overcome the resistance to chemotherapy agents. Many clinical trials are underway to develop newer agents to target chemo-resistant cancer cells. Some of them are briefly described as follows.

## Salmonella typhimurium A1-R

Nanoparticle albumin bound (nabpaclitaxel) is a chemotherapeutic option for several cancers. *Salmonella typhimurium* A1-R induces cytotoxicity, thus potentially enhancing the effect of chemotherapeutic agent with which it is administered like nabpaclitaxel. It increases the levels of inflammatory markers interleukin1-β (IL-1β) and tumor necrosis factor-alpha (TNF-α) inside the tumor, thereby killing the tumor cells. *S. typhimurium* A1-R can be combined with nabpaclitaxel to overcome paclitaxel resistance. This can be further studied in clinical trials as a potential option to overcome resistance in cervical cancer [28].

## Erlotinib

Erlotinib is an inhibitor of epidermal growth factor receptor tyrosine kinase (EGFR-TKI). In vitro and in vivo studies have shown that erlotinib can be potentially utilized to sensitize the tumor cells to chemotherapeutic drugs as shown in Fig. 2. It targets cancer stem cells (CSCs) which possess high regenerative potential. The CSCs have many mechanisms by which they evade the effects of chemotherapeutic agents like increased DNA repair and decreased apoptosis. The EGFR pathway is involved in the pathogenesis of many cancers, and its constitutive activation leads to downstream sequential cascade that results in the unchecked proliferation of cancer cells. Inhibition of this pathway is the basis of action of many chemotherapy agents. Erlotinib can inhibit IL6 and can potentially overcome resistance to paclitaxel in tumor cells [29].

## Inhibitors of PI3K/AKT/mTOR pathway

The phosphatidylinositol-3-kinase (PI3K)/Akt and the mammalian target of rapamycin (mTOR) signaling pathways play an integral role(s) in tumorigenesis as depicted in Fig. 2. It is involved in cellular proliferation, growth, and survival. Inhibitors of this pathway can be potentially combined with nabpaclitaxel to increase the sensitivity of resistant tumor cells. Large clinical trials are underway [30, 31].

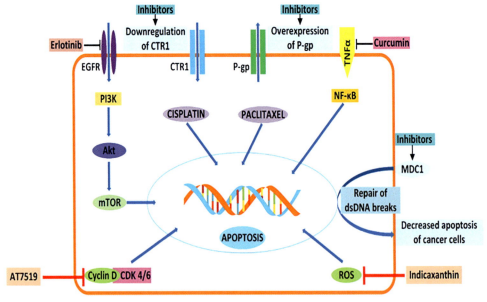

FIG. 2    Mechanisms of resistance to cisplatin and paclitaxel and ways to overcome resistance. Some of the potential targets for inhibitors that are shown can be utilized to chemosensitize cervical tumor cells.

# MicroRNAs

MicroRNAs are small RNA molecules that act on messenger RNAs (mRNA) and modulate gene expression. They are involved in many biological as well as pathological processes. Minute changes in microRNAs can occur inside the tumor cells, making them resistant to chemotherapeutic agents. Many such alterations have been identified in cancer cells. MicroRNAs control a variety of different cellular pathways. An important example is cytochrome P (CYP)-450 family of cytochrome. MicroRNAs regulate the gene expression of some of these enzymes as well, and their downregulation can lead to overexpression of these proteins, leading to downstream cellular effects. One such microRNA called miR-375 is overexpressed as a result of paclitaxel therapy, thus leading to decreased paclitaxel cellular sensitivity. It targets E-cadherin which plays an important role in cell-to-cell adhesion and keeping cellular growth in check. Suppression of E-cadherin can lead to cellular proliferation and unchecked growth. This pathway can be targeted to improve sensitivity of cervical cancer cells and overcoming chemoresistance [32–34].

# Cyclin-dependent kinases

Cyclin-dependent kinases (CDK) also play important role(s) in cell cycle and cellular proliferation. The CDK inhibitors can be used in combination with chemotherapeutic agents to overcome drug resistance. One such inhibitor being studied in clinical trials is AT7519, which can lead to apoptosis particularly in paclitaxel-resistant cervical cancer cells and also in colon cancer cells that are resistant to 5-FU [35].

# P-glycoprotein

P-glycoprotein (P-gp) is a membrane transporter pump that leads to efflux of anticancer drugs from tumor cells, as shown in Fig. 2. Overexpression of p-glycoprotein is one of the major mechanisms involved in multidrug resistance (MDR). Over the past years, significant efforts have been made toward the development of inhibitors of P-glycoprotein. Despite that, there have not been many promising developments. One important factor is that P-gp is present in many normal cells of the body as well, so inhibitors of this molecule can have much potential toxicity. This is one of the avenues that can be explored further [36].

# Mediator of DNA damage checkpoint 1

The human papilloma virus (HPV) has been implicated in the development and progression of cervical cancer. At the molecular level, HPV encodes for a subset of specific proteins that activate the pathways ataxia telangiectasia-mutated serine/threonine kinase (ATM) and checkpoint kinase 2 (CHK2). This pathway becomes upregulated as a result of DNA damage. As a result, it has been a target for molecular genomics because the downregulation of this pathway would inhibit further replication of HPV.

The key protein that is activated as a part of this pathway is mediator of the DNA damage checkpoint 1 (MDC1). The MDC1 is required in order to mount a DNA damage repair (DDR) response as shown in Fig. 2. The ability to selectively target the checkpoint regulator MDC1 presents the possibility to circumvent cisplatin drug resistance. Downregulation of MDC1 cell lines was tested with notable increased efficacy of cisplatin in these MDC1-silenced cell lines. Cervical cancer cell lines with high expression of MDC1 resulted in a lack of response to cisplatin. The mechanism for this finding is explained by downstream regulation of yH2AX phosphorylation which influences the DDR proteins to aid in the repair of damaged chromatin. In cell lines with silenced MDC1, yH2AX was unable to be upregulated sufficiently to recruit the DNA damage repair proteins, allowing cisplatin to have a greater effect on depleting cervical cancer cells.

Understanding the ATM and CHK2 pathways is fundamental to oncology research aimed at preventing and circumventing the cisplatin drug resistance. In addition to cisplatin, these cell-specific pathways are an area for target of novel therapies. This study has demonstrated that certain checkpoint regulators such as MDC1 are potential targets for improving the efficacy of current chemotherapeutic agents such as cisplatin [37].

## Phytochemicals and their role in decreasing chemoresistance

Many newer agents are being developed to overcome drug resistance in cancer treatment. These agents can play pivotal roles in the management of cancers in the coming future. Among these newer agents are some natural compounds or phytochemicals as well. Phytochemicals are natural plant-derived compounds like tea polyphenols, curcumin, resveratrol, and flavonoids that possess healing properties. In vivo studies have shown that a phytochemical compound called curcumin has the ability to sensitize cervical tumor cells to platinum-based chemotherapeutic agents like paclitaxel. The mechanism(s) of action of curcumin involves downregulation of several important pathways such as the nuclear factor kappa-B (NF-κB) and mitogen-activated protein kinase (MAPK) that play pivotal roles in cellular proliferation and survival outcomes. Curcumin inhibits the activation of these pathways and enhances the effects of paclitaxel, which also helps in minimizing toxicities. In addition, decreased activity of tumor-suppressor gene p53 is commonly involved in carcinogenesis. This pathway involves oncoproteins E6 and E7. Studies have showed that curcumin can inhibit the expression of these oncoproteins. Phytochemicals can be used as an adjuvant therapy to overcome drug resistance [38–40].

Nanophytochemicals [41] are also being studied as potential anticancer agents. Whether they can help in sensitizing cancer cells to conventional chemotherapy agents remains to be determined. The limiting factor for using phytochemicals in fighting cancer is mainly their low bioavailability, necessitating larger quantities for any significant effect. Many studies have shown that combination of phytochemicals and chemotherapeutic agents can be one modality to overcome drug resistance for many gynecologic malignancies. In one study [42], photochemical Indicaxanthin was combined with cisplatin against cervical cancer cells, and it showed favorable results by increasing the cellular apoptosis via reactive oxygen species and oxidative stress, as shown in Fig. 2. Another study done by Jakubowicz-Gil et al. reported that a flavonoid called quercetin enhances the effects of cisplatin on cervical cancer cells by sensitizing them to apoptosis [43]. Similarly, many phytochemicals studied in combination with paclitaxel show promising results, e.g., apigenin which belongs to the class of flavonoids as well [44].

## Conclusions and future perspectives

Cervical cancer was first described in 400 BCE by a Greek physician Hippocrates. It would then take until the 1970s for the HPV to be detected. Over 100 different HPV genotypes have been sequenced, with at least 12 known to cause cervical cancer. In the United States, it is estimated that there will be 13,800 new cases of invasive cervical cancer diagnosed in the year 2020 and 4290 women will succumb to the disease. Cisplatin is a first-line chemotherapeutic agent that has been used since the early 1970s for the treatment of advanced-stage or recurrent cervical cancer. Chemotherapy is usually combined with radiation therapy for better clinical outcomes. With the increasing use of chemotherapeutic agents, drug resistance concerns are emerging and have posed significant problems for oncologists in recent years. Mechanisms by which resistance occurs include DNA repair, decreased drug uptake by cancerous cells, and inactivation of the drug. Understanding the mechanisms by which resistance occurs allows for novel therapies to be developed to overcome this problem. Newer studies and clinical trials are needed to develop agents acting on specific signaling pathways to sensitize tumor cells to commonly used chemotherapeutic agents. One such avenue for exploration is field of phytochemicals. Combination of phytochemicals and chemotherapeutic drugs has been a new ray of hope for many malignancies including gynecological cancers. The future is promising with multiple trials currently underway with the aim to ensure cervical cancer can continue to be treatable and offer patients the most effective therapy with better qualityoflife and affordability.

## Author contributions

Dr. Anum Jalil primarily conceived the idea and subsequently all the coauthors have diligently contributed to the development and preparation of this research chapter, including the literature search, concept organization, data interpretation and writings, and approval for publication.

## Financial disclosures

None to disclose.

# References

[1] Chibwesha CJ, Stringer JSA. Cervical cancer as a global concern: contributions of the dual epidemics of HPV and HIV. JAMA 2019;322(16):1558–60. https://doi.org/10.1001/jama.2019.16176.

[2] Campos NG, Sharma M, Clark A, et al. The health and economic impact of scaling cervical cancer prevention in 50 low- and lower-middle-income countries. Int J Gynaecol Obstet 2017;138(Suppl. 1):47–56. https://doi.org/10.1002/ijgo.12184.

[3] Bray F, Ferlay J, Soerjomataram I, Siegel RL, Torre LA, Jemal A. Global cancer statistics 2018: GLOBOCAN estimates of incidence and mortality worldwide for 36 cancers in 185 countries. CA Cancer J Clin 2018;68(6): 394–424. https://doi.org/10.3322/caac.21492.

[4] American Cancer Society. Cancer facts & figures 2018. Atlanta: American Cancer Society; 2018.

[5] Walboomers JM, Jacobs MV, Manos MM, et al. Human papillomavirus is a necessary cause of invasive cervical cancer worldwide. J Pathol 1999;189(1):12–9. https://doi.org/10.1002/(SICI)1096-9896(199909)189:1<12:: AID-PATH431>3.0.CO;2-F.

[6] de Sanjose S, Quint WG, Alemany L, et al. Human papillomavirus genotype attribution in invasive cervical cancer: a retrospective cross-sectional worldwide study. Lancet Oncol 2010;11(11):1048–56. https://doi.org/10.1016/S1470-2045(10)70230-8.

[7] Hildesheim A, Schiffman M, Bromley C, et al. Human papillomavirus type 16 variants and risk of cervical cancer. J Natl Cancer Inst 2001;93(4):315–8. https://doi.org/10.1093/jnci/93.4.315.

[8] Wang K, Zhou B, Zhang J, et al. Association of signal transducer and activator of transcription 3 gene polymorphisms with cervical cancer in Chinese women. DNA Cell Biol 2011;30(11):931–6. https://doi.org/10.1089/dna.2010.1179.

[9] Schiffman M, Castle PE, Jeronimo J, Rodriguez AC, Wacholder S. Human papillomavirus and cervical cancer. Lancet 2007;370(9590):890–907. https://doi.org/10.1016/S0140-6736(07)61416-0.

[10] Manhart LE, Holmes KK, Koutsky LA, et al. Human papillomavirus infection among sexually active young women in the United States: implications for developing a vaccination strategy. Sex Transm Dis 2006;33 (8):502–8. https://doi.org/10.1097/01.olq.0000204545.89516.0a.

[11] Workowski KA, Bolan GA. Centers for disease control and prevention. sexually transmitted diseases treatment guidelines, 2015 [published correction appears in MMWR Recomm Rep. 2015 Aug 28;64(33):924]. MMWR Recomm Rep 2015;64(RR-03):1–137.

[12] Kjaer SK, de Villiers EM, Cağlayan H, et al. Human papillomavirus, herpes simplex virus and other potential risk factors for cervical cancer in a high-risk area (Greenland) and a low-risk area (Denmark)-a second look. Br J Cancer 1993;67(4):830–7. https://doi.org/10.1038/bjc.1993.152.

[13] Kayes O, Ahmed HU, Arya M, Minhas S. Molecular and genetic pathways in penile cancer. Lancet Oncol 2007;8 (5):420–9. https://doi.org/10.1016/S1470-2045(07)70137-7.

[14] Annede P, Gouy S, Haie-Meder C, Morice P, Chargari C. Place of radiotherapy and surgery in the treatment of cervical cancer patients. Cancer Radiother 2019;23(6–7):737–44. https://doi.org/10.1016/j.canrad.2019.07.151.

[15] Landoni F, Maneo A, Colombo A, et al. Randomised study of radical surgery versus radiotherapy for stage Ib-IIa cervical cancer. Lancet 1997;350(9077):535–40. https://doi.org/10.1016/S0140-6736(97)02250-2.

[16] Kumar L, Harish P, Malik PS, Khurana S. Chemotherapy and targeted therapy in the management of cervical cancer. Curr Probl Cancer 2018;42(2):120–8. https://doi.org/10.1016/j.currproblcancer.2018.01.016.

[17] Koh WJ, Abu-Rustum NR, Bean S, et al. Cervical cancer, version 3.2019, NCCN clinical practice guidelines in oncology. J Natl Compr Canc Netw 2019;17(1):64–84. https://doi.org/10.6004/jnccn.2019.0001.

[18] Kim YS, Shin SS, Nam JH, et al. Prospective randomized comparison of monthly fluorouracil and cisplatin versus weekly cisplatin concurrent with pelvic radiotherapy and high-dose rate brachytherapy for locally advanced cervical cancer. Gynecol Oncol 2008;108(1):195–200. https://doi.org/10.1016/j.ygyno.2007.09.022.

[19] Cerrotta A, Gardan G, Cavina R, et al. Concurrent radiotherapy and weekly paclitaxel for locally advanced or recurrent squamous cell carcinoma of the uterine cervix. A pilot study with intensification of dose. Eur J Gynaecol Oncol 2002;23(2):115–9.

[20] Vordermark D. Radiotherapy of cervical cancer. Oncol Res Treat 2016;39(9):516–20. https://doi.org/10.1159/000448902.

[21] Cohen PA, Jhingran A, Oaknin A, Denny L. Cervical cancer. Lancet 2019;393(10167):169–82. https://doi.org/10.1016/S0140-6736(18)32470-X.

[22] Tewari KS, Sill MW, Long 3rd HJ, et al. Improved survival with bevacizumab in advanced cervical cancer [published correction appears in N Engl J Med. 2017 Aug 17;377(7):702]. N Engl J Med 2014;370(8): 734–43. https://doi.org/10.1056/NEJMoa1309748.

[23] Zhu H, Luo H, Zhang W, Shen Z, Hu X, Zhu X. Molecular mechanisms of cisplatin resistance in cervical cancer. Drug Des Devel Ther 2016;10:1885–95. https://doi.org/10.2147/DDDT.S106412.

[24] Basu A, Krishnamurthy S. Cellular responses to Cisplatin-induced DNA damage. J Nucleic Acids 2010;2010:201367. https://doi.org/10.4061/2010/201367.

[25] Amable L. Cisplatin resistance and opportunities for precision medicine. Pharmacol Res 2016;106: 27–36. https://doi.org/10.1016/j.phrs.2016.01.001.

[26] Siddik ZH. Cisplatin: mode of cytotoxic action and molecular basis of resistance. Oncogene 2003;22(47): 7265–79. https://doi.org/10.1038/sj.onc.1206933.

[27] Chang X, Monitto CL, Demokan S, et al. Identification of hypermethylated genes associated with cisplatin resistance in human cancers. Cancer Res 2010;70(7):2870–9. https://doi.org/10.1158/0008-5472.CAN-09-3427.

[28] Miyake K, Murata T, Murakami T, et al. Tumor-targeting *Salmonella typhimurium* A1-R overcomes nab-paclitaxel resistance in a cervical cancer PDOX mouse model. Arch Gynecol Obstet 2019;299(6):1683–90. https://doi.org/10.1007/s00404-019-05147-3.

[29] Lv Y, Cang W, Li Q, et al. Erlotinib overcomes paclitaxel-resistant cancer stem cells by blocking the EGFR-CREB/GRβ-IL-6 axis in MUC1-positive cervical cancer. Oncogenesis 2019;8(12):70. https://doi.org/10.1038/s41389-019-0179-2.

[30] Liu JJ, Ho JY, Lee HW, et al. Inhibition of phosphatidylinositol 3-kinase (PI3K) signaling synergistically potentiates antitumor efficacy of paclitaxel and overcomes paclitaxel-mediated resistance in cervical cancer. Int J Mol Sci 2019;20(14):3383. https://doi.org/10.3390/ijms20143383.

[31] Sharma P, Abramson VG, O'Dea A, Pathak HB, Pessetto ZY, Wang YY, et al. Clinical and biomarker results from phase I/II study of PI3K inhibitor BYL 719 (alpelisib) plus nab-paclitaxel in HER2-negative metastatic breast cancer. J Clin Oncol 2018;36:1018. https://doi.org/10.1200/JCO.2018.36.15_suppl.1018.

[32] Lu TX, Rothenberg ME. MicroRNA. J Allergy Clin Immunol 2018;141(4):1202–7. https://doi.org/10.1016/j.jaci.2017.08.034.

[33] Geretto M, Pulliero A, Rosano C, Zhabayeva D, Bersimbaev R, Izzotti A. Resistance to cancer chemotherapeutic drugs is determined by pivotal microRNA regulators. Am J Cancer Res 2017;7(6):1350–71.

[34] Shen Y, Zhou J, Li Y, et al. miR-375 mediated acquired chemo-resistance in cervical cancer by facilitating EMT. PLoS ONE 2014;9(10):e109299. https://doi.org/10.1371/journal.pone.0109299.

[35] Xi C, Wang L, Yu J, Ye H, Cao L, Gong Z. Inhibition of cyclin-dependent kinases by AT7519 is effective to overcome chemoresistance in colon and cervical cancer. Biochem Biophys Res Commun 2019;513(3):589–93. https://doi.org/10.1016/j.bbrc.2019.04.014.

[36] Waghray D, Zhang Q. Inhibit or evade multidrug resistance P-glycoprotein in cancer treatment. J Med Chem 2018;61(12):5108–21. https://doi.org/10.1021/acs.jmedchem.7b01457.

[37] Singh N, Bhakuni R, Chhabria D, Kirubakaran S. MDC1 depletion promotes cisplatin induced cell death in cervical cancer cells. BMC Res Notes 2020;13(1):146. https://doi.org/10.1186/s13104-020-04996-5.

[38] Farrand L, Oh SW, Song YS, Tsang BK. Phytochemicals: a multitargeted approach to gynecologic cancer therapy. Biomed Res Int 2014;2014:890141. https://doi.org/10.1155/2014/890141.

[39] Sreekanth CN, Bava SV, Sreekumar E, Anto RJ. Molecular evidences for the chemosensitizing efficacy of liposomal curcumin in paclitaxel chemotherapy in mouse models of cervical cancer. Oncogene 2011;30(28):3139–52. https://doi.org/10.1038/onc.2011.23.

[40] Roy M, Mukherjee S. Reversal of resistance towards cisplatin by curcumin in cervical cancer cells. Asian Pac J Cancer Prev 2014;15(3):1403–10. https://doi.org/10.7314/apjcp.2014.15.3.1403.

[41] Yadav N, Parveen S, Banerjee M. Potential of nano-phytochemicals in cervical cancer therapy. Clin Chim Acta 2020;505:60–72. https://doi.org/10.1016/j.cca.2020.01.035.

[42] Allegra M, D'Anneo A, Frazzitta A, et al. The phytochemical indicaxanthin synergistically enhances cisplatin-induced apoptosis in HeLa cells via oxidative stress-dependent p53/p21$^{\text{waf1}}$ axis. Biomolecules 2020;10(7): E994. https://doi.org/10.3390/biom10070994.

[43] Jakubowicz-Gil J, Paduch R, Piersiak T, Głowniak K, Gawron A, Kandefer-Szerszeń M. The effect of quercetin on pro-apoptotic activity of cisplatin in HeLa cells. Biochem Pharmacol 2005;69(9):1343–50. https://doi.org/10.1016/j.bcp.2005.01.022.

[44] Xu Y, Xin Y, Diao Y, et al. Synergistic effects of apigenin and paclitaxel on apoptosis of cancer cells. PLoS ONE 2011;6(12):e29169. https://doi.org/10.1371/journal.pone.0029169.

# Immune response, inflammation pathway gene polymorphisms, and the risk of cervical cancer

*Henu Kumar Verma[a], Batoul Farran[b], and Lakkakula V.K.S. Bhaskar[c]*

[a]Institute of Endocrinology and Oncology, Naples, Italy [b]Department of Hematology & Medical Oncology, Winship Cancer Institute, Emory University, Atlanta, GA, United States [c]Department of Zoology, Guru Ghasidas Vishwavidyalaya, Bilaspur, Chhattisgarh, India

## Abstract

Numerous studies have provided novel insights into the potential role of genetic polymorphisms as prognostic markers in gynecologic cancers, particularly cervical cancer. Cervical cancer is the second most common and primary cause of cancer-related deaths in developing countries. Deep research into single-nucleotide polymorphisms (SNPs) may help to explain potential differences in distinct cancers. Candidate SNPs may be involved in immune cell regulation, cell-cycle control, DNA damage and repair, cancer metabolism, and apoptosis. In the present chapter, we have summarized important findings of genetic association studies exploring the role of glutathione-*S*-transferase (GST), interleukins (ILs), p53 codon 72, tumor necrosis factor A gene (TNFA), human leukocyte antigen (HLA), Fas gene polymorphism, murine double-minute two homologs (MDM2), and interferons-γ (IFNG) gene polymorphism with cervical cancer. Genetic association studies will require further validation to provide direction for future research leading to better clinical outcomes for patients with cervical cancer.

## Abbreviations

| | |
|---|---|
| CDC | Centers for Disease Control and Prevention (US) |
| GC | gynecologic cancer |
| GST | glutathione-*S*-transferase |
| GSTO | glutathione-*S*-transferase omega |
| HLA | human leukocyte antigen |
| HPV | human papillomavirus |

R. Basha, S. Ahmad (eds.)
*Overcoming Drug Resistance in Gynecologic Cancers*
ISBN 978-0-12-824299-5
https://doi.org/10.1016/B978-0-12-824299-5.00011-3

**HRM**    high-resolution melting analysis
**IFNG**   interferons-γ gene
**ILs**    interleukins
**JNK**    Jun N-terminal kinase
**MDM2**   murine double-minute two homologs
**NF-κB**  nuclear factor κB
**SICL**   squamous intraepithelial cervical lesions
**SNPs**   single-nucleotide polymorphisms
**TNFA**   tumor necrosis factor A gene

## Conflict of interest

No potential conflicts of interest were disclosed.

## Introduction

Gynecologic cancer is any type of cancer that begins in the reproductive organs of a woman and is the main cause of morbidity and mortality. According to the data from the US Centers for Disease Control and Prevention (CDC), approximately 94,000 women were diagnosed with gynecologic cancers per year from 2012 to 2016 in the United States. Previous family history, overweight, age, and human papillomavirus (HPV) have been identified as major risk factors for developing gynecologic cancers [1]. Mainly three types of gynecologic cancers have been highly observed in women, with first being endometrial, second ovarian, and third cervical cancer. Frequency of gynecologic cancers among women varies depending on their ethnicity and the type of cancer. Despite significant advances in the first-line chemotherapy, mortality rates for all gynecologic cancers, including cervical cancer, are higher due to recurrence of the disease and remain the second most common cancer in women worldwide [2–4].

The uterine cervix and the ovary are the primary hormone-responsive tissues of the female reproductive system. The uterine cervix is highly responsive to estrogen; it is estimated that over 99% of the patients with cervical cancer are positive for HPV infections and associated with the progression of the severity of cancer [5, 6].

In the United States, endometrial cancer is the most common gynecologic cancer, with around 65,620 new cases and 13,940 deaths estimated in the year 2020. The least common type is cervical cancer (6230 new cases and 1450 deaths estimated for the year 2020) [7]. The Global Cancer Observatory 2018 database showed that there were almost 570,000 cases and 311,000 deaths from cervical cancer in the year 2018 [8]. In addition, approximately 22,240 new cases were diagnosed in the same year, and 14,070 deaths from ovarian cancer were reported [9].

Among all gynecologic cancers, cervical cancer is the major cause of death in developing countries because in the initial stages, none of the clinical tests can diagnose or identify cervical cancer when treatment can be most successful. Since there is no way to know or detect any cancer, it is very important to identify the risk factors for gynecologic cancers associated with biomarkers during their early symptoms and prior to tumor development.

Although a number of pharmacogenomics studies have suggested newer insights into the genetic variant and its association with environmental factors leading to tumorigenesis [10–12], this genetic predisposition accounts for approximately 5%–10% of gynecologic cancer cases in populations, and it may be of more public health significance [13]. To date, there has been no systematic review to guide future cervical cancer research. Nevertheless, there is an urgent need for a new growing number of case-control studies of specific genetic markers that may have a role in cancer susceptibility. In this chapter, we have focused on cervical cancer and summarized the following important points: (i) the role of genetic polymorphism in cervical cancer, (ii) identifying possible candidate biomarkers and early detection pathways for cervical cancer, and (iii) the current state of research and future directions.

## Single-nucleotide polymorphisms in susceptibility to cervical tumors

Single-nucleotide polymorphisms (SNPs) are substitutions of a single nucleotide at a specific position in the human genome. The current focus of cancer research endeavors has been SNPs targeting to explore the variation in different cancer susceptibilities. More than 335 million SNPs, which are functionally "silent" and do not alter the protein expression of a gene, can be found in the promoters, exons, introns, as well as 5'- and 3'-UTRs of the human genome [14]. Various studies have identified links between a specific SNP and cell-cycle control, DNA damage repair, metabolism, and immunity, as well as genetic susceptibility to cervical cancer development [15–21]. From a clinical perspective, it is essential to understand the role of SNPs in regulating several mechanisms required for the molecular pathogenesis of different cancers, including cervical cancer (Fig. 1).

Therefore, single-base substitutions have a specific role in increasing the risk of cancer and vary depending on their location. Distantly located SNPs from the defined gene regions decrease or increase the gene transcription levels through long-range cis-acting elements [22]. In addition, the association of the SNPs with a clinical effect is not necessarily due to a causal relationship. Some SNPs are present at the same location of a gene that has been inherited from a group of alleles known as "haplotype." In population genetics, the word "linkage disequilibrium" is characterized as a nonrandom combination of two or more loci in the general population. Subsequently, the correlation of SNPs and their role in raising the risk of tumorigenesis may be simple since they combine the disequilibrium with another polymorphism in the same chromosomal area. Moreover, these SNPs are not accurate in interpreting cancer susceptibility as a highly multifactorial disease, and a specific SNP of a particular gene is neither mandatory nor necessary for cancer development, as the distribution of certain SNPs may differ on the basis of ethnic or racial background.

## Immune regulator gene

The HPV infection is known to be the main etiologic agent of cervical cancer. Previous studies have shown that most cases of cervical cancer develop into preinvasive lesions than invasive cancer through chronic HPV infection in deeper tissues and may spread to other parts of the body (metastasize) [23]. However, the precise molecular mechanisms that contribute to HPV-induced cervical cancer are not well established. Several immunological factors, particularly innate and adaptive immune pathways, are responsible for triggering the immune response and the progression of HPV infection [24].

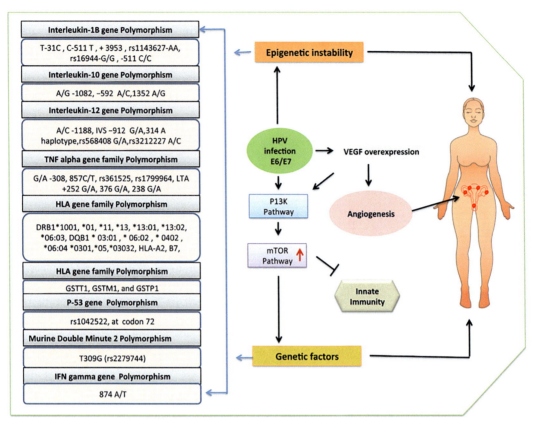

FIG. 1   Gynecologic cancers-associated genetic polymorphisms. Survival pathway (PI3K-Akt) induces genetic instability and variations in genetic factors that influence gynecologic cancer progression. PI3K-Akt pathway and angiogenesis are induced by VEGF. PI3K stimulated mTOR pathway that leads to inhibition of innate immunity in cervical cancer.

Furthermore, it has been found that genetic polymorphisms have a causal relationship to HPV infection, in particular polymorphisms such as glutathione-S-transferase (GST), interleukins (ILs), p53 codon 72, tumor necrosis factor A gene (TNFA), human leukocyte antigen genes (HLA), Fas gene polymorphism, murine double-minute homologs (MDM2) gene, and interferons-γ gene (IFNG) have been shown to be associated with HPV infection and cancer progression [25].

## Interleukin-1B gene

Interleukin-1 (IL-1) is the first mediator for innate immunity and inflammation. IL-1 has a large family of cytokines consisting of IL-1 alpha, IL-1 beta, and IL-1 receptor antagonists. It is understood that host genetic factors have a vital role to play in cervical cancer [26]. Among them, the IL-1β gene, encoding IL-1beta cytokine, contains several SNPs. IL-1β is a proinflammatory cytokine produced by monocytes. A number of studies have shown that

IL-1β plays a significant role in chronic inflammation and increased the risk of developing cervical cancer [27, 28]. Findings also indicate that the mean plasma level of IL-1β in cervical cancer is substantially elevated relative to the healthy controls' plasma levels. More than 75% cases were also associated with a 2.49-fold elevated risk of cervical cancer. Furthermore, patients carrying a T/T genotype in C-511T variant were significantly higher in cervical cancer in Egyptian women [29].

Qian et al. [30] found that mean plasma levels of IL-1β were significantly higher in patients with cervical cancer relative to controls ($P < .0002$) and were associated with a 1.74-fold increased risk. In addition to IL-1β T-31C TC/CC and C-511 T CT/TT, polymorphism was significantly correlated with an increased risk of cervical cancer among Chinese women [30]. Singh et al. [31] found that IL-1β (-511C/T) polymorphism in cases was significantly higher than controls ($P = .012$) and was prone to cervical cancer [31]. Sobti et al. [32] have found that $IL-1\beta + 3953$ polymorphism is significantly associated with an increased risk of cervical cancer [32]. The MassARRAY analysis of 267 patients with cervical cancer recently revealed that IL-1β rs1143627-AA and rs16944-GG were associated with an increased risk of cervical cancer in the Chinese Uygur population [33]. Although Kang et al. [34] reported a contradictory outcome of the association between IL-1β-511 polymorphism and cervical cancer risk in the Korean population [34], a number of case-control studies have analyzed the role of IL-1β polymorphisms and the progression of cervical cancer. However, the results are not consistent, and further studies with a larger sample size are desirable for further validation.

## Interleukin-10 gene

Interleukin-10 (IL-10) is a Th2-type cytokine located on chromosome 1, with antiinflammatory properties, involved in the development of various cancers [35]. Numerous studies have identified higher levels of Th2 and lower levels of Th1 cytokines in premalignant cervical legions [36, 37]. The initial study investigated 77 patients with higher genotype frequency of −1082 A/G polymorphism of IL-10, documenting that they were significantly associated with the development of cervical cancer [38]. Shrestha et al. [39] demonstrated that IL-10 GCC haplotype was correlated with higher rates of IL-10 production, and these polymorphisms influenced the pathogenesis of HPV and cervical cancer progression [39].

High-resolution melting analysis (HRM) in 171 HPV-positive Brazilian women has shown that 1082 A/G polymorphism is significantly associated with the susceptibility to cervical cancer progression [40]. Gao et al. [41] observed that cervical cancer patients with HPV16 infection had a significantly higher frequency of genotype AA and allele A. In addition, this multivariate analysis showed a positive association of −1082 loci and the TNM stages, vascular infiltration, and lymph node metastases in Chinese women [41]. Another study indicated that IL-10 −1082 G alleles were significantly associated with 29.8% higher production of interleukin-10 levels in Japanese cervical cancer patients [42]. Furthermore, Torres-Poveda et al. [43] found that −592 polymorphism is associated with an increased risk of squamous intraepithelial cervical lesions (SICL), but no association at loci −819, −1082, and −1352 in Mexican women [43]. Accordingly, based on all these cumulative data, the role of IL-10 polymorphism in the progression and development of cervical cancer requires further studies for clearer evidence.

## Interleukin-12 gene

Interleukin 12 (IL-12) is a cytokine produced by dendritic cells, neutrophils, and macrophages in response to antigenic stimulation. IL-12 is a disulfide-linked heterodimeric cytokine consisting of a 40-kD heavy-chain and a 35-kD light-chain subunit [44]. The first clinical trial findings in 34 women with cervical cancer showed that IL-12 therapy was significantly linked to the enrichment of lymphoproliferative immune responses to HPV16 E4, E6, and E7 peptides but not in survival [45]. Tamandani et al. [46] have shown that IL-12β (1188 A/C) registers a 1.8% increase in the risk of cervical cancer compared to the healthy controls. In addition to passive AC and AC+CC smokers, genotypes were associated with a 2.8% elevated risk compared to the nonsmokers [46].

The antitumor activity of the IL-12 gene was documented in an in vitro HPV16-positive tumor model, where it was shown that the IL-12 gene promotes the activation of the cellular immune response through the expression of a cytokine profile of the type Th1 [47]. A study found that the presence of functional 1188 (A/C) IL-12B gene polymorphism is associated with cervical cancer women in Brazil [48]. Han et al. [49] showed that the allelic frequencies of the GG genotype of IVS2 −912 G/A gene polymorphism and +314 A haplotype of IL-12B gene are associated with an increased risk of cervical cancer as compared with the healthy controls in Korean women [49]. Similarly, Chen et al. [50] found that IL-12A rs568408 GA/AA and IL-12B rs3212227 AC/CC genetic polymorphism significantly increased the risk of 404 Chinese cervical cancer women [51]. Recently, Zhang et al. [51] have shown that the retroviral gene transfer of IL-12 in transgenic murine tumor models has increased the antitumor efficacy and low circulating IL-12 serum levels, indicating that this strategy can be used for cellular therapy [51].

## Interferons-γ gene

Interferon-gamma (IFN-γ) encoded by the IFNG gene at 2q33 is a soluble cytokine member of the type II class of interferons and is critical for the innate and adaptive immunity against viral infections [52]. A major study reported that the genotype frequency of IFN-γ genotype and HPV status are significantly correlated with cervical cancer susceptibility [53]. Kim et al. [54] reported that IFN-γ decreased the HPV-16 E6 and E7 transcription levels in cervical cancer cell lines, while IFN-γ increased the level of these transcripts in SiHa cells and showed conflicting clinical outcomes [54]. Tamandani et al. [55] reported that IFN-γ +874 gene polymorphisms AT and AA + AT increased the risk of cervical cancer, and passive smokers were highly susceptible to the increased risk as compared to the healthy controls in a north Indian population [55]. Similarly, Gangwar et al. [56] demonstrated that allele A of IFN-γ +874 gene polymorphism was associated with a 1.54-fold increase in the risk of cervical cancer, while other tobacco users modulated risk severity [56]. Ma et al. [57] found that higher methylation rates and lower mRNA expression of the IFN-γ gene were detected in cervical cancer tissues.

In addition, hypermethylation of the IFN-γ gene may be correlated with cervical cancer tumorigenesis [57]. While most studies reported IFN-γ gene polymorphism as a risk factor, some studies documented the exception of genotype and cervical cancer outcomes. For example, Zidi et al. [58] showed that IFN-γ +874 is a protective factor for cervical cancer among 122

Tunisian women [58]. A similar study in a Chinese Han Women population found that IFN-γ +874 A/T polymorphisms have no significant role in the progression of cervical cancer [59]. The 458 cases of cervical cancer of +874 (A/T) IFN polymorphism showed a significant difference between the African groups compared to the mixed population group. Furthermore, it seems that differences between the ethnicity and gene polymorphism in IFN-γ cytokines do not influence the development of invasive cervical cancer [60].

## Tumor necrosis factor a gene

Tumor necrosis factor-alpha (TNF-α) encoded by the TNFA gene is a pleiotropic inflammatory cytokine involved in the control of HPV infection. TNF-α plays a dual role in the induction and progression of tumors through the activation of various signaling pathways, including nuclear factor κB (NF-κB) and c-Jun N-terminal kinase (JNK) [61]. Studies reported in peer-reviewed literature show that the cervicovaginal wash fluid in 33 patients with cervical cancer has a significantly higher level of TNF-α [28]. In cervical cancer cases, most of the studies have focused on the −308 G/A (r1800629) promoter gene polymorphism because it has a direct effect on TNF-α gene regulation [62]. Zuo et al. [63] and Wang et al. [64] have shown that TNF-α −857 C/T and −308 G/A promoter gene polymorphisms are significantly associated with an increased risk of cervical cancer in Southwest Chinese women [63, 64]. Li et al. [65] found that patients with cervical cancer had a high risk of infection with papillomavirus and allele frequencies of rs361525, rs1800629, and rs1799964. TNF-α gene polymorphism was significantly associated with cervical cancer. Hence, it can potentially be used for diagnostic purposes [65].

Rotar et al. [66] have shown that there is a risk of invasive cervical cancer in the presence of A allele at-308 TNF-α [66]. Kohaar et al. [67] noted that there was a significant association between TNFA-308 G/A and LTA+252 A/G polymorphisms and cervical cancer, as well as a functional correlation between TNF-α-308 G/A and elevated plasma levels in Indian populations [67]. Du et al. [68] documented that TNF-α-308 AA genotypes may increase the susceptibility to cervical cancer by altering the immune response of an individual [68]. Furthermore, Nieves-Ramirez et al. [69] have shown that TNF-α −376 genotype G/A has a 2.48-fold higher risk of cervical cancer as compared to the healthy controls in Mexican women [69]. In addition, the recently published peer-reviewed metaanalysis of 27 studies showed that −308 G/A polymorphism is significantly associated with both the Asians and Caucasians, and −238 G/A polymorphism is associated with Asian cervical cancer patients [70].

While most published research studies point to an association of TNF-α gene polymorphisms with cervical cancer, a few reports have also documented contradictory observations. For instance, a study by Wang et al. [71] contradicted previous findings and revealed that the Chinese population displays no significant correlations between rs1800629 polymorphism and high-risk HPV infection [71]. Similarly, Govan et al. [72] showed no association between the TNF-α −308 polymorphism and cervical cancer risk among South Africans [72]. Nevertheless, since the distribution of the TNF-α gene polymorphism in different population groups was surprisingly diverse, these findings indicate that the rates of TNF-α development could be influenced by the ethnic inequality.

# Human leukocyte antigen genes

Human leukocyte antigen (HLA) is essential to the presence of viral antigens on the surface of tumor cells and promotes tumorigenesis [73, 74]. Several genetic factors play a vital role in the susceptibility of cervical cancer, including HLA polymorphisms, which have been assumed to be involved in cervical cancer pathogenesis through their role in the immunological management of HPV infection. Among them, Madeleine et al. [75] have shown that HLA DRB1*1001, DRB1*1101, and DQB1*0301 increased the risks of squamous cell cervical cancer in 315 women [75]. A previous study has also stated that the carriers of allele 184 at the MICA locus near the genes HLA-DQ and HLA-DR were strongly correlated with cervical cancer development [76].

Hosono et al. [77] showed that HLA-A*0206 incidence is significantly lower in such cases and that there is an inverse association of A*0206 with cervical cancer risk in Japanese women [77]. Safaeian et al. [78] found that HLA B*07:02 increased the risk and DRB1*13:01, DRB1*13:02, *06:03, and DQB1DQB1*06:04 decreased the risk of cervical cancer [78]. Recently, Chambuso et al. [79] found that HLA class IIDQB1*03:01 and DQB1*06:02 alleles had a protective role in cervical cancer in the HIV/HPV coinfected women in South Africa [79]. Ivansson et al. [80] reported that the HLA class II DQB1 alleles *0301, *0402, and *0602 increased the susceptibility to cancer, while *0501 and *0603 increased the susceptibility to cervical cancer [80]. Ghaderi et al. [81] documented that HLA II DR15- and DQ6-positive haplotype women have a 3.73-fold risk of cervical cancer [81]. Davidson et al. [82] showed a significant association in the frequencies of HLA-A2, HLA-B7, HLA-DRB1*01, HLA-DRB1*11, HLA-DRB1*13, HLA-DQB1*05, and HLA-DQB1*03032 with British cervical cancer women [82]. In addition, a large sample size metaanalysis showed a positive association between HLA-DPB1 at 03:01 alleles and the risk of cervical cancer, and HLA-DPB1 at 04:02 and HLA-DP at 3117027 G alleles reduced the risk of cervical cancer [83]. Besides, a recently peer-reviewed published study documented that HLA-A*30:01 and HLA-A*33:03 alleles are protective, while HLA-A*01:01 serves as the risk factor for cervical cancer in China [84].

# Glutathione-S-transferase gene polymorphisms

Glutathione-S-transferase (GST) enzyme plays a key role in protecting the cells and tissues from oxidative stress damage. The lack of polymorphic variants in the genes *GSTT1*, *GSTM1*, and *GSTP1* can reduce the functional activity of glutathione enzymes and increase the clinical severity of many diseases, including cancer [85]. A study showed an increase in the concentration of GSH after chemotherapy in cervical cancer patients, which may be vital for predicting the efficacy of the treatment [86]. Abbas et al. [87] reported that patients with GSTM1 in combination with TT1-null and TP1 (AG + GG) had a higher survival benefit for cervical cancer [87]. Later, it was shown that there is no significant relationship between the GST gene polymorphism and the treatment response to cervical cancer [88]. Based on the cumulative data, most recent studies are from the Asian populations, and the GST gene polymorphism has been shown to be associated with cervical cancer. As reported by Phuthong et al. [89], GSTP1 gene polymorphism influences the risk of cervical cancer among Northeast

Thai women in combination with high-risk HPV infection [89]. Zamani et al. [90] found that glutathione-S-transferase omega (GSTO) 1 plays a crucial role in the HPV-related cervical cancer susceptibility [90]. Thakre et al. [91] demonstrated that polymorphisms GSTM1 and GSTT1 play a crucial role in the development of cervical cancer and that tobacco chewing and alcohol use are rather not associated with the risk of disease [91].

Furthermore, Tacca et al. [92] documented that GSTM1- and GSTT1-null polymorphisms were not associated with cervical cancer risk, but GSTT1 gene polymorphism was associated with poor prognosis [92]. Recently, some metaanalyses have shown inconsistent results, including GSTM1-null genotype, which may contribute to the development of cervical cancer in the Chinese populations [93]. The GSTM1- and GSTT1-null polymorphism showed increased susceptibility to cervical cancer, and no association between the GSTP1 polymorphism and risk of cervical lesions was observed [94, 95]. Another report found that GSTM1-null genotype and GSTM1/GSTT1 were the risk factors for cervical cancer, and GSTT1 was not associated with the risk of cervical cancer [96].

## p53 codon polymorphism

Tumor protein p53 (TP53) is known as the guardian of the genome, a well-characterized tumor-suppressor gene that is involved in several signaling pathways, including apoptotic cell stress responses, transcription control, and cell-cycle arrest prosurvival response [97]. Mutation in the p53 gene not only disrupts the normal cell function, but also has the ability to develop and promote tumorigenesis. [98]. A study reported that SNPs (rs1042522) at codon 72 of the p53 gene in exon four results in the expression of either arginine (CGC) or proline (CCC) residues are associated with susceptibility for the development of cancer [99]. Jiang et al. [100] have shown that the genotype of Arg/Arg has been significantly associated with an increased risk of cervical cancer in the southern Chinese population [100]. Dokianakis and Spandidos [101] reported that p53 Arg/Arg may be a potential risk factor for the tumorigenesis in cervix [101]. Although some studies have found that there are no associations of codon 72 of the p53 gene polymorphism and risk of cervical cancer in women from the United Kingdom, India, United States, Thailand, and South Africa [102–108].

In addition, a pooled analysis of 7946 cases from 49 different worldwide studies recommended that there should be no association between cervical cancer and TP53 codon 72 polymorphism [109]. However, this global study suggested that the results of almost all of the peer-reviewed published studies are inconsistent and that larger sample size studies with detailed methodology are needed for a clearer picture.

## Murine double minute 2SNPs

The murine double minute 2 (MDM2) overexpression is associated with the development and progression of multiple cancers [110, 111]. Single-nucleotide polymorphism (SNP) in the first intronic region known as SNP309 (rs2279744) was shown to influence the MDM2 expression via p53 pathway. Several studies have documented that in MDM2 SNP309 gene polymorphism, replacing TT with GG increases the risk of cervical cancer. Singhal et al. [112]

have shown that MDM2 polymorphisms have a 3.5-fold higher risk of susceptibility for cervical cancer [112]. Zhuo et al. [113] suggested that the GG alleles of MDM2 T309G polymorphism have a higher risk for cervical cancer in the Asian populations [113].

Similarly, Jiang et al. [114] have shown that GG alleles may be a risk factor for cervical cancer in the Chinese population [114]. Recently, a study found that the TG genotype was associated with lower rates of risk for cervical cancer in the Philippine population [115]. Furthermore, a large sample size metaanalysis has also revealed that MDM2 T309G polymorphism is a risk factor for cervical cancer [116]. However, some studies are inconsistent with the previous results and have reported that the GG alleles of MDM2 T309G polymorphism have no significant association with the progression or development of cervical lesions [117–119].

## Fas gene promoter polymorphism

FAS is an apoptotic cell surface receptor located in chromosome 10q24.1 and plays a vital role in the physiological process that regulates normal homeostasis, altering any related gene to the apoptosis pathway likely to contribute to cancer pathogenesis [120]. The FAS gene regulates a number of genetic variants, among which −670 A/G, rs1800682, and FASL-844 T gene polymorphisms were found to be associated with cervical cancer risk [121, 122]. Ueda et al. [123] have found that the G allele in FAS −670 A/G polymorphism increases the risk of cervical cancer [123]. Kang et al. [124] found that the FAS −1377 G/A polymorphism may be a risk factor for lymph node metastases in Korean cervical cancer patients [124]. Lai et al. [53] reported that FASL −844 C is associated with a combination of FAS −1377 A/−670 A haplotype in cervical cancer. However, the study was inconsistent with other results/reports and showed that FAS-SNPs were not related to cervical cancer [125]. Similarly, a metaanalysis of 10 FAS 670 genotyping studies showed that there was no association between FAS-SNPs and cervical cancer risk [126].

## Conclusions and future perspectives

Cervical cancer is one of the most common gynecologic malignancies among women in the developing countries. Although previous studies have resulted in functional gene polymorphism and cervical cancer risk, there has been a lack of consistency globally. Still, it is a vital target for future research that focuses on the prevention and early detection of the disease, since all individuals have risk factors related to their health and lifestyle activities that could lead to the development of cervical cancer. Although some other risk factors are also responsible for the development of cervical cancer, they need to be identified at an early stage. Seen from this perspective, the identification of a precise association between the genetic variants and cervical cancer as documented by the peer-reviewed published studies highlights a distinctive link between the immune response gene polymorphisms of GST gene isoforms, HLA gene, ILs, p53 codon 72, Fas gene, MDM2 gene, IFNG gene, HPV infection, and cancer progression.

In this chapter, we summarized a broad picture of the host genetic causes of HPV infection and the progression of cervical cancer. We noted that the association between genetic polymorphisms and cervical cancer among limited populations was due to heterogeneity in the study design, small sample sizes, ethnic diversity, and the lack of risk stratification elements (e.g., nutrition, alcohol intake, smoking, and other environmental exposures). However, it is not possible yet to make a definitive statement focusing on two or three genes that cause severity or protect from a multifaceted disorder such as cervical cancer. Although this remains a major challenge, it would be of great interest to identify critical interactions between the genetic variants and cancer risk, which could be used as valuable clinical marker for the diagnosis and management of cervical cancer.

# References

[1] U.S. Cancer Statistics Data Briefs. Gynecologic cancer incidence, United States—2012–2016. CDC; September 2019.

[2] Ushijima K. Treatment for recurrent ovarian cancer-at first relapse. J Oncol 2010;2010:497429. https://doi.org/10.1155/2010/497429.

[3] Pectasides D, Pectasides E, Economopoulos T. Systemic therapy in metastatic or recurrent endometrial cancer. Cancer Treat Rev 2007;33(2):177–90. https://doi.org/10.1016/j.ctrv.2006.10.007.

[4] Tsuda N, Watari H, Ushijima K. Chemotherapy and molecular targeting therapy for recurrent cervical cancer. Chin J Cancer Res 2016;28(2):241–53. https://doi.org/10.21147/j.issn.1000-9604.2016.02.14.

[5] Walboomers JM, Jacobs MV, Manos MM, Bosch FX, Kummer JA, Shah KV, Snijders PJ, Peto J, Meijer CJ, Muñoz N. Human papillomavirus is a necessary cause of invasive cervical cancer worldwide. J Pathol 1999;189(1):12–9. https://doi.org/10.1002/(sici)1096-9896(199909)189:1<12::aid-path431>3.0.co;2-f.

[6] Cadenas C, Bolt HM. Estrogen receptors in human disease. Arch Toxicol 2012;86(10):1489–90. https://doi.org/10.1007/s00204-012-0928-x.

[7] Constantine GD, Kessler G, Graham S, Goldstein SR. Increased incidence of endometrial cancer following the women's health initiative: an assessment of risk factors. J Womens Health (Larchmt) 2019;28(2):237–43. https://doi.org/10.1089/jwh.2018.6956.

[8] Arbyn M, Weiderpass E, Bruni L, de Sanjose S, Saraiya M, Ferlay J, Bray F. Estimates of incidence and mortality of cervical cancer in 2018: a worldwide analysis. Lancet Glob Health 2020;8(2):e191–203. https://doi.org/10.1016/S2214-109X(19)30482-6.

[9] Torre LA, Trabert B, DeSantis CE, Miller KD, Samimi G, Runowicz CD, Gaudet MM, Jemal A, Siegel RL. Ovarian cancer statistics, 2018. CA Cancer J Clin 2018;68(4):284–96. https://doi.org/10.3322/caac.21456.

[10] Gibbon S. Calibrating cancer risk, uncertainty and environments: genetics and their contexts in southern Brazil. Biosocieties 2018;13(4):761–79. https://doi.org/10.1057/s41292-017-0095-7.

[11] Radford C, Prince A, Lewis K, Pal T. Factors which impact the delivery of genetic risk assessment services focused on inherited cancer genomics: expanding the role and reach of certified genetics professionals. J Genet Couns 2014;23(4):522–30. https://doi.org/10.1007/s10897-013-9668-1.

[12] Pomerantz MM, Freedman ML. The genetics of cancer risk. Cancer J 2011;17(6):416–22. https://doi.org/10.1097/PPO.0b013e31823e5387.

[13] Holman LL, Lu KH. Genetic risk and gynecologic cancers. Hematol Oncol Clin N Am 2012;26(1):13–29. https://doi.org/10.1016/j.hoc.2011.11.003.

[14] Robert F, Pelletier J. Exploring the impact of single-nucleotide polymorphisms on translation. Front Genet 2018;9:507. https://doi.org/10.3389/fgene.2018.00507.

[15] Apu MNH, Rashed AZM, Bashar T, Rahman MM, Mostaid MS. TP53 genetic polymorphisms and susceptibility to cervical cancer in Bangladeshi women: a case-control study. Mol Biol Rep 2020. https://doi.org/10.1007/s11033-020-05523-2.

[16] Wongpratate M, Ishida W, Phuthong S, Natphopsuk S, Ishida T. Genetic polymorphisms of the human cytochrome P450 1A1 (CYP1A1) and cervical cancer susceptibility among northeast Thai women. Asian Pac J Cancer Prev 2020;21(1):243–8. https://doi.org/10.31557/APJCP.2020.21.1.243.

[17] Kuguyo O, Tsikai N, Thomford NE, Magwali T, Madziyire MG, Nhachi CFB, Matimba A, Dandara C. Genetic susceptibility for cervical cancer in African populations: what are the host genetic drivers? OMICS 2018;22(7):468–83. https://doi.org/10.1089/omi.2018.0075.

[18] Li X, Yin G, Li J, Wu A, Yuan Z, Liang J, Sun Q. The correlation between TNF-alpha promoter gene polymorphism and genetic susceptibility to cervical cancer. Technol Cancer Res Treat 2018;17. https://doi.org/10.1177/1533033818782793, 1533033818782793.

[19] Zidi S, Stayoussef M, Zouidi F, Benali S, Gazouani E, Mezlini A, Yacoubi-Loueslati B. Tumor necrosis factor alpha (-238/-308) and TNFRII-VNTR (-322) polymorphisms as genetic biomarkers of susceptibility to develop cervical cancer among Tunisians. Pathol Oncol Res 2015;21(2):339–45. https://doi.org/10.1007/s12253-014-9826-2.

[20] Ma D, Hovey RL, Zhang Z, Fye S, Huettner PC, Borecki IB, Rader JS. Genetic variations in EGFR and ERBB4 increase susceptibility to cervical cancer. Gynecol Oncol 2013;131(2):445–50. https://doi.org/10.1016/j.ygyno.2013.07.113.

[21] Chen D, Gyllensten U. Systematic investigation of contribution of genetic variation in the HLA-DP region to cervical cancer susceptibility. Carcinogenesis 2014;35(8):1765–9. https://doi.org/10.1093/carcin/bgu096.

[22] Kleinjan DA, van Heyningen V. Long-range control of gene expression: emerging mechanisms and disruption in disease. Am J Hum Genet 2005;76(1):8–32. https://doi.org/10.1086/426833.

[23] Lepique AP, Rabachini T, Villa LL. HPV vaccination: the beginning of the end of cervical cancer?—a review. Mem Inst Oswaldo Cruz 2009;104(1):1–10. https://doi.org/10.1590/s0074-02762009000100001.

[24] Gheit T, Cornet I, Clifford GM, Iftner T, Munk C, Tommasino M, Kjaer SK. Risks for persistence and progression by human papillomavirus type 16 variant lineages among a population-based sample of Danish women. Cancer Epidemiol Biomarkers Prev 2011;20(7):1315–21. https://doi.org/10.1158/1055-9965.EPI-10-1187.

[25] Nunobiki O, Ueda M, Toji E, Yamamoto M, Akashi K, Sato N, Izuma S, Torii K, Tanaka I, Okamoto Y, Noda S. Genetic polymorphism of cancer susceptibility genes and HPV infection in cervical carcinogenesis. Patholog Res Int 2011;2011:364069. https://doi.org/10.4061/2011/364069.

[26] Dinarello CA. Overview of the IL-1 family in innate inflammation and acquired immunity. Immunol Rev 2018;281(1):8–27. https://doi.org/10.1111/imr.12621.

[27] Behbakht K, Friedman J, Heimler I, Aroutcheva A, Simoes J, Faro S. Role of the vaginal microbiological ecosystem and cytokine profile in the promotion of cervical dysplasia: a case-control study. Infect Dis Obstet Gynecol 2002;10(4):181–6. https://doi.org/10.1155/S1064744902000200.

[28] Tjiong MY, van der Vange N, ter Schegget JS, Burger MP, ten Kate FW, Out TA. Cytokines in cervicovaginal washing fluid from patients with cervical neoplasia. Cytokine 2001;14(6):357–60. https://doi.org/10.1006/cyto.2001.0909.

[29] Al-Tahhan MA, Etewa RL, El Behery MM. Association between circulating interleukin-1 beta (IL-1beta) levels and IL-1beta C-511T polymorphism with cervical cancer risk in Egyptian women. Mol Cell Biochem 2011;353(1–2):159–65. https://doi.org/10.1007/s11010-011-0782-9.

[30] Qian N, Chen X, Han S, Qiang F, Jin G, Zhou X, Dong J, Wang X, Shen H, Hu Z. Circulating IL-1beta levels, polymorphisms of IL-1B, and risk of cervical cancer in Chinese women. J Cancer Res Clin Oncol 2010;136(5):709–16. https://doi.org/10.1007/s00432-009-0710-5.

[31] Singh H, Sachan R, Goel H, Mittal B. Genetic variants of interleukin-1RN and interleukin-1beta genes and risk of cervical cancer. BJOG 2008;115(5):633–8. https://doi.org/10.1111/j.1471-0528.2007.01655.x.

[32] Sobti RC, Kordi Tamandani DM, Shekari M, Kaur P, Malekzadeh K, Suri V. Interleukin 1 beta gene polymorphism and risk of cervical cancer. Int J Gynaecol Obstet 2008;101(1):47–52. https://doi.org/10.1016/j.ijgo.2007.10.014.

[33] Wang L, Zhao W, Hong J, Niu F, Li J, Zhang S, Jin T. Association between IL1B gene and cervical cancer susceptibility in Chinese Uygur population: a case-control study. Mol Genet Genomic Med 2019;7(8). https://doi.org/10.1002/mgg3.779, e779.

[34] Kang S, Kim JW, Park NH, Song YS, Park SY, Kang SB, Lee HP. Interleukin-1 beta-511 polymorphism and risk of cervical cancer. J Korean Med Sci 2007;22(1):110–3. https://doi.org/10.3346/jkms.2007.22.1.110.

[35] Mocellin S, Marincola F, Rossi CR, Nitti D, Lise M. The multifaceted relationship between IL-10 and adaptive immunity: putting together the pieces of a puzzle. Cytokine Growth Factor Rev 2004;15(1):61–76. https://doi.org/10.1016/j.cytogfr.2003.11.001.

[36] Clerici M, Merola M, Ferrario E, Trabattoni D, Villa ML, Stefanon B, Venzon DJ, Shearer GM, De Palo G, Clerici E. Cytokine production patterns in cervical intraepithelial neoplasia: association with human papillomavirus infection. J Natl Cancer Inst 1997;89(3):245–50. https://doi.org/10.1093/jnci/89.3.245.

[37] Mota F, Rayment N, Chong S, Singer A, Chain B. The antigen-presenting environment in normal and human papillomavirus (HPV)-related premalignant cervical epithelium. Clin Exp Immunol 1999;116(1):33–40. https://doi.org/10.1046/j.1365-2249.1999.00826.x.

[38] Stanczuk GA, Sibanda EN, Perrey C, Chirara M, Pravica V, Hutchinson IV, Tswana SA. Cancer of the uterine cervix may be significantly associated with a gene polymorphism coding for increased IL-10 production. Int J Cancer 2001;94(6):792–4. https://doi.org/10.1002/ijc.1543.

[39] Shrestha S, Wang C, Aissani B, Wilson CM, Tang J, Kaslow RA. Interleukin-10 gene (*IL10*) polymorphisms and human papillomavirus clearance among immunosuppressed adolescents. Cancer Epidemiol Biomarkers Prev 2007;16(8):1626–32. https://doi.org/10.1158/1055-9965.epi-06-0881.

[40] Chagas BS, Gurgel AP, da Cruz HL, Amaral CM, Cardoso MV, da Silva Neto JC, da Silva LA, de Albuquerque EM, Muniz MT, de Freitas AC. An interleukin-10 gene polymorphism associated with the development of cervical lesions in women infected with human papillomavirus and using oral contraceptives. Infect Genet Evol 2013;19:32–7. https://doi.org/10.1016/j.meegid.2013.06.016.

[41] Shifang Gao JZ, Wu Y, Xu L, Wang H. Correlation between 1082G/a gene polymorphism of interleukin 10 promoter and cervical carcinoma. Int J Clin Exp Pathol 2016;9(4):4539–44.

[42] Matsumoto K, Oki A, Satoh T, Okada S, Minaguchi T, Onuki M, Ochi H, Nakao S, Sakurai M, Abe A, Hamada H, Yoshikawa H. Interleukin-10 -1082 gene polymorphism and susceptibility to cervical cancer among Japanese women. Jpn J Clin Oncol 2010;40(11):1113–6. https://doi.org/10.1093/jjco/hyq094.

[43] Torres-Poveda K, Burguete-Garcia AI, Cruz M, Martinez-Nava GA, Bahena-Roman M, Ortiz-Flores E, Ramirez-Gonzalez A, Lopez-Estrada G, Delgado-Romero K, Madrid-Marina V. The SNP at −592 of human IL-10 gene is associated with serum IL-10 levels and increased risk for human papillomavirus cervical lesion development. Infect Agent Cancer 2012;7(1):32. https://doi.org/10.1186/1750-9378-7-32.

[44] Lasek W, Zagożdżon R, Jakobisiak M. Interleukin 12: still a promising candidate for tumor immunotherapy? Cancer Immunol Immunother 2014;63(5):419–35. https://doi.org/10.1007/s00262-014-1523-1.

[45] Wadler S, Levy D, Frederickson HL, Falkson CI, Wang Y, Weller E, Burk R, Ho G, Kadish AS. A phase II trial of interleukin-12 in patients with advanced cervical cancer: clinical and immunologic correlates. Eastern cooperative oncology group study E1E96. Gynecol Oncol 2004;92(3):957–64. https://doi.org/10.1016/j.ygyno.2003.12.022.

[46] Tamandani DM, Shekari M, Suri V. Interleukin-12 gene polymorphism and cervical cancer risk. Am J Clin Oncol 2009;32(5):524–8. https://doi.org/10.1097/COC.0b013e318192519a.

[47] García Paz F, Madrid Marina V, Morales Ortega A, Santander González A, Peralta Zaragoza O, Burguete García A, Torres Poveda K, Moreno J, Alcocer González J, Hernandez Marquez E, Bermúdez Morales V. The relationship between the antitumor effect of the IL-12 gene therapy and the expression of Th1 cytokines in an HPV16-positive murine tumor model. Mediators Inflamm 2014;2014:510846. https://doi.org/10.1155/2014/510846.

[48] do Carmo Vasconcelos de Carvalho V, de Macedo JL, de Lima CA, da Conceicao Gomes de Lima M, de Andrade Heraclio S, Amorim M, de Mascena Diniz Maia M, Porto AL, de Souza PR. IFN-gamma and IL-12B polymorphisms in women with cervical intraepithelial neoplasia caused by human papillomavirus. Mol Biol Rep 2012;39(7):7627–34. https://doi.org/10.1007/s11033-012-1597-9.

[49] Han SS, Cho EY, Lee TS, Kim JW, Park NH, Song YS, Kim JG, Lee HP, Kang SB. Interleukin-12 p40 gene (IL12B) polymorphisms and the risk of cervical cancer in Korean women. Eur J Obstet Gynecol Reprod Biol 2008;140(1):71–5. https://doi.org/10.1016/j.ejogrb.2008.02.007.

[50] Chen X, Han S, Wang S, Zhou X, Zhang M, Dong J, Shi X, Qian N, Wang X, Wei Q, Shen H, Hu Z. Interactions of IL-12A and IL-12B polymorphisms on the risk of cervical cancer in Chinese women. Clin Cancer Res 2009;15(1):400–5. https://doi.org/10.1158/1078-0432.CCR-08-1829.

[51] Zhang L, Davies JS, Serna C, Yu Z, Restifo NP, Rosenberg SA, Morgan RA, Hinrichs CM. Enhanced efficacy and limited systemic cytokine exposure with membrane-anchored interleukin-12 T-cell therapy in murine tumor models. J Immunother Cancer 2020;8(1). https://doi.org/10.1136/jitc-2019-000210.

[52] Schroder K, Hertzog PJ, Ravasi T, Hume DA. Interferon-gamma: an overview of signals, mechanisms and functions. J Leukoc Biol 2004;75(2):163–89. https://doi.org/10.1189/jlb.0603252.

[53] Lai HC, Chang CC, Lin YW, Chen SF, Yu MH, Nieh S, Chu TW, Chu TY. Genetic polymorphism of the interferon-gamma gene in cervical carcinogenesis. Int J Cancer 2005;113(5):712–8. https://doi.org/10.1002/ijc.20637.

[54] Kim KY, Blatt L, Taylor MW. The effects of interferon on the expression of human papillomavirus oncogenes. J Gen Virol 2000;81(Pt 3):695–700. https://doi.org/10.1099/0022-1317-81-3-695.

[55] Kordi Tamandani MK, Sobti RC, Shekari M, Mukesh M, Suri V. Expression and polimorphism of IFN-gamma gene in patients with cervical cancer. Exp Oncol 2008;30(3):224–9.

[56] Gangwar R, Pandey S, Mittal RD. Association of interferon-gamma +874A polymorphism with the risk of developing cervical cancer in north-Indian population. BJOG 2009;116(12):1671–7. https://doi.org/10.1111/j.1471-0528.2009.02307.x.

[57] Ma D, Jiang C, Hu X, Liu H, Li Q, Li T, Yang Y, Li O. Methylation patterns of the IFN-γ gene in cervical cancer tissues. Sci Rep 2014;4:6331. https://doi.org/10.1038/srep06331.

[58] Zidi S, Benothmen Y, Sghaier I, Ghazoueni E, Mezlini A, Slimen B, Yacoubi-Loueslati B. Association of IL10-1082 and IFN-γ+874 polymorphisms with cervical cancer among Tunisian women. ISRN Genet 2014;2014:706516. https://doi.org/10.1155/2014/706516.

[59] Yuan Y, Fan J-L, Yao F-L, Wang K-T, Yu Y, Carlson J, Li M. Association study of single-nucleotide polymorphisms of STAT2/STAT3/IFN-γ genes in cervical cancer in Southern Chinese Han women. Asian Pac J Cancer Prev 2015;16(8):3117–20.

[60] Govan V, Carrara H, Sachs J, Hoffman M, Stanczuk G, Williamson A-L. Ethnic differences in allelic distribution of IFN-g in south African women but no link with cervical cancer. J Carcinog 2003;2(1):3. https://doi.org/10.1186/1477-3163-2-3.

[61] Wang X, Lin Y. Tumor necrosis factor and cancer, buddies or foes? Acta Pharmacol Sin 2008;29(11):1275–88. https://doi.org/10.1111/j.1745-7254.2008.00889.x.

[62] Wilson AG, Symons JA, McDowell TL, McDevitt HO, Duff GW. Effects of a polymorphism in the human tumor necrosis factor alpha promoter on transcriptional activation. Proc Natl Acad Sci U S A 1997;94(7):3195–9. https://doi.org/10.1073/pnas.94.7.3195.

[63] Zuo F, Liang W, Ouyang Y, Li W, Lv M, Wang G, Ding M, Wang B, Zhao S, Liu J, Jiang Z, Li M. Association of TNF-α gene promoter polymorphisms with susceptibility of cervical cancer in Southwest China. Lab Med 2011;42(5):287–90. https://doi.org/10.1309/lm532dspduxirjvn.

[64] Wang L, Ma K, Wang Z, Mou Y, Ma L, Guo Y. Association between tumor necrosis factor alpha rs1800629 polymorphism and risk of cervical cancer. Int J Clin Exp Med 2015;8(2):2108–17.

[65] Li X, Yin G, Li J, Wu A, Yuan Z, Liang J, Sun Q. The correlation between TNF-α promoter gene polymorphism and genetic susceptibility to cervical cancer. Technol Cancer Res Treat 2018;17. https://doi.org/10.1177/1533033818782793, 1533033818782793.

[66] Rotar IC, Muresan D, Radu P, Petrisor F, Apostol S, Mariana T, Butuza C, Stamatian F. TNF-alpha 308 G/A polymorphism and cervical intraepithelial neoplasia. Anticancer Res 2014;34(1):373–8.

[67] Kohaar ISH, Kumar A, Singhal P, Choudhury SR, Das BC, Bharadwaj M. Impact of haplotype TNF-LTA locus with susceptibility to cervical cancer in Indian population. Obstet Gynecol 2014. https://doi.org/10.5171/2014.831817.

[68] Du GH, Wang JK, Richards JR, Wang JJ. Genetic polymorphisms in tumor necrosis factor alpha and interleukin-10 are associated with an increased risk of cervical cancer. Int Immunopharmacol 2019;66:154–61. https://doi.org/10.1016/j.intimp.2018.11.015.

[69] Nieves-Ramirez ME, Partida-Rodriguez O, Alegre-Crespo PE, Tapia-Lugo Mdel C, Perez-Rodriguez ME. Characterization of single-nucleotide polymorphisms in the tumor necrosis factor alpha promoter region and in lymphotoxin alpha in squamous intraepithelial lesions, precursors of cervical cancer. Transl Oncol 2011;4(6):336–44. https://doi.org/10.1593/tlo.11226.

[70] Wang Y, Yang J, Huang J, Tian Z. Tumor necrosis factor-α polymorphisms and cervical cancer: evidence from a meta-analysis. Gynecol Obstet Invest 2020;85(2):153–8. https://doi.org/10.1159/000502955.

[71] Wang N, Yin D, Zhang S, Wei H, Wang S, Zhang Y, Lu Y, Dai S, Li W, Zhang Q. TNF-alpha rs1800629 polymorphism is not associated with HPV infection or cervical cancer in the Chinese population. PLoS One 2012;7(9). https://doi.org/10.1371/journal.pone.0045246, e45246.

[72] Govan VA, Constant D, Hoffman M, Williamson AL. The allelic distribution of −308 tumor necrosis factor-alpha gene polymorphism in South African women with cervical cancer and control women. BMC Cancer 2006;6:24. https://doi.org/10.1186/1471-2407-6-24.

[73] Campoli M, Ferrone S. HLA antigen changes in malignant cells: epigenetic mechanisms and biologic significance. Oncogene 2008;27(45):5869–85. https://doi.org/10.1038/onc.2008.273.

[74] Chang CC, Campoli M, Ferrone S. Classical and nonclassical HLA class I antigen and NK cell-activating ligand changes in malignant cells: current challenges and future directions. Adv Cancer Res 2005;93:189–234. https://doi.org/10.1016/S0065-230X(05)93006-6.

[75] Madeleine MM, Brumback B, Cushing-Haugen KL, Schwartz SM, Daling JR, Smith AG, Nelson JL, Porter P, Shera KA, McDougall JK, Galloway DA. Human leukocyte antigen class II and cervical cancer risk: a population-based study. J Infect Dis 2002;186(11):1565–74. https://doi.org/10.1086/345285.

[76] Zoodsma M, Nolte IM, Schipper M, Oosterom E, van der Steege G, de Vries EGE, te Meerman GJ, van der Zee AGJ. Analysis of the entire HLA region in susceptibility for cervical cancer: a comprehensive study. J Med Genet 2005;42(8):e49. https://doi.org/10.1136/jmg.2005.031351.

[77] Hosono S, Kawase T, Matsuo K, Watanabe M, Kajiyama H, Hirose K, Suzuki T, Kidokoro K, Ito H, Nakanishi T, Yatabe Y, Hamajima N, Kikkawa F, Tajima K, Tanaka H. HLA-A alleles and the risk of cervical squamous cell carcinoma in Japanese women. J Epidemiol 2010;20(4):295–301. https://doi.org/10.2188/jea.je20090155.

[78] Safaeian M, Johnson LG, Yu K, Wang SS, Gravitt PE, Hansen JA, Carrington M, Schwartz SM, Gao X, Hildesheim A, Madeleine MM. Human leukocyte antigen class I and II alleles and cervical adenocarcinoma. Front Oncol 2014;4:119. https://doi.org/10.3389/fonc.2014.00119.

[79] Chambuso R, Ramesar R, Kaambo E, Denny L, Passmore JA, Williamson AL, Gray CM. Human leukocyte antigen (HLA) class II -DRB1 and -DQB1 alleles and the association with cervical cancer in HIV/HPV co-infected women in South Africa. J Cancer 2019;10(10):2145–52. https://doi.org/10.7150/jca.25600.

[80] Ivansson EL, Magnusson JJ, Magnusson PK, Erlich HA, Gyllensten UB. MHC loci affecting cervical cancer risk: distinguishing the effects of HLA-DQB1 and non-HLA genes TNF, LTA, TAP1 and TAP2. Genes Immun 2008;9 (7):613–23. https://doi.org/10.1038/gene.2008.58.

[81] Ghaderi M, Wallin KL, Wiklund F, Zake LN, Hallmans G, Lenner P, Dillner J, Sanjeevi CB. Risk of invasive cervical cancer associated with polymorphic HLA DR/DQ haplotypes. Int J Cancer 2002;100(6):698–701. https://doi.org/10.1002/ijc.10551.

[82] Davidson EJ, Davidson JA, Sterling JC, Baldwin PJ, Kitchener HC, Stern PL. Association between human leukocyte antigen polymorphism and human papillomavirus 16-positive vulval intraepithelial neoplasia in British women. Cancer Res 2003;63(2):400–3.

[83] Cheng L, Guo Y, Zhan S, Xia P. Association between HLA-DP gene polymorphisms and cervical cancer risk: a meta-analysis. Biomed Res Int 2018;2018:7301595. https://doi.org/10.1155/2018/7301595.

[84] Alifu M, Hu Y-H, Dong T, Wang R-Z. HLA-A*30:01 and HLA-A*33:03 are the protective alleles while HLA-A*01:01 serves as the susceptible gene for cervical cancer patients in Xinjiang, China. J Cancer Res Ther 2018;14(6):1266–72. https://doi.org/10.4103/0973-1482.199430.

[85] Allocati N, Masulli M, Di Ilio C, Federici L. Glutathione transferases: substrates, inhibitors and pro-drugs in cancer and neurodegenerative diseases. Oncogenesis 2018;7(1):8. https://doi.org/10.1038/s41389-017-0025-3.

[86] Daukantiene L, Kazbariene B, Valuckas KP, Didziapetriene J, Krikstaponiene A, Aleknavicius E. The significance of reduced glutathione and glutathione S-transferase during chemoradiotherapy of locally advanced cervical cancer. Medicina (Kaunas) 2014;50(4):222–9. https://doi.org/10.1016/j.medici.2014.09.005.

[87] Abbas M, Kushwaha VS, Srivastava K, Banerjee M. Glutathione S-transferase gene polymorphisms and treatment outcome in cervical cancer patients under concomitant chemoradiation. PLoS One 2015;10(11). https://doi.org/10.1371/journal.pone.0142501, e0142501.

[88] Abbas M, Kushwaha VS, Srivastava K, Raza ST, Banerjee M. Impact of GSTM1, GSTT1 and GSTP1 genes polymorphisms on clinical toxicities and response to concomitant chemoradiotherapy in cervical cancer. Br J Biomed Sci 2018;75(4):169–74. https://doi.org/10.1080/09674845.2018.1482734.

[89] Phuthong S, Settheetham-Ishida W, Natphopsuk S, Ishida T. Genetic polymorphism of the glutathione S-transferase Pi 1 (GSTP1) and susceptibility to cervical cancer in human papilloma virus infected northeastern Thai women. Asian Pac J Cancer Prev 2018;19(2):381–5. https://doi.org/10.22034/APJCP.2018.19.2.381.

[90] Zamani S, Sohrabi A, Rahnamaye-Farzami M, Hosseini SM. Glutathione S-transferase omega gene polymorphism as a biomarker for human papilloma virus and cervical cancer in Iranian women. J Turk Ger Gynecol Assoc 2018;19(4):193–200. https://doi.org/10.4274/jtgga.2018.0056.

[91] Tulsi Rani Thakre AS, Mitra M. Investigation on glutathione-S-transferase M1 and T1 gene polymorphisms as risk factor in cervical cancer. Res J Pharm Tech 2016;9(12):2295–300. https://doi.org/10.5958/0974-360X.2016.00462.5.

[92] Tacca ALM, Lopes AK, Vilanova-Costa CAST, Silva AMTC, Costa SHN, Nogueira NA, Ramos JEP, Ribeiro AA, Saddi VA. Null polymorphisms in GSTT1 and GSTM1 genes and their associations with smoking and cervical cancer. Genet Mol Res 2019;18(1):GMR18067.

[93] Sun P, Song WQ. GSTM1 null genotype and susceptibility to cervical cancer in the Chinese population: an updated meta-analysis. J Cancer Res Ther 2016;12(2):712–5. https://doi.org/10.4103/0973-1482.154004.

[94] Tian S, Yang X, Zhang L, Zhao J, Pei M, Yu Y, Yang T. Polymorphic variants conferring genetic risk to cervical lesions support GSTs as important associated loci. Medicine (Baltimore) 2019;98(41):e17487.

[95] Zhao E, Hu K, Zhao Y. Associations of the glutathione S-transferase P1 Ile105Val genetic polymorphism with gynecological cancer susceptibility: a meta-analysis. Oncotarget 2017;8(25):41734–9.

[96] Wang D, Wang B, Zhai JX, Liu DW, Sun GG. Glutathione S-transferase M1 and T1 polymorphisms and cervical cancer risk: a meta-analysis. Neoplasma 2011;58(4):352–9. https://doi.org/10.4149/neo_2011_04_352.

[97] Ko LJ, Prives C. p53: puzzle and paradigm. Genes Dev 1996;10(9):1054–72. https://doi.org/10.1101/gad.10.9.1054.

[98] Brosh R, Rotter V. When mutants gain new powers: news from the mutant p53 field. Nat Rev Cancer 2009;9(10):701–13. https://doi.org/10.1038/nrc2693.

[99] Lin H-Y, Huang C-H, Wu W-J, Chang L-C, Lung F-W. TP53 codon 72 gene polymorphism paradox in associated with various carcinoma incidences, invasiveness and chemotherapy responses. Int J Biomed Sci 2008;4(4):248–54.

[100] Jiang P, Liu J, Zeng X, Li W, Tang J. Association of TP53 codon 72 polymorphism with cervical cancer risk in Chinese women. Cancer Genet Cytogenet 2010;197(2):174–8. https://doi.org/10.1016/j.cancergencyto.2009.11.011.

[101] Dokianakis DN, Spandidos DA. P53 codon 72 polymorphism as a risk factor in the development of HPV-associated cervical cancer. Mol Cell Biol Res Commun 2000;3(2):111–4. https://doi.org/10.1006/mcbr.2000.0196.

[102] Rosenthal AN, Ryan A, Al-Jehani RM, Storey A, Harwood CA, Jacobs IJ. p53 codon 72 polymorphism and risk of cervical cancer in UK. Lancet 1998;352(9131):871–2. https://doi.org/10.1016/S0140-6736(98)07357-7.

[103] Bansal A, Das P, Kannan S, Mahantshetty U, Mulherkar R. Effect of p53 codon 72 polymorphism on the survival outcome in advanced stage cervical cancer patients in India. Indian J Med Res 2016;144(3):359–65. https://doi.org/10.4103/0971-5916.198685.

[104] Malcolm EK, Baber GB, Boyd JC, Stoler MH. Polymorphism at codon 72 of p53 is not associated with cervical cancer risk. Mod Pathol 2000;13(4):373–8. https://doi.org/10.1038/modpathol.3880061.

[105] Ojeda JM, Ampuero S, Rojas P, Prado R, Allende JE, Barton SA, Chakraborty R, Rothhammer F. p53 codon 72 polymorphism and risk of cervical cancer. Biol Res 2003;36(2):279–83. https://doi.org/10.4067/s0716-97602003000200017.

[106] Settheetham-Ishida W, Singto Y, Yuenyao P, Tassaneeyakul W, Kanjanavirojkul N, Ishida T. Contribution of epigenetic risk factors but not p53 codon 72 polymorphism to the development of cervical cancer in Northeastern Thailand. Cancer Lett 2004;210(2):205–11. https://doi.org/10.1016/j.canlet.2004.03.039.

[107] Pegoraro R, Moodley J, Naiker S, Lanning P, Rom L. The p53 codon 72 polymorphism in black south African women and the risk of cervical cancer. BJOG 2000;107(9):1164–5. https://doi.org/10.1111/j.1471-0528.2000.tb11118.x.

[108] Brenna SMF, IDCGd S, Zeferino LC, Pereira J, Martinez EZ, Syrjänen KJ. Prevalence of codon 72 P53 polymorphism in Brazilian women with cervix cancer. Genet Mol Biol 2004;27:496–9.

[109] Klug SJ, Ressing M, Koenig J, Abba MC, Agorastos T, Brenna SM, Ciotti M, Das BR, Del Mistro A, Dybikowska A, Giuliano AR, Gudleviciene Z, Gyllensten U, Haws AL, Helland A, Herrington CS, Hildesheim A, Humbey O, Jee SH, Kim JW, Madeleine MM, Menczer J, Ngan HY, Nishikawa A, Niwa Y, Pegoraro R, Pillai MR, Ranzani G, Rezza G, Rosenthal AN, Roychoudhury S, Saranath D, Schmitt VM, Sengupta S, Settheetham-Ishida W, Shirasawa H, Snijders PJ, Stoler MH, Suarez-Rincon AE, Szarka K, Tachezy R, Ueda M, van der Zee AG, von Knebel DM, Wu MT, Yamashita T, Zehbe I, Blettner M. TP53 codon 72 polymorphism and cervical cancer: a pooled analysis of individual data from 49 studies. Lancet Oncol 2009;10(8):772–84. https://doi.org/10.1016/S1470-2045(09)70187-1.

[110] Hav M, Libbrecht L, Ferdinande L, Pattyn P, Laurent S, Peeters M, Praet M, Pauwels P. MDM2 gene amplification and protein expressions in colon carcinoma: is targeting MDM2 a new therapeutic option? Virchows Arch 2011;458(2):197–203. https://doi.org/10.1007/s00428-010-1012-7.

[111] Chen X, Sturgis EM, Lei D, Dahlstrom K, Wei Q, Li G. Human papillomavirus seropositivity synergizes with MDM2 variants to increase the risk of oral squamous cell carcinoma. Cancer Res 2010;70(18):7199–208. https://doi.org/10.1158/0008-5472.CAN-09-4733.

[112] Singhal P, Hussain S, Thakur N, Batra S, Salhan S, Bhambani S, Bharadwaj M. Association of MDM2 and p53 polymorphisms with the advancement of cervical carcinoma. DNA Cell Biol 2013;32(1):19–27.

[113] Zhuo X, Ren J, Li D, Wu Y, Zhou Q. MDM2 SNP309 variation increases cervical cancer risk among Asians. Tumor Biol 2014;35(6):5331–7. https://doi.org/10.1007/s13277-014-1695-5.

[114] Jiang P, Liu J, Zeng X, Li W, Tang J. MDM2 gene promoter polymorphism and risk of cervical cancer in Chinese population. In: 2010 4th international conference on bioinformatics and biomedical engineering, 18–20 June 2010; 2010. p. 1–4. https://doi.org/10.1109/icbbe.2010.5516487.

[115] Tantengco OAG, Nakura Y, Yoshimura M, Llamas-Clark EF, Yanagihara I. Association of PIK3CA and MDM2 SNP309 with cervical squamous cell carcinoma in a Philippine population. Asian Pac J Cancer Prev 2019;20 (7):2103–7. https://doi.org/10.31557/apjcp.2019.20.7.2103.

[116] Zhang J, Zhang Y, Zhang Z. Association of rs2279744 and rs117039649 promoter polymorphism with the risk of gynecological cancer: a meta-analysis of case–control studies. Medicine (Baltimore) 2018;97(2).

[117] Amaral CMM, Cetkovská K, Gurgel APAD, Cardoso MV, Chagas BS, Paiva Júnior SSL, de Lima RCP, Silva-Neto JC, LAF S, MTC M, Balbino VQ, Freitas AC. MDM2 polymorphism associated with the development of cervical lesions in women infected with human papillomavirus and using of oral contraceptives. Infect Agent Cancer 2014;9(1):24. https://doi.org/10.1186/1750-9378-9-24.

[118] Meissner Rde V, Barbosa RN, Fernandes JV, Galvao TM, Galvao AF, Oliveira GH. No association between SNP309 promoter polymorphism in the MDM2 and cervical cancer in a study from northeastern Brazil. Cancer Detect Prev 2007;31(5):371–4. https://doi.org/10.1016/j.cdp.2007.09.001.

[119] Hu X, Zhang Z, Ma D, Huettner PC, Massad LS, Nguyen L, Borecki I, Rader JS. TP53, MDM2, NQO1, and susceptibility to cervical cancer. Cancer Epidemiol Biomarkers Prev 2010;19(3):755–61. https://doi.org/10.1158/1055-9965.epi-09-0886.

[120] Gerl R, Vaux DL. Apoptosis in the development and treatment of cancer. Carcinogenesis 2005;26(2):263–70. https://doi.org/10.1093/carcin/bgh283.

[121] Sun T, Zhou Y, Li H, Han X, Shi Y, Wang L, Miao X, Tan W, Zhao D, Zhang X, Guo Y, Lin D. FASL -844C polymorphism is associated with increased activation-induced T cell death and risk of cervical cancer. J Exp Med 2005;202(7):967–74. https://doi.org/10.1084/jem.20050707.

[122] Zucchi F, da Silva ID, Ribalta JC, de Souza NC, Speck NM, Girão MJ, Brenna SM, Syrjänen KJ. Fas/CD95 promoter polymorphism gene and its relationship with cervical carcinoma. Eur J Gynaecol Oncol 2009;30(2):142–4.

[123] Ueda M, Terai Y, Kanda K, Kanemura M, Takehara M, Yamaguchi H, Nishiyama K, Yasuda M, Ueki M. Fas gene promoter −670 polymorphism in gynecological cancer. Int J Gynecol Cancer 2006;16(Suppl. 1):179–82. https://doi.org/10.1111/j.1525-1438.2006.00505.x.

[124] Kang S, Dong SM, Seo SS, Kim JW, Park SY. FAS -1377 G/A polymorphism and the risk of lymph node metastasis in cervical cancer. Cancer Genet Cytogenet 2008;180(1):1–5. https://doi.org/10.1016/j.cancergencyto.2007.09.002.

[125] Chatterjee K, Engelmark M, Gyllensten U, Dandara C, van der Merwe L, Galal U, Hoffman M, Williamson AL. Fas and FasL gene polymorphisms are not associated with cervical cancer but differ among black and mixed-ancestry South Africans. BMC Res Notes 2009;2:238. https://doi.org/10.1186/1756-0500-2-238.

[126] Huang Q, Wang J, Hu Y. FAS-670 gene polymorphism and cervical carcinogenesis risk: a meta-analysis. Biomed Rep 2013;1(1):889–94. https://doi.org/10.3892/br.2013.159.

# Overcoming chemotherapy resistance in endometrial cancer

*Thomas A. Paterniti[a], Evan A. Schrader[b], Aditi Talkad[c],
Kasey Shepp[c], Jesse Wayson[a], Alexandra M. Poch[a], and
Sarfraz Ahmad[d]*

[a]Department of Obstetrics and Gynecology, Augusta University Medical Center, Augusta, GA,
United States [b]Department of Obstetrics and Gynecology, Atrium Health Carolinas Medical
Center, Charlotte, NC, United States [c]Medical College of Georgia, Augusta, GA, United States
[d]Gynecologic Oncology, AdventHealth Cancer Institute, Florida State University, University of
Central Florida, Orlando, FL, United States

## Abstract

Endometrial cancer (EC) is the 6th most common and the 14th most deadly cancer among women worldwide; however, outcomes remain poor for women with advanced-stage or recurrent disease. Substantial research is currently being conducted on molecular mechanisms of endometrial carcinogenesis with the goal of creating new therapeutic strategies for overcoming chemotherapy resistance. Initially classified as either type I or type II tumors, analysis of The Cancer Genome Atlas has led to a reorganization of EC into four groups based on molecular characteristics, with studies ongoing to determine the prognostic and therapeutic significance of this schema. The role of Wnt signaling in EC has been demonstrated in multiple studies and is believed to mediate the proliferative influence of estrogen on endometrial tissue. Current therapies in phase I/II clinical trials for solid tumors include several small-molecule inhibitors of the canonical Wnt pathway, a Porcupine inhibitor, a Wnt5A mimetic, an ROR1 antibody, and an ROR1 CART-cell therapy. Alterations in PI3K-AKT-mTOR signaling, especially via *PTEN* mutations, are common in type 1 EC, making this pathway an attractive target for molecular therapy. Temsirolimus, everolimus, and ridaforolimus have all been evaluated in phase I/II clinical trials, where they have proven most effective in treatment-naïve patients and as adjuncts to front-line therapy. Second-generation mTORC1/2 inhibitors, PI3K inhibitors, dual mTOR/PI3K inhibitors, and AKT inhibitors are also in early-phase clinical trials. Metformin has shown an intriguing ability to reverse hyperplasia in conjunction with progestogens and to prevent the recurrence of AEH and EC, making it an attractive option for patients desiring fertility preservation. Due to the significance of angiogenic growth factors in EC, several agents targeting multiple upstream receptor tyrosine kinases within the EGFR/FGF-Ras-Raf-MEK-ERK signaling pathway are currently in clinical trials, including brivanib, nintedanib, dovitinib, lenvatinib, and ponatinib. Of these, dovitinib and lenvatinib have shown the most promising results. Hypoxia and triggered

R. Basha, S. Ahmad (eds.)
*Overcoming Drug Resistance in Gynecologic Cancers*
ISBN 978-0-12-824299-5
https://doi.org/10.1016/B978-0-12-824299-5.00012-5

225

angiogenesis are critical, but understudied, components of tumorigenesis in EC. Bevacizumab is the best-studied antiangiogenic agent, with moderate activity seen when used as a single agent and more promising results obtained by combining it with front-line therapies. Going forward, robust molecular screening tools are needed for both improved patient selection and identification of prognostic biomarkers, and further study with more selective agents is warranted to parse out the isolated impact of tyrosine kinase inhibition and antiangiogenic agents in recurrent EC.

## Abbreviations

| | |
|---|---|
| AEH | atypical complex endometrial hyperplasia |
| AKT | protein kinase B |
| AMPK | adenosine monophosphate-activated protein kinase |
| APC | adenomatous polyposis coli |
| ARID1A | AT-rich interactive domain-containing protein 1A |
| CART cell | chimeric antigen receptorT cell |
| CNH | copy-number high |
| CNL | copy-number low |
| CR | complete response |
| DFI | disease-free interval |
| DKKs | Dickkopf protein |
| DSS | disease-specific survival |
| E2 | estradiol |
| EC | endometrial cancer |
| EGF | epidermal growth factor |
| EGFR | epidermal growth factor receptor |
| ER | estrogen receptor |
| ERK | extracellular signal-related kinase |
| FGF | fibroblast growth factor |
| FGFR | fibroblast growth factor receptor |
| FIH | factor inhibiting hypoxia-inducible factor |
| GβL/LST8 | G-protein β-subunit-like protein/lethal with SEC13 protein 8 |
| GSK-3β | glycogen synthase kinase 3β |
| GTPase | guanosine triphosphatase |
| HER2/neu | human epidermal growth factor receptor 2 |
| HIF | hypoxia-inducible factors |
| HSGAG | heparin sulfate glycosaminoglycan |
| IGF-1 | insulin-like growth factor 1 |
| IHC | immunohistochemistry |
| KRAS | Kirsten rat sarcoma |
| LKB1 | serine-threonine liver kinase B1 |
| MAPK | mitogen-activated protein kinase |
| MEK | mitogen-activated protein kinase/extracellular signal-related kinase kinase |
| MPA | medroxyprogesterone acetate |
| MSI | microsatellite instability |
| MSS | microsatellite stable |
| mTOR | mammalian target of rapamycin |
| mTORC | mammalian target of rapamycin complex |
| ORR | overall response rate |
| OS | overall survival |
| PARP | poly-ADP ribose polymerase |
| PD-1 | programmed cell death protein 1 |
| PDGF | platelet-derived growth factor |
| PD-L1 | programmed cell death ligand 1 |

| PFS | progression-free survival |
| PHD | propyl hydroxylase |
| PI3K | phosphoinositide 3-kinase |
| PKB | protein kinase B |
| PKC | protein kinase C |
| PLC-$\gamma$ | phospholipase C-$\gamma$ |
| POLE | DNA polymerase epsilon |
| PR | partial response |
| PTEN | phosphatase and homolog |
| Raf | rapidly accelerated fibrosarcoma |
| Ras | rat sarcoma |
| ROR | receptor tyrosine kinase-like orphan receptor |
| SD | stable disease |
| SFRP | secreted frizzled-related proteins |
| TCF/LEF | T-cell factor/lymphoid enhancer factor |
| TIL | tumor-infiltrating lymphocytes |
| VEGF | vascular endothelial growth factor |
| VEGFR | vascular endothelial growth factor receptor |
| VHL | Von Hippel-Lindau protein |
| Wnt | wingless-type integration |

## Conflict of interest

No potential conflicts of interest were disclosed.

# Introduction

## Chemotherapy resistance in endometrial cancer

Endometrial cancer (EC) is the 6th most common and the 14th most deadly cancer among women worldwide, accounting for 5% of the global cancer incidence and 2% of the global cancer mortality among women per year [1]. The 5-year relative survival rate for all stages of EC in the United States has remained steady at 81%, with an overall survival rate of 95% in those with locally confined disease, 69% in those with regional spread, and 16% in those with distant metastases [2]. Incidence rates of EC are higher in more economically developed countries than in less-developed countries, likely due to a variety of lifestyle factors, and global mortality rates vary widely based on access to high-quality health care within each country [1].

The landmark trials that led to currently accepted first-line therapies in EC have been reviewed extensively [3]. Multiagent therapy is preferred over single agents where tolerated, with the National Comprehensive Cancer Network recommending carboplatin/paclitaxel, cisplatin/doxorubicin, cisplatin/doxorubicin/paclitaxel, or carboplatin/docetaxel as first-line agents for recurrent or metastatic disease, or for high-risk histologies (clear cell, serous, carcinosarcoma, and undifferentiated) [3]. The role of radiation in advanced disease is controversial, and outcomes remain poor for women who present with either advanced-stage or recurrent disease, independent of the treatment modality chosen [3]. Treatment options are limited for recurrent or progressive disease, with no therapies considered standard [3].

In light of the poor outcomes associated with progressive and recurrent EC, substantial research is being conducted on molecular mechanisms of endometrial carcinogenesis in the hope that a better understanding of the pathogenesis of disease at a molecular level will lead to new therapeutic strategies for overcoming chemotherapy resistance [3]. Ultimately, risk stratification using molecular markers in addition to traditional pathological criteria is needed in order to better predict the risk of local and systemic recurrence, and to understand how specific mutations or combinations of molecular features may lend themselves to targeted therapies [3, 4].

## Classification of endometrial cancer

Endometrial cancer has historically been classified into two broad groups on the basis of histological, epidemiologic, and molecular factors [5]. Within this schema, type I carcinomas represent approximately 80%–90% of cases, are typically low-grade, endometrioid, diploid, and hormone receptor-positive, and carry a good prognosis [5]. These tumors typically show phosphatase and homolog gene (*PTEN*) mutations, are associated with states of hyperestrogenism, and arise in a background of hyperproliferative endometrium [5, 6]. Type II carcinomas, by contrast, are typically high-grade, nonendometrioid, aneuploid, and hormone receptor-negative, and carry a significantly worse prognosis [5]. These tumors, which comprise the remaining 10%–20% of EC, are typically *TP53*-mutated, are less associated with hyperestrogenism, and classically arise in a background of atrophic endometrium [5, 6]. Subsequent efforts have been made to further define characteristic genetic alterations in type I and type II EC using molecular profiling techniques [6]. This classification system is not without limitations, however, as 20% of type I cancers relapse, whereas 50% of type II cancers do not [6]. Furthermore, 15%–20% of type I cancers are high-grade lesions, and their place within this dualistic model remains controversial [7]. It is becoming increasingly clear that EC exhibits a range of distinct genetic and molecular features, and widespread efforts have been undertaken to reclassify this disease for both prognostic and therapeutic reasons [6].

Recent analysis of The Cancer Genome Atlas (TCGA) focusing on endometrioid and serous carcinomas has led to a proposed reorganization of EC into four distinct groups, each with idiosyncratic mutation rates and patterns (Table 1): (1) DNA polymerase epsilon (*POLE*) ultramutated; (2) Microsatellite instability (MSI) hypermutated; (3) Copy-number low (CNL); and (4) Copy-number high (CNH) [14]. Following publication of this landmark study, attempts were made in earnest to correlate this reclassification system with clinical outcomes [6]. A cohort study of 50 patients with high-grade endometrioid EC showed a 100% disease-specific survival (DSS) among *POLE*-ultramutated tumors, 82% in the MSI-hypermutated group, 77.8% in the CNL group, and 42.9% in the CNH group [15]. The $P$ value ($P = 0.065$) for this study did not reach statistical significance, but was nevertheless suggestive [15]. Five-year recurrence-free survival (RFS) has also been correlated with molecular subtype, with 93% reported for *POLE*-ultramutated tumors, 95% in the MSI-hypermutated group, 52% in the CNL group, and 42% in the CNH group [1]. Several studies have used outcomes in CNL tumors (*TP53* wild-type group) as a basis for generating hazard ratios for the other molecular subgroups and have demonstrated a strong correlation between molecular subtype and overall survival (OS), RFS, and progression-free survival (PFS) [16].

## Therapeutic implications of molecular classification of endometrial cancer

Multiple studies are underway to determine the prognostic and therapeutic significance of this reclassification of EC by molecular characteristics [3]. The *POLE*-ultramutated group represents 7% of tumors, is characterized by increased mutations in the *POLE* exonuclease

TABLE 1  Molecular alterations in endometrial cancer subtypes [6, 8–13].

| Histology or molecular subtype | Incidence | TP53 mutation | PI3K alterations | KRAS alteration | ERBB alteration | FGFR alteration | Wnt/β-catenin mutation | ARID1A mutation | FBXW7 mutation |
|---|---|---|---|---|---|---|---|---|---|
| Endometrioid | 68% | 11.4% | PTEN mutation (75%–85%) PIK3CA mutation (50%–60%) PIK3R1 mutation (40%–50%) | 20%–30% | None | FGFR2 mutation (12%) | CTNNB1 mutation (25%) | 35%–40% | |
| Serous | 10% | >90% | PTEN mutation (11%) PIK3CA mutation (35%) PIK3CA amplification (45%) PIK3R1 mutation (12%) | 3% | ERBB2 amplification (25%–30%) | FGFR2 mutation (5%) Frequent FGFR1 and FGFR3 amplification | CTNNB1 mutation (3%) | | 20% of undifferentiated endometrial carcinoma |
| Clear cell | 3% | 35% | PTEN loss (80%) PIK3CA mutation (18%) | 0% | ERBB2 amplification (16%) ERBB2 mutation (12%) | | | 25% | |
| Carcinosarcoma | 5% | 60%–90% | PTEN mutation (19%) PIK3CA mutation (35%) PIK3R1 mutation (14%) | 17% | ERBB2 amplification (13%–20%) ERBB3 amplification or mutation (13%) | FGFR3 amplification (20%) | | 25% | 35%–40% |

*Continued*

**TABLE 1** Molecular alterations in endometrial cancer subtypes [6, 8–13]—cont'd

| Histology or molecular subtype | Incidence | TP53 mutation | PI3K alterations | KRAS alteration | ERBB alteration | FGFR alteration | Wnt/β-catenin mutation | ARID1A mutation | FBXW7 mutation |
|---|---|---|---|---|---|---|---|---|---|
| POLE ultramutated | 7% | 35% | PTEN mutation (94%) PIK3CA mutation (71%) PIK3R1 mutation (65%) | 53% | | | | | 76% | 82% |
| MSI hypermutated | 28% | 8% | PTEN mutation (87.7%) PIK3CA mutation (53.8%) PIK3R1 mutation (41.5%) | 35.4% | | FGFR2 mutation (13.8%) | CTNNB1 mutation (20%) CCND1 (12.3%) | 36.9% | <10% |
| Copy-number low | 39% | 1% | PTEN mutation (76.7%) PIK3CA mutation (53.3%) PIK3R1 mutation (33.3%) | 15.6% | | FGFR2 mutation (13.3%) | CTNNB1 mutation (52%) | 42.2% | <10% |
| Copy-number high | 26% | 91.7% | PTEN mutation (<10%) PIK3CA mutation (46.7%) PIK3R1 mutation (13.3%) | <10% | | | | <10% | 22% |

ARID1A: AT-rich interactive domain-containing protein 1A; CTNNB1: Catenin beta-1; ErbB: aka epidermal growth factor receptor aka HER; FBXW7: F-box/WD repeat-containing protein 7; FGFR: fibroblast growth factor receptor; KRAS: Kirsten rat sarcoma; MSI: microsatellite instability; PIK3CA: phosphatidylinositol-4,5-bisphospate 3-kinase, catalytic subunit alpha; PIK3R1: phosphatidylinositol 3-kinase regulatory subunit alpha; POLE: DNA polymerase epsilon; PTEN: phosphatase and homolog; TP53: tumor protein P53 aka p53; Wnt: wingless-type integration.

domain, and is associated with improved PFS [8]. Importantly, these tumors display increased numbers of tumor-infiltrating lymphocytes (TIL), suggesting that they may be good candidates for immune checkpoint inhibition [17, 18]. Tumors exhibiting mismatch repair deficiencies are characterized by MSI and are associated with Lynch syndrome [8]. This subgroup of EC accounts for 28% of tumors and is also believed to be highly susceptible to immune checkpoint inhibition due to its association with high neoantigen loads, a large number of TILs, and increased expression of both programmed cell death ligand 1 (PD-L1) and programmed cell death protein 1 (PD-1), a cell surface receptor that represses Th1 cytotoxic immune responses [8, 17, 19, 20]. Microsatellite-stable (MSS) CNL EC accounts for 39% of tumors and displays increased progesterone receptor expression, suggesting heightened responsiveness to hormonal therapy [8]. CNH EC accounts for 26% of tumors and contains nearly all of the serous and one-quarter of the grade three endometrioid tumors [8]. Most of these tumors feature *TP53*, *FBXW7*, and *PPP2R1A* mutations, which were previously only reported in uterine serous carcinomas, suggesting a potential increased responsiveness to treatment approaches currently utilized in serous EC [8]. PORTEC-4a, a randomized trial comparing standard versus molecular subtype-driven recommendations for adjuvant radiotherapy in early-stage EC, is currently ongoing [8, 21].

Pembrolizumab is the first site-agnostic molecular marker-driven therapy approved for EC, particularly for those with tumor MSI or abnormal immunohistochemical (IHC) findings [3]. Several other targeted therapies hold promise, including Phosphoinositide 3-kinase (PI3K) inhibitors for PI3K-Protein Kinase B (PKB, also known as AKT)-mammalian target of rapamycin (mTOR) signaling pathway mutations, hormonal therapy for estrogen receptor (ER)/progesterone receptor mutations, mTOR inhibitors for PI3K-AKT-mTOR pathway and *PTEN* mutations, enhancer of zeste homolog 2 inhibitors for AT-rich interactive domain-containing protein 1A (*ARID1A*) mutations, fibroblast growth factor receptor (FGFR) inhibitors for FGFR2 pathway mutations, and poly-ADP ribose polymerase (PARP) inhibitors for *PTEN*, homologous repair deficiency, and *ARID1A* mutations [3]. Multiple studies are investigating combinations of immunotherapy, radiation, and chemotherapy in the treatment of EC, with the goal of inducing a proinflammatory state [4]. Increased sensitivity to immune checkpoint blockade is presumed with this approach as well [4]. Bevacizumab, a vascular endothelial growth factor (VEGF) inhibitor with widespread use in ovarian cancer, has demonstrated a single-agent activity in EC, phase II trials have shown synergistic activity with chemotherapy, and numerous trials are ongoing to assess its effectiveness in recurrent EC [3, 22]. Ultimately, although many promising advances have been made in the molecular characterization of EC, high-quality studies and clinical trials are needed to both correlate molecular markers with clinical outcomes and further the development and implementation of targeted therapies [3].

## Wnt/β-catenin signaling pathway

## Introduction

Wingless-type integration site (Wnt) signaling is a cellular pathway consisting of an evolutionarily conserved family of 19 ligands and 10 receptors that act at the cell surface to regulate diverse cellular behaviors (Fig. 1) [23, 24]. In the absence of Wnt signaling, unstimulated

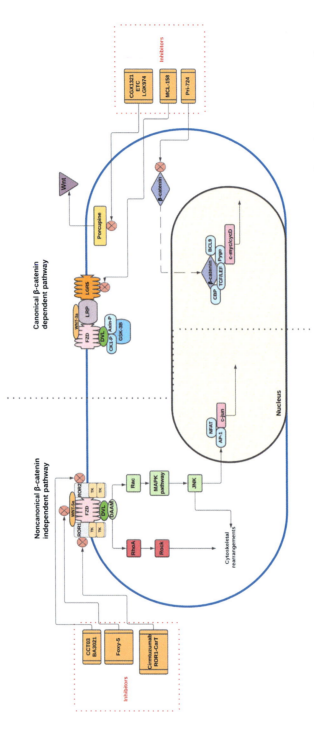

**FIG. 1**    Wnt/β-catenin signaling and current therapies targeting this pathway. Inhibitors of the canonical β-catenin-dependent pathway are shown on the right, including the Porcupine inhibitors CGX1321, ETC, and LGK974; the LGR5 inhibitor MCL-158; and the direct β-catenin inhibitor Pri-274. Inhibitors of the noncanonical β-catenin-independent pathway are shown on the left, including the ROR1 inhibitor cirmtuzumab and an ROR1 CART-cell therapy: the ROR2 inhibitors CCT03 and BA3021; and the Wnt5A inhibitor Foxy-5.

cells limit β-catenin levels through phosphorylation, subsequent ubiquitination, and degradation [23–25]. When a Wnt ligand binds its frizzled receptor (FZD), a signaling cascade is initiated that destabilizes the degradation complex and allows unphosphorylated β-catenin to accumulate in the nucleus, where it acts as a cofactor in the transcription of genes determining cell fate and proliferation [25]. Wnt signaling involves two distinct pathways: a canonical β-catenin-dependent pathway and several noncanonical β-catenin-independent pathways, with considerable crosstalk and feedback seen between these two [24, 25]. Canonical β-catenin-dependent signaling is understood to regulate cell proliferation and survival, and is mediated through nuclear accumulation of β-catenin [23, 24]. Noncanonical β-catenin-independent Wnt signaling is linked to cell differentiation, polarity, and migration, and can be further subdivided into Wnt/calcium signaling and planar cell polarity signaling, which is achieved via activation of the tyrosine kinase receptors [receptor tyrosine kinase-like orphan receptor (ROR)1 and ROR2] [23, 24, 26]. Canonical Wnt signaling is inhibited by several antagonists through both direct and indirect means: Wnt inhibitory factor-1 and secreted frizzled-related proteins (SFRP) act by directly binding Wnt molecules, whereas Dickkopf proteins (DKKs) act through indirect binding to the receptor complex [25].

The role of Wnt signaling in embryonic development and tissue homeostasis has long been recognized [24]. Defective Wnt/β-catenin signaling was first described in colorectal cancer, where activating mutations of the canonical Wnt signaling pathway have been demonstrated in over 90% of cases [23, 25]. Wnt pathway activation is most commonly mediated through mutational inactivation of adenomatous polyposis coli (APC); however, some colorectal cancers exhibit constitutive β-catenin transcriptional activity even in the absence of an APC mutation [25]. Ultimately, either scenario leads to increased tumorigenesis and cancer progression [25]. Wnt signaling has been explored in several other solid tumors, including melanoma, osteosarcoma, prostate, breast, liver, lung, ovarian, and several other gastrointestinal cancers, and the results of these findings have been reviewed extensively [25]. Noncanonical β-catenin-independent signaling may at times antagonize β-catenin-dependent signaling; however, recent studies have revealed a more complex and nuanced understanding of Wnt signaling that eschews the dichotomous and antagonistic relationship between the two pathways in favor of a Wnt network theory, which features significant crosstalk between the two pathways [23, 24, 26]. It is now clear that aberrant Wnt signaling does not exclusively promote tumor development, but may instead regulate nearly every aspect of carcinogenesis both positively and negatively, ranging from tumor initiation to metastatic spread [23, 24].

Wnt/β-catenin signaling is mediated through several transcription factors and target genes [25]. T-cell factor/lymphoid enhancer factor (TCF/LEF) transcription factors are multifunctional proteins that use their sequence-specific, DNA-binding, and context-dependent interactions to mediate Wnt signaling by repressing target gene expression when Wnt signals are absent and by recruiting β-catenin to target genes for activation when these signals are present [25]. Several target genes have been identified as transcriptionally activated by β-catenin signaling via TCF/LEF transcription factors [25]. These include genes relevant to cell cycling and growth (e.g., c-Myc, c-Jun, cyclin D1, and gastrin), others involved in cell survival (e.g., inhibitor of DNA binding-2 and MDR1), and still others implicated in tumor invasion and metastasis (e.g., matrilysin and VEGF) [25]. Axin and APC are the most well-known suppressors of the canonical Wnt/β-catenin pathway, and their inactivation has been shown to result in neoplastic transformation of affected cells through enhanced expression of immunoglobulin transcription factor-2 [27]. This role of Axin makes it, along with the oncogenes *myc* and cyclin D, one of the most important target genes related to Wnt-mediated carcinogenesis [25].

## Wnt/β-catenin signaling in physiologic endometrium

In the postpubertal female, the endometrium undergoes cyclic structural remodeling in response to hormonal changes, with proliferation of glands and stroma under the influence of estrogen (proliferative phase) and subsequent decidualization (secretory phase) in response to elevated progesterone levels following ovulation [24]. It is now well documented that most elements of the Wnt signaling pathway are expressed in the endometrium and that this pathway mediates the tissue response seen to fluctuating hormone levels [28]. During the proliferative phase, Wnt signaling is increased in response to estrogen, while the Wnt inhibitor Dickkopf-1 is downregulated [29]. As progesterone levels rise during the secretory phase, progesterone induces Dickkopf-1 with an associated decrease in Wnt signaling [29]. With the onset of menopause, levels of both estrogen and progesterone fall, with an accompanying decrease in Wnt signaling [24]. In the presence of obesity and hyperestrogenic states, induction of Wnt signaling mediates endometrial proliferation and hyperplasia, although this signaling mechanism does not appear sufficient on its own to trigger neoplastic transformation [24].

## Wnt/β-catenin signaling in endometrial hyperplasia and cancer

The role of Wnt signaling in endometrial hyperplasia and cancer has been consistently demonstrated in multiple studies over the course of two decades, and the results of these have been reviewed extensively [24]. Briefly, Fukuchi et al. were the first to demonstrate accumulation of β-catenin in 38% of EC, and an analysis of the TCGA dataset demonstrated β-catenin mutations in 36.6% of nonultramutated and 52% of MSS/CNL EC [14, 30]. Many of the studies examining the role of Wnt signaling in endometrial carcinogenesis have been conducted on relatively small cohorts from single sites with a focus on endometrioid EC and with wide variations in methodology, yielding estimates of activating β-catenin mutations in EC that range from 12% to 31% [24]. Liu et al. in a validation study comparing the findings of the TCGA dataset with a large independent cohort demonstrated overexpression of multiple Wnt pathway components, including Wnt5A, Frizzled-10, TCF7, and LEF1, and correlated these with decreased OS [31, 32]. The role of Wnt7A in endometrial hyperplasia and malignancy remains unclear, with one IHC study correlating increased levels of Wnt7A with hyperplasia and tumorigenesis and another showing the opposite [31, 33]. This incongruity may be explained by differences in antibodies used to detect Wnt expression in various IHC studies, as unopposed estrogen-driven tumorigenesis should theoretically result in increased levels of Wnt7A [24].

Several studies, most of which are small, heterogeneous cohort studies examining primarily endometrioid EC, have investigated the role of Wnt inhibitors in endometrial tumorigenesis [24, 34]. Cellular expression of Dickkopf1 and Dickkopf3 has been shown in some studies to be significantly lower in EC than in benign endometrium, with correlation to high-grade disease, extrauterine spread, and a poor prognosis [34, 35]. Other studies have shown increased levels of serum Dickkopf1 and Dickkopf3 in endometrial and cervical cancer patients compared to healthy subjects [36]. The reason for these conflicting results remains elusive, and the mechanism underlying altered regulation of DKKS in EC is unknown, but may be due to epigenetic silencing [25]. Similar conflicting patterns of expression have been described for SFRP, although recent studies have consistently correlated downregulation of SFRP with increased Wnt expression in endometrial hyperplasia and cancer [25]. Abu-Jawdeh et al. were the first to demonstrate the expression of SFRP4 in the stroma of proliferative endometrium, finding marked upregulation of its mRNA in hyperplastic and malignant endometrial stromal cells and concluding that it was hormonally regulated [37]. Other authors have shown decreased expression of various SFRPs in endometrial hyperplasia and cancer, finding downregulation of SFRP1 and SFRP4 in high-grade disease, in both low-grade endometrial stromal sarcoma and undifferentiated endometrial sarcoma, and in MSI as compared to MSS tumors [35, 38, 39]. Downregulation of SFRP was also associated with locoregional and distant disease recurrence, suggesting a therapeutic role for targeting Wnt signaling in EC [35]. Several studies have suggested epigenetic factors as the mechanism behind the changes observed, with promoter methylation of SFRP1, SFRP2, and SFRP5 and demethylation of SFRP4 seen in EC, and with corresponding upregulation of the Wnt target fibroblast growth factor (FGF) 18 [39, 40]. In each case, decreased expression of Wnt inhibitors is consistent with increased Wnt signaling and subsequent pathologic endometrial proliferation.

Given the known association between type I EC and a hyperestrogenic state, as well as the previously demonstrated interplay between estrogen, progesterone, and the Wnt/β-catenin signaling pathway in the normal endometrial cycle, it is no surprise that a connection has been suggested between states of hormonal dysregulation, altered Wnt/β-catenin signaling, and pathologic endometrial proliferation [25]. Estrogen has been shown in multiple studies to exert its stimulatory effects on Wnt signaling through inhibition of glycogen synthase kinase 3 (GSK-3β), which in turn results in increased stability and nuclear translocation of β-catenin [41]. B-catenin then regulates transcription by both binding to TCF transcription factors and associating with estrogen receptor α (ERα) [41]. Other studies have shown estrogen-mediated upregulation of the Wnt signaling target insulin-like growth factor-1 (IGF-1) with corresponding progesterone-mediated downregulation [42]. In addition to its action though GSK-3β inhibition and IGF-1 upregulation, estrogen in a murine model was shown to upregulate canonical Wnt signaling independent of estrogen receptors through upregulation of Wnt4, Wnt5A, and Frizzled2, and through downregulation of Wnt7A, with upregulation of the Wnt inhibitor SFRP2 circumventing the effects of estrogen on uterine epithelial cell growth [43].

The ability of progesterone to counter the proliferative influence of estrogen on the endometrium may also be mediated through the Wnt/β-catenin signaling pathway, as several studies have shown increased Wnt signaling with progesterone receptor knockdown or inhibition [25, 44, 45]. Whereas mifepristone, a progesterone receptor antagonist, has been shown to mediate its effects through upregulation of Wnt5A, progesterone seems to act through both inhibition of GSK-3β and upregulation of the Wnt inhibitors Dickkopf-1 and Forkhead box protein O1, with increased levels of progesterone inducing Wnt inhibition and knockout of these Wnt inhibitors circumventing progesterone-mediated inhibition of Wnt signaling [25, 28, 44]. These findings strongly suggest that Wnt signaling is the downstream mediator of both the proliferative effects of estrogen and the antiproliferative effects of progesterone on pathologic endometrial proliferation, and further suggest that modulation of Wnt signaling may prove to be a potent target for novel therapeutic agents in the treatment of EC [25].

Several studies have suggested crosstalk between the Wnt/β-catenin signaling pathway and several other important signaling pathways [2]. Wnt ligands are one of many upstream factors regulating mTOR signaling through the canonical Wnt pathway, whereas overexpression of Gli1, a transcriptional factor and target gene of the hedgehog signaling pathway, has been positively correlated with β-catenin nuclear immunoreactivity in both atypical complex endometrial hyperplasia (AEH) and EC [25, 46]. Such interactions hold promise for potentially targeting multiple pathways in order to achieve a synergistic therapeutic response in the treatment of EC [25]. In vitro studies have suggested a correlation between the expression of Wnt5A, mediator of the noncanonical Wnt signaling pathway, and protein kinase Cα (PKCα), which is associated with enhanced vasculogenic capacity, motility, and invasiveness of ovarian cancer cells [47]. Furthermore, upregulation of Wnt5A is associated with increased expression of PI3K and Snail, indicating significant crosstalk between the noncanonical Wnt and the PI3K-AKT-mTOR pathways [47].

Most research conducted on Wnt signaling has focused on the canonical β-catenin-dependent pathway, and relatively little is known about the role of the noncanonical β-catenin-independent pathway in endometrial carcinogenesis [24]. Bui et al. found downregulation of Wnt4 mRNA in four EC cell lines compared to benign controls as well as possible downregulation of Wnt2, Wnt3, and Wnt5A mRNA [48]. Wnt4 and Wnt5A both act on the noncanonical pathway, with Wnt5A overexpression previously seen in cervical, breast, and ovarian cancers [24]. Wnt5A has generated particular interest because it is capable of signaling via multiple Wnt pathways and because it acts as both a promoter and a suppressor in different cancers [49]. It has also been associated with the initial development and progression of ovarian cancer, especially through induction of vasculogenic mimicry and promotion of the epithelial-to-mesenchymal transition [47]. Wnt5A was shown to do this via noncanonical signaling, with possible interplay with the PKCα pathway [47]. Most data on Wnt5A have come from ovarian cancer and melanoma, where knockdown of Wnt5A has been shown to reduce proliferation, migration, and invasion, and to promote cell-cycle arrest and apoptosis [50].

Wnt5A binds a family of receptors called receptor tyrosine kinase-like orphan receptors (RORs), a class of receptors initially labeled "orphan" before their ligand was discovered, but now clearly identified as Wnt receptors (Fig. 1) [24]. These RORs function in developmental processes such as skeletal and neuronal development, but more importantly regulate cell movement and polarity, and are associated with aggressive disease and a poor prognosis in a number of cancers [24]. Limited studies have examined ROR expression in EC, with one finding ROR1 expression in 28/29 (96%) uterine cancer cell lines and no ROR1 expression in benign counterparts, and others correlating ROR1 expression with higher stage and increased tumor volume [51]. Investigations into ROR2 expression in EC are very limited [24]. One study examined ROR2 expression in leiomyosarcoma and demonstrated reduced tumor mass with ROR2 knockdown and worse clinical outcomes with increased ROR2 expression, suggesting that ROR2 may serve as a useful prognostic indicator in soft-tissue sarcomas and may represent a novel therapeutic target [52]. Another study using results extrapolated from ovarian cancer demonstrated that CD55 regulation via ROR2/c-Jun N-terminal kinase signaling may provide an opportunity to target key signaling pathways in endometrioid tumors [53]. A third study looked at ROR1 and ROR2 expression more broadly in endometrioid EC and found that increased ROR1 expression correlated with worse overall survival, while increased ROR2 expression correlated with better survival, concluding that the two likely play opposite roles in EC, with ROR1 promoting tumor progression and ROR2 acting as a tumor suppressor [54]. The role of ROR1 and ROR2 in several other cancers has been reviewed extensively, and both may ultimately prove to be promising therapeutic targets in EC [24].

## Therapies targeting Wnt/β-catenin signaling currently under investigation

Given the limited availability of treatment options for women who have failed traditional cytotoxic chemotherapy for EC, as well as increasing recognition of Wnt signaling as a mediator of endometrial carcinogenesis, there is a clear need for therapies that can target this signaling pathway at multiple levels [25]. Several existing drugs and natural compounds are believed to exert influence on the Wnt/β-catenin pathway, including NSAIDs such as aspirin, sulindac, and celecoxib; vitamins such as retinoids; and polyphenols [25]. A firm understanding of the role that these compounds play in modulation of Wnt signaling is, however, currently lacking, limiting their therapeutic usefulness in the treatment of EC [25].

Despite an increased understanding of Wnt signaling and promising results seen from targeting various components within the pathway in preclinical studies, few clinical trials of immunotherapy agents targeting the Wnt signaling pathway are currently underway [24, 26]. Documented challenges in conducting trials on Wnt-modulating therapies include poor recruitment, withdrawn funding, dosage adjustments, and more broadly a need for improved understanding of the interplay between various pathway components on endometrial carcinogenesis [24]. For example, DKN-01, a Dickkopf1 inhibitor, is currently in a clinical trial; however, since reduced expression of Dickkopf1 has been correlated with EC progression and metastasis, the rationale for including EC patients in this trial is unclear and potentially dangerous [24]. Several small-molecule inhibitors of the canonical Wnt pathway are currently in phase I/II clinical trials for breast, colon, and pancreatic cancers (Fig. 1), and extension to patients with endometrioid EC could prove beneficial; however, potential activating effects on noncanonical signaling with inhibition of the canonical pathway may produce unintended effects [24]. Inhibition of Porcupine protein, which processes Wnt ligands prior to secretion, is the subject of another clinical trial, with the goal of broad Wnt inhibition, including ligands in both the canonical and noncanonical signaling pathways (Fig. 1) [24]. Similar caution is advised in this scenario, as it is unknown what the net effect of inhibiting all Wnt ligands may have on endometrial tumorigenesis [24].

There is only one drug involving noncanonical Wnt signaling currently under investigation, Foxy-5, a Wnt5A mimetic, which has only been tested in metastatic prostate cancer in murine models (Fig. 1) [55]. Cirmtuzumab, an ROR1 antibody, is currently in phase I/II clinical trials for chronic lymphoid leukemia and breast cancer, as is a basket design study for an ROR1 chimeric antigen receptor (CAR)T-cell therapy for ROR1-positive breast, lung, and hematologic cancer (Fig. 1) [24]. Caution is advised in the application of these therapies to EC, as increased expression of Wnt5A and ROR receptors has been associated with tumor progression in some cancers, such as breast, and ROR1 and ROR2 have been shown to exert opposing influence on endometrial carcinogenesis [24].

Ultimately, further development of agents targeting the Wnt/β-catenin signaling pathway in EC are limited by a poor understanding of various biomarkers in the pathway, by challenges in developing treatments that effectively target specific molecules within it, and by a lack of sufficient knowledge regarding potential side effects that may result from targeting various ligands and receptors [24]. The potential for targeting this pathway with immunotherapeutic agents is clearly promising; however, the role of the canonical and noncanonical signaling pathways needs to be further elucidated in endometrial carcinogenesis before safe and effective targeted therapeutic agents can be developed and brought to clinical trial [24].

## PI3K-AKT-mTOR signaling pathway

### Introduction

*PTEN* is a tumor-suppressor gene whose primary role is the downregulation of the PI3K-AKT-mTOR signaling pathway, which is itself responsible for governing multiple cellular functions, including cell replication, cell-cell adhesion, cell migration, and apoptosis [56]. PTEN acts via dephosphorylation of phosphatidylinositol (3,4,5)-triphosphate, which is essential to the phosphorylation and subsequent activation of AKT by phosphoinositide-dependent kinase-1, and is in this way antagonistic to the activity of PI3K, which also activates AKT (Fig. 2) [57]. Thus, PTEN acts as a tumor suppressor, and any loss of function results in activation of the PI3K-AKT-mTOR pathway with resultant increased cell proliferation, angiogenesis, ability to invade, and propensity for metastasis [57]. *PTEN* mutations and associated PI3K-AKT-mTOR pathway activation have also been linked to mechanisms of DNA double-strand break repair that occurs via homologous recombination [58].

mTOR, a member of the PI3K-AKT family, is a serine-threonine kinase that responds to a variety of growth factors, nutrients, and other signals to regulate several metabolic and macromolecular processes within the cell, including cell survival and proliferation, angiogenesis, and various aspects of protein synthesis [57]. Once activated, mTOR forms two separate protein complexes, mammalian target of rapamycin complex 1 (mTORC1) and mTORC2, which themselves regulate distinct cellular functions through control of mRNA translation [57]. mTORC1 is a protein complex consisting of mTOR, regulatory-associated protein of mTOR, G-protein β-subunit-like protein (GβL)/lethal with SEC13 protein 8 (LST8), and proline-rich AKT substrate 40 kDA [57]. The regulatory-associated protein of mTOR recruits substrates to mTORC1 for phosphorylation, leading to downstream activation of ribosomal initiation and elongation factors, which ultimately promotes mRNA translation [57]. mTORC2 is another protein complex containing GβL, rapamycin-insensitive companion of mTOR, and mammalian stress-activated protein kinase-interacting protein, whose activation by growth factors regulates cytoskeleton organization and leads to AKT activation, which in turn upregulates mTORC1 and protein synthesis [57]. mTORC2 is typically insensitive to allosteric inhibition by mTOR rapalog inhibitors such as sirolimus, but is effectively inhibited by small-molecule catalytic mTOR inhibitors [57].

### PI3K-AKT-mTOR signaling in physiologic endometrium

The role of PI3K-AKT-mTOR signaling in physiologic endometrium has not been clearly established, but investigations have primarily focused on the role of this pathway in the

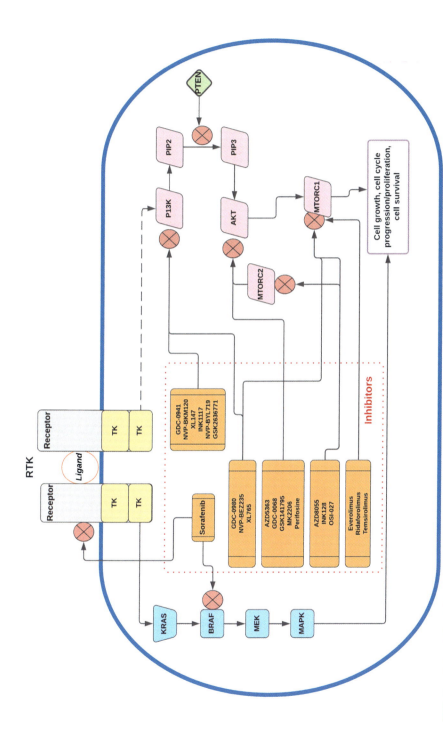

FIG. 2 PI3K-AKT-mTOR signaling and current therapies targeting this pathway. Illustrated are the BRAF and receptor tyrosine kinase (VEGF, PDGFR) inhibitor sorafenib; the MTORC1 inhibitors everolimus, ridaforolimus, and temsirolimus; the pan-PI3K inhibitors GDC-0941, NVP-BKM120, and XL147, along with the isoform-specific PI3K inhibitors INK1117, NVP-BYL719, and GSK2636771; the mTOR/PI3K inhibitors GDC-0980 (apitolisib), NVP-BEZ235, and XL765 (voxtalisib); the combined mTORC1/2 inhibitors AZD8055, INK128, and OSI-027; and the AKT inhibitors AZD5363, GDC-0068, GSK141795, MK2206, and Perifosine.

nongenomic mechanism of estrogen signaling [59]. In the traditional genomic mechanism of estrogen signaling, estrogen binds to ER, which in turn dimerizes, translocates to the cell nucleus, and binds estrogen response elements to promote gene transcription [59]. Estrogen has, however, also shown the ability to rapidly influence intracellular signaling pathways apart from de novo gene transcription [60]. Nongenomic estrogen-mediated activation has been described for the protein kinase A (PKA), mitogen-activated protein kinase (MAPK), and the PI3K pathways, with cell membrane-localized estrogen receptors implicated in the activation of MAPK and PI3K signaling [61]. Estradiol has been shown in vitro to mediate PI3K signaling in physiologic endometrium via increased phosphorylation of AKT, with resultant proliferative and antiapoptotic effects [59]. Other studies have suggested that estrogen may downregulate PTEN, a potent antiproliferative factor, via increased phosphorylation, whereas progesterone may upregulate it via decreased phosphorylation [62]. IGF-1 is suggested as an activator of AKT via PI3K in human endometrium, and AKT expression has been demonstrated in decidualized stromal cells in both murine and human models; however, there is little evidence of cyclical expression of AKT in physiologic endometrium or for induction of a decidualized reaction in human endometrial stromal cells apart from a direct cell-to-cell interaction with trophoblasts [59]. AKT signaling in response to estradiol has also been shown to protect endometrial cells from apoptosis, triggering proliferation, whereas inhibition of AKT is associated with caspase-3 activation and apoptosis [63]. Finally, downregulation of AKT activity has been shown to represent a critical step in the initiation of decidualization in human endometrial stromal cells [64]. Ultimately, further study is needed to clarify the role of PI3K-AKT-mTOR signaling in physiologic endometrium.

## PI3K-AKT-mTOR signaling in endometrial hyperplasia and cancer

Common genetic and molecular alterations in type I EC have been reviewed extensively, but in short involve *PTEN*, *PIK3CA*, β-catenin, Kirsten rat sarcoma (KRAS), microsatellite instability (MSI), serine-threonine liver kinase B1 (LKB1), E-cadherin, and *TP53* (Table 1) [57, 65]. *PTEN* loss or mutation represents the most common alteration seen in type I EC, occurring in 70%–83% of cases, although it is frequently mutated in type II cancers as well (Table 1) [65]. The presence of *PTEN* mutations in endometrial hyperplasia both with and without atypia suggests its alteration as an early step in endometrial carcinogenesis [65]. By promoting endometrial proliferation, estrogen is thought to provide a survival advantage to *PTEN*-mutant clones that have evaded usual proapoptotic, antiproliferative checkpoints, whereas progesterone, in opposing the proliferative effects of estrogen, is thought to decrease the competitive advantage of *PTEN*-mutant cells, promoting their involution [66]. Indeed, in murine models, 100% of mice with *PTEN* heterozygosity develop endometrial hyperplasia, and 20% develop EC [67]. PTEN also downregulates cellular production of IGF-1, which is typically increased in EC, and in this way antagonizes the effects of estrogen [42]. Although *PTEN* loss has been consistently associated with increased mTOR activity in EC, *PTEN*-independent overexpression of mTOR is rare and may not correlate with disease progression, overall survival, or sensitivity to mTOR inhibition [65].

The *PIK3CA* gene, which encodes the α catalytic subunit of PI3K, is upregulated in EC, with isolated *PIK3CA* mutations seen in 41%–52% of type I and 33%–38% of type II cancers, *PIK3R1* mutations seen in 21%–43% of type I and 12%–17% of type II cancers, and concurrent *PTEN* and *PIK3CA* mutations seen in 26% of all EC [65, 68]. *PIK3CA* mutations are additionally associated with endometrioid histology, higher stage, and increased lymphovascular invasion [69]. AKT mutations are rare and are only described in 2% of type I EC [65].

*KRAS*, a protooncogene that encodes a signaling plasma membrane guanosine triphosphatase (GTPase) involved in cell growth and differentiation, is mutated with constitutive activity in 10%–30% of type I EC and in 16% of hyperplastic lesions, leading to upregulation of mTOR activity and associated genomic instability [65]. LKB1, also known as serine/threonine kinase 11, activates AMP-activated protein kinase (AMPK), itself a potent inhibitor of mTOR [70]. Mutations of the *LKB1* gene in animal models are associated with aggressive features in EC, and human studies have shown that approximately 20% of endometrial tumors are LKB1-negative and up to 65% possess a functionally deficient version of this protein [71].

mTOR activation in EC has been studied in murine models, tissue samples, and clinical studies, and the findings of these investigations have been reviewed extensively [57, 65]. In short, increased mTOR activity is associated with advanced stage, and increased phosphorylation of eukaryotic translation initiation factor 4E-binding protein, a key component of mTOR signaling, has been shown to occur more frequently in high-grade endometrial tumors and to be associated with a worse prognosis [72]. Other studies, however, have not shown a significant correlation between PI3K-AKT-mTOR activity and either survival, stage, grade, or lymphovascular spread [73].

## Therapies targeting PI3K-AKT-mTOR signaling currently under investigation

### mTOR inhibitors

In principle, the well-documented overactivity of PI3K-AKT-mTOR signaling in EC should make this pathway a prime target for molecular therapy [57]. There are four categories of PI3K-AKT-mTOR signaling pathway inhibitors: mTOR inhibitors, PI3K inhibitors, dual mTOR/PI3K inhibitors, and AKT inhibitors [65]. Metformin also inhibits the PI3K-AKT-mTOR signaling pathway via AMPK activation, and is currently being evaluated in several clinical trials (Fig. 2) [65]. There are two broad classes of mTOR inhibitors: so-called "first-generation" mTORC1 inhibitors called rapalogs, which include sirolimus (rapamycin), temsirolimus, everolimus, and ridaforolimus, and which primarily show activity in type I EC; and newer catalytic inhibitors that directly inhibit mTORC1 and mTORC2, and which include the compounds AZD8055, INK128, and OSI-027 [57, 65]. Temsirolimus, everolimus, and ridaforolimus are all analogs of sirolimus that allosterically inhibit mTORC1 through a mechanism that is poorly understood [57].

Multiple preclinical studies have investigated the impact of rapalogs on mTOR pathway proteins, cellular responses to mTOR inhibitors both in vitro and in murine models, and the role that small-molecule mTOR kinase inhibitors play in EC cell lines [57, 65]. Ultimately, such studies have suggested that rapalogs are more effective as cytostatic rather than cytotoxic agents, and that their principal utility may lie in potentiating the effects of standard front-line chemotherapeutic agents such as cisplatin and paclitaxel [57]. Cell culture studies have shown that single-agent rapalogs have antiproliferative properties in a dose-dependent manner, and models in mice predisposed to hyperplasia have found that rapalogs reduce the incidence of hyperplastic lesions, reduce tumor burden, and slow progression of disease [57]. Tissue culture studies have yielded mixed results on the relationship between *PTEN* mutation status and rapamycin sensitivity [57].

Multiple phase I/II clinical trials have examined oral and parenteral mTOR inhibitors in advanced and recurrent EC, with variable responses seen [57]. One phase II study ($n = 45$) of ridaforolimus in patients with recurrent or progressive EC showed an 11% partial response (PR) rate, an 18% stable disease (SD) rate, and a 6-month PFS of 18%, while another ($n = 34$) similar study showed an 8.8% PR rate and a 52.9% SD rate [74, 75]. Everolimus was evaluated in a phase II study ($n = 35$) in recurrent EC with endometrioid histology and showed a 21% rate of stable disease at 20 weeks [76]. The best response to rapalogs has been seen in treatment-naïve patients, as a phase II study ($n = 50$) comparing single-agent temsirolimus in chemotherapy-naïve versus previously treated patients with recurrent or metastatic EC showed a 14% PR rate and a 69% SD rate in the chemotherapy-naïve group, compared to a 4% PR rate and a 48% SD rate in previously treated patients, with no correlation seen between *PTEN* status and response rates [77]. Finally, a large phase II trial ($n = 349$) randomized patients with chemotherapy-naïve stage III–IVA or with recurrent or stage IVB EC to either paclitaxel, carboplatin, and bevacizumab (Arm 1); paclitaxel, carboplatin, and temsirolimus (Arm 2); or ixabepilone, carboplatin, and bevacizumab (Arm 3); and used the paclitaxel and carboplatin arm of GOG209 as a historical control [78]. This study found similar response rates among all arms (60%, 55%, and 53%) and did not find a difference in PFS among any arms compared to the historical control, however did find a significantly increased OS in arm 1 [78]. Tuberous sclerosis complex 2 mutations appeared to be predictive of response to temsirolimus [78]. Due to the potential for loss of negative regulation of PI3K-AKT-mTOR signaling with exclusive mTORC1 inhibitors, second-generation drugs that inhibit both mTORC1 and mTORC2 have been developed, of which AZD8055, INK128, and OSI-027 are currently in early-stage clinical trials [65].

## PI3K inhibitors

There are two classes of PI3K inhibitors: pan-PI3K inhibitors, including GDC-0941, NVP-BKM120, and XL147; and isoform-specific PI3K inhibitors, with INK1117 and NVP-BYL719 acting on p110α and GSK2636771 acting on P110β (Fig. 2) [65]. The pan-PI3K inhibitors GDC-0941 and NVP-BKM120 have each shown activity in a variety of cancer cell lines in preclinical studies, with NVP-BKM showing increased activity in cells with *PIK3CA* mutations and GDC-0941 showing increased activity in murine models with *FGFR2*-mutated endometrial cells [79, 80]. The pan-PI3K inhibitors GDC-0941, NVP-BKM120, and XL147 are all in phase I/II clinical trials for patients with advanced solid tumors, while NVP-BKM120 and XL147 are in phase II trials for patients with EC [65, 80]. The isoform-specific PI3K inhibitors were created in order to mitigate the side effect profile of pan-PI3K inhibitors, and several are currently in preclinical and early-phase clinical studies [65]. The p110α-selective inhibitors INK1117 and NVP-BYL719 have shown activity in tumor cell lines with *PIK3CA* mutations and are currently in early-phase clinical trials [65]. INK1117 has much less activity in *PTEN*-deficient tumors, which rely on p110β for PI3K signaling, prompting the development of p110β-selective inhibitors such as GSK2636771 [65]. Dual p110α- and p110β-selective inhibitors are also currently in development, with early studies indicating that the success of isoform-specific inhibitors will depend on the ability to determine the *PIK3CA* and *PTEN* status of the targeted tumor [65].

### Dual mTOR/PI3K inhibitors

Dual mTOR/PI3K inhibitors include GDC-0980 (apitolisib), NVP-BEZ235, and XL765 (voxtalisib) (Fig. 2) [65]. These compete with the ATP-binding cleft of both class I PI3Ks and mTORC1/2, and should in theory more completely suppress the PI3K-AKT-mTOR signaling pathway than previously designed inhibitors [65]. GDC-0980 (apitolisib) and NVP-BEZ235 have both shown activity against several cancer cell lines in preclinical studies, with NVP-BEZ235 showing the greatest activity in EC cell lines containing *PIK3CA* or *PTEN* mutations in both in vitro and murine models, and NVP-BEZ235 producing similar, but not better, results than everolimus in xenograft studies [81–83].

GDC-0980 (apitolisib) and NVP-BEZ235 have both shown activity against advanced solid tumors in phase I clinical trials, while several phase II trials of XL765 (voxtalisib) have demonstrated an acceptable safety profile, but modest or limited efficacy in solid tumors [84–87]. A phase II trial of GDC-0980 (apitolisib) ($n=56$) showed only a 6% ORR, a median PFS of 3.5 months, and a median OS of 15.7 months, and dose reductions were required for 39% of patients due to drug toxicity [88]. Molecular profiling showed *PIK3CA*, *PTEN*, or *AKT1* mutations in 57% of the study population, all three patients with a confirmed response had at least one mutation in the PI3K-AKT-mTOR pathway, and it was suggested that patients with these mutations may have derived increased benefit from the drug [88].

### AKT inhibitors

Despite the rarity of AKT mutations, increased AKT signaling is commonly seen in EC, prompting the development of AKT inhibitors that either compete for an ATP-binding site, or allosterically inhibit AKT (Fig. 2) [65]. One drawback to AKT inhibition is the potential for increased compensatory signaling via AKT-independent activators of PI3K, and indeed there has been some indirect, albeit no clinical, evidence for this possibility [65]. The allosteric AKT inhibitors perifosine and MK2206 demonstrated antitumor activity in preclinical studies, where perifosine was shown to induce apoptosis in human EC cell lines more effectively than everolimus and the EGFR inhibitor gefitinib; however, it showed very little antitumor activity on soft-tissue sarcomas in a phase II trial [89, 90]. MK2206, ARQ 092, AZD5363, GDC-0068, and GSK141795 are in early-stage clinical trials for advanced solid tumors, while MK2206 is undergoing a separate phase II trial in EC [65].

### Metformin

Metformin regulates PI3K-AKT-mTOR signaling though activation of the mTOR inhibitor AMPK and through inhibition of insulin receptor substrate 1 (Fig. 2) [91]. Metformin has also been shown to reduce PD-L1 expression [92]. Multiple studies have evaluated the ability of metformin to alter various proliferative markers both in vitro and in clinical studies, and these results have been reviewed extensively [93]. Metformin has demonstrated in case reports, case series, and cohort studies, its ability to reverse AEH in conjunction with progestogens when progestogens alone have been unsuccessful [93]. One randomized-controlled trial ($n=42$) found that addition of metformin to megestrol was more effective than megestrol alone at inducing regression of simple endometrial hyperplasia at 12 weeks, while another randomized-controlled trial ($n = 43$) comparing metformin and megestrol head-to-head in patients with either simple hyperplasia, AEH, or EC showed that metformin alone was superior to megestrol at resolving endometrial hyperplasia [94, 95].

A phase II study ($n = 17$) of patients with AEH or FIGO stage 1 EC given medroxyprogesterone acetate (MPA) and metformin showed an 81% CR rate, a 14% PR rate, a 3-year RFS of 89%, and a relapse rate of 10% [96]. Compared to a previous similar study ($n = 45$) that administered only MPA, addition of metformin increased the CR rate from 55% to 68% in EC, and from 82% to 94% in AEH, and reduced the recurrence rate from 57% to 15% in EC, and from 38% to 6% in AEH [96, 97]. In a randomized-controlled trial ($n = 150$) of patients with EC or AEH who were randomized to either MPA only or MPA plus metformin, the addition of metformin to MPA improved the 16-week response rate from 20.4% to 39.6% in AEH, but no significant change was seen in EC [98]. The benefit of metformin in this study was still seen in nonobese and insulin-sensitive women [98].

A recent metaanalysis found that metformin use was not associated with a lower risk of EC in either diabetics or nondiabetics; however, its use was associated with improved OS and a reduced recurrence risk in EC, including a significant survival advantage among diabetics [99]. A large phase III randomized-controlled trial ($n = 88$) that used Ki-67 as a surrogate marker of endometrial proliferation in endometrial tumor cells did not show any benefit to giving patients with either AEH or EC metformin for 1 to 5 weeks prior to hysterectomy [100]. Finally, in a randomized-controlled trial ($n = 102$) of women with breast cancer comparing tamoxifen use only to tamoxifen use plus metformin, metformin was shown to reduce average endometrial thickness from a baseline of 2.5 mm to 2.3 mm, which was considered statistically significant, although the clinical significance of this reduction is questionable [101].

PI3K-AKT-mTOR signaling alterations are heavily implicated in EC, and many therapies targeting this pathway are currently undergoing clinical trials [65]. Preclinical data have offered useful guidance into how these agents may be most effectively used and have suggested which patients may derive the greatest benefit from them; however, correlation in well-designed clinic trials is needed to confirm these preliminary findings [65]. mTOR inhibitors have shown the greatest activity in treatment-naïve patients and as adjunctive therapies to front-line agents [65]. Crosstalk and negative feedback among other molecular signaling pathways remains a challenge in targeting the PI3K-AKT-mTOR pathway [65]. Ultimately, this challenge may be overcome by combining agents targeting multiple pathways; however, well-designed clinical trials are needed to confirm this hypothesis [65].

## EGFR/FGF-Ras-Raf-MEK-ERK signaling pathway

### Introduction

The fibroblast growth factor (FGF) family is a diverse group of five fibroblast growth factor receptors (FGFRs) and 22 FGF ligands that transmits intracellular and extracellular signals in either an intracrine, paracrine, or endocrine manner [102–104]. Each FGFR consists of an extracellular ligand-binding domain, a transmembrane domain, and an intracellular tyrosine kinase domain (Fig. 3) [103]. When small polypeptide FGF ligands in the extracellular matrix bind one of the extracellular FGFRs, it dimerizes, activating the associated intracellular

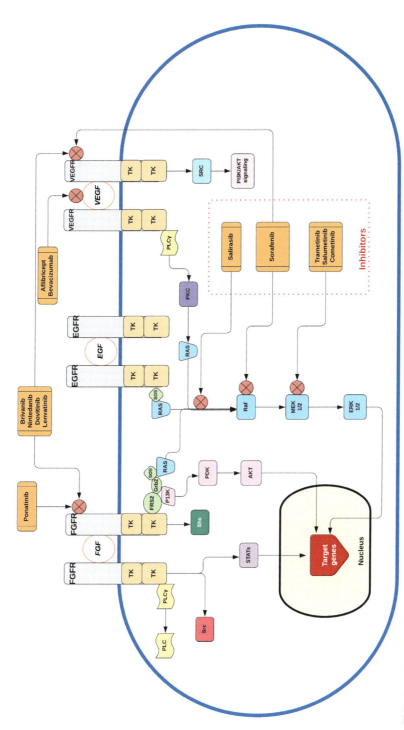

FIG. 3  EGFR/FGF-Ras-Raf-MEK-ERK signaling, VEGF signaling, and current therapies targeting these pathways. Illustrated are the multikinase inhibitors brivanib, nintedanib, dovitinib, and lenvatinib; the FGFR inhibitor ponatinib; the humanized monoclonal antibody against VEGF bevacizumab and the VEGF sequestering agent aflibercept; the Ras inhibitor salirasib; the BRAF and VEGF (and PDGFR) inhibitor sorafenib; and the MEK inhibitors trametinib, selumetinib, and cobimetinib.

tyrosine kinase domains via transautophosphorylation [103, 105]. Heparin sulfate glycosami-noglycan (HSGAG) binds both the FGF and FGFR components of this dimer, promoting protein-protein interactions between ligand and receptor, stabilizing the FGF against degra-dation, acting as a reservoir for ligands, and modulating the maximal distance of ligand dif-fusion [105]. Differences in HSGAG affinity to various FGFs appear to account for some of the divergence seen in activity among them [105]. FGF-FGFR interactions are also regulated by alternative splicing of FGFR mRNA and by tissue-specific expression of various ligands, re-ceptors, and cell-surface proteins [102]. Ligand specificity of FGFRs produced through alter-native splicing results in a IIIb isoform preferentially expressed in epithelial cells that binds FGF3, FGF7, FGF10, and FGF22 exclusively, and a IIIc isoform preferentially expressed in mesenchymal cells that binds FGF1 exclusively [104]. Ligand expression is typically cell specific, such that physiologic stimulation of these receptors tends to occur in a paracrine rather than an autocrine fashion [104].

Ligand binding of the FGFR leads to autophosphorylation and production of phosphotyrosine residues, which then act as binding sites for intracellular proteins, leading to the activation of four possible signaling cascades [104]: (1) Mitogen-activated protein kin-ase (MAPK) pathway activation begins when activated FGFR phosphorylates FGFR substrate 2 and recruits growth factor receptor-bound 2, which in turn activates rat sarcoma (Ras) and subsequently the Ras/MAPK pathway [102]. This pathway promotes cell proliferation through apoptosis signaling and regulation of the cell cycle through increased transcription factor exp-ression and modulation of the p53 pathway [106]; (2) The PI3K-AKT-dependent signaling pathway is also activated by growth factor receptor-bound 2 and results in cell growth and pro-liferation, as well as initiation of antiapoptotic factors [102]; (3) Phospholipase C-γ (PLC-γ), activated independently from FGFR substrate 2, activates protein kinase C (PKC), which incr-eases MAPK signaling [102]; (4) Signal transducer and activator of transcription dimers trans-locate to the nucleus to regulate gene transcription [104]. This complex signaling pathway is activated by several upstream signals, including genetic alterations, growth factors, cytokines, interleukin, and mitogens [106]. Regulation of FGF signaling occurs through negative feedback mechanisms, which include downregulation of receptors via ubiquitination and upregulation of negative regulators [102]. These include MAPK phosphatase 3, Sprouty proteins, and similar expression to FGF proteins, which all work at several points to attenuate the signaling cascade [102].

The human epidermal growth factor receptor (EGFR) family belongs to the receptor tyro-sine kinase ErbB-2 family, also known as human epidermal growth factor receptor 2 (*HER2/neu*), which like FGFR contains an extracellular domain, a hydrophobic transmembrane seg-ment, and an intracellular segment with a protein kinase domain [107]. Similar to FGFR, bind-ing of growth factors to EGFR induces a conformational change that leads to dimerization and subsequent activation of several potential downstream signaling pathways, including the PI3K-AKT pathway, the Ras-Raf-MEK-ERK pathway, and the PLC-γ pathway [107]. Overactivation of either FGFR or EGFR signaling leads to increased activity of the Ras GTPase, which is mutationally activated or overexpressed in a wide variety of cancers [108]. Ras in turn activates rapidly accelerated fibrosarcoma (Raf) serine-threonine kinases, which leads to downstream activation of one of four widely studied MAPK cascades, all of which play a key role in the regulation of cell proliferation, survival, and differentiation [108]. Specifically, Raf activates the MAPK/extracellular signal-related kinase (ERK) kinase

(MEK)1/2 protein kinases, which in turn activate ERK1/2 [108]. Activated ERKs phosphorylate and regulate a variety of substrates, most of which are nuclear proteins, although some are found in the cytoplasm and other organelles, ultimately resulting in altered gene expression [108].

## EGFR/FGF-Ras-Raf-MEK-ERK signaling in physiologic endometrium

The EGFR/FGF-Ras-Raf-MEK-ERK pathway is primarily involved in embryogenesis, tissue homeostasis and repair, wound healing, and inflammation [102]. FGFR pathway activation promotes cell proliferation through the MAPK cascade, and antiapoptosis through the PI3K/AKT cascade, although crosstalk is seen between these pathways, and FGF signaling may actually inhibit tumorigenesis in some contexts [109]. During the physiologic endometrial cycle, the endometrium undergoes continuous remodeling through three phases: the follicular phase (day 0 to day 13), ovulation (day 14), and the luteal phase (day 15 to day 28). This process is governed by a complex interplay between ovarian hormones and various growth factors and signaling cascades initiated within the endometrial stroma and epithelial cells [110]. During the follicular phase, increased mitotic activity throughout the endometrial glandular epithelium and stroma thickens the functional layer for implantation [110]. This process is governed by the interaction of estradiol (E2) with estrogen receptor alpha (ERα) within the stroma, and is augmented by feedback pathways and paracrine signaling in the endometrial epithelium [110]. E2 binds nuclear ER in the cytoplasm, translocates to the nucleus, and acts as a transcription factor by directly binding to estrogen response elements on estrogen-responsive genes [110]. The result of E2-nucleur ER-mediated transcription is the upregulation of genes promoting cell-cycle progression from G1 to S, as well as induction of IGF-1 and MAPK pathway-related genes, which regulate endometrial proliferation in both an ER-dependent and ER-independent manner [110, 111].

In the most well characterized of these nongenomic regulatory mechanisms, cytosolic E2-ERα complexes bind the transmembrane portion of IGF receptor, resulting in downstream activation of both the Ras-MAPK and PI3K-AKT pathways [112]. Activation of the Ras-MAPK pathway triggers a kinase cascade that enhances the activity of available transcription factors and may phosphorylate nuclear ER, which in turn translocates to the nucleus and initiates transcription of MAPK-related genes in an E2-independent manner [110]. Endometrial proliferation may also be induced in an E2-independent manner through the action of ligand-activated G protein-coupled estrogen receptor 1, which stimulates adenylate cyclase to activate the protein kinase A pathway via production of cyclic adenosine monophosphate, and activates EGFR, resulting in downstream signaling through the MAPK and PI3K pathways, with subsequent proliferation-associated gene expression [64, 113]. ER-mediated activation of the PI3K-AKT and Ras-MAPK pathways also upregulates Wnt/β-catenin signaling via inhibition of GSK-3β [114].

Following ovulation, decidualization occurs, whereby pituitary hormones and progesterone induce differentiation of the functional layer in anticipation of blastocyst implantation [110]. The precise mechanism of this process is not well established; however, it is believed that membrane progesterone receptor-induced inflow of intracellular calcium ions activates MAPK signaling cascades [115]. When implantation does not occur, withdrawal of progesterone in decidualized stromal cells results in endometrial breakdown and shedding through a mechanism that is also poorly understood [110]. Menstruation is followed by regeneration of the functionalis layer, with angiogenesis and restoration of its vascular support [110]. The process of glandular epithelial cell proliferation and migration is poorly understood, but endometrial inflammation as well as platelet-rich plasma is believed to play a significant role in this process [116]. Platelets in this milieu release several growth factors and cytokines, including VEGF, transforming growth factor β, platelet-derived growth factor (PDGF), FGF, and IGF-1, among others, resulting in angiogenesis, cell proliferation, and migration [117]. The p38 MAPK and AKT signaling pathways are also believed to play a critical role in endometrial stem cell-mediated stromal repair [118].

## EGFR/FGF-Ras-Raf-MEK-ERK signaling in endometrial hyperplasia and cancer

KRAS mutations are seen in 10%–30% of endometrial hyperplasia and endometrioid EC, suggesting an association with early tumor progression, although it should be noted that KRAS mutations have never been directly associated with tumor stage, grade, depth of myometrial invasion, age, or clinical outcome in endometrioid EC [119, 120]. BRAF encodes a Raf protein downstream of Ras, and somatic BRAF mutations are common in low-grade serous ovarian carcinoma, where they provide an alternative pathway for Ras signaling [121]. One study reported an increased correlation between BRAF mutations and endometrial hyperplasia and carcinoma; however, attempts to replicate these findings have proven unfruitful, and it is doubtful whether there is an increased prevalence of BRAF mutations in EC or a firm association between BRAF mutations and other frequent molecular alterations, such as KRAS or PTEN mutations, or MSI [121, 122]. Recent studies have suggested that inactivation of Ras association domain-containing protein 1A by promoter hypermethylation may lead to increased EGFR/FGF-Ras-Raf-MEK-ERK activity, and elevated levels of the kinase suppressor of Ras, a protein associated with increased EGFR/FGF-Ras-Raf-MEK-ERK activity, have also been found in EC and could play a role in modulation of the apoptotic response [123, 124].

FGF1 and FGF2 are the main drivers of angiogenesis in physiologic endometrium, promoting endothelial cell proliferation, migration, vessel formation, and wound healing [102]. FGF2 mutations are seen in up to 16% of EC and typically result in mutated receptors that are either constitutively active, or that demonstrate increased ligand-biding affinity or decreased required ligand specificity, ultimately leading to increased cell proliferation, migration, and premature differentiation [104]. FGFR2 mutations are commonly found in tumors with MSI, have been implicated in resistance to FGFR inhibitors, and are associated with a shorter disease-free interval (DFI) and decreased OS [125]. In contrast, FGFR2 inhibition in preclinical endometrial cell lines has been associated with decreased proliferation and even cell death [126]. Ultimately, these findings, while preliminary, are promising, and suggest that identification and inhibition of FGFR2 in EC may be of both prognostic and therapeutic utility [104].

FGFR2 mutations are typically mutually exclusive with KRAS mutations; however, PTEN mutations occur in 77% of endometrial tumors with FGFR2 mutations, suggesting a potential benefit to combination therapy [104]. This hypothesis is supported by the synergistic activity seen with a combination of the FGFR inhibitor ponatinib and the mTOR inhibitor ridaforolimus, and FGFR inhibition has been previously shown to synergize with other chemotherapeutic agents as well [127]. Amplification of *FGFR1* and *FGFR3* has also been identified in EC, and both are notably associated with a worse prognosis [14]. In addition to alterations in FGFRs, multiple studies have correlated increased expression of *FGF1* and *FGF2* with hyperplastic and malignant endometrial tissue, and with increased stage, tumor grade, and myometrial invasion [104].

Angiogenic growth factors are highly expressed in EC, and increased levels of VEGF expression have been previously associated with advanced stage, high tumor grade, deep myometrial invasion, lymphovascular invasion, lymph node metastases, and poor prognosis [104, 128]. FGFRs play a significant role in angiogenesis within the tumor microenvironment, and in vitro and xenograft studies have both demonstrated an indirect synergistic effect with combined VEGF and PDGF pathway inhibition [102, 129]. FGFs may promote endothelial cell tumor angiogenesis via both autocrine and paracrine mechanisms, and FGF overexpression has been correlated with increased tumor size, microvessel tumor density, and decreased survival [102, 130]. Resistance to VEGF pathway inhibition occurs through a process of angiogenic escape, whereby tumors employ alternative angiogenic pathways, leading to the hypothesis that targeting the FGF pathway, either alone or in combination with the VEGF pathway, may prove synergistic in vivo [131].

## Therapies targeting EGFR/FGF-Ras-Raf-MEK-ERK signaling currently under investigation

Due to the significance of angiogenic pathways in EC and the notable overlap between tyrosine kinase receptors upstream of these, clinical trials have investigated a number of therapeutic agents targeting multiple growth factors, including VEGF, FGF, and PDGF (Fig. 3) [104]. These include the tyrosine kinase inhibitors sunitinib and sorafenib, the immunomodulatory drug thalidomide, the VEGF monoclonal antibody bevacizumab, and the VEGF-sequestering agent aflibercept [104]. Most of these agents have shown only modest activity in EC with response rates of 4%–18% and a median survival of 5.9–19.4 months [104]. New research has focused on targeting the FGF pathway, both with single-agent therapy or combined with anti-VEGF therapies [104]. Several FGFR-targeted therapies are undergoing evaluation in preclinical and clinical trials, including brivanib, nintedanib, dovitinib, lenvatinib, and ponatinib [104].

Brivanib is a tyrosine kinase inhibitor targeting VEGF receptor (VEGFR)2/3 and FGFR1/2 that underwent a phase II trial ($n=43$) for recurrent or persistent EC (Fig. 3) [132]. It was well tolerated and showed an overall 19% response rate with a median PFS of 3.3 months and a median OS of 10.7 months; however, only three patients had FGFR2 mutations, limiting a robust analysis of its role in treating patients with FGFR pathway mutations [132]. Nintedanib is a PDGF receptor α/β, FGFR1–3, and VEGFR1–3 inhibitor that underwent a phase II trial ($n=32$) for patients with recurrent or persistent EC [133]. This agent demonstrated an overall response rate of 9% with a median PFS of 3.3 months and a median OS of 10.1 months [133]. Dovitinib is an FGFR1–3, VEGFR1–3, and PDGF receptor inhibitor that underwent a phase II trial ($n=53$) as second-line therapy in patients with advanced or metastatic EC, or both [134]. Patients in this trial were prospectively screened for *FGFR2* mutation status, the first study to do so, with 22 patients having an *FGFR2*-mutated tumor and 31 having an *FGFR2*-nonmutated tumor [134]. Among patients with an *FGFR2* mutation, 5% achieved a partial response, 59% achieved stable disease, median PFS was 4.1 months, and median OS was 20.2 months, whereas among those without *FGFR2*-mutated tumors, 16% achieved a partial response, 35% achieved stable disease, median PFS was 2.7 months, and median OS was 9.3 months [134].

Lenvatinib is a multikinase inhibitor targeting VEGFR1–2, FGFR1–4, PDGF receptor, RET, and c-Kit that underwent a phase I dose-escalation trial ($n = 77$) in patients with advanced solid tumors and showed a 25% response rate, prompting an ongoing phase II trial ($n = 133$) in patients with advanced or progressive EC following platinum-based therapy (Fig. 3) [135, 136]. This agent has shown a 14% response rate as assessed by independent investigators (with a 22% response rate as assessed by study investigators), a median PFS of 5.4 months, and a median OS of 10.6 months [136, 137]. Lower baseline levels of angiopoietin-2 correlated with improved overall response rate (ORR), median PFS, and median OS, with various additional cytokine and angiogenic factors also correlated with survival [136, 137]. MAPK and PI3K signaling pathway mutations were associated with Lenvatinib resistance, and patients with a *PIK3CA* mutation showed a trend toward worse OS that did not reach statistical significance [137]. Ponatinib is a pan-FGFR inhibitor targeting FGFR1–4 that has shown promising results in murine, xenograft, and in vitro models with additional suggestion of synergistic activity with the mTOR inhibitor ridaforolimus [127]. Ponatinib was slated for clinical trial; however, this study was withdrawn prior to recruitment, and there are no active studies currently investigating this agent in the treatment of EC [104, 127].

Two MEK inhibitors, trametinib and cobimetinib, have undergone preliminary studies in EC (Fig. 3) [138, 139]. A clinical trial studying the combination of trametinib and an AKT inhibitor in EC demonstrated intolerable levels of toxicity and was discontinued after a lower, more tolerable dose was found to be inefficacious [139]. Additionally, in vitro studies of *KRAS*-mutated EC showed an increase in ER signaling with MEK inhibition, suggesting that addition of a MEK inhibitor may be necessary to achieve an adequate response to antiestrogen therapies in *KRAS*-mutated EC [138].

Current evidence indicates that FGFR and VEGF pathway inhibitors have similar activity in vivo, and that drugs targeting the FGF pathway may provide alternative or adjunctive treatment options in patients with recurrent or persistent EC [104]. Preclinical studies suggest that improved patient selection is needed in order to quantify the efficacy of these agents; however, with *FGFR* mutations occurring in <20% of endometrial tumors, this has proven to be a challenge [104]. Additionally, robust molecular screening tools are needed for both improved patient selection and for identification of specific biomarkers which may ultimately prove to be of prognostic utility [104]. Finally, further study with more selective agents is needed to parse out the specific impact of FGFR inhibition and to better determine the feasibility of these agents for the treatment of chemotherapy resistance in EC [104].

# HIF-1 signaling pathway

## Introduction

Hypoxia and triggered angiogenesis are key components of tumorigenesis, as tumor growth usually outstrips the rate of its own oxygen supply, creating a hypoxic environment

[106, 140]. Hypoxia serves as a state of physiologic stress for the tumor itself, inducing the expression of regulatory factors that are protective against cell necrosis and apoptosis [140, 141]. Of these, hypoxia-inducible factors (HIF) predominate, modifying the tumor microenvironment, and allowing tumor cells to gain independence from external regulatory stimuli in order to facilitate migration and metastasis to less hypoxic areas of the body [142, 143]. HIF-1 is the most studied of the HIFs, it serves as a prognosticator of tumor cell response to chemotherapy, and it has itself become a target of several molecular therapies in both in vitro and in vivo studies [144, 145].

HIF-1 is composed of two subunits, HIF-1α and HIF-1β, which are needed in a 1:1 concentration for assembly of functional HIF-1 (Fig. 4) [142]. Whereas HIF-1β is constitutively expressed, the intracellular concentration of HIF-1α is regulated by hypoxic conditions via two mechanisms [142]. First, HIF-1α is hydroxylated by propyl hydroxylases (PHD) in normoxic conditions, then polyubiquitinated via a protein complex in which the Von Hippel-Lindau protein (VHL) participates, before finally undergoing proteolysis in proteasomes [141, 142]. PHDs are oxygen dependent and unable to hydroxylate HIF-1α under hypoxic conditions, preventing recognition by VHL and allowing HIF-1α to escape from degradation and to combine with the already-present HIF-1β to form active HIF-1 [141, 142]. A second mechanism involving oxygen-dependent asparaginyl hydroxylase factor inhibiting HIF-1 (FIH), which inhibits the HIF-1 pathway via a separate mechanism, is also inactivated in the presence of hypoxia (Fig. 4) [142]. Alterations to both PHDs and FIHs under anoxic conditions trigger increased production of HIF-1, which then binds to HIF-1 response elements on promoter regions and induces the transcription of various genes responsible for cellular adaptation mechanisms, including pH regulation, glucose transportation and metabolism, angiogenesis, nitric oxide synthesis, and erythropoiesis [141, 142, 146].

HIF-1 may inhibit cell-cycle progression and regulate apoptosis, with the ultimate goal of preserving energy and allowing the cell to survive hypoxic conditions [140, 142]. HIF-1 also modifies the local T-cell immune response, allowing for tumor immune-escape, and may contribute to tumor cell invasion and metastasis [142, 147]. Stabilization of HIF-1α leading to constitutive HIF-1 activity in normoxic conditions has also been reported, typically in association with viral oncogenes, with genetic alterations in PHD, or with VHL mutations, as in Von Hippel-Lindau disease [141]. Several other molecular pathways, including the PI3K-AKT-mTOR and the EGFR/FGF-Ras-Raf-MEK-ERK signaling pathways, also increase HIF-1 activity by stabilizing HIF-1α [141].

## HIF-1 signaling in physiologic endometrium

HIF-1α was not initially thought to play a significant role in the physiologic endometrial cycle, but has since been shown to contribute to endometrial tissue repair [140]. HIF-1 is maximally secreted when progesterone levels fall during the late secretory phase and menstruation, it acts as a coactivator of estrogen-induced VEGF production, and it mediates IL-8 production, all of which are critical for endometrial tissue repair following menstruation [140]. HIF-1α has also demonstrated activity as a coactivator of estrogen-induced VEGF production, and inhibition of PI3K activity has been shown in animal models to block both HIF-1α activation and recruitment of ER to the VEGF promoter [140, 148]. Two explanations are

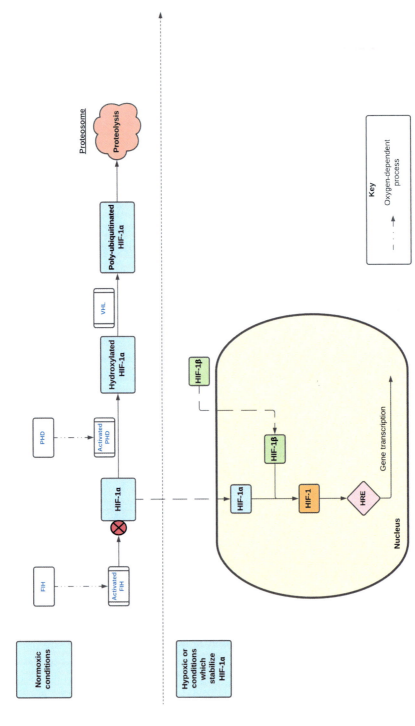

FIG. 4 HIF-1 signaling. Normoxic conditions are depicted above, where levels of HIF-1α are suppressed by proteolysis under the control of FIHs and PHDs. Hypoxic conditions are depicted below, where HIF-1α suppression is removed, allowing it to accumulate and subsequently combine with constitutively active HIF-1β to generate HIF-1, which then binds response elements on promoter regions, inducing the transcription of various genes responsible for mechanisms of cellular adaptation.

proposed for this: it may be that upon binding to the VEGF promoter, ERα is unable to activate the VEGF promoter on its own; alternatively, HIF-1α may play a role in recruiting ERα to the VEGF promoter [140, 148]. Additionally, since PI3K inhibition has been shown to inhibit HIF-1α and ERα activation and recruitment, it has been hypothesized that PI3K pathway overactivity (as is often seen in the setting of EC) may increase HIF-1α activity [140]. Thus, angiogenesis secondary to HIF-1α and estrogen-induced VEGF production provide a plausible mechanism for both normal endometrial repair following menstruation as well as tumorigenesis in EC [140].

## HIF-1 signaling in endometrial hyperplasia and cancer

The primary role of HIF-1 in EC stems from its proangiogenic properties as outlined earlier [140]. In short, EC is a hypoxic tumor affecting the epithelial layer of the endometrium [106, 140]. As the tumor grows, the resulting hypoxic microenvironment triggers increased stabilization of HIF-1α, which induces VEGF expression with the resultant formation of a large, intertumoral network of new blood vessels [106]. These increase blood flow and nutrient availability to the tumor and facilitate elimination of metabolic products, both of which promote adaptation to the hypoxic environment and allow for further tumor growth and spread [106].

Not only does HIF-1α trigger angiogenesis during myometrial invasion by mediating the production of VEGF in myofibroblasts, but increased HIF-1α expression has also been shown in animal models to act synergistically with the prostaglandin E receptor 4 pathway to promote endometrial tumorigenesis [149, 150]. HIF-1α additionally regulates the cell cycle via induction of cell-cycle inhibitors, inhibition of proliferative cell-cycle regulators, and modulation of apoptosis [142, 151]. It remains unclear whether this interaction is a component of a larger network of hypoxia-induced factors, and the so-called "hypoxia signature" seen in breast and ovarian cancers has yet to be demonstrated in EC [140, 151]. HIF-1α polymorphisms seen in EC are associated with increased tumor microvascular density and possibly with increased tumor growth; however, it is unclear whether such polymorphisms themselves predispose to EC [152, 153]. Mutations in *PTEN*, a regulator of the PI3K-AKT-mTOR pathway, may lead to increased HIF-1α activity and increased tumorigenesis [140]. Conversely, inhibition of the PI3K-AKT-mTOR pathway may sensitize EC to radiation therapy via downstream inhibition of HIF-1α and VEGF [154].

Despite increasing interest in hypoxia-induced tumorigenesis, there is a paucity of literature on its impact in EC [140]. Nuclear staining of HIF-1α is seen to gradually increase throughout the progression from normal endometrium to AEH to EC, with a corresponding increase seen in both immediate downstream products such as carbonic anhydrase 9 and Glut 1 transporters and in tissue microvascular density [140]. HIF-1α expression has also been correlated with an increased angiopoietin-1/angiopoietin-2 ratio and with increased IL-8 production [155]. Although HIF-1α is more abundant in type II carcinomas, statistically significant correlations have only been made in type I carcinomas, where HIF-1α expression is associated with higher histologic grade, myometrial invasion, adnexal or vascular invasion, and increased clinical stage [140, 156]. The prognostic significance of HIF-1α-associated necrosis is controversial due to significant heterogeneity in the studies that have considered this issue [140].

## Therapies targeting HIF-1 signaling currently under investigation

Due to the prominent role of VEGF in EC and its involvement across several pathways in addition to HIF-1α, multiple targets of this pathway have been studied as potential therapeutic agents [157]. Bevacizumab is a humanized monoclonal antibody that binds VEGF-A and neutralizes it [157]. It was approved by the Food and Drug Administration in 2003 and is commonly used as an adjunctive and maintenance therapy in ovarian cancer; however, its use in EC remains relatively understudied [157].

Several phase II trials have shown moderate ORR to single-agent bevacizumab in recurrent or persistent EC, with more promising results seen by combining it with front-line therapies [157]. One phase II trial ($n = 56$) examining the efficacy and safety of single-agent bevacizumab in patients with recurrent or persistent EC showed one complete response (CR), six PR, an ORR of 13.5%, and no significant toxicity, while another phase II trial ($n = 49$) considering bevacizumab with temsirolimus in recurrent or persistent EC showed an ORR of 24.5%, a median PFS of 5.6 months, and a median OS of 16.9 months [22, 158]. A large phase II trial ($n = 349$) randomized patients with either chemotherapy-naïve stage III–IVA EC, recurrent EC, or stage IVB EC to one of three treatment arms: paclitaxel, carboplatin, and bevacizumab (Arm 1); paclitaxel, carboplatin, and temsirolimus (Arm 2); or ixabepilone, carboplatin, and bevacizumab (Arm 3); and used the paclitaxel and carboplatin arm of GOG209 as a historical control [78]. Of these three regimens, paclitaxel, carboplatin, and bevacizumab (Arm 1) was the only one that showed a significant improvement in OS compared to historical controls [78, 159]. Finally, in the MITO-END 2 study, a randomized phase II trial ($n = 108$), the addition of bevacizumab to carboplatin and paclitaxel in advanced stage, recurrent EC increased ORR from 54.3% to 72.7% and increased PFS from 8.7 to 13.0 months [160]. There was, however, an increase in cardiovascular events seen in the group given bevacizumab [160].

Multiple studies have suggested a role for antiangiogenic agents in both inhibiting cancer stem cells that are resistant to radiation and in sensitizing tumors to radiotherapy by interfering with upregulation of VEGF, which malignant cells may use to overcome radiation damage to cell vasculature [161, 162]. Bevacizumab was evaluated in a phase II trial ($n = 15$) to determine its safety and activity with concurrent external beam radiation in recurrent EC, with findings that included a 1-, 2-, and 3-year PFS of 80%, 80%, and 67%, respectively, a 1-, 2-, and 3-year OS of 93%, 80%, and 80%, respectively, no recurrences in the vaginal/pelvic region, and acceptable toxicity rates [157, 163]. Another phase II trial ($n = 30$) of patients undergoing intensity-modulated radiation therapy with concurrent cisplatin and bevacizumab with optional vaginal brachytherapy, followed by four additional cycles of carboplatin and paclitaxel, showed a 2-year DFS of 79.1%, a 2-year estimated OS of 96.7%, and no within-field pelvic-only failures [164]. There is significant overlap between therapies targeting VEGF inhibitors and those targeting the EGFR/FGF-Ras-Raf-MEK-ERK signaling pathway, and several other therapies currently under investigation with activity on both pathways are described in more detail in the relevant section discussed earlier.

## Conclusions and future perspectives

Traditionally, stage and histology have determined the treatment for EC, with regimens of carboplatin/paclitaxel, cisplatin/doxorubicin, cisplatin/doxorubicin/paclitaxel, or carboplatin/docetaxel recommended as first-line agents for recurrent or metastatic disease, or for high-risk histologies. Translational research is rapidly progressing in EC, however, and has shown that tumors with a similar histology and stage may in fact have very different underlying molecular and genomic profiles. As a result of the chemotherapy resistance ultimately seen in advanced-stage patients, novel therapeutic agents based on genetic profiling are being actively developed and brought to clinical trial. This new approach has led to approval of the first tissue-agnostic immune checkpoint inhibitor, pembrolizumab (anti-PD-1), for solid tumors with defective mismatch repair, and it is estimated that this drug may benefit up to 20%–30% of patients with EC. Other genomic changes and molecular markers, including hormone receptor status, continue to be investigated, with the hope of identifying effective targeted therapies for patients with advanced, recurrent, and chemotherapy-resistant disease. Single-agent targeted therapies have consistently shown modest efficacy, with more promising results seen when novel therapeutics are added to front-line therapy. The principal challenge facing the next generation of clinical trials will be determining how best to utilize targeted therapies and identifying what biomarkers may aid in more accurately selecting patients who may benefit from them.

## Author contributions

All of the authors have diligently contributed to the conception, development, and preparation of this manuscript, including the literature search, concept organization, and data interpretation. All of the authors have read and approved the final draft.

## Financial disclosures

None to disclose.

## References

[1] Torre LA, Islami F, Siegel RL, Ward EM, Jemal A. Global cancer in women: burden and trends. Cancer Epidemiol Biomarkers Prev 2017;26(4):444–57. https://doi.org/10.1158/1055-9965.EPI-16-0858.

[2] American Cancer Society. Cancer facts & figures 2018. Atlanta: American Cancer Society; 2018.

[3] Arend RC, Jones BA, Martinez A, Goodfellow P. Endometrial cancer: molecular markers and management of advanced stage dise ase. Gynecol Oncol 2018;150(3):569–80. https://doi.org/10.1016/j.ygyno.2018.05.015.

[4] Cowan M, Strauss JB, Barber EL, Matei D. Updates on adjuvant chemotherapy and radiation therapy for endometrial cancer. Curr Opin Obstet Gynecol 2019;31(1):31–7. https://doi.org/10.1097/GCO.0000000000000506.

[5] Bokhman JV. Two pathogenetic types of endometrial carcinoma. Gynecol Oncol 1983;15(1):10–7. https://doi.org/10.1016/0090-8258(83)90111-7.

[6] Morice P, Leary A, Creutzberg C, Abu-Rustum N, Darai E. Endometrial cancer. Lancet 2016;387(10023):1094–108. https://doi.org/10.1016/S0140-6736(15)00130-0.

[7] Brinton LA, Felix AS, McMeekin DS, et al. Etiologic heterogeneity in endometrial cancer: evidence from a gynecologic oncology group trial. Gynecol Oncol 2013;129(2):277–84. https://doi.org/10.1016/j.ygyno.2013.02.023.

[8] Mitamura T, Dong P, Ihira K, Kudo M, Watari H. Molecular-targeted therapies and precision medicine for endometrial cancer. Jpn J Clin Oncol 2019;49(2):108–20. https://doi.org/10.1093/jjco/hyy159. Feb 1 PMID: 30423148.

[9] Henley SJ, Miller JW, Dowling NF, Benard VB, Richardson LC. Uterine Cancer Incidence and Mortality—United States, 1999–2016. MMWR Morb Mortal Wkly Rep 2018;67(48):1333–8. https://doi.org/10.15585/mmwr.mm6748a1. Dec 7 PMID: 30521505; PMCID: PMC6329484.

[10] Hamilton CA, Cheung MK, Osann K, Chen L, Teng NN, Longacre TA, Powell MA, Hendrickson MR, Kapp DS, Chan JK. Uterine papillary serous and clear cell carcinomas predict for poorer survival compared to grade 3 endometrioid corpus cancers. Br J Cancer 2006;94(5):642–6. https://doi.org/10.1038/sj.bjc.6603012. Mar 13 PMID: 16495918; PMCID: PMC2361201.

[11] Hong B, Le Gallo M, Bell DW. The mutational landscape of endometrial cancer. Curr Opin Genet Dev 2015;30: 25–31. https://doi.org/10.1016/j.gde.2014.12.004. Feb Epub 2015 Jan 23. PMID: 25622247; PMCID: PMC4476916.

[12] Iijima M, Banno K, Okawa R, Yanokura M, Iida M, Takeda T, Kunitomi-Irie H, Adachi M, Nakamura K, Umene K, Nogami Y, Masuda K, Tominaga E, Aoki D. Genome-wide analysis of gynecologic cancer: the Cancer Genome Atlas in ovarian and endometrial cancer. Oncol Lett 2017;13(3):1063–70. https://doi.org/10.3892/ol.2017.5582. Mar Epub 2017 Jan 10. PMID: 28454214; PMCID: PMC5403284.

[13] Schultheis AM, Martelotto LG, De Filippo MR, Piscuglio S, Ng CK, Hussein YR, Reis-Filho JS, Soslow RA, Weigelt B. TP53 mutational spectrum in endometrioid and serous endometrial cancers. Int J Gynecol Pathol 2016;35(4):289–300. https://doi.org/10.1097/PGP.0000000000000243. Jul PMID: 26556035; PMCID: PMC5087968.

[14] Cancer Genome Atlas Research Network, Kandoth C, Schultz N, Cherniack AD, Akbani R, Liu Y, Shen H, Robertson AG, Pashtan I, Shen R, Benz CC, Yau C, Laird PW, Ding L, Zhang W, Mills GB, Kucherlapati R, Mardis ER, Levine DA. Integrated genomic characterization of endometrial carcinoma. Nature 2013;497(7447):67–73. https://doi.org/10.1038/nature12113. May 2 Erratum in: Nature. 2013 Aug 8;500(7461):242. PMID: 23636398; PMCID: PMC3704730.

[15] Piulats JM, Guerra E, Gil-Martín M, Roman-Canal B, Gatius S, Sanz-Pamplona R, Velasco A, Vidal A, Matias-Guiu X. Molecular approaches for classifying endometrial carcinoma. Gynecol Oncol 2017;145(1):200–7. https://doi.org/10.1016/j.ygyno.2016.12.015. Apr Epub 2016 Dec 29. PMID: 28040204.

[16] Talhouk A, McConechy MK, Leung S, Yang W, Lum A, Senz J, Boyd N, Pike J, Anglesio M, Kwon JS, Karnezis AN, Huntsman DG, Gilks CB, McAlpine JN. Confirmation of ProMisE: a simple, genomics-based clinical classifier for endometrial cancer. Cancer 2017;123(5):802–13. https://doi.org/10.1002/cncr.30496. Mar 1 Epub 2017 Jan 6. PMID: 28061006.

[17] Eggink FA, Van Gool IC, Leary A, Pollock PM, Crosbie EJ, Mileshkin L, Jordanova ES, Adam J, Freeman-Mills L, Church DN, Creutzberg CL, De Bruyn M, Nijman HW, Bosse T. Immunological profiling of molecularly classified high-risk endometrial cancers identifies *POLE*-mutant and microsatellite unstable carcinomas as candidates for checkpoint inhibition. Oncoimmunology 2016;6(2):e1264565. https://doi.org/10.1080/2162402X.2016.1264565. Dec 9. PMID: 28344870; PMCID: PMC5353925.

[18] van Gool IC, Eggink FA, Freeman-Mills L, Stelloo E, Marchi E, de Bruyn M, Palles C, Nout RA, de Kroon CD, Osse EM, Klenerman P, Creutzberg CL, Tomlinson IP, Smit VT, Nijman HW, Bosse T, Church DN. POLE proof-reading mutations elicit an antitumor immune response in endometrial cancer. Clin Cancer Res 2015;21 (14):3347–55. https://doi.org/10.1158/1078-0432.CCR-15-0057. Epub 2015 Apr 15. PMID: 25878334; PMCID: PMC4627582.

[19] Howitt BE, Shukla SA, Sholl LM, Ritterhouse LL, Watkins JC, Rodig S, Stover E, Strickland KC, D'Andrea AD, Wu CJ, Matulonis UA, Konstantinopoulos PA. Association of polymerase e-mutated and microsatellite-instable endometrial cancers with neoantigen load, number of tumor-infiltrating lymphocytes, and expression of PD-1 and PD-L1. JAMA Oncol 2015;1(9):1319–23. https://doi.org/10.1001/jamaoncol.2015.2151. PMID: 26181000.

[20] Liu J, Liu Y, Wang W, Wang C, Che Y. Expression of immune checkpoint molecules in endometrial carcinoma. Exp Ther Med 2015;10(5):1947–52. https://doi.org/10.3892/etm.2015.2714. Epub 2015 Aug 28. PMID: 26640578; PMCID: PMC4665362.

[21] van den Heerik ASVM, Horeweg N, Nout RA, Lutgens LCHW, van der Steen-Banasik EM, Westerveld GH, van den Berg HA, Slot A, Koppe FLA, Kommoss S, JWM M, Nowee ME, Bijmolt S, Cibula D, Stam TC, Jurgenliemk-Schulz IM, Snyers A, Hamann M, Zwanenburg AG, VLMA C, Vandecasteele K, Gillham C, Chargari C, Verhoeven-Adema KW, Putter H, van den Hout WB, Wortman BG, Nijman HW, Bosse T, Creutzberg CL. PORTEC-4a: international randomized trial of molecular profile-based adjuvant treatment for women with high-intermediate risk endometrial cancer. Int J Gynecol Cancer 2020. https://doi.org/10.1136/ijgc-2020-001929. Oct 12. Epub ahead of print. PMID: 33046573: ijgc-2020-001929.

[22] Aghajanian C, Sill MW, Darcy KM, Greer B, McMeekin DS, Rose PG, Rotmensch J, Barnes MN, Hanjani P, Leslie KK. Phase II trial of bevacizumab in recurrent or persistent endometrial cancer: a Gynecologic Oncology Group study. J Clin Oncol 2011;29(16):2259–65. https://doi.org/10.1200/JCO.2010.32.6397. Jun 1 Epub 2011 May 2. PMID: 21537039; PMCID: PMC3107744.

[23] Anastas JN. Functional crosstalk between WNT signaling and tyrosine kinase signaling in cancer. Semin Oncol 2015;42(6):820–31. December Semin Oncol. 2016;43(4):526 https://doi.org/10.1053/j.seminoncol.2016.06.010.

[24] Coopes A, Henry CE, Llamosas E, Ford CE. An update of Wnt signalling in endometrial cancer and its potential as a therapeutic target [published online ahead of print, 2018 Aug 9]. Endocr Relat Cancer 2018. https://doi.org/10.1530/ERC-18-0112, ERC-18-0112.

[25] Dellinger TH, Planutis K, Tewari KS, Holcombe RF. Role of canonical Wnt signaling in endometrial carcinogenesis. Expert Rev Anticancer Ther 2012;12(1):51–62. https://doi.org/10.1586/era.11.194.

[26] Ford CE, Henry C, Llamosas E, Djordjevic A, Hacker N. Wnt signalling in gynaecological cancers: a future target for personalised medicine? Gynecol Oncol 2016;140(2):345–51. https://doi.org/10.1016/j.ygyno.2015.09.085.

[27] Kolligs FT, Nieman MT, Winer I, Hu G, Van Mater D, Feng Y, Smith IM, Wu R, Zhai Y, Cho KR, Fearon ER. ITF-2, a downstream target of the Wnt/TCF pathway, is activated in human cancers with beta-catenin defects and promotes neoplastic transformation. Cancer Cell 2002 Mar;1(2):145–55. https://doi.org/10.1016/s1535-6108 (02)00035-1. PMID: 12086873.

[28] Wang Y, Hanifi-Moghaddam P, Hanekamp EE, Kloosterboer HJ, Franken P, Veldscholte J, van Doorn HC, Ewing PC, Kim JJ, Grootegoed JA, Burger CW, Fodde R, Blok LJ. Progesterone inhibition of Wnt/beta-catenin signaling in normal endometrium and endometrial cancer. Clin Cancer Res 2009;15(18):5784–93. https://doi.org/10.1158/1078-0432.CCR-09-0814. Sep 15 Epub 2009 Sep 8. PMID: 19737954.

[29] Tulac S, Overgaard MT, Hamilton AE, Jumbe NL, Suchanek E, Giudice LC. Dickkopf-1, an inhibitor of Wnt signaling, is regulated by progesterone in human endometrial stromal cells. J Clin Endocrinol Metab 2006;91(4):1453–61. https://doi.org/10.1210/jc.2005-0769. Apr Epub 2006 Jan 31. PMID: 16449346.

[30] Fukuchi T, Sakamoto M, Tsuda H, Maruyama K, Nozawa S, Hirohashi S. Beta-catenin mutation in carcinoma of the uterine endometrium. Cancer Res 1998;58(16):3526–8.

[31] Liu Y, Meng F, Xu Y, et al. Overexpression of Wnt7a is associated with tumor progression and unfavorable prognosis in endometrial cancer. Int J Gynecol Cancer 2013;23(2):304–11. https://doi.org/10.1097/IGC.0b0 13e31827c7708.

[32] Liu Y, Patel L, Mills GB, et al. Clinical significance of CTNNB1 mutation and Wnt pathway activation in endometrioid endometrial carcinoma. J Natl Cancer Inst 2014;106(9), dju245. Published 2014 Sep 10 https://doi.org/10.1093/jnci/dju245.

[33] Peng C, et al. Expression and prognostic significance of wnt7a in human endometrial carcinoma. Obstet Gynecol Int 2012;2012:134962. https://doi.org/10.1155/2012/134962.

[34] Dellinger T, Planutis K, Gatcliffe T, DiSaia P, Monk B, Holcombe R. Wnt inhibitor expression as a predictor of progression in patients with endometrial carcinoma: a potential role as prognostic biomarker. Gynecol Oncol 2010;116(3):S86. Suppl. 1.

[35] Eskander RN, Ali S, Dellinger T, Lankes HA, Randall LM, Ramirez NC, Monk BJ, Walker JL, Eisenhauer E, Hoang BH. Expression patterns of the Wnt pathway inhibitors Dickkopf3 and secreted frizzled-related proteins 1 and 4 in endometrial endometrioid adenocarcinoma: an NRG Oncology/Gynecologic Oncology Group Study. Int J Gynecol Cancer 2016;26(1):125–32. https://doi.org/10.1097/IGC.0000000000000563. Jan PMID: 26397159; PMCID: PMC5061499.

[36] Jiang T, Wang S, Huang L, Zhang S. Clinical significance of serum DKK-1 in patients with gynecological cancer. Int J Gynecol Cancer 2009;19(7):1177–81. https://doi.org/10.1111/IGC.0b013e31819d8b2d. Oct PMID: 19820386.

[37] Abu-Jawdeh G, Comella N, Tomita Y, Brown LF, Tognazzi K, Sokol SY, Kocher O. Differential expression of frpHE: a novel human stromal protein of the secreted frizzled gene family, during the endometrial cycle and malignancy. Lab Invest 1999;79(4):439–47. Apr PMID: 10211996 (web archive link).

[38] Hrzenjak A, Tippl M, Kremser ML, Strohmeier B, Guelly C, Neumeister D, Lax S, Moinfar F, Tabrizi AD, Isadi-Moud N, Zatloukal K, Denk H. Inverse correlation of secreted frizzled-related protein 4 and beta-catenin expression in endometrial stromal sarcomas. J Pathol 2004;204(1):19–27. https://doi.org/10.1002/path.1616. Sep PMID: 15307134.

[39] Risinger JI, Maxwell GL, Chandramouli GV, Aprelikova O, Litzi T, Umar A, Berchuck A, Barrett JC. Gene expression profiling of microsatellite unstable and microsatellite stable endometrial cancers indicates distinct pathways of aberrant signaling. Cancer Res 2005;65(12):5031–7. https://doi.org/10.1158/0008-5472.CAN-04-0850. Jun 15 PMID: 15958545.

[40] Di Domenico M, Santoro A, Ricciardi C, Iaccarino M, Iaccarino S, Freda M, Feola A, Sanguedolce F, Losito S, Pasquali D, Di Spiezio SA, Bifulco G, Nappi C, Bufo P, Guida M, De Rosa G, Abbruzzese A, Caraglia M,

Pannone G. Epigenetic fingerprint in endometrial carcinogenesis: the hypothesis of a uterine field cancerization. Cancer Biol Ther 2011;12(5):447–57. https://doi.org/10.4161/cbt.12.5.15963. Sep 1 Epub 2011 Sep 1. PMID: 21709441.

[41] Cardona-Gomez P, Perez M, Avila J, Garcia-Segura LM, Wandosell F. Estradiol inhibits GSK3 and regulates interaction of estrogen receptors, GSK3, and beta-catenin in the hippocampus. Mol Cell Neurosci 2004;25(3): 363–73. https://doi.org/10.1016/j.mcn.2003.10.008. Mar PMID: 15033165.

[42] McCampbell AS, Broaddus RR, Loose DS, Davies PJ. Overexpression of the insulin-like growth factor I receptor and activation of the AKT pathway in hyperplastic endometrium. Clin Cancer Res 2006;12(21):6373–8. https://doi.org/10.1158/1078-0432.CCR-06-0912. Nov 1 PMID: 17085648.

[43] Katayama S, Ashizawa K, Fukuhara T, Hiroyasu M, Tsuzuki Y, Tatemoto H, Nakada T, Nagai K. Differential expression patterns of Wnt and beta-catenin/TCF target genes in the uterus of immature female rats exposed to 17alpha-ethynyl estradiol. Toxicol Sci 2006;91(2):419–30. https://doi.org/10.1093/toxsci/kfj167. Epub 2006 Mar 21. Jun PMID: 16551644.

[44] Saegusa M, Hamano M, Kuwata T, Yoshida T, Hashimura M, Akino F, Watanabe J, Kuramoto H, Okayasu I. Up-regulation and nuclear localization of beta-catenin in endometrial carcinoma in response to progesterone therapy. Cancer Sci 2003;94(1):103–11. https://doi.org/10.1111/j.1349-7006.2003.tb01360.x. Jan PMID: 12708483.

[45] Satterfield MC, Song G, Hayashi K, Bazer FW, Spencer TE. Progesterone regulation of the endometrial WNT system in the ovine uterus. Reprod Fertil Dev 2008;20(8):935–46. https://doi.org/10.1071/rd08069. PMID: 19007558.

[46] Liao X, Siu MK, Au CW, Chan QK, Chan HY, Wong ES, Ip PP, Ngan HY, Cheung AN. Aberrant activation of hedgehog signaling pathway contributes to endometrial carcinogenesis through beta-catenin. Mod Pathol 2009;22(6):839–47. https://doi.org/10.1038/modpathol.2009.45. Epub 2009 Mar 27. Jun PMID: 19329935.

[47] Qi H, Sun B, Zhao X, Du J, Gu Q, Liu Y, Cheng R, Dong X. Wnt5a promotes vasculogenic mimicry and epithelial-mesenchymal transition via protein kinase Cα in epithelial ovarian cancer. Oncol Rep 2014;32(2):771–9. https://doi.org/10.3892/or.2014.3229. Epub 2014 Jun 4. Aug PMID: 24898696.

[48] Bui TD, Zhang L, Rees MC, Bicknell R, Harris AL. Expression and hormone regulation of Wnt2, 3, 4, 5a, 7a, 7b and 10b in normal human endometrium and endometrial carcinoma. Br J Cancer 1997;75(8):1131–6. https://doi.org/10.1038/bjc.1997.195.

[49] McDonald SL, Silver A. The opposing roles of Wnt-5a in cancer. Br J Cancer 2009;101(2):209–14. https://doi.org/10.1038/sj.bjc.6605174. PMID: 19603030; PMCID: PMC2720208.

[50] Chen S, Wang J, Gou WF, Xiu YL, Zheng HC, Zong ZH, Takano Y, Zhao Y. The involvement of RhoA and Wnt-5a in the tumorigenesis and progression of ovarian epithelial carcinoma. Int J Mol Sci 2013;14(12):24187–99. https://doi.org/10.3390/ijms141224187. Dec 12 PMID: 24351810; PMCID: PMC3876104.

[51] Zhang H, Yan X, Ke J, Zhang Y, Dai C, Zhu M, Jiang F, Wan X. ROR1 promotes the proliferation of endometrial cancer cells. Int J Clin Exp Pathol 2017;10(10):10603–10. Oct 1 PMID: 31966402; PMCID: PMC6965769.

[52] Edris B, Espinosa I, Mühlenberg T, Mikels A, Lee CH, Steigen SE, Zhu S, Montgomery KD, Lazar AJ, Lev D, Fletcher JA, Beck AH, West RB, Nusse R, van de Rijn M. ROR2 is a novel prognostic biomarker and a potential therapeutic target in leiomyosarcoma and gastrointestinal stromal tumour. J Pathol 2012;227(2):223–33. https://doi.org/10.1002/path.3986. Jun Epub 2012 Feb 17. PMID: 22294416; PMCID: PMC3992120.

[53] Saygin C, Wiechert A, Rao VS, Alluri R, Connor E, Thiagarajan PS, Hale JS, Li Y, Chumakova A, Jarrar A, Parker Y, Lindner DJ, Nagaraj AB, Kim JJ, DiFeo A, Abdul-Karim FW, Michener C, Rose PG, DeBernardo R, Mahdi H, McCrae KR, Lin F, Lathia JD, Reizes O. CD55 regulates self-renewal and cisplatin resistance in endometrioid tumors. J Exp Med 2017;214(9):2715–32. https://doi.org/10.1084/jem.20170438. Sep 4 Epub 2017 Aug 24. PMID: 28838952; PMCID: PMC5584126.

[54] Henry CE, Llamosas E, Daniels B, Coopes A, Tang K, Ford CE. ROR1 and ROR2 play distinct and opposing roles in endometrial cancer. Gynecol Oncol 2018;148(3):576–84. https://doi.org/10.1016/j.ygyno.2018.01.025. Epub 2018 Feb 1. Mar PMID: 29395309.

[55] Canesin G, Evans-Axelsson S, Hellsten R, Krzyzanowska A, Prasad CP, Bjartell A, Andersson T. Treatment with the WNT5A-mimicking peptide Foxy-5 effectively reduces the metastatic spread of WNT5A-low prostate cancer cells in an orthotopic mouse model. PLoS One 2017;12(9). https://doi.org/10.1371/journal.pone.0184418, e0184418. Sep 8 PMID: 28886116; PMCID: PMC5590932.

[56] Creasman WT, Odicino F, Maisonneuve P, Quinn MA, Beller U, Benedet JL, Heintz AP, Ngan HY, Pecorelli S. Carcinoma of the corpus uteri. FIGO 26th annual report on the results of treatment in gynecological cancer.

Int J Gynaecol Obstet 2006;95(Suppl. 1):S105–43. https://doi.org/10.1016/S0020-7292(06)60031-3. Nov PMID: 17161155.

[57] Korets SB, Czok S, Blank SV, Curtin JP, Schneider RJ. Targeting the mTOR/4E-BP pathway in endometrial cancer. Clin Cancer Res 2011;17(24):7518–28. https://doi.org/10.1158/1078-0432.CCR-11-1664. Epub 2011 Dec 5. Dec 15 PMID: 22142830.

[58] Philip CA, Laskov I, Beauchamp MC, Marques M, Amin O, Bitharas J, Kessous R, Kogan L, Baloch T, Gotlieb WH, Yasmeen A. Inhibition of PI3K-AKT-mTOR pathway sensitizes endometrial cancer cell lines to PARP inhibitors. BMC Cancer 2017;17(1):638. https://doi.org/10.1186/s12885-017-3639-0. Sep 8 PMID: 28886696; PMCID: PMC5591502.

[59] Guzeloglu-Kayisli O, Kayisli UA, Luleci G, Arici A. In vivo and in vitro regulation of Akt activation in human endometrial cells is estrogen dependent. Biol Reprod 2004;71(3):714–21. https://doi.org/10.1095/biolreprod.104.027235. Epub 2004 Apr 28. Sep PMID: 15115729.

[60] Falkenstein E, Tillmann HC, Christ M, Feuring M, Wehling M. Multiple actions of steroid hormones—a focus on rapid, nongenomic effects. Pharmacol Rev 2000;52(4):513–56. PMID: 11121509.

[61] Levin ER. Cell localization, physiology, and nongenomic actions of estrogen receptors. J Appl Physiol (1985) 2001;91(4):1860–7. https://doi.org/10.1152/jappl.2001.91.4.1860. Oct PMID: 11568173.

[62] Guzeloglu-Kayisli O, Kayisli UA, Al-Rejjal R, Zheng W, Luleci G, Arici A. Regulation of PTEN (phosphatase and tensin homolog deleted on chromosome 10) expression by estradiol and progesterone in human endometrium. J Clin Endocrinol Metab 2003;88(10):5017–26. https://doi.org/10.1210/jc.2003-030414. Oct PMID: 14557489.

[63] Dery MC, Leblanc V, Shooner C, Asselin E. Regulation of Akt expression and phosphorylation by 17beta-estradiol in the rat uterus during estrous cycle. Reprod Biol Endocrinol 2003;1:47. https://doi.org/10.1186/1477-7827-1-47. Jun 12 PMID: 12816542; PMCID: PMC161822.

[64] Fabi F, Grenier K, Parent S, Adam P, Tardif L, Leblanc V, Asselin E. Regulation of the PI3K/Akt pathway during decidualization of endometrial stromal cells. PLoS One 2017;12(5). https://doi.org/10.1371/journal.pone.0177387, e0177387. May 5 PMID: 28475617; PMCID: PMC5419658.

[65] Slomovitz BM, Coleman RL. The PI3K/AKT/mTOR pathway as a therapeutic target in endometrial cancer. Clin Cancer Res 2012;18(21):5856–64. https://doi.org/10.1158/1078-0432.CCR-12-0662. Epub 2012 Oct 18. Nov 1 PMID: 23082003.

[66] Zheng W, Baker HE, Mutter GL. Involution of PTEN-null endometrial glands with progestin therapy. Gynecol Oncol 2004;92(3):1008–13. https://doi.org/10.1016/j.ygyno.2003.11.026. Mar PMID: 14984979.

[67] Stambolic V, Tsao MS, Macpherson D, Suzuki A, Chapman WB, Mak TW. High incidence of breast and endometrial neoplasia resembling human Cowden syndrome in pten+/− mice. Cancer Res 2000;60(13):3605–11. Jul 1 PMID: 10910075.

[68] Oda K, Stokoe D, Taketani Y, McCormick F. High frequency of coexistent mutations of PIK3CA and PTEN genes in endometrial carcinoma. Cancer Res 2005;65(23):10669–73. https://doi.org/10.1158/0008-5472.CAN-05-2620. Dec 1 PMID: 16322209.

[69] Catasus L, Gallardo A, Cuatrecasas M, Prat J. PIK3CA mutations in the kinase domain (exon 20) of uterine endometrial adenocarcinomas are associated with adverse prognostic parameters. Mod Pathol 2008;21(2):131–9. https://doi.org/10.1038/modpathol.3800992. Epub 2007 Dec 14. Feb PMID: 18084252.

[70] Woods A, Johnstone SR, Dickerson K, Leiper FC, Fryer LG, Neumann D, Schlattner U, Wallimann T, Carlson M, Carling D. LKB1 is the upstream kinase in the AMP-activated protein kinase cascade. Curr Biol 2003;13(22):2004–8. https://doi.org/10.1016/j.cub.2003.10.031. Nov 11 PMID: 14614828.

[71] Contreras CM, Gurumurthy S, Haynie JM, Shirley LJ, Akbay EA, Wingo SN, Schorge JO, Broaddus RR, Wong KK, Bardeesy N, Castrillon DH. Loss of Lkb1 provokes highly invasive endometrial adenocarcinomas. Cancer Res 2008;68(3):759–66. https://doi.org/10.1158/0008-5472.CAN-07-5014. Feb 1 PMID: 18245476.

[72] Darb-Esfahani S, Faggad A, Noske A, Weichert W, Buckendahl AC, Müller B, Budczies J, Röske A, Dietel M, Denkert C. Phospho-mTOR and phospho-4EBP1 in endometrial adenocarcinoma: association with stage and grade in vivo and link with response to rapamycin treatment in vitro. J Cancer Res Clin Oncol 2009;135(7):933–41. https://doi.org/10.1007/s00432-008-0529-5. Epub 2008 Dec 24. Jul PMID: 19107520.

[73] Mori N, Kyo S, Sakaguchi J, Mizumoto Y, Ohno S, Maida Y, Hashimoto M, Takakura M, Inoue M. Concomitant activation of AKT with extracellular-regulated kinase 1/2 occurs independently of PTEN or PIK3CA mutations in endometrial cancer and may be associated with favorable prognosis. Cancer Sci 2007;98(12):1881–8. https://doi.org/10.1111/j.1349-7006.2007.00630.x. Epub 2007 Oct 9. Dec PMID: 17924977.

[74] Colombo N, McMeekin DS, Schwartz PE, Sessa C, Gehrig PA, Holloway R, Braly P, Matei D, Morosky A, Dodion PF, Einstein MH, Haluska F. Ridaforolimus as a single agent in advanced endometrial cancer: results of a single-arm, phase 2 trial. Br J Cancer 2013;108(5):1021–6. https://doi.org/10.1038/bjc.2013.59. Mar 19 Epub 2013 Feb 12. PMID: 23403817; PMCID: PMC3619076.

[75] Tsoref D, Welch S, Lau S, Biagi J, Tonkin K, Martin LA, Ellard S, Ghatage P, Elit L, Mackay HJ, Allo G, Tsao MS, Kamel-Reid S, Eisenhauer EA, Oza AM. Phase II study of oral ridaforolimus in women with recurrent or metastatic endometrial cancer. Gynecol Oncol 2014;135(2):184–9. https://doi.org/10.1016/j.ygyno.2014.06.033. Epub 2014 Aug 28. Nov PMID: 25173583.

[76] Slomovitz BM, Lu KH, Johnston T, Coleman RL, Munsell M, Broaddus RR, Walker C, Ramondetta LM, Burke TW, Gershenson DM, Wolf J. A phase 2 study of the oral mammalian target of rapamycin inhibitor, everolimus, in patients with recurrent endometrial carcinoma. Cancer 2010;116(23):5415–9. https://doi.org/10.1002/cncr.25515. Dec 1 Epub 2010 Aug 2. PMID: 20681032; PMCID: PMC5120730.

[77] Oza AM, Elit L, Tsao MS, Kamel-Reid S, Biagi J, Provencher DM, Gotlieb WH, Hoskins PJ, Ghatage P, Tonkin KS, Mackay HJ, Mazurka J, Sederias J, Ivy P, Dancey JE, Eisenhauer EA. Phase II study of temsirolimus in women with recurrent or metastatic endometrial cancer: a trial of the NCIC Clinical Trials Group. J Clin Oncol 2011;29(24):3278–85. https://doi.org/10.1200/JCO.2010.34.1578. Aug 20 Epub 2011 Jul 25. PMID: 21788564; PMCID: PMC3158598.

[78] Aghajanian C, Filiaci V, Dizon DS, Carlson JW, Powell MA, Secord AA, Tewari KS, Bender DP, O'Malley DM, Stuckey A, Gao J, Dao F, Soslow RA, Lankes HA, Moore K, Levine DA. A phase II study of frontline paclitaxel/carboplatin/bevacizumab, paclitaxel/carboplatin/temsirolimus, or ixabepilone/carboplatin/bevacizumab in advanced/recurrent endometrial cancer. Gynecol Oncol 2018;150(2):274–81. https://doi.org/10.1016/j.ygyno.2018.05.018. Aug Epub 2018 May 24. PMID: 29804638; PMCID: PMC6179372.

[79] Folkes AJ, Ahmadi K, Alderton WK, Alix S, Baker SJ, Box G, Chuckowree IS, Clarke PA, Depledge P, Eccles SA, Friedman LS, Hayes A, Hancox TC, Kugendradas A, Lensun L, Moore P, Olivero AG, Pang J, Patel S, Pergl-Wilson GH, Raynaud FI, Robson A, Saghir N, Salphati L, Sohal S, Ultsch MH, Valenti M, Wallweber HJ, Wan NC, Wiesmann C, Workman P, Zhyvoloup A, Zvelebil MJ, Shuttleworth SJ. The identification of 2-(1H-indazol-4-yl)-6-(4-methanesulfonyl-piperazin-1-ylmethyl)-4-morpholin-4-yl-thieno[3,2-d]pyrimidine (GDC-0941) as a potent, selective, orally bioavailable inhibitor of class I PI3 kinase for the treatment of cancer. J Med Chem 2008;51(18):5522–32. https://doi.org/10.1021/jm800295d. Sep 25 PMID: 18754654.

[80] Maira SM, Pecchi S, Huang A, Burger M, Knapp M, Sterker D, Schnell C, Guthy D, Nagel T, Wiesmann M, Brachmann S, Fritsch C, Dorsch M, Chène P, Shoemaker K, De Pover A, Menezes D, Martiny-Baron G, Fabbro D, Wilson CJ, Schlegel R, Hofmann F, García-Echeverría C, Sellers WR, Voliva CF. Identification and characterization of NVP-BKM120, an orally available pan-class I PI3-kinase inhibitor. Mol Cancer Ther 2012;11(2):317–28. https://doi.org/10.1158/1535-7163.MCT-11-0474. Epub 2011 Dec 21. Feb PMID: 22188813.

[81] Maira SM, Stauffer F, Brueggen J, Furet P, Schnell C, Fritsch C, Brachmann S, Chène P, De Pover A, Schoemaker K, Fabbro D, Gabriel D, Simonen M, Murphy L, Finan P, Sellers W, García-Echeverría C. Identification and characterization of NVP-BEZ235, a new orally available dual phosphatidylinositol 3-kinase/mammalian target of rapamycin inhibitor with potent in vivo antitumor activity. Mol Cancer Ther 2008;7(7):1851–63. https://doi.org/10.1158/1535-7163.MCT-08-0017. Epub 2008 Jul 7. Jul PMID: 18606717.

[82] Shoji K, Oda K, Kashiyama T, Ikeda Y, Nakagawa S, Sone K, Miyamoto Y, Hiraike H, Tanikawa M, Miyasaka A, Koso T, Matsumoto Y, Wada-Hiraike O, Kawana K, Kuramoto H, McCormick F, Aburatani H, Yano T, Kozuma S, Taketani Y. Genotype-dependent efficacy of a dual PI3K/mTOR inhibitor, NVP-BEZ235, and an mTOR inhibitor, RAD001, in endometrial carcinomas. PLoS One 2012;7(5):e37431. https://doi.org/10.1371/journal.pone.0037431. Epub 2012 May 25. PMID: 22662154; PMCID: PMC3360787.

[83] Wallin JJ, Edgar KA, Guan J, Berry M, Prior WW, Lee L, Lesnick JD, Lewis C, Nonomiya J, Pang J, Salphati L, Olivero AG, Sutherlin DP, O'Brien C, Spoerke JM, Patel S, Lensun L, Kassees R, Ross L, Lackner MR, Sampath D, Belvin M, Friedman LS. GDC-0980 is a novel class I PI3K/mTOR kinase inhibitor with robust activity in cancer models driven by the PI3K pathway. Mol Cancer Ther 2011;10(12):2426–36. https://doi.org/10.1158/1535-7163.MCT-11-0446. Epub 2011 Oct 13. Dec PMID: 21998291.

[84] Brown JR, Hamadani M, Hayslip J, Janssens A, Wagner-Johnston N, Ottmann O, Arnason J, Tilly H, Millenson M, Offner F, Gabrail NY, Ganguly S, Ailawadhi S, Kasar S, Kater AP, Doorduijn JK, Gao L, Lager JJ, Wu B, Egile C, Kersten MJ. Voxtalisib (XL765) in patients with relapsed or refractory non-Hodgkin lymphoma or chronic lymphocytic leukaemia: an open-label, phase 2 trial. Lancet Haematol 2018;5(4):e170–80. https://doi.org/10.1016/S2352-3026(18)30030-9. Apr Epub 2018 Mar 14. Erratum in: Lancet Haematol. 2018 Jun;5(6):e240. PMID: 29550382; PMCID: PMC7029813.

[85] Burris H, Rodon J, Sharma S, Herbst RS, Tabernero J, Infante JR, et al. First-in-human phase I study of the oral PI3K inhibitor BEZ235 in patients (pts) with advanced solid tumors [abstract]. J Clin Oncol 2010;28. Abstract nr 3005.

[86] Papadopoulos KP, Egile C, Ruiz-Soto R, Jiang J, Shi W, Bentzien F, Rasco D, Abrisqueta P, Vose JM, Tabernero J. Efficacy, safety, pharmacokinetics and pharmacodynamics of SAR245409 (voxtalisib, XL765), an orally administered phosphoinositide 3-kinase/mammalian target of rapamycin inhibitor: a phase 1 expansion cohort in patients with relapsed or refractory lymphoma. Leuk Lymphoma 2015;56(6):1763–70. https://doi.org/10.3109/10428194.2014.974040. Epub 2014 Nov 19. Jun PMID: 25300944.

[87] Wen PY, Omuro A, Ahluwalia MS, Fathallah-Shaykh HM, Mohile N, Lager JJ, Laird AD, Tang J, Jiang J, Egile C, Cloughesy TF. Phase I dose-escalation study of the PI3K/mTOR inhibitor voxtalisib (SAR245409, XL765) plus temozolomide with or without radiotherapy in patients with high-grade glioma. Neuro Oncol 2015;17 (9):1275–83. https://doi.org/10.1093/neuonc/nov083. Sep Epub 2015 May 26. PMID: 26019185; PMCID: PMC4588757.

[88] Makker V, Recio FO, Ma L, Matulonis UA, Lauchle JO, Parmar H, Gilbert HN, Ware JA, Zhu R, Lu S, Huw LY, Wang Y, Koeppen H, Spoerke JM, Lackner MR, Aghajanian CA. A multicenter, single-arm, open-label, phase 2 study of apitolisib (GDC-0980) for the treatment of recurrent or persistent endometrial carcinoma (MAGGIE study). Cancer 2016;122(22):3519–28. https://doi.org/10.1002/cncr.30286. Nov 15 Epub 2016 Sep 7. PMID: 27603005; PMCID: PMC5600677.

[89] Hirai H, Sootome H, Nakatsuru Y, Miyama K, Taguchi S, Tsujioka K, Ueno Y, Hatch H, Majumder PK, Pan BS, Kotani H. MK-2206, an allosteric Akt inhibitor, enhances antitumor efficacy by standard chemotherapeutic agents or molecular targeted drugs in vitro and in vivo. Mol Cancer Ther 2010;9(7):1956–67. https://doi.org/10.1158/1535-7163.MCT-09-1012. Epub 2010 Jun 22. Jul PMID: 20571069.

[90] Knowling M, Blackstein M, Tozer R, Bramwell V, Dancey J, Dore N, Matthews S, Eisenhauer E. A phase II study of perifosine (D-21226) in patients with previously untreated metastatic or locally advanced soft tissue sarcoma: a National Cancer Institute of Canada Clinical Trials Group trial. Invest New Drugs 2006;24(5):435–9. https://doi.org/10.1007/s10637-006-6406-7. Sep PMID: 16528419.

[91] Dowling RJ, Zakikhani M, Fantus IG, Pollak M, Sonenberg N. Metformin inhibits mammalian target of rapamycin-dependent translation initiation in breast cancer cells. Cancer Res 2007;67(22):10804–12. https://doi.org/10.1158/0008-5472.CAN-07-2310. Nov 15 PMID: 18006825.

[92] Xue J, Li L, Li N, Li F, Qin X, Li T, Liu M. Metformin suppresses cancer cell growth in endometrial carcinoma by inhibiting PD-L1. Eur J Pharmacol 2019;859:172541. https://doi.org/10.1016/j.ejphar.2019.172541. Epub 2019 Jul 15. Sep 15 PMID: 31319067.

[93] Meireles CG, Pereira SA, Valadares LP, Rêgo DF, Simeoni LA, Guerra ENS, Lofrano-Porto A. Effects of metformin on endometrial cancer: systematic review and meta-analysis. Gynecol Oncol 2017;147(1):167–80. https://doi.org/10.1016/j.ygyno.2017.07.120. Epub 2017 Jul 29. Oct PMID: 28760367.

[94] Sharifzadeh F, Aminimoghaddam S, Kashanian M, Fazaeli M, Sheikhansari N. A comparison between the effects of metformin and megestrol on simple endometrial hyperplasia. Gynecol Endocrinol 2017;33(2):152–5. https://doi.org/10.1080/09513590.2016.1223285. Epub 2016 Sep 30. Feb PMID: 27690687.

[95] Tabrizi AD, Melli MS, Foroughi M, Ghojazadeh M, Bidadi S. Antiproliferative effect of metformin on the endometrium—a clinical trial. Asian Pac J Cancer Prev 2014;15(23):10067–70. https://doi.org/10.7314/apjcp.2014.15.23.10067. PMID: 25556427.

[96] Mitsuhashi A, Sato Y, Kiyokawa T, Koshizaka M, Hanaoka H, Shozu M. Phase II study of medroxyprogesterone acetate plus metformin as a fertility-sparing treatment for atypical endometrial hyperplasia and endometrial cancer. Ann Oncol 2016;27(2):262–6. https://doi.org/10.1093/annonc/mdv539. Epub 2015 Nov 16. Feb PMID: 26578736.

[97] Ushijima K, Yahata H, Yoshikawa H, Konishi I, Yasugi T, Saito T, Nakanishi T, Sasaki H, Saji F, Iwasaka T, Hatae M, Kodama S, Saito T, Terakawa N, Yaegashi N, Hiura M, Sakamoto A, Tsuda H, Fukunaga M, Kamura T. Multicenter phase II study of fertility-sparing treatment with medroxyprogesterone acetate for endometrial carcinoma and atypical hyperplasia in young women. J Clin Oncol 2007;25(19):2798–803. https://doi.org/10.1200/JCO.2006.08.8344. Jul 1 PMID: 17602085.

[98] Yang BY, Gulinazi Y, Du Y, Ning CC, Cheng YL, Shan WW, Luo XZ, Zhang HW, Zhu Q, Ma FH, Liu J, Sun L, Yu M, Guan J, Chen XJ. Metformin plus megestrol acetate compared with megestrol acetate alone as fertility-sparing treatment in patients with atypical endometrial hyperplasia and well-differentiated endometrial cancer: a randomised controlled trial. BJOG 2020;127(7):848–57. https://doi.org/10.1111/1471-0528.16108. Epub 2020 Feb 16. Jun PMID: 31961463.

[99] Chu D, Wu J, Wang K, Zhao M, Wang C, Li L, Guo R. Effect of metformin use on the risk and prognosis of endometrial cancer: a systematic review and meta-analysis. BMC Cancer 2018;18(1):438. https://doi.org/10.1186/s12885-018-4334-5. Apr 18 PMID: 29669520; PMCID: PMC5907461.

[100] Kitson SJ, Maskell Z, Sivalingam VN, Allen JL, Ali S, Burns S, Gilmour K, Latheef R, Slade RJ, Pemberton PW, Shaw J, Ryder WD, Kitchener HC, Crosbie EJ. PRE-surgical metformin in uterine malignancy (PREMIUM): a multi-center, randomized double-blind, placebo-controlled phase III trial. Clin Cancer Res 2019;25(8):2424–32. https://doi.org/10.1158/1078-0432.CCR-18-3339. Apr 15 Epub 2018 Dec 18. PMID: 30563932; PMCID: PMC6586555.

[101] Davis SR, Robinson PJ, Jane F, White S, Brown KA, Piessens S, Edwards A, McNeilage J, Woinarski J, Chipman M, Bell RJ. The benefits of adding metformin to tamoxifen to protect the endometrium—a randomized placebo-controlled trial. Clin Endocrinol (Oxf) 2018;89(5):605–12. https://doi.org/10.1111/cen.13830. Epub 2018 Sep 9. Nov PMID: 30107043.

[102] Dieci MV, Arnedos M, Andre F, Soria JC. Fibroblast growth factor receptor inhibitors as a cancer treatment: from a biologic rationale to medical perspectives. Cancer Discov 2013;3(3):264–79. https://doi.org/10.1158/2159-8290.CD-12-0362. Epub 2013 Feb 15. Mar PMID: 23418312.

[103] Itoh N, Ornitz DM. Fibroblast growth factors: from molecular evolution to roles in development, metabolism and disease. J Biochem 2011;149(2):121–30. https://doi.org/10.1093/jb/mvq121. Feb Epub 2010 Oct 12. PMID: 20940169; PMCID: PMC3106964.

[104] Winterhoff B, Konecny GE. Targeting fibroblast growth factor pathways in endometrial cancer. Curr Probl Cancer 2017;41(1):37–47. https://doi.org/10.1016/j.currproblcancer.2016.11.002. Epub 2016 Nov 11. Jan-Feb PMID: 28041631.

[105] Beenken A, Mohammadi M. The FGF family: biology, pathophysiology and therapy. Nat Rev Drug Discov 2009;8(3):235–53. https://doi.org/10.1038/nrd2792. Mar PMID: 19247306; PMCID: PMC3684054.

[106] Chiu HC, Li CJ, Yiang GT, Tsai AP, Wu MY. Epithelial to mesenchymal transition and cell biology of molecular regulation in endometrial carcinogenesis. J Clin Med 2019;8(4):439. https://doi.org/10.3390/jcm8040439. Mar 30 PMID: 30935077; PMCID: PMC6518354.

[107] Roskoski Jr R. The ErbB/HER family of protein-tyrosine kinases and cancer. Pharmacol Res 2014;79:34–74. https://doi.org/10.1016/j.phrs.2013.11.002. Epub 2013 Nov 20. Jan PMID: 24269963.

[108] Roberts PJ, Der CJ. Targeting the Raf-MEK-ERK mitogen-activated protein kinase cascade for the treatment of cancer. Oncogene 2007;26(22):3291–310. https://doi.org/10.1038/sj.onc.1210422. May 14 PMID: 17496923.

[109] Turner N, Grose R. Fibroblast growth factor signalling: from development to cancer. Nat Rev Cancer 2010;10(2):116–29. https://doi.org/10.1038/nrc2780. Feb PMID: 20094046.

[110] Makieva S, Giacomini E, Ottolina J, Sanchez AM, Papaleo E, Viganò P. Inside the endometrial cell signaling subway: mind the gap(s). Int J Mol Sci 2018;19(9):2477. https://doi.org/10.3390/ijms19092477. Aug 21 PMID: 30134622; PMCID: PMC6164241.

[111] Hewitt SC, Li Y, Li L, Korach KS. Estrogen-mediated regulation of Igf1 transcription and uterine growth involves direct binding of estrogen receptor alpha to estrogen-responsive elements. J Biol Chem 2010;285(4):2676–85. https://doi.org/10.1074/jbc.M109.043471. Jan 22 Epub 2009 Nov 17. PMID: 19920132; PMCID: PMC2807324.

[112] Kahlert S, Nuedling S, van Eickels M, Vetter H, Meyer R, Grohe C. Estrogen receptor alpha rapidly activates the IGF-1 receptor pathway. J Biol Chem 2000;275(24):18447–53. https://doi.org/10.1074/jbc.M910345199. Jun 16 PMID: 10749889.

[113] Jacenik D, Cygankiewicz AI, Krajewska WM. The G protein-coupled estrogen receptor as a modulator of neoplastic transformation. Mol Cell Endocrinol 2016;429:10–8. https://doi.org/10.1016/j.mce.2016.04.011. Epub 2016 Apr 21. Jul 5 PMID: 27107933.

[114] Wang Y, van der Zee M, Fodde R, Blok LJ. Wnt/B-catenin and sex hormone signaling in endometrial homeostasis and cancer. Oncotarget 2010;1(7):674–84. https://doi.org/10.18632/oncotarget.101007. Nov PMID: 21317462; PMCID: PMC3248134.

[115] Karteris E, Zervou S, Pang Y, Dong J, Hillhouse EW, Randeva HS, Thomas P. Progesterone signaling in human myometrium through two novel membrane G protein-coupled receptors: potential role in functional progesterone withdrawal at term. Mol Endocrinol 2006;20(7):1519–34. https://doi.org/10.1210/me.2005-0243. Epub 2006 Feb 16. Jul PMID: 16484338.

[116] Kaitu'u-Lino TJ, Morison NB, Salamonsen LA. Neutrophil depletion retards endometrial repair in a mouse model. Cell Tissue Res 2007;328(1):197–206. https://doi.org/10.1007/s00441-006-0358-2. Epub 2006 Dec 22. Apr PMID: 17186309.

[117] Gargett CE, Chan RW, Schwab KE. Hormone and growth factor signaling in endometrial renewal: role of stem/progenitor cells. Mol Cell Endocrinol 2008;288(1–2):22–9. https://doi.org/10.1016/j.mce.2008.02.026. Epub 2008 Mar 4. Jun 25 PMID: 18403104.

[118] Zhu H, Jiang Y, Pan Y, Shi L, Zhang S. Human menstrual blood-derived stem cells promote the repair of impaired endometrial stromal cells by activating the p38 MAPK and AKT signaling pathways. Reprod Biol 2018;18(3):274–81. https://doi.org/10.1016/j.repbio.2018.06.003. Epub 2018 Jun 23. Sep PMID: 29941287.

[119] Alexander-Sefre F, Salvesen HB, Ryan A, Singh N, Akslen LA, MacDonald N, Wilbanks G, Jacobs IJ. Molecular assessment of depth of myometrial invasion in stage I endometrial cancer: a model based on K-ras mutation analysis. Gynecol Oncol 2003;91(1):218–25. https://doi.org/10.1016/s0090-8258(03)00505-5. Oct PMID: 14529685.

[120] Ito K, Watanabe K, Nasim S, Sasano H, Sato S, Yajima A, Silverberg SG, Garrett CT. K-ras point mutations in endometrial carcinoma: effect on outcome is dependent on age of patient. Gynecol Oncol 1996;63(2):238–46. https://doi.org/10.1006/gyno.1996.0313. Nov PMID: 8910634.

[121] Moreno-Bueno G, Sanchez-Estevez C, Palacios J, Hardisson D, Shiozawa T. Low frequency of BRAF mutations in endometrial and in cervical carcinomas. Clin Cancer Res 2006;12(12):3865. author reply 3865-6 https://doi.org/10.1158/1078-0432.CCR-06-0284. Jun 15 PMID: 16778116.

[122] Feng YZ, Shiozawa T, Miyamoto T, Kashima H, Kurai M, Suzuki A, Konishi I. BRAF mutation in endometrial carcinoma and hyperplasia: correlation with KRAS and p53 mutations and mismatch repair protein expression. Clin Cancer Res 2005;11(17):6133–8. https://doi.org/10.1158/1078-0432.CCR-04-2670. Sep 1 PMID: 16144912.

[123] Pallarés J, Velasco A, Eritja N, Santacana M, Dolcet X, Cuatrecasas M, Palomar-Asenjo V, Catasús L, Prat J, Matias-Guiu X. Promoter hypermethylation and reduced expression of RASSF1A are frequent molecular alterations of endometrial carcinoma. Mod Pathol 2008;21(6):691–9. https://doi.org/10.1038/modpathol.2008.38. Epub 2008 May 9. Jun PMID: 18469797.

[124] Yeramian A, Moreno-Bueno G, Dolcet X, Catasus L, Abal M, Colas E, Reventos J, Palacios J, Prat J, Matias-Guiu X. Endometrial carcinoma: molecular alterations involved in tumor development and progression. Oncogene 2013;32(4):403–13. https://doi.org/10.1038/onc.2012.76. Epub 2012 Mar 19. Jan 24 PMID: 22430211.

[125] Byron SA, Chen H, Wortmann A, Loch D, Gartside MG, Dehkhoda F, Blais SP, Neubert TA, Mohammadi M, Pollock PM. The N550K/H mutations in FGFR2 confer differential resistance to PD173074, dovitinib, and ponatinib ATP-competitive inhibitors. Neoplasia 2013;15(8):975–88. https://doi.org/10.1593/neo.121106. Aug PMID: 23908597; PMCID: PMC3730048.

[126] Konecny GE, Kolarova T, O'Brien NA, Winterhoff B, Yang G, Qi J, Qi Z, Venkatesan N, Ayala R, Luo T, Finn RS, Kristof J, Galderisi C, Porta DG, Anderson L, Shi MM, Yovine A, Slamon DJ. Activity of the fibroblast growth factor receptor inhibitors dovitinib (TKI258) and NVP-BGJ398 in human endometrial cancer cells. Mol Cancer Ther 2013;12(5):632–42. https://doi.org/10.1158/1535-7163.MCT-12-0999. Epub 2013 Feb 26. May PMID: 23443805.

[127] Gozgit JM, Squillace RM, Wongchenko MJ, Miller D, Wardwell S, Mohemmad Q, Narasimhan NI, Wang F, Clackson T, Rivera VM. Combined targeting of FGFR2 and mTOR by ponatinib and ridaforolimus results in synergistic antitumor activity in FGFR2 mutant endometrial cancer models. Cancer Chemother Pharmacol 2013;71(5):1315–23. https://doi.org/10.1007/s00280-013-2131-z. Epub 2013 Mar 7. May PMID: 23468082.

[128] Kamat AA, Merritt WM, Coffey D, Lin YG, Patel PR, Broaddus R, Nugent E, Han LY, Landen Jr CN, Spannuth WA, Lu C, Coleman RL, Gershenson DM, Sood AK. Clinical and biological significance of vascular endothelial growth factor in endometrial cancer. Clin Cancer Res 2007;13(24):7487–95. https://doi.org/10.1158/1078-0432.CCR-07-1017. Dec 15 PMID: 18094433.

[129] Pepper MS, Ferrara N, Orci L, Montesano R. Potent synergism between vascular endothelial growth factor and basic fibroblast growth factor in the induction of angiogenesis in vitro. Biochem Biophys Res Commun 1992;189(2):824–31. https://doi.org/10.1016/0006-291x(92)92277-5. Dec 15 PMID: 1281999.

[130] Birrer MJ, Johnson ME, Hao K, Wong KK, Park DC, Bell A, Welch WR, Berkowitz RS, Mok SC. Whole genome oligonucleotide-based array comparative genomic hybridization analysis identified fibroblast growth factor 1 as a prognostic marker for advanced-stage serous ovarian adenocarcinomas. J Clin Oncol 2007;25(16):2281–7. https://doi.org/10.1200/JCO.2006.09.0795. Jun 1 Erratum in: J Clin Oncol. 2007 Jul 20;25(21):3184. PMID: 17538174.

[131] Casanovas O, Hicklin DJ, Bergers G, Hanahan D. Drug resistance by evasion of antiangiogenic targeting of VEGF signaling in late-stage pancreatic islet tumors. Cancer Cell 2005;8(4):299–309. https://doi.org/10.1016/j.ccr.2005.09.005. Oct PMID: 16226705.

[132] Powell MA, Sill MW, Goodfellow PJ, Benbrook DM, Lankes HA, Leslie KK, Jeske Y, Mannel RS, Spillman MA, Lee PS, Hoffman JS, McMeekin DS, Pollock PM. A phase II trial of brivanib in recurrent or persistent endometrial cancer: an NRG Oncology/Gynecologic Oncology Group Study. Gynecol Oncol 2014;135(1):38–43. https://doi.org/10.1016/j.ygyno.2014.07.083. Oct Epub 2014 Jul 11. PMID: 25019571; PMCID: PMC4278402.

[133] Dizon DS, Sill MW, Schilder JM, McGonigle KF, Rahman Z, Miller DS, Mutch DG, Leslie KK. A phase II evaluation of nintedanib (BIBF-1120) in the treatment of recurrent or persistent endometrial cancer: an NRG Oncology/Gynecologic Oncology Group Study. Gynecol Oncol 2014;135(3):441–5. https://doi.org/10.1016/j.ygyno.2014.10.001. Dec Epub 2014 Oct 13. PMID: 25312396; PMCID: PMC4373614.

[134] Konecny GE, Finkler N, Garcia AA, Lorusso D, Lee PS, Rocconi RP, Fong PC, Squires M, Mishra K, Upalawanna A, Wang Y, Kristeleit R. Second-line dovitinib (TKI258) in patients with FGFR2-mutated or FGFR2-non-mutated advanced or metastatic endometrial cancer: a non-randomised, open-label, two-group, two-stage, phase 2 study. Lancet Oncol 2015;16(6):686–94. https://doi.org/10.1016/S1470-2045(15)70159-2. Epub 2015 May 13. Jun PMID: 25981814.

[135] Hong DS, Kurzrock R, Wheler JJ, Naing A, Falchook GS, Fu S, Kim KB, Davies MA, Nguyen LM, George GC, Xu L, Shumaker R, Ren M, Mink J, Bedell C, Andresen C, Sachdev P, O'Brien JP, Nemunaitis J. Phase I dose-escalation study of the multikinase inhibitor lenvatinib in patients with advanced solid tumors and in an expanded cohort of patients with melanoma. Clin Cancer Res 2015;21(21):4801–10. https://doi.org/10.1158/1078-0432.CCR-14-3063. Nov 1 Epub 2015 Jul 13. PMID: 26169970; PMCID: PMC4840931.

[136] Vergote I, Powell MA, Teneriello MG, Miller DS, Garcia AA, Mikheeva ON, Bidzinski M, Cebotaru CL, Dutcus CE, Ren M, Kadowaki T, Funahashi Y, Penson RT. Second-line lenvatinib in patients with recurrent endometrial cancer. Gynecol Oncol 2020;156(3):575–82. https://doi.org/10.1016/j.ygyno.2019.12.039. Epub 2020 Jan 17. Mar PMID: 31955859.

[137] Funahashi Y, Penson RT, Powell MA, et al. Analysis of a plasma biomarker and tumor genetic alterations from a phase II trial of Lenvatinib in patients with advanced endometrial cancer. J Clin Oncol 2013;31(suppl) [ASCO Annual Meeting Abstracts, Abstract 5591].

[138] Ring KL, Yates MS, Schmandt R, Onstad M, Zhang Q, Celestino J, Kwan SY, Lu KH. Endometrial cancers with activating KRas mutations have activated estrogen signaling and paradoxical response to MEK inhibition. Int J Gynecol Cancer 2017;27(5):854–62. https://doi.org/10.1097/IGC.0000000000000960. Jun PMID: 28498246; PMCID: PMC5438270.

[139] Westin SN, Sill MW, Coleman RL, Waggoner S, Moore KN, Mathews CA, Martin LP, Modesitt SC, Lee S, Ju Z, Mills GB, Schilder RJ, Fracasso PM, Birrer MJ, Aghajanian C. Safety lead-in of the MEK inhibitor trametinib in combination with GSK2141795, an AKT inhibitor, in patients with recurrent endometrial cancer: An NRG Oncology/GOG study. Gynecol Oncol 2019;155(3):420–8. https://doi.org/10.1016/j.ygyno.2019.09.024. Dec Epub 2019 Oct 15. PMID: 31623857; PMCID: PMC6922584.

[140] Dousias V, Vrekoussis T, Navrozoglou I, Paschopoulos M, Stefos T, Makrigiannakis A, Jeschke U. Hypoxia-induced factor-1α in endometrial carcinoma: a mini-review of current evidence. Histol Histopathol 2012;27(10):1247–53. https://doi.org/10.14670/HH-27.1247. Oct PMID: 22936443.

[141] Seeber LM, Horrée N, Vooijs MA, Heintz AP, van der Wall E, Verheijen RH, van Diest PJ. The role of hypoxia inducible factor-1alpha in gynecological cancer. Crit Rev Oncol Hematol 2011;78(3):173–84. https://doi.org/10.1016/j.critrevonc.2010.05.003. Epub 2010 Jun 7. Jun PMID: 20627616.

[142] Ruan K, Song G, Ouyang G. Role of hypoxia in the hallmarks of human cancer. J Cell Biochem 2009;107(6):1053–62. https://doi.org/10.1002/jcb.22214. Aug 15 PMID: 19479945.

[143] Semenza GL. Life with oxygen. Science 2007;318(5847):62–4. https://doi.org/10.1126/science.1147949. Oct 5 PMID: 17916722.

[144] DeClerck K, Elble RC. The role of hypoxia and acidosis in promoting metastasis and resistance to chemotherapy. Front Biosci 2010;15:213–25. https://doi.org/10.2741/3616. Jan 1 PMID: 20036816.

[145] Seeber LM, Zweemer RP, Verheijen RH, van Diest PJ. Hypoxia-inducible factor-1 as a therapeutic target in endometrial cancer management. Obstet Gynecol Int 2010;2010:580971. https://doi.org/10.1155/2010/580971. Epub 2010 Feb 14. PMID: 20169098; PMCID: PMC2821774.

[146] Semenza GL. Regulation of metabolism by hypoxia-inducible factor 1. Cold Spring Harb Symp Quant Biol 2011;76:347–53. https://doi.org/10.1101/sqb.2011.76.010678. Epub 2011 Jul 22 21785006.

[147] Gort EH, Groot AJ, van der Wall E, van Diest PJ, Vooijs MA. Hypoxic regulation of metastasis via hypoxia-inducible factors. Curr Mol Med 2008;8(1):60–7. https://doi.org/10.2174/156652408783565568. Feb PMID: 18289014.

[148] Kazi AA, Koos RD. Estrogen-induced activation of hypoxia-inducible factor-1alpha, vascular endothelial growth factor expression, and edema in the uterus are mediated by the phosphatidylinositol 3-kinase/Akt pathway. Endocrinology 2007;148(5):2363–74. https://doi.org/10.1210/en.2006-1394. Epub 2007 Feb 1. May PMID: 17272396.

[149] Catalano RD, Wilson MR, Boddy SC, McKinlay AT, Sales KJ, Jabbour HN. Hypoxia and prostaglandin E receptor 4 signalling pathways synergise to promote endometrial adenocarcinoma cell proliferation and tumour growth. PLoS One 2011;6(5):e19209. https://doi.org/10.1371/journal.pone.0019209. May 12 PMID: 21589857; PMCID: PMC3093383.

[150] Orimo A, Tomioka Y, Shimizu Y, Sato M, Oigawa S, Kamata K, Nogi Y, Inoue S, Takahashi M, Hata T, Muramatsu M. Cancer-associated myofibroblasts possess various factors to promote endometrial tumor progression. Clin Cancer Res 2001;7(10):3097–105. Oct PMID: 11595701.

[151] Chi JT, Wang Z, Nuyten DS, Rodriguez EH, Schaner ME, Salim A, Wang Y, Kristensen GB, Helland A, Børresen-Dale AL, Giaccia A, Longaker MT, Hastie T, Yang GP, van de Vijver MJ, Brown PO. Gene expression programs in response to hypoxia: cell type specificity and prognostic significance in human cancers. PLoS Med 2006;3(3): e47. https://doi.org/10.1371/journal.pmed.0030047. Mar PMID: 16417408; PMCID: PMC1334226.

[152] Horrée N, Groot AJ, van Hattem WA, Heintz AP, Vooijs M, van Diest PJ. HIF-1A gene mutations associated with higher microvessel density in endometrial carcinomas. Histopathology 2008;52(5):637–9. https://doi.org/10.1111/j.1365-2559.2008.02991.x. Apr PMID: 18370960.

[153] Konac E, Onen HI, Metindir J, Alp E, Biri AA, Ekmekci A. An investigation of relationships between hypoxia-inducible factor-1 alpha gene polymorphisms and ovarian, cervical and endometrial cancers. Cancer Detect Prev 2007;31(2):102–9. https://doi.org/10.1016/j.cdp.2007.01.001. Epub 2007 Apr 6. PMID: 17418979.

[154] Miyasaka A, Oda K, Ikeda Y, Sone K, Fukuda T, Inaba K, Makii C, Enomoto A, Hosoya N, Tanikawa M, Uehara Y, Arimoto T, Kuramoto H, Wada-Hiraike O, Miyagawa K, Yano T, Kawana K, Osuga Y, Fujii T. PI3K/mTOR pathway inhibition overcomes radioresistance via suppression of the HIF1-α/VEGF pathway in endometrial cancer. Gynecol Oncol 2015;138(1):174–80. https://doi.org/10.1016/j.ygyno.2015.04.015. Epub 2015 Apr 22. Jul PMID: 25913131.

[155] Fujimoto J, Sato E, Alam SM, Jahan I, Toyoki H, Hong BL, Sakaguchi H, Tamaya T. Plausible linkage of hypoxia-inducible factor (HIF) in uterine endometrial cancers. Oncology 2006;71(1–2):95–101. https://doi.org/10.1159/000100477. Epub 2007 Mar 6. PMID: 17341889.

[156] Pansare V, Munkarah AR, Schimp V, Haitham Arabi M, Saed GM, Morris RT, Ali-Fehmi R. Increased expression of hypoxia-inducible factor 1alpha in type I and type II endometrial carcinomas. Mod Pathol 2007;20(1): 35–43. https://doi.org/10.1038/modpathol.3800718. Epub 2006 Nov 10. Jan PMID: 17099695.

[157] Papa A, Zaccarelli E, Caruso D, Vici P, Benedetti Panici P, Tomao F. Targeting angiogenesis in endometrial cancer—new agents for tailored treatments. Expert Opin Investig Drugs 2016;25(1):31–49. https://doi.org/10.1517/13543784.2016.1116517. Epub 2015 Dec 19. PMID: 26560489.

[158] Alvarez EA, Brady WE, Walker JL, Rotmensch J, Zhou XC, Kendrick JE, Yamada SD, Schilder JM, Cohn DE, Harrison CR, Moore KN, Aghajanian C. Phase II trial of combination bevacizumab and temsirolimus in the treatment of recurrent or persistent endometrial carcinoma: a Gynecologic Oncology Group study. Gynecol Oncol 2013;129(1):22–7. https://doi.org/10.1016/j.ygyno.2012.12.022. Epub 2012 Dec 20. Apr PMID: 23262204.

[159] Simpkins F, Drake R, Escobar PF, Nutter B, Rasool N, Rose PG. A phase II trial of paclitaxel, carboplatin, and bevacizumab in advanced and recurrent endometrial carcinoma (EMCA). Gynecol Oncol 2015;136(2):240–5. https://doi.org/10.1016/j.ygyno.2014.12.004. Epub 2014 Dec 6. Feb PMID: 25485782.

[160] Lorusso D, Ferrandina G, Colombo N, Pignata S, Pietragalla A, Sonetto C, Pisano C, Lapresa MT, Savarese A, Tagliaferri P, Lombardi D, Cinieri S, Breda E, Sabatucci I, Sabbatini R, Conte C, Cecere SC, Maltese G, Scambia G. Carboplatin-paclitaxel compared to carboplatin-paclitaxel-bevacizumab in advanced or recurrent endometrial cancer: MITO END-2—a randomized phase II trial. Gynecol Oncol 2019;155(3):406–12. https://doi.org/10.1016/j.ygyno.2019.10.013. Epub 2019 Oct 31. Dec PMID: 31677820.

[161] Gorski DH, Beckett MA, Jaskowiak NT, Calvin DP, Mauceri HJ, Salloum RM, Seetharam S, Koons A, Hari DM, Kufe DW, Weichselbaum RR. Blockage of the vascular endothelial growth factor stress response increases the antitumor effects of ionizing radiation. Cancer Res 1999;59(14):3374–8. Jul 15 PMID: 10416597 (web archive link).

[162] Lee CG, Heijn M, di Tomaso E, Griffon-Etienne G, Ancukiewicz M, Koike C, Park KR, Ferrara N, Jain RK, Suit HD, Boucher Y. Anti-vascular endothelial growth factor treatment augments tumor radiation response under normoxic or hypoxic conditions. Cancer Res 2000;60(19):5565–70. Oct 1 PMID: 11034104.

[163] Viswanathan AN, Lee H, Berkowitz R, Berlin S, Campos S, Feltmate C, Horowitz N, Muto M, Sadow CA, Matulonis U. A prospective feasibility study of radiation and concurrent bevacizumab for recurrent endometrial cancer. Gynecol Oncol 2014;132(1):55–60. https://doi.org/10.1016/j.ygyno.2013.10.031. Epub 2013 Nov 4. Jan PMID: 24201015.

[164] Viswanathan AN, Moughan J, Miller BE, Xiao Y, Jhingran A, Portelance L, Bosch WR, Matulonis UA, Horowitz NS, Mannel RS, Souhami L, Erickson BA, Winter KA, Small Jr W, Gaffney DK. NRG Oncology/RTOG 0921: a phase 2 study of postoperative intensity-modulated radiotherapy with concurrent cisplatin and bevacizumab followed by carboplatin and paclitaxel for patients with endometrial cancer. Cancer 2015;121(13):2156–63. https://doi.org/10.1002/cncr.29337. Jul 1 Epub 2015 Apr 6. PMID: 25847373; PMCID: PMC4685031.

# Chemoresistance in uterine cancer: Mechanisms of resistance and current therapies

*Abeer Arain[a], Ibrahim N. Muhsen[b], Ala Abudayyeh[c], and Maen Abdelrahim[a]*

[a]Houston Methodist Cancer Center, Houston Methodist Hospital, Houston, TX, United States
[b]Department of Medicine, Houston Methodist Hospital, Houston, TX, United States [c]Section of Nephrology, MD Anderson Cancer Center, Houston, TX, United States

## Abstract

Uterine cancer is one of the most common gynecological cancers, with significant morbidity and mortality. Many patients will fail treatment or have a recurrence, with resistance to chemotherapy being one of the main reasons for this worse outcome. Current chemotherapeutic modalities in uterine cancer include platinum-based and taxane-based chemotherapy; however, resistance to chemotherapy is a challenge that limits the efficacy of these agents in many patients. In this chapter, we discussed the current understanding of chemoresistance mechanisms in uterine cancers and the current and future approaches to overcome the resistance to chemotherapeutic agents. Primary molecular mechanisms of chemoresistance to platinum-based and taxane-based chemotherapy in uterine cancer involve DNA repair mechanisms, efflux pumps, and survival pathways (e.g., PI3K/AKT pathway and MAPK pathway). Multiple targeted drugs have been investigated, primarily in combination with chemotherapy, including AKT inhibitors, PARP inhibitors, mTOR inhibitors, and EFGR inhibitors. More studies will be needed to optimize the use of these drugs. Additionally, multiple mechanisms and targets are being currently investigated, including microRNA, leptin signaling pathway, and others.

## Abbreviations

| | |
|---|---|
| **BAX** | BCL2-associated X |
| **BCL2** | B-cell lymphoma 2 |
| **cJNT** | c-Jun N-terminal kinase |
| **CR** | complete response |
| **ER** | estrogen receptors |
| **ERE** | estrogen response element |

R. Basha, S. Ahmad (eds.)
*Overcoming Drug Resistance in Gynecologic Cancers*
ISBN 978-0-12-824299-5
https://doi.org/10.1016/B978-0-12-824299-5.00013-7

| **ERK** | extracellular-signal-regulated kinase |
| **GRP** | G protein-coupled receptor |
| **MAPK** | microtubule-associated protein kinase |
| **MDM2** | mouse double minute 2 |
| **MDR** | multidrug resistance |
| **miRNA** | microRNA |
| **MMR** | mismatch repair |
| **MSI** | microsatellite instability |
| **mTOR** | mammalian target of rapamycin |
| **NILCO** | Notch, interleukin-1, and leptin outcome |
| **ORR** | objective response rate |
| **PARP** | poly-ADP ribose polymerase |
| **PI3K** | phosphatidylinositol 3-kinase |
| **TP** | tumor protein |
| **XIAP** | X-linked inhibitor of apoptosis protein |

## Conflict of interest

No potential conflicts of interest were disclosed.

## Endometrial cancer

Gynecologic cancers account for about 10% of the new cancers and cancer deaths among women, both in North America and in Europe [1, 2]. Uterine (endometrial) cancer is one of the most frequent gynecologic cancers, and in most of the cases, it is diagnosed early, leading to a better outcome [2]. Most of the cancers in the uterus occur in the endometrial lining (about 95%) known as endometrial cancers [3]. Endometrial cancer is a malignancy with an overall good prognosis; however, about 25% of the population is diagnosed at an advanced stage, which is defined by the presence of invasive primary tumor and the presence of distant metastases [4]. Unfortunately, among the patients diagnosed with advanced-stage disease or recurrent endometrial cancer, the prognosis is unfavorable (<20%) despite the use of aggressive chemotherapies and immunotherapies [3].

In the United States, it is estimated that about 65,520 new cases and 12,590 deaths from endometrial occurred in 2020 [5]. The average age at the time of diagnosis is about 60 years [5]. Traditionally, there are two types of endometrial cancers, viz., type I tumors are estrogen-dependent endometrioid adenocarcinoma, and type II tumors are nonestrogen-dependent cancers of the endometrium [6]. Type II tumors are considered to be highly aggressive variant of endometrial cancers and are the ones that usually show recurrence and death from endometrial cancer [6]. One of the common reasons of treatment failure or recurrence in endometrial cancers is the development of resistance to the chemotherapy by the cancer cells [7]. Additionally, lack of efficacy of many currently available chemotherapy regimens (e.g., lack of selectivity) has contributed to the disease recurrence [8].

## Current treatments of endometrial cancer

The initial treatment strategy for endometrial cancer usually includes surgical removal of the primary tumor when still localized [9]. Further treatments are decided based on the stage of endometrial cancer with the aim of eradicating local and metastatic disease [3, 9]. The current guidelines recommend that patients with advanced-stage endometrial cancers and the ones with poor prognostic factors should receive chemotherapy [10]. Poor prognosis includes patients with stage IIIa or IIIb or any histology, patients with stage IA with myometrium invasion, and patients with stage IB, II, or IIIA with serous or clear cell histology [10]. While chemotherapy is the mainstay of therapeutic management after surgery, radiation and hormone therapies are the other two potential methods of elimination of the remaining cancer cells, common in most gynecologic cancers such as endometrial and ovarian cancers [3]. In endometrial cancer, radiation therapy is used most frequently in the initial stages of the disease and not as much for the advanced-stage cases [3, 11].

Different chemotherapy regimens are used in gynecologic cancers including platinum compounds (e.g., cisplatin or carboplatin), taxanes (e.g., paclitaxel or docetaxel), and doxorubicin [4]. Among the chemotherapeutic agents, cisplatin is one of the most effective platinum-based cytotoxic chemotherapy for endometrial cancer [12]. It works by interfering with the cell division, thus halting the proliferation of cancer cells and causing DNA damage resulting in programmed cell death of the malignant cells [12]. Taxanes work by microtubule polymerization, inhibit mitosis, and induce apoptosis [13]. Doxorubicin is an anthracycline that inhibits topoisomerase II by stabilizing its complex and by the generation of free radicals that lead to the death of cancer cells [14]. Other chemotherapeutic agents that are utilized in the management of endometrial cancers include topotecan (topoisomeraseI inhibitor), cyclophosphamide (an alkylating agent), and gemcitabine (a nucleoside analog) [15]. These agents are generally used in combination with taxane-or platinum-based regimens [15]. The overall response rate of gynecologic cancers with chemotherapy is good, ranging from 45% to 70% in most cases [16]. However, there is a risk of disease relapse and primary tumor becoming resistant to chemotherapeutic agents, contributing to a relatively lower survival rate [15–17].

## Mechanisms of chemoresistance in endometrial cancer

Chemoresistance in endometrial cancers is responsible for up to 90% of all the treatment failures, leading to increased mortality [18]. This is seen commonly in advanced-stage endometrial cancers. Multiple mechanisms were reported in the peer-reviewed literature, including increase in the efflux pumps causing drug ejection and limiting the effect of chemotherapeutic compounds that target mitosis arrest [3]. Cancer genes can also get modified and can influence the efficiency of repair proteins [3, 9]. They can also harbor diverse survival pathways and decrease the level of tumor-suppressor genes [18]. In the following section, we describe some of the most common mechanisms of chemoresistance in endometrial cancers.

## DNA repair mechanisms

### Mismatch repair

Mismatch repair (MMR) is a mechanism by which cells can repair unmatched and/or mismatched DNA fragments which increases cellular proliferation [19]. Platinum compounds act on this pathway leading to apoptosis, thus preventing cellular proliferation. However, MMR deficiency (attributed to loss of MLH1 gene) is proposed as a possible mechanism by which cells become resistant to chemotherapy [19]. On the other hand, other studies reported no association between the presence of MMR deficiency in gynecologic cancers, such as ovarian cancer, and the resistance to platinum compounds [20]. Lack of consistent results can be explained by other possible mechanisms of resistance that might confound results [21]. In endometrial cancer studies, the chemoresistance secondary to the MMR deficiency was seen when the HEC59 cell line was used [22, 23]. Moreover, in endometrial cancer, the MMR deficiency is associated with microsatellite instability (MSI) leading to lower efficacy and resistance of platinum compounds [24, 25].

### BRCA1 *and* BRCA2 *genes (homologous recombination)*

The two genes, *BRCA1* and *BRCA2*, are involved in the repair mechanism of the DNA via homologous recombination of double-stranded breaks in the DNA [26]. The two genes are known for increasing the risk of many cancers, including breast and ovarian cancers, and are among one of the most frequent hereditary mutations associated with familial cancers [27, 28]. The downregulation of *BRCA1* gene is frequently observed in ovarian cancers (about 70% of the times) and in some endometrial cancers also [29]. In endometrial cancer, mutations in *BRCA1* and *BRCA2* genes are associated with the increased risk, and it is particularly seen among women taking tamoxifen (an estrogen modulator) [30]. The role of these two genes has also been studied in the area of chemoresistance [31, 32]. In ovarian cancer, BRCA1 downregulation leads to sensitivity toward platinum compounds and resistance toward taxane agents [33, 34].

## Efflux pumps

One of the reasons of resistance against the drug paclitaxel is through the overexpression of multidrug resistance (MDR) gene *MDR1* [24, 35]. This resistance is explained by the presence of increased number of efflux pump P-glycoprotein (P-gp) [35, 36]. This eliminates the presence of the drug when used in endometrial or ovarian cancers [36]. When it comes to resistance toward the platinum compounds such as cisplatin or carboplatin, an increased number of copper pumps plays an important role [37, 38]. Here, the chemoresistance is acquired via increased protein level of copper-transporting ATPases (ATP 7A and 7B) [38]. In the clinical settings, the presence of ATP7B is considered as a marker of chemoresistance and an overall unfavorable outcome in patients with ovarian and endometrial cancers [39, 40].

## Survival pathways

### PI3K/AKT pathway

The phosphatidylinositol 3-kinase (PI3K)/protein kinase B (AKT).pathway is one of the common mutated or hyperactive pathways in certain gynecologic cancers, such as endometrial and ovarian carcinomas [41]. The pathway is one of the major signaling cascades and shows a high frequency of alteration in both ovarian (>40%) and endometrial (>90%) cancers [41].

The alterations are not only involved in the tumorigenesis but also responsible for chemoresistance [42]. The PI3K pathway lies on the cellular membrane, and under the stimulation of growth hormones, it causes the phosphorylation of phosphatidylinositol 4,5-bisphosphate (PIP2) to phosphatidylinositol 3,4,5-trisphosphate (PIP3) [42, 43]. Once phosphorylated, PIP3 leads to the downstream pathways such as phosphoinositide-dependent kinase (PDK)-1 and AKT kinases which are involved in cellular growth. Mutation of the PI3K pathway or its subunits is a frequent cause of chemoresistance in gynecologic cancers [43, 44].

Chemoresistance against platinum-containing compounds, taxane drugs, and anthracycline such as doxorubicin is explained by the downstreaming of PI3K and AKT isoforms [45, 46]. Among the AKT isoforms, only AKT1 and AKT2 are accountable for resistance against paclitaxel and cisplatin, while all the isoforms of AKT cause resistance against doxorubicin in endometrial cancer. Phosphatase and tensin homolog (PTEN) also plays a role in downregulating the PI3K pathway via dephosphorylating PIP3 to PIP2 [46, 47]. It is a tumor-suppressor phosphatase that causes negative impact on the PI3K pathway leading to drug resistance. In endometrial cancer, PTEN mutations are expressed in the high percentage (up to 65% or more). The mutations of PTEN leading to its downregulation and inactivity further lead to an increase resistance against the platinum compounds [48]. X-linked inhibitor of apoptosis protein (XIAP) is an apoptosis inhibitor involved in the PI3K/AKT pathway and promotes the activity of AKT by increasing its interaction with PI3K [48]. This overall process leads to chemoresistance against cisplatin and doxorubicin in endometrial cancer and against taxane compounds in ovarian cancer. Tumor protein (TP)-53 also plays an active role in chemoresistance by inhibiting the PI3K activity by binding to PIK3CA gene promoters. This eventually leads to the inhibition of the AKT pathway. In order to overcome the chemoresistance caused by p53, wild-type P53 is needed for the sensitization of cancer cells to chemotherapy agents [49].

## MAPK pathway

Microtubule-associated protein kinase (MAPK) is another molecular pathway of significance in gynecologic cancers. It includes the series of kinase pathways leading to cell growth, survival, and apoptosis [50]. The stimuli for these pathways can be genotoxic stress or growth factors [50]. The role of MAPK is important when it comes to RAS oncogene, which is frequently deregulated in certain cancers including endometrial cancer [51]. The MAPK activation requires various cascades including extracellular-signal-regulated kinase (ERK)1/ERK2, c-Jun N-terminal kinase (JNK)/stress-activated protein kinase (SAPK), and p-38 [7, 52]. RAS oncogene activates RAF, which then leads to the activation of ERK1/ERK2 cascade leading to cell division and survival. Certain chemotherapeutic agents also stimulate the ERK1/ERK2 pathway via phosphorylation and play roles in apoptosis and survival of the cells [53]. P-38 and JNK/SAPK are also stimulated by certain chemotherapeutic agents and genotoxic stressors leading to inflammation, cell growth, and apoptosis [54]. The role of MAPK and its interconnection with PI3K has been studied in detail in ovarian cancer where the two pathways influence each other [52, 53]. Studies have shown that the chemoresistance from AKT2 in ovarian tumors occurs when the platinum compounds are not able to activate p38 and JNK pathways to induce apoptosis and cell death [55]. In endometrial cancer, the role of MAPK and chemoresistance has not been studied in detail, but studies hypothesize that the observed effect of MAPK pathways in endometrial cancer would be relatable to ovarian cancer models [55, 56].

### Estrogen receptors in endometrial cancer

The concept of estrogen receptors (ER α/β) overexpression leading to cell proliferation and tumor growth has been well studied in gynecologic cancers such as ovarian and endometrial cancers [57, 58]. Estrogen can follow certain different paths, it can either bind with the estrogen receptors on the plasma membrane or can then interact with other receptors such as insulin-like growth factor (IGF)-1 and ErbB2 [59]. Estrogen can also bind to its receptor, undergo dimerization and translocation followed by binding to ERE (estrogen response element), and act as a transcription factor [60]. Estrogen can also show a direct binding to G protein-coupled receptor 30 (GRP30) receptor, without interacting the estrogen receptors [60]. Studies have shown that these interactions of estrogen can influence the pathways of P3K and MAPK, leading to chemoresistance [61]. In endometrial cancer, estrogen can directly activate GRP78, resulting in chemoresistance against cisplatin and paclitaxel via inhibition of cancer cell apoptosis [3, 62, 63] (Figs. 1 and 2).

### P53

P53 (also known as TP53) is a well-known tumor-suppressor gene involved in DNA repair, cell growth control, and apoptosis [64]. Structurally, P53 is a tetramer and is regulated by mouse double minute 2 homolog (MDM2) [65]. P53 is activated as a result of diverse stimuli including DNA damage, hypoxia, and oncogene activation in gynecological tumors. P53 is frequently found to be mutated, about 25%–85% in the endometrial tumors [3, 66–68].

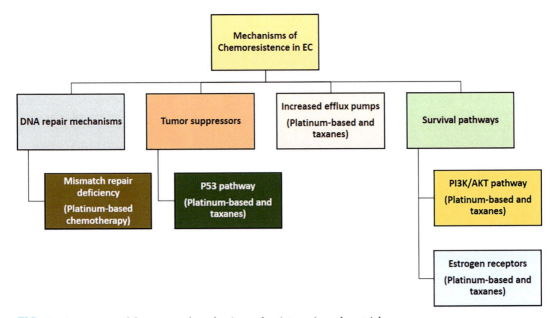

FIG. 1    A summary of the proposed mechanisms of resistance in endometrial cancer.

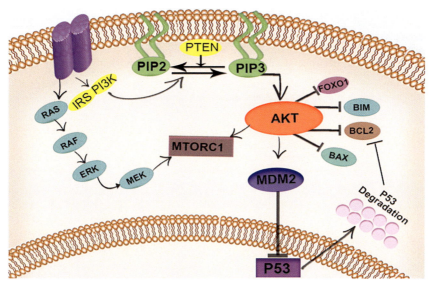

**FIG. 2** Simplified representation of major signaling pathways that can be altered and induce carcinogenesis and resistance in endometrial cancer. Intracellular signaling of the PI3K/AKT pathway starts by the activation of tyrosine kinase receptors leading to phosphorylation of PIP2 to PIP3 through IRS-PI3K, which activates AKT. Functions of AKT include the activation of mTORC1 through the inhibition of TSC2 (not shown in the figure), thus promoting cell proliferation, and also the inhibition of proapoptotic proteins (BCL-2, BAX, and BAM), in addition to P53 degradation. The MAPK pathway is another pathway that leads to increased cellular proliferation via multiple intracellular pathways including mTORC1.

Chemoresistance from P53 is via P3K pathway alterations [69]. One of the mechanisms of P53 inactivation is by an increase expression of mitochondrial B-cell lymphoma 2 (BCL2). BCL2 is an antiapoptotic protein thatis responsible for chemoresistance against several platinum compounds in endometrial and ovarian cancers [70–72] (Table 1).

## Overcoming chemoresistance-targeted therapies

## Targeting the repair mechanisms

### PARP inhibitors

In order to target the chemoresistance in *BRCA1/2* gene-mutated cancers, the role of poly-ADP ribose polymerase (PARP) protein inhibition is important [73]. The idea is to inhibit PARP and decrease the cells' ability to repair the single-strand breaks. The accumulation of single-strand breaks eventually leads to double-strand breaks in the DNA [74]. Normally, the cells are able to repair the breaks; but in the *BRCA* gene-mutated cells, this process further leads to chromosomal instability. The mutation of *BRCA1/2* genes results in the defective homologous recombination. Pairing this homologous recombination with PARP inhibition leads to the destruction of cancer cells [74, 75]. The treatment of PARP inhibitors (though not necessarily to be used only with BRCA1/2 mutations) becomes effective when used in

TABLE 1　Targets and examples of potential drugs to overcome resistance.

| Targets | Examples of potential drugs/therapy | Mechanism(s) of action | Comments |
| --- | --- | --- | --- |
| PARP | Olaparib | Inhibition of poly-ADP ribose polymerase (PARP) | Possible efficacy based on PTEN status |
| AKT | MK-2206 Capivasertib Perifosine | Inhibition of AKT through multiple methods including pleckstrin homology domain targeting (e.g., perifosine), ATP-competitive inhibitors (e.g., capivasertib), and allosteric manner (e.g., MK-2206) | Largely preclinical and early phase trials Possible role in recurrent endometrial cancer |
| mTORC1 | Temsirolimus | Binding to intracellular protein (FKBP-12) creating a complex that inhibits mTOR activity | Multiple phase II trials in chemotherapy-naïve patients with recurrent endometrial cancer showing possible efficacy but increased toxicity |
| EGFR | Erlotinib Cetuximab | Inhibition of epidermal growth factor receptor (EGFR) Potential drugs include chimeric monoclonal antibodies (e.g., cetuximab) and tyrosine kinase inhibitors (e.g., erlotinib) | Possible efficacy in phase II trials for erlotinib Only preclinical evidence for cetuximab |

the BRCA1/2-mutated environment of the cancer cells [76]. Some of the PARP inhibitors (olaparib, niraparib, veliparib, and rucaparib) developed and recently approved are used in gynecologic cancers [73, 75, 77].

The role of PARP inhibitors is well-established in the management of BRCA1/2-mutated ovarian cancers [3, 75]. Many phase II and III trials have shown an improvement in progression-free survival (PFS) with the use of olaparib in ovarian cancer [75]. When used in the relapsed setting regardless of the BRCA status, olaparib has shown decreased progression of the disease when used in the longterm (PFS of 8.4 months compared with 4.8 months of placebo) [75]. The long-term use of other PARP inhibitors in ovarian cancer is under various current clinical trials [78].

The BRCA gene mutations are also seen in endometrial cancers, but the role of PARP inhibition is not well studied in this area [79]. Given the high mutation rates of PTEN in endometrial cancer, olaparib is being studied to target PTEN considering the role of PTEN in maintaining genomic stability and homologous recombination [79–82].

## Targeting PI3K/AKT and MAPK

In preclinical studies, PI3K inhibitors have shown some considerable success rates with mice models and in some phase I trials of ovarian cancers [3]. The results have been promising and prompting the need of more clinical trials [3, 83]. In endometrial cancers, pilaralisib has been used in phase II studies, but the objective response rate (ORR) and complete response

(CR) rate seen were minimal. Also, the molecular association between the PI3K pathways and the presence of PTEN has not been wellstudied [3, 83].

### AKT pathway

As noted earlier, AKT is a downstream pathway of PI3K and is a potential target to overcome chemoresistance in gynecologic cancers [84]. Some of the AKT-specific inhibitors being developed currently in the research include MK-2206, Perifosine, and AZD5363 and are in preclinical to phase II trial phase [85–87]. The AKT-targeting agent, MK-2206, has been studied in the recurrent endometrial cancer settings [88, 89]. The study included 39 patients, and PI3KCA was checked—27 wild-type and 9 true-mutated patients were identified, and MK-2206 was studied in them [3, 86]. The results, however, showed that only one patient from each group showed partial response (about 30% reduction of the tumor) [3]. Similar results have been noticed when AZD5363 was used in a phase I trial and only two patients achieved a partial response [87]. The results thus are limiting the current use of AKT-inhibiting agent for the single or combination use and necessitating the need of more clinical trials in this exciting area of clinical research [87].

The molecular network of PI3 kinase is vast, and more drugs targeting additional keyplayers in this area are needed [90]. Some studies have suggested the role of *KRAS* mutation in indicating the resistance to PI3K inhibitors, providing the area for more translational research [90]. PI3K mutation and inhibition has also been associated with homologous recombination, and the novel idea of adding PARP inhibition can provide better pathways for the development of more targeting agents [90].

### Targeting mTOR

Mammalian target of rapamycin (mTOR) is associated with two major complexes including mTORC1 and MTORC2 [91]. mTOR is a complex and is associated with cell growth and survival. mTOR is a downstream pathway in the cascade of PI3K/AKT pathways [91]. It is found to be frequently mutated in patients with endometrial cancer and can be a potential targeting pathway in chemoresistance [92]. Temsirolimus (inhibitor of only mTORC1) has shown modest activity in chemo-resistant ovarian cancers. In one phase II trial, temsirolimus was used in 54 women with chemo-resistant ovarian cancers, and 9 women showed partial response [3]. In endometrial cancer, a phase II trial studied temsirolimus with hormonal therapy with the idea that the mTOR pathway may be involved in hormonal therapy resistance [93]. Fifty patients received temsirolimus in the study, and 29 received prior chemotherapy. The remaining patients did not receive any prior chemotherapy. The overall response rate was not much different between the two groups (24% ORR in chemotherapy arm and 19% in without chemotherapy arm) [3, 93]. The study also highlighted that the combination of megestrol acetate with temsirolimus was associated with increased toxicity and did not show much efficacy benefit [93]. Similar results have been noticed in another phase II trial with temsirolimus, which is another mTOR inhibitor [94]. Thus, given the modest results of the earlier-discussed trials and the need for improvement in targeted therapy, the overall pathway of mTOR inhibition along with targeting PTEN mutations in endometrial cancer requires further investigations [3, 94–96].

## Additional concepts, targets, and mechanisms

### MicroRNA-135a

Recently, the role of microRNAs (miRNA) has been studied by some investigators, highlighting its role in tumorigenesis and progression in various cancers including endometrial cancers [97]. miRNA is currently under investigations regarding its ability to promote cell proliferation, migration, and invasion of endometrial cells [97]. A study was performed on the endometrial cancer cell lines with an altered expression of miR-135a [98]. The study found that upregulation of microRNA leads to proliferation, invasion, and migration of endometrial cancer cells [99]. Apoptosis induced by cisplatin was also inhibited by miRNA via regulation of BCL2 and BCL2-associated X protein (BAX) expression [99]. Given its crucial role in tumorigenesis and chemoresistance, the study indicated that miRNA can act as a potential targeting biomarker to predict the chemotherapy response and prognosis in endometrial cancer [100, 101].

### The role of leptin

Another novel concept in the study of chemoresistance in uterine/endometrial cancer highlights the role of leptin [102]. Obesity is a well-established risk factor in endometrial cancer. Leptin is a hormone with increased levels found in obese individuals [102]. The role of leptin-induced signaling crosstalk (e.g., Notch, interleukin-1, and leptin outcome (NILCO)) has been wellstudied in breast cancer and has been found to have a role in the progression of the disease [103]. Recently published experimental data on leptin have indicated potential role(s) of leptin in inducing Notch in endometrial cancer [102, 104]. The study suggested that leptin increases the expression of NILCO and promotes cellular proliferation and disease progression [105]. Increased expression of NILCO was found to reduce the cytotoxic effects of paclitaxel on endometrial cancer cells, thus linking the progression of endometrial cancer and chemoresistance in obese patients [102, 106]. Newer research is underway developing the antagonist of leptin signaling pathway called IONP-LPrA2 (leptin antagonist linked to iron oxide nanoparticles), which is showing promising results in sensitizing endometrial cancer cells of paclitaxel, thus emerging as a promising new drug [107–109].

### Targeting aurora-A

Another novel concept under study is about the role of Aurora-A, which is a serine-threonine kinase [110]. It is found to be overexpressed on the endometrial cancer cells and has also been linked with an overall reduced survival [110, 111]. Recent study from China has focused on the role of Aurora-A overexpression in endometrial cancer cell lines (viz., ISHIKAWA and HEC-1B cells) [111]. The study found that the presence of Aurora-A kinase promoted cellular proliferation and induced chemoresistance of cisplatin and paclitaxel agents [111]. As noted earlier, recent peer-reviewed literature is highlighting the activation of the AKT-mTOR pathway by Aurora-A, further increasing the chemoresistance [112]. The sensitivity of endometrial cancer cells to the AKT/mTOR pathway is controlled through

Aurora-A, and targeted therapy against the overexpression of Aurora-A can potentiate the overcome on chemoresistance [111–113].

In summary, the overall complexity of endometrial cancer and the involvement of several molecular pathways leading to resistance to chemotherapeutic agents currently require more thorough research to develop multitargeted therapies.

# References

[1] Ferlay J, Steliarova-Foucher E, Lortet-Tieulent J, Rosso S, Coebergh JW, Comber H, et al. Cancer incidence and mortality patterns in Europe: estimates for 40 countries in 2012. Eur J Cancer 2013;49(6):1374–403.

[2] Siegel R, Naishadham D, Jemal A. Cancer statistics, 2013. CA Cancer J Clin 2013;63(1):11–30.

[3] Brasseur K, Gevry N, Asselin E. Chemoresistance and targeted therapies in ovarian and endometrial cancers. Oncotarget 2017;8(3):4008–42.

[4] Plataniotis G, Castiglione M, Group EGW. Endometrial cancer: ESMO clinical practice guidelines for diagnosis, treatment and follow-up. Ann Oncol 2010;21(Suppl. 5):v41–5.

[5] Van Nyen T, Moiola CP, Colas E, Annibali D, Amant F. Modeling endometrial cancer: past, present, and future. Int J Mol Sci 2018;19(8).

[6] Felix AS, Weissfeld JL, Stone RA, Bowser R, Chivukula M, Edwards RP, et al. Factors associated with type I and type II endometrial cancer. Cancer Causes Control 2010;21(11):1851–6.

[7] Patch AM, Christie EL, Etemadmoghadam D, Garsed DW, George J, Fereday S, et al. Whole-genome characterization of chemoresistant ovarian cancer. Nature 2015;521(7553):489–94.

[8] Bast Jr RC, Hennessy B, Mills GB. The biology of ovarian cancer: new opportunities for translation. Nat Rev Cancer 2009;9(6):415–28.

[9] Agarwal R, Kaye SB. Ovarian cancer: strategies for overcoming resistance to chemotherapy. Nat Rev Cancer 2003;3(7):502–16.

[10] Koh WJ, Abu-Rustum NR, Bean S, Bradley K, Campos SM, Cho KR, et al. Uterine neoplasms, version 1.2018, NCCN clinical practice guidelines in oncology. J Natl Compr Canc Netw 2018;16(2):170–99.

[11] Sjoquist KM, Martyn J, Edmondson RJ, Friedlander ML. The role of hormonal therapy in gynecological cancers-current status and future directions. Int J Gynecol Cancer 2011;21(7):1328–33.

[12] Florea AM, Busselberg D. Cisplatin as an anti-tumor drug: cellular mechanisms of activity, drug resistance and induced side effects. Cancers (Basel) 2011;3(1):1351–71.

[13] Orr GA, Verdier-Pinard P, McDaid H, Horwitz SB. Mechanisms of taxol resistance related to microtubules. Oncogene 2003;22(47):7280–95.

[14] Jordan MA, Wilson L. Microtubules as a target for anticancer drugs. Nat Rev Cancer 2004;4(4):253–65.

[15] Thorn CF, Oshiro C, Marsh S, Hernandez-Boussard T, McLeod H, Klein TE, et al. Doxorubicin pathways: pharmacodynamics and adverse effects. Pharmacogenet Genomics 2011;21(7):440–6.

[16] Boere IA, van der Burg ME. Review of dose-intense platinum and/or paclitaxel containing chemotherapy in advanced and recurrent epithelial ovarian cancer. Curr Pharm Des 2012;18(25):3741–53.

[17] Yap TA, Carden CP, Kaye SB. Beyond chemotherapy: targeted therapies in ovarian cancer. Nat Rev Cancer 2009;9(3):167–81.

[18] Longley DB, Johnston PG. Molecular mechanisms of drug resistance. J Pathol 2005;205(2):275–92.

[19] Kunkel TA, Erie DA. DNA mismatch repair. Annu Rev Biochem 2005;74:681–710.

[20] Helleman J, van Staveren IL, Dinjens WN, van Kuijk PF, Ritstier K, Ewing PC, et al. Mismatch repair and treatment resistance in ovarian cancer. BMC Cancer 2006;6:201.

[21] Mesquita B, Veiga I, Pereira D, Tavares A, Pinto IM, Pinto C, et al. No significant role for beta tubulin mutations and mismatch repair defects in ovarian cancer resistance to paclitaxel/cisplatin. BMC Cancer 2005;5:101.

[22] Fink D, Nebel S, Aebi S, Zheng H, Cenni B, Nehme A, et al. The role of DNA mismatch repair in platinum drug resistance. Cancer Res 1996;56(21):4881–6.

[23] Lin X, Howell SB. Effect of loss of DNA mismatch repair on development of topotecan-, gemcitabine-, and paclitaxel-resistant variants after exposure to cisplatin. Mol Pharmacol 1999;56(2):390–5.

[24] Esteller M, Catasus L, Matias-Guiu X, Mutter GL, Prat J, Baylin SB, et al. hMLH1 promoter hypermethylation is an early event in human endometrial tumorigenesis. Am J Pathol 1999;155(5):1767–72.

[25] Kanaya T, Kyo S, Maida Y, Yatabe N, Tanaka M, Nakamura M, et al. Frequent hypermethylation of MLH1 promoter in normal endometrium of patients with endometrial cancers. Oncogene 2003;22(15):2352–60.

[26] Welcsh PL, King MC. BRCA1 and BRCA2 and the genetics of breast and ovarian cancer. Hum Mol Genet 2001;10 (7):705–13.

[27] Miki Y, Swensen J, Shattuck-Eidens D, Futreal PA, Harshman K, Tavtigian S, et al. A strong candidate for the breast and ovarian cancer susceptibility gene BRCA1. Science 1994;266(5182):66–71.

[28] Brose MS, Rebbeck TR, Calzone KA, Stopfer JE, Nathanson KL, Weber BL. Cancer risk estimates for BRCA1 mutation carriers identified in a risk evaluation program. J Natl Cancer Inst 2002;94(18):1365–72.

[29] Pal T, Permuth-Wey J, Betts JA, Krischer JP, Fiorica J, Arango H, et al. BRCA1 and BRCA2 mutations account for a large proportion of ovarian carcinoma cases. Cancer 2005;104(12):2807–16.

[30] King MC, Marks JH, Mandell JB, New York Breast Cancer Study G. Breast and ovarian cancer risks due to inherited mutations in BRCA1 and BRCA2. Science 2003;302(5645):643–6.

[31] Segev Y, Iqbal J, Lubinski J, Gronwald J, Lynch HT, Moller P, et al. The incidence of endometrial cancer in women with BRCA1 and BRCA2 mutations: an international prospective cohort study. Gynecol Oncol 2013;130 (1):127–31.

[32] Beiner ME, Finch A, Rosen B, Lubinski J, Moller P, Ghadirian P, et al. The risk of endometrial cancer in women with BRCA1 and BRCA2 mutations. A prospective study. Gynecol Oncol 2007;104(1):7–10.

[33] Quinn JE, James CR, Stewart GE, Mulligan JM, White P, Chang GK, et al. BRCA1 mRNA expression levels predict for overall survival in ovarian cancer after chemotherapy. Clin Cancer Res 2007;13(24):7413–20.

[34] Husain A, He G, Venkatraman ES, Spriggs DR. BRCA1 up-regulation is associated with repair-mediated resistance to cis-diamminedichloroplatinum(II). Cancer Res 1998;58(6):1120–3.

[35] Schneider J, Efferth T, Centeno MM, Mattern J, Rodriguez-Escudero FJ, Volm M. High rate of expression of multidrug resistance-associated P-glycoprotein in human endometrial carcinoma and normal endometrial tissue. Eur J Cancer 1993;29A(4):554–8.

[36] Holzmayer TA, Hilsenbeck S, Von Hoff DD, Roninson IB. Clinical correlates of MDR1 (P-glycoprotein) gene expression in ovarian and small-cell lung carcinomas. J Natl Cancer Inst 1992;84(19):1486–91.

[37] Katano K, Kondo A, Safaei R, Holzer A, Samimi G, Mishima M, et al. Acquisition of resistance to cisplatin is accompanied by changes in the cellular pharmacology of copper. Cancer Res 2002;62(22):6559–65.

[38] Nakayama K, Kanzaki A, Ogawa K, Miyazaki K, Neamati N, Takebayashi Y. Copper-transporting P-type adenosine triphosphatase (ATP7B) as a cisplatin based chemoresistance marker in ovarian carcinoma: comparative analysis with expression of MDR1, MRP1, MRP2, LRP and BCRP. Int J Cancer 2002;101(5):488–95.

[39] Samimi G, Safaei R, Katano K, Holzer AK, Rochdi M, Tomioka M, et al. Increased expression of the copper efflux transporter ATP7A mediates resistance to cisplatin, carboplatin, and oxaliplatin in ovarian cancer cells. Clin Cancer Res 2004;10(14):4661–9.

[40] Komatsu M, Sumizawa T, Mutoh M, Chen ZS, Terada K, Furukawa T, et al. Copper-transporting P-type adenosine triphosphatase (ATP7B) is associated with cisplatin resistance. Cancer Res 2000;60(5):1312–6.

[41] Oda K, Stokoe D, Taketani Y, McCormick F. High frequency of coexistent mutations of PIK3CA and PTEN genes in endometrial carcinoma. Cancer Res 2005;65(23):10669–73.

[42] Cerami E, Gao J, Dogrusoz U, Gross BE, Sumer SO, Aksoy BA, et al. The cBio cancer genomics portal: an open platform for exploring multidimensional cancer genomics data. Cancer Discov 2012;2(5):401–4.

[43] Fabi F, Asselin E. Expression, activation, and role of AKT isoforms in the uterus. Reproduction 2014;148(5): R85–95.

[44] Lee S, Choi EJ, Jin C, Kim DH. Activation of PI3K/Akt pathway by PTEN reduction and PIK3CA mRNA amplification contributes to cisplatin resistance in an ovarian cancer cell line. Gynecol Oncol 2005;97(1):26–34.

[45] Mitsuuchi Y, Johnson SW, Selvakumaran M, Williams SJ, Hamilton TC, Testa JR. The phosphatidylinositol 3-kinase/AKT signal transduction pathway plays a critical role in the expression of p21WAF1/CIP1/SDI1 induced by cisplatin and paclitaxel. Cancer Res 2000;60(19):5390–4.

[46] Gagnon V, Van Themsche C, Turner S, Leblanc V, Asselin E. Akt and XIAP regulate the sensitivity of human uterine cancer cells to cisplatin, doxorubicin and taxol. Apoptosis 2008;13(2):259–71.

[47] Fraser M, Leung BM, Yan X, Dan HC, Cheng JQ, Tsang BK. p53 is a determinant of X-linked inhibitor of apoptosis protein/Akt-mediated chemoresistance in human ovarian cancer cells. Cancer Res 2003;63(21):7081–8.

[48] Asselin E, Mills GB, Tsang BK. XIAP regulates Akt activity and caspase-3-dependent cleavage during cisplatin-induced apoptosis in human ovarian epithelial cancer cells. Cancer Res 2001;61(5):1862–8.

[49] Yang X, Fraser M, Moll UM, Basak A, Tsang BK. Akt-mediated cisplatin resistance in ovarian cancer: modulation of p53 action on caspase-dependent mitochondrial death pathway. Cancer Res 2006;66(6):3126–36.

[50] Roberts PJ, Der CJ. Targeting the Raf-MEK-ERK mitogen-activated protein kinase cascade for the treatment of cancer. Oncogene 2007;26(22):3291–310.

[51] Pearson G, Robinson F, Beers Gibson T, Xu BE, Karandikar M, Berman K, et al. Mitogen-activated protein (MAP) kinase pathways: regulation and physiological functions. Endocr Rev 2001;22(2):153–83.

[52] Cagnol S, Chambard JC. ERK and cell death: mechanisms of ERK-induced cell death—apoptosis, autophagy and senescence. FEBS J 2010;277(1):2–21.

[53] Mansouri A, Ridgway LD, Korapati AL, Zhang Q, Tian L, Wang Y, et al. Sustained activation of JNK/p38 MAPK pathways in response to cisplatin leads to Fas ligand induction and cell death in ovarian carcinoma cells. J Biol Chem 2003;278(21):19245–56.

[54] Mansouri A, Zhang Q, Ridgway LD, Tian L, Claret FX. Cisplatin resistance in an ovarian carcinoma is associated with a defect in programmed cell death control through XIAP regulation. Oncol Res 2003;13(6–10):399–404.

[55] Villedieu M, Deslandes E, Duval M, Heron JF, Gauduchon P, Poulain L. Acquisition of chemoresistance following discontinuous exposures to cisplatin is associated in ovarian carcinoma cells with progressive alteration of FAK, ERK and p38 activation in response to treatment. Gynecol Oncol 2006;101(3):507–19.

[56] Chan DW, Liu VW, Tsao GS, Yao KM, Furukawa T, Chan KK, et al. Loss of MKP3 mediated by oxidative stress enhances tumorigenicity and chemoresistance of ovarian cancer cells. Carcinogenesis 2008;29(9):1742–50.

[57] Pearce ST, Jordan VC. The biological role of estrogen receptors alpha and beta in cancer. Crit Rev Oncol Hematol 2004;50(1):3–22.

[58] Wray S, Noble K. Sex hormones and excitation-contraction coupling in the uterus: the effects of oestrous and hormones. J Neuroendocrinol 2008;20(4):451–61.

[59] Dauvois S, White R, Parker MG. The antiestrogen ICI 182780 disrupts estrogen receptor nucleocytoplasmic shuttling. J Cell Sci 1993;106(Pt. 4):1377–88.

[60] Bjornstrom L, Sjoberg M. Mechanisms of estrogen receptor signaling: convergence of genomic and nongenomic actions on target genes. Mol Endocrinol 2005;19(4):833–42.

[61] Won YS, Lee SJ, Yeo SG, Park DC. Effects of female sex hormones on clusterin expression and paclitaxel resistance in endometrial cancer cell lines. Int J Med Sci 2012;9(1):86–92.

[62] Luvsandagva B, Nakamura K, Kitahara Y, Aoki H, Murata T, Ikeda S, et al. GRP78 induced by estrogen plays a role in the chemosensitivity of endometrial cancer. Gynecol Oncol 2012;126(1):132–9.

[63] Mabuchi S, Ohmichi M, Kimura A, Nishio Y, Arimoto-Ishida E, Yada-Hashimoto N, et al. Estrogen inhibits paclitaxel-induced apoptosis via the phosphorylation of apoptosis signal-regulating kinase 1 in human ovarian cancer cell lines. Endocrinology 2004;145(1):49–58.

[64] Bieging KT, Mello SS, Attardi LD. Unravelling mechanisms of p53-mediated tumour suppression. Nat Rev Cancer 2014;14(5):359–70.

[65] Lane DP. Cancer. p53, guardian of the genome. Nature 1992;358(6381):15–6.

[66] Garg K, Leitao Jr MM, Wynveen CA, Sica GL, Shia J, Shi W, et al. p53 overexpression in morphologically ambiguous endometrial carcinomas correlates with adverse clinical outcomes. Mod Pathol 2010;23(1):80–92.

[67] Leung EL, Fraser M, Fiscus RR, Tsang BK. Cisplatin alters nitric oxide synthase levels in human ovarian cancer cells: involvement in p53 regulation and cisplatin resistance. Br J Cancer 2008;98(11):1803–9.

[68] Sato S, Kigawa J, Minagawa Y, Okada M, Shimada M, Takahashi M, et al. Chemosensitivity and p53-dependent apoptosis in epithelial ovarian carcinoma. Cancer 1999;86(7):1307–13.

[69] Jones NA, Turner J, McIlwrath AJ, Brown R, Dive C. Cisplatin- and paclitaxel-induced apoptosis of ovarian carcinoma cells and the relationship between bax and bak up-regulation and the functional status of p53. Mol Pharmacol 1998;53(5):819–26.

[70] Vandenput I, Capoen A, Coenegrachts L, Verbist G, Moerman P, Vergote I, et al. Expression of ERCC1, p53, and class III beta-tubulin do not reveal chemoresistance in endometrial cancer: results from an immunohistochemical study. Int J Gynecol Cancer 2011;21(6):1071–7.

[71] Sultana H, Kigawa J, Kanamori Y, Itamochi H, Oishi T, Sato S, et al. Chemosensitivity and p53-Bax pathway-mediated apoptosis in patients with uterine cervical cancer. Ann Oncol 2003;14(2):214–9.

[72] Rouette A, Parent S, Girouard J, Leblanc V, Asselin E. Cisplatin increases B-cell-lymphoma-2 expression via activation of protein kinase C and Akt2 in endometrial cancer cells. Int J Cancer 2012;130(8):1755–67.

[73] Tutt AN, Lord CJ, McCabe N, Farmer H, Turner N, Martin NM, et al. Exploiting the DNA repair defect in BRCA mutant cells in the design of new therapeutic strategies for cancer. Cold Spring Harb Symp Quant Biol 2005;70:139–48.

[74] Audeh MW, Carmichael J, Penson RT, Friedlander M, Powell B, Bell-McGuinn KM, et al. Oral poly(ADP-ribose) polymerase inhibitor olaparib in patients with BRCA1 or BRCA2 mutations and recurrent ovarian cancer: a proof-of-concept trial. Lancet 2010;376(9737):245–51.

[75] Ledermann J, Harter P, Gourley C, Friedlander M, Vergote I, Rustin G, et al. Olaparib maintenance therapy in platinum-sensitive relapsed ovarian cancer. N Engl J Med 2012;366(15):1382–92.

[76] Miyasaka A, Oda K, Ikeda Y, Wada-Hiraike O, Kashiyama T, Enomoto A, et al. Anti-tumor activity of olaparib, a poly (ADP-ribose) polymerase (PARP) inhibitor, in cultured endometrial carcinoma cells. BMC Cancer 2014;14:179.

[77] Mendes-Pereira AM, Martin SA, Brough R, McCarthy A, Taylor JR, Kim JS, et al. Synthetic lethal targeting of PTEN mutant cells with PARP inhibitors. EMBO Mol Med 2009;1(6–7):315–22.

[78] Gelmon KA, Tischkowitz M, Mackay H, Swenerton K, Robidoux A, Tonkin K, et al. Olaparib in patients with recurrent high-grade serous or poorly differentiated ovarian carcinoma or triple-negative breast cancer: a phase 2, multicentre, open-label, non-randomised study. Lancet Oncol 2011;12(9):852–61.

[79] Kaufman B, Shapira-Frommer R, Schmutzler RK, Audeh MW, Friedlander M, Balmana J, et al. Olaparib monotherapy in patients with advanced cancer and a germline BRCA1/2 mutation. J Clin Oncol 2015;33 (3):244–50.

[80] Lee JM, Hays JL, Annunziata CM, Noonan AM, Minasian L, Zujewski JA, et al. Phase I/Ib study of olaparib and carboplatin in BRCA1 or BRCA2 mutation-associated breast or ovarian cancer with biomarker analyses. J Natl Cancer Inst 2014;106(6), dju089.

[81] Yin Y, Shen WH. PTEN: a new guardian of the genome. Oncogene 2008;27(41):5443–53.

[82] Forster MD, Dedes KJ, Sandhu S, Frentzas S, Kristeleit R, Ashworth A, et al. Treatment with olaparib in a patient with PTEN-deficient endometrioid endometrial cancer. Nat Rev Clin Oncol 2011;8(5):302–6.

[83] Ohta T, Ohmichi M, Hayasaka T, Mabuchi S, Saitoh M, Kawagoe J, et al. Inhibition of phosphatidylinositol 3-kinase increases efficacy of cisplatin in in vivo ovarian cancer models. Endocrinology 2006;147(4):1761–9.

[84] Juvekar A, Burga LN, Hu H, Lunsford EP, Ibrahim YH, Balmana J, et al. Combining a PI3K inhibitor with a PARP inhibitor provides an effective therapy for BRCA1-related breast cancer. Cancer Discov 2012;2(11):1048–63.

[85] Matulonis U, Vergote I, Backes F, Martin LP, McMeekin S, Birrer M, et al. Phase II study of the PI3K inhibitor pilaralisib (SAR245408; XL147) in patients with advanced or recurrent endometrial carcinoma. Gynecol Oncol 2015;136(2):246–53.

[86] Lin YH, Chen BY, Lai WT, Wu SF, Guh JH, Cheng AL, et al. The Akt inhibitor MK-2206 enhances the cytotoxicity of paclitaxel (Taxol) and cisplatin in ovarian cancer cells. Naunyn Schmiedebergs Arch Pharmacol 2015;388 (1):19–31.

[87] Fu S, Hennessy BT, Ng CS, Ju Z, Coombes KR, Wolf JK, et al. Perifosine plus docetaxel in patients with platinum and taxane resistant or refractory high-grade epithelial ovarian cancer. Gynecol Oncol 2012;126(1):47–53.

[88] Pant A, Lee II, Lu Z, Rueda BR, Schink J, Kim JJ. Inhibition of AKT with the orally active allosteric AKT inhibitor, MK-2206, sensitizes endometrial cancer cells to progestin. PLoS ONE 2012;7(7), e41593.

[89] Engel JB, Honig A, Schonhals T, Weidler C, Hausler S, Krockenberger M, et al. Perifosine inhibits growth of human experimental endometrial cancers by blockade of AKT phosphorylation. Eur J Obstet Gynecol Reprod Biol 2008;141(1):64–9.

[90] Ibrahim YH, Garcia-Garcia C, Serra V, He L, Torres-Lockhart K, Prat A, et al. PI3K inhibition impairs BRCA1/2 expression and sensitizes BRCA-proficient triple-negative breast cancer to PARP inhibition. Cancer Discov 2012;2(11):1036–47.

[91] Wullschleger S, Loewith R, Hall MN. TOR signaling in growth and metabolism. Cell 2006;124(3):471–84.

[92] Behbakht K, Sill MW, Darcy KM, Rubin SC, Mannel RS, Waggoner S, et al. Phase II trial of the mTOR inhibitor, temsirolimus and evaluation of circulating tumor cells and tumor biomarkers in persistent and recurrent epithelial ovarian and primary peritoneal malignancies: a Gynecologic Oncology Group study. Gynecol Oncol 2011;123(1):19–26.

[93] Fleming GF, Filiaci VL, Marzullo B, Zaino RJ, Davidson SA, Pearl M, et al. Temsirolimus with or without megestrol acetate and tamoxifen for endometrial cancer: a gynecologic oncology group study. Gynecol Oncol 2014;132(3):585–92.

[94] Oza AM, Elit L, Tsao MS, Kamel-Reid S, Biagi J, Provencher DM, et al. Phase II study of temsirolimus in women with recurrent or metastatic endometrial cancer: a trial of the NCIC Clinical Trials Group. J Clin Oncol 2011;29 (24):3278–85.

[95] Mabuchi S, Altomare DA, Connolly DC, Klein-Szanto A, Litwin S, Hoelzle MK, et al. RAD001 (Everolimus) delays tumor onset and progression in a transgenic mouse model of ovarian cancer. Cancer Res 2007;67(6):2408–13.

[96] Slomovitz BM, Lu KH, Johnston T, Coleman RL, Munsell M, Broaddus RR, et al. A phase 2 study of the oral mammalian target of rapamycin inhibitor, everolimus, in patients with recurrent endometrial carcinoma. Cancer 2010;116(23):5415–9.

[97] Jurcevic S, Olsson B, Klinga-Levan K. MicroRNA expression in human endometrial adenocarcinoma. Cancer Cell Int 2014;14(1):88.

[98] Ludwig N, Leidinger P, Becker K, Backes C, Fehlmann T, Pallasch C, et al. Distribution of miRNA expression across human tissues. Nucleic Acids Res 2016;44(8):3865–77.

[99] Yanokura M, Banno K, Kobayashi Y, Kisu I, Ueki A, Ono A, et al. MicroRNA and endometrial cancer: roles of small RNAs in human tumors and clinical applications (review). Oncol Lett 2010;1(6):935–40.

[100] Baranwal S, Alahari SK. miRNA control of tumor cell invasion and metastasis. Int J Cancer 2010;126(6):1283–90.

[101] Gong B, Yue Y, Wang R, Zhang Y, Jin Q, Zhou X. Overexpression of microRNA-194 suppresses the epithelial-mesenchymal transition in targeting stem cell transcription factor Sox3 in endometrial carcinoma stem cells. Tumour Biol 2017;39(6). https://doi.org/10.1177/1010428317706217.

[102] Daley-Brown D, Oprea-Iles G, Vann KT, Lanier V, Lee R, Candelaria PV, et al. Type II endometrial cancer overexpresses NILCO: a preliminary evaluation. Dis Markers 2017;2017:8248175.

[103] Lipsey CC, Harbuzariu A, Daley-Brown D, Gonzalez-Perez RR. Oncogenic role of leptin and Notch interleukin-1 leptin crosstalk outcome in cancer. World J Methodol 2016;6(1):43–55.

[104] Abdullah LN, Chow EK. Mechanisms of chemoresistance in cancer stem cells. Clin Transl Med 2013;2(1):3.

[105] Gonzalez RR, Leary K, Petrozza JC, Leavis PC. Leptin regulation of the interleukin-1 system in human endometrial cells. Mol Hum Reprod 2003;9(3):151–8.

[106] Jonusiene V, Sasnauskiene A, Lachej N, Kanopiene D, Dabkeviciene D, Sasnauskiene S, et al. Down-regulated expression of Notch signaling molecules in human endometrial cancer. Med Oncol 2013;30(1):438.

[107] Daley-Brown D, Harbuzariu A, Kurian AA, Oprea-Ilies G, Gonzalez-Perez RR. Leptin-induced Notch and IL-1 signaling crosstalk in endometrial adenocarcinoma is associated with invasiveness and chemoresistance. World J Clin Oncol 2019;10(6):222–33.

[108] Cobellis L, Caprio F, Trabucco E, Mastrogiacomo A, Coppola G, Manente L, et al. The pattern of expression of Notch protein members in normal and pathological endometrium. J Anat 2008;213(4):464–72.

[109] Mitsuhashi Y, Horiuchi A, Miyamoto T, Kashima H, Suzuki A, Shiozawa T. Prognostic significance of Notch signalling molecules and their involvement in the invasiveness of endometrial carcinoma cells. Histopathology 2012;60(5):826–37.

[110] Damodaran AP, Vaufrey L, Gavard O, Prigent C. Aurora a kinase is a priority pharmaceutical target for the treatment of cancers. Trends Pharmacol Sci 2017;38(8):687–700.

[111] Wu J, Cheng Z, Xu X, Fu J, Wang K, Liu T, et al. Aurora-a induces chemoresistance through activation of the AKT/mTOR pathway in endometrial cancer. Front Oncol 2019;9:422.

[112] Umene K, Yanokura M, Banno K, Irie H, Adachi M, Iida M, et al. Aurora kinase A has a significant role as a therapeutic target and clinical biomarker in endometrial cancer. Int J Oncol 2015;46(4):1498–506.

[113] Yang H, He L, Kruk P, Nicosia SV, Cheng JQ. Aurora-A induces cell survival and chemoresistance by activation of Akt through a p53-dependent manner in ovarian cancer cells. Int J Cancer 2006;119(10):2304–12.

# Ovarian cancer: Targeted therapies and mechanisms of resistance

*Deepika Sarvepalli[a], Mamoon Ur Rashid[b], Hammad Zafar[c], Sundas Jehanzeb[d], Effa Zahid[e], and Sarfraz Ahmad[f]*

[a]Guntur Medical College, Guntur, Andhra Pradesh, India [b]Department of Gastroenterology, Cleveland Clinic, Weston, FL, United States [c]University of Iowa Hospitals & Clinics, Iowa City, IA, United States [d]Khyber Medical College, Peshawar, Pakistan [e]Services Institute of Medical Sciences, Lahore, Pakistan [f]Gynecologic Oncology, AdventHealth Cancer Institute, Florida State University, University of Central Florida, Orlando, FL, United States

## Abstract

Ovarian cancer is one of the leading causes of cancer-related mortality in women. The treatment of ovarian cancer consists of surgical removal of vast majority of the tumor mass followed by chemotherapy, hormone therapy, and/or radiotherapy of the remaining mass and metastases. A variety of chemotherapeutic drugs have been used in gynecologic cancers either alone or in combination therapy such as cisplatin, carboplatin, paclitaxel, docetaxel, doxorubicin, cyclophosphamide, gemcitabine, topotecan, and vinorelbine. Even though the combination therapy has a 70% response rate, tumor relapse and subsequent chemoresistance are frequently observed, eventually resulting in treatment failure and mortality in majority of the patients. Chemoresistance can be intrinsic or acquired through various cellular modifications, and drugs targeting these systems can overcome the chemoresistance and hopefully extend patient survival. Targeted therapy is a relatively new treatment approach where the drugs attack the oncogenic pathways in the malignant cells. Targeted therapy is designed to selectively target the molecules and mechanisms that cause chemoresistance and eliminate them so that chemotherapeutic drugs can work to their full potential of antitumor activity. Since these pathways are associated with chemoresistance, targeting them could help sensitize the malignant cells to standard chemotherapy drugs. Some of the drugs include efflux pump inhibitor, tariquidar, tyrosine kinase inhibitors such as erlotinib and gefitinib, and PARP inhibitors such as olaparib, veliparib, and p53 inhibitors. This research chapter discusses in detail about the current targets and related drugs for better management of patients with ovarian cancer and improved quality of life.

## Abbreviations

| | |
|---|---|
| AI | aromatase inhibitor |
| AKT | protein kinase B |

R. Basha, S. Ahmad (eds.)
*Overcoming Drug Resistance in Gynecologic Cancers*
ISBN 978-0-12-824299-5
https://doi.org/10.1016/B978-0-12-824299-5.00014-9

ATP        adenosine triphosphate
BCL-2      B-cell lymphoma 2
BRCA       breast cancer gene
CBR        clinical benefit rate
CR         complete response
EGFR       epithelial growth factor receptor
EMT        epithelial-mesenchymal transition
EOC        epithelial ovarian cancer
ER         estrogen receptor
ErbB2      erythroblastic oncogene B
ERK        extracellular signal-regulated kinase
HGSOC      high-grade serous ovarian cancer
IGF        insulin growth factor
JNK        c-Jun N-terminal kinase
KRAS       Kirsten rat sarcoma 2 viral oncogene homolog
MAPK       mitogen-activated protein kinase
MDR        multidrug resistance
MEK        MAPK enzymes kinase
MMR        mismatch repair
mTOR       mammalian target of rapamycin
Mut        mutation
NER        nucleotide excision repair
OC         ovarian cancer
ORR        objective response rate
OS         overall survival
PAR        prostate apoptosis response
PARP       poly ADP ribose polymerase
PDK1       3-phosphoinositide-dependent protein kinase-1
PFS        progression-free survival
P-gp       P-glycoprotein
PI3K       phosphatidylinositol 3-kinase
PIP        prolactin-induced protein
PR         partial response
PTEN       phosphatase and tensin homolog
RTK        receptor tyrosine kinase
SD         stable disease
TK         tyrosine kinase
TKI        tyrosine kinase inhibitor
VEGFR      vascular endothelial growth factor receptor
WT         wild type
XIAP       X-linked inhibitor of apoptosis protein

## Conflict of interest

No potential conflicts of interest were disclosed.

# Introduction

Ovarian cancer, the top five leading cause of cancer-related mortality among women, is a fatal gynecologic cancer with a 5-year survival of about 46% [1–6]. With a current incidence rate of 10.3 per 100,000 women and a mortality rate of 6.8 per 100,000 women, over 21,750 women are estimated to be newly diagnosed, and 13,940 will be deceased in the United States alone in 2020, a number that has only changed slightly after over 30 years of research [5]. There are various histologic types of ovarian cancer, of which the major variant is the one with epithelial origin. Cancer chemoresistance, intrinsic or acquired, is one of the prominent causes of mortality in ovarian cancer patients. The disease is hard to detect in the early stages because of lack of symptoms, and most cases (75%) are presented to the clinic only at late stages in the course of the disease progression [4].

The treatment of ovarian cancer consists of surgical removal of vast majority of the tumor mass followed by chemotherapy, hormone therapy, and/or radiotherapy of the remaining mass and metastases [7]. Interestingly, the residual tumor size postsurgery is a good predictor of the patient survival in metastatic ovarian cancers. Since most gynecologic tumors are diagnosed relatively late, radiation therapy is reserved for recurrent tumors and patients intolerant to chemotherapy and those with contraindications to surgical removal of the tumor [8, 9]. Hormone therapy is also less effective in ovarian cancer, since most tumors develop mutations and become hormone-independent in the early stages of the treatment [10, 11]. So, chemotherapy is most often opted in patients for treating metastatic tumors after the initial surgical removal of the tumor mass.

A variety of chemotherapeutic drugs have been used in gynecologic cancers such as cisplatin, carboplatin, paclitaxel, docetaxel, and doxorubicin [4, 12, 13] (Table 1). Platinum compounds cause DNA damage by forming platinum-DNA adducts, leading to DNA replication inhibition and subsequent cell death [34]. Taxanes target microtubule polymerization, inhibiting mitosis and thus inducing apoptosis [35, 36]. Doxorubicin deters the topoisomerase-II activity ultimately leading to cell death [37]. Few other drugs mostly used in combination chemotherapy along with platinum drugs include cyclophosphamide, gemcitabine, topotecan, and vinorelbine [4, 38]. Even though the combination therapy has a 70% response rate, tumor relapse and subsequent chemoresistance are frequently observed, eventually resulting in a low survival rate [4, 38, 39]. Generally, chemotherapy is a very effective initial treatment, although the response declines after a few phases resulting in tumor recurrence in around 80%–85% cases [6].

In >90% of the patients with advanced malignancy, cancer chemoresistance is a significant cause of treatment failure and unfortunately death [40]. Chemoresistance can be intrinsic or acquired through various cellular modifications, and drugs targeting these systems can overcome the resistance and hopefully extend patient survival. Targeted therapy is a relatively new treatment approach where the drugs attack the oncogenic pathways in the malignant cells. Since these pathways are associated with chemoresistance, targeting them could help sensitize the malignant cells to standard chemotherapy drugs. The various mechanisms of chemoresistance in ovarian cancer are shown in the Table 1.

This research chapter focuses on ovarian cancer chemoresistance and the emerging therapies that have great potentials in mitigating and overcoming it. The various mechanisms of overcoming chemoresistance and the drugs currently used and are being tested in clinical trials in ovarian cancer are discussed below.

## Targeting efflux pumps

P-glycoprotein (P-gp) is an efflux pump protein that helps in the elimination of chemotherapeutic drugs out of the cells making them less efficient in various malignancies including ovarian and endometrial cancers [7]. Studies have shown that overexpression of the

TABLE 1    Chemotherapy drugs and their mechanisms of actions in ovarian cancer.

| Name of the drugs | Mechanism of resistance | Reference(s) |
|---|---|---|
| *Targeting efflux pumps* | | |
| Platinum drugs | Upregulation of ATP7A/ATP7B | [14–17] |
| Doxorubicin and taxane drugs | Upregulation of P-glycoprotein | [18, 19] |
| *Targeting DNA repair mechanisms* | | |
| Platinum drugs | Upregulation of NER | [20] |
| Taxane drugs | Downregulation of BRCA1/2 | [21] |
| Platinum drugs | Downregulation of MMR | [22] |
| *Targeting signaling pathways* | | |
| Platinum drugs | Upregulation of PI3K | [23] |
| | Downregulation of JNK | [24] |
| | Downregulation of PTEN | [23] |
| | Upregulation of AKT | [25] |
| | Downregulation of p-38 | [24] |
| | Upregulation of XIAP | [26, 27] |
| | Upregulation of ERK1/2 | [28] |
| | Downregulation of mut P53 | [29] |
| | Upregulation of BCL-2 | [30] |
| | Upregulation of ErbB2 | [31] |
| Taxane drugs | Upregulation of AKT | [25] |
| | Upregulation of ErbB2 | [31] |
| | Downregulation of mut P53 | [29] |
| | Downregulation of PAR-4 | [32] |
| | Upregulation of XIAP | [27] |
| | Upregulation of EGFR | [33] |
| Doxorubicin | Upregulation of AKT | [25] |

multidrug-resistance (MDR) gene *MDR1* helps in the upregulation of P-gp thus resulting in chemoresistance [18, 19]. Tariquidar (XR9576) is a potent P-gp inhibitor, which in combination with other chemotherapeutic drugs such as docetaxel has shown minimal adverse effects [41]. Studies have shown that Tariquidar when combined with other drugs such as doxorubicin and paclitaxel resulted in effective reversal of the chemoresistance in ovarian cancer cells [41]. Moreover, it showed potential in increasing the effectiveness of doxorubicin and paclitaxel against cancer cells in ovarian tumor biopsies. However, further clinical studies are required to assess the effectiveness and safety of Tariquidar in different types of malignancies.

## Targeting repair mechanisms

### Poly-ADP-ribose polymerase inhibitors

Studies have shown that mutations in *BRCA-1/2* genes cause chemoresistance in many gynecologic cancers, primarily in ovarian cancers [42]. In the cancer cells, BRCA mutations induce a defect in homologous recombination repair of breaks in the DNA strands. This change is the basis for development of the enzyme poly-ADP-ribose polymerase (PARP) inhibitors in ovarian tumors. PARP is the protein involved in DNA repair process. The PARP inhibitors act by inhibiting DNA repair of single-strand breaks in the cancer cells with BRCA mutations ultimately resulting in cell death. Studies have shown that PARP inhibitors are more effective in patients with BRCA mutations; however, it is not a prerequisite for initiating therapy in patients who does not have BRCA mutations [42]. Some of the PARP inhibitors used in the management of ovarian cancer patients include the following:

(i) *Olaparib*: In patients with high-grade ovarian cancer, Olaparib has demonstrated positive results in phase II clinical studies in terms of progression-free survival (PFS) and objective response rate (ORR) [43–46]. During an initial study of Olaparib in patients with recurrent and advanced cancer of ovary, 11 out of 33 people showed a positive ORR [43]. The results of the study prompted for a large-scale randomized Olaparib-placebo study comprising 265 patients with platinum-sensitive ovarian cancer relapse. Olaparib showed significant results in terms of PFS (8.4 months in Olaparib vs 4.8 months in the placebo group). Moreover, the results were positive irrespective of the BRCA status of the patients [44]. In another clinical trial, Olaparib was tested in patients with advanced-stage ovarian tumors. The ORR was improved irrespective of the BRCA status including 41% of people with BRCA mutations and 24% without mutations [45]. In a phase II trial consisting of 193 patients, Olaparib demonstrated good ORR in patients with platinum-resistant ovarian cancer as well. Recent studies have tested the efficiency of Olaparib in ovarian malignancies as a combination chemotherapy [46]. In a phase II trial comprising 162 women with advanced-stage platinum-sensitive ovarian cancer, patients received either a combination therapy with olaparib and paclitaxel and carboplatin or chemotherapy alone. The combination therapy showed significant results in terms of PFS (12.2 months in combination group vs 9.6 months in chemotherapy alone) [47]. Another phase II trial involving olaparib in combination with Cediranib showed positive results when compared with Olaparib alone in terms of median PFS of 9 months in the combination group vs 17.7 months in Olaparib alone group [48].

(ii) *Veliparib*: Veliparib is another PARP inhibitor that was tested as a monotherapy in BRCA-positive ovarian cancer. In a phase II trial comprising 50 BRCA-positive ovarian cancer patients, people were treated with veliparib irrespective of the platinum sensitivity status. Veliparib showed promising results in terms of ORR, which was 26% for all the patients and was 20% for the platinum-resistant group and 35% in the sensitive group [49]. Another study assessed the impact of veliparib in combination with cyclophosphamide in BRCA-positive ovarian cancer patients; however, the results were insignificant in terms of ORR and PFS [50].

Another PARP inhibitor, **Niraparib**, a dual inhibitor of PARP-1 and PARP-2, was tested in phase I trials as a monotherapy in BRCA-positive ovarian cancer; but the results were insignificant [51]. **Iniparib** is another PARP inhibitor that showed positive results in relapsing ovarian malignancies when given along with carboplatin and gemcitabine. In the phase II trial, the ORR was significantly improved in the Iniparib group (71%) compared to those who were treated with carboplatin and gemcitabine only (47% ORR). Moreover, the study outcomes were independent on the BRCA status of the patients [52]. Interestingly, when Iniparib was tested as a combination therapy in other gynecologic cancers as well such as uterine carcinosarcoma, the results were not significant due to the limited patient population [53].

(iii) *Rucaparib*: In BRCA- and PTEN-positive cancer cell lines, in vitro studies of Rucaparib in combination with platinum-based chemotherapeutic drugs such as topotecan, carboplatin, doxorubicin, paclitaxel, or gemcitabine have shown promising results in terms of synergistic response and accelerated apoptosis of cancer cells [54]. A phase I trial of rucaparib in advanced-stage solid tumors showed positive results in terms of clinical benefit rate (CBR), which was 86% particularly in ovarian cancer and peritoneal cancer cell lines [55]. A phase II trial of Rucaparib in patients with BRCA-positive advanced-stage ovarian and breast cancers showed good response (CR, PR, or SD for more than 12 weeks) when the drug was administered continuously compared to the oral administration, particularly in ovarian cancer patients [56]. One of the advantages of PARP inhibitors is their relatively low toxicity profile, which makes them ideal for the combination chemotherapy.

## Targeting mismatch repair deficiency

Mismatch repair (MMR), coded by *MLH1* gene, is a mechanism of identifying a mismatched DNA sequence and reconstructing the proper functional DNA [57]. In ovarian cancer, hypermethylation of *MLH1* gene is thought to be the source of MMR deficiency. Studies have shown that administration of platinum drugs results in downregulation of MMR activity thus resulting in decreased efficiency of cisplatin and other related drugs, ultimately leading to chemoresistance [22]. However, few other studies refute the correlation between the platinum compounds and MMR deficiency [58, 59]. Studies have shown that when decitabine, a methylating agent, was added to platinum compounds and epirubicin in ovarian cancer cell lines, it resulted in an increase in sensitivity to chemotherapy agents [60]. Moreover, the re-expression of *MLH1* gene in the tumor cells was also noticed. In a phase I study, 10 patients with chemo-resistant ovarian cancer were given a combination of carboplatin and decitabine. The trial results were encouraging in terms of clinical outcomes. Moreover, demethylation of other cancer genes such as *BRCA1* and *HOXA11* was also observed [61]. Later, the same team conducted a phase II study where 17 patients pretreated with carboplatin were given decitabine. The treatment resulted in an improvement in PFS (10.2 months) via demethylation of cancer-causing genes such as *HOXA10*, *HOXA11*, *MLH1*, and *RASSF1A* [62]. The positive results of the study show that incorporation of decitabine in MMR deficient cancers can be beneficial in overcoming chemoresistance.

# Targeting PI3K/AKT pathway

In >40% of ovarian cancers, various components of phosphatidylinositol 3-kinase (PI3K)/ protein kinase B (AKT) pathway are commonly upregulated and mutated thus resulting in chemoresistance and cancer forming cells [63]. PIK3 is a cell membrane kinase enzyme that converts PIP2 into PIP3, which further activates other kinase enzymes such as PDK1, and AKT plays role in peptide formation and cell growth. Protein kinase B, also known as AKT, is involved in a variety of cellular processes such as cell growth and apoptosis. In ovarian cancer, mutations of the AKT isoforms (AKT1 and AKT2) are involved in cellular chemoresistance against taxane drugs and platinum drugs [25]. Phosphatase and tensin homolog (PTEN), a protein encoded by PTEN gene, is involved in the development of various cancers [23]. It converts PIP3 to PIP2, an inverse effect of the PI3K. X-linked inhibitor of apoptosis protein (XIAP), associated with *XIAP* gene, plays a major role in inhibiting the cell death. XIAP is also involved in the chemoresistance against cisplatin and taxane compounds [26, 27].

## PI3K inhibitors

(i) *Buparlisib*: Buparlisib is a pan-class I PI3K inhibitor. In vivo studies in the BRCA-1-mutated mice showed a significant synergistic effect when buparlisib was combined with a PARP inhibitor drug, olaparib [64]. Based on the in vitro study results, a phase I study involving 34 women with high-grade serous ovarian cancer (HGSOC) was given the combination therapy of olaparib and buparlisib at various dosages. Interestingly, 26 out of 34 women had mutations of BRCA1/2 [65]. The study reported good clinical response in the majority of the people, which encouraged and paved the way to further clinical trials.

(ii) *BYL719*: It is a PI3K inhibitor, which has been tested in various advanced-stage solid tumors both alone, and in combination with MAPK enzyme kinase (MEK) inhibitor, Binimetinib, during phase I clinical trials. One of the studies reported that the combination therapy showed PR in patients with Kirsten rat sarcoma 2 viral oncogene homolog KRAS mutations [66]. The results were interesting since tumors with KRAS mutations are frequently associated with minimal clinical response to other drugs such as mammalian target of rapamycin (mTOR) inhibitors and PI3K inhibitors when used alone [67].

(iii) *NVP-BEZ235*: It is a molecule that has the capability to inhibit both PI3K and mTOR. During the in vitro studies in transgenic mice with the KRAS and PTEN mutations, treatment with NVP-BEZ235 showed good results in terms of prolonged survival [68].

## AKT inhibitors

In vitro studies on ovarian and endometrial cancer cell lines with some of the AKT inhibitors such as MK-2206, Perifosine, AZD5363, and GSK2141795 produced promising results, which led to inducing phase I clinical trials. These drugs were tested both alone and in combination with other drugs such as MEK1 inhibitors in both platinum-sensitive and platinum-resistant tumors. However, they showed limited clinical activity in humans [69, 70].

## Targeting mToR

Mammalian target of rapamycin (mTOR) is a subcomponent of complexes mTORC1/C2, which plays a vital role in various cellular processes such as cell growth and differentiation, and protein synthesis. Since mTOR is placed distal to the PI3K-AKT pathway, it is a potential target for targeted chemotherapy in various gynecologic cancers.

**(i)** *Temsirolimus*: Temsirolimus is a water-soluble molecule extracted from rapamycin. Studies have reported its capability to inhibit tumor spread and growth [71, 72]. A phase II study was conducted on 54 women with advanced ovarian cancer where patients received Temsirolimus therapy. The study reported only minimal response in patients with therapy [71]. Later, Temsirolimus was given in combination with bevacizumab in 31 women with advanced-stage ovarian malignancies including both chemo-resistant and chemo-sensitive cancers. The combination approach showed some clinical response with only nine patients reporting SD and three with PR [72].

**(ii)** *Everolimus*: Everolimus (RAD-001) is another mTORC1-selective mTOR inhibitor. Experiments on 58 transgenic mice with serous ovarian cancer with clinical presentation of ascites and peritoneal metastases have shown that when everolimus was given, good clinical response was observed in terms of tumor size (84% decrease in size of tumor) and peritoneal spread (only 21% reported in treatment group vs 74% in placebo) [73]. Later, everolimus was tested alone, and along with cisplatin in ovarian tumor cell lines. Interestingly, the combination therapy showed good response in terms of increased tumor cell death [74]. Further investigations are needed to assess the response in humans.

## Targeting epithelial growth factor receptors

### Tyrosine kinase inhibitors

The receptor tyrosine kinases (RTKs) are transmembrane glycoproteins that comprise of an extracellular domain for ligand attachment, a transmembrane domain and an intracellular tyrosine kinase motif [75–77]. The extracellular domain of the RTK helps in the identification of various subfamilies of the kinases. Binding of the corresponding ligands to the RTKs results in their activation via phosphorylation of tyrosine molecules and through intracellular signaling proteins [78]. The activated RTKs play major role(s) in managing many cellular processes such as proliferation, differentiation, adhesion, migration, and survival [79]. When a bivalent ligand binds to two receptor molecules, it forms a dimer, which results in the activation of the kinases. The activation of kinases is dependent on two key steps including the augmentation of catalytic activity intrinsically and the formation of binding sites to recruit downstream signaling proteins, both of which are dependent on tyrosine autophosphorylation [80]. RTKs are classified into 21 groups, such as the vascular endothelial growth factor receptor (VEGFR), epidermal growth factor receptor (EGFR), fibroblast growth factor receptor (FGFR), and platelet-derived growth factor receptor (PDGFR) families [81]. In tumor tissues, cellular processes such cell growth and differentiation are stimulated via RTK dimerization. Studies have

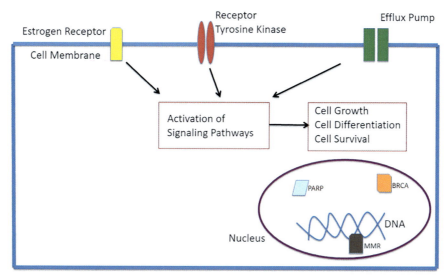

FIG. 1   Overview of some of the most common chemoresistance mechanisms in ovarian cancer. *Abbreviations:* PARP, poly ADP ribose polymerase; BRCA, breast cancer gene; MMR, mismatch repair.

shown that monoclonal antibodies can inhibit the activation and overexpression of kinases in the cancer cells. An overview of some of the most common chemoresistance mechanisms in ovarian cancer is shown in Fig. 1.

(i) *Geftinib*: The potential of geftinib as a single-agent therapy in the treatment of both ovarian and endometrial cancers was tested in phase II trials. Interestingly, geftinib showed beneficial clinical response in terms of PFS in endometrial cancer but not in ovarian cancer [82, 83]. In another phase II study, geftinib in combination with tamoxifen was tested in patients with platinum-resistant ovarian cancer; however, the drug combination did not produce any significant results [84]. Later, geftinib was tested in combination with paclitaxel and carboplatin against a variety of cancers including ovarian, peritoneal cavity, and tubal cancers, irrespective of the platinum-sensitivity status. The combination was effective in platinum-sensitive cancers with an ORR of 61.9% and CBR of 81% compared to the platinum-resistant malignancy cases (ORR 19.2% and CBR 69.2%) [85]. Similarly, in a different phase I/II trial, when geftinib was combined with oxaliplatin and vinorelbine in patients with metastatic ovarian cancer, the treatment was more effective in platinum-sensitive group (ORR 90% in sensitive group vs 23.8% in resistant group) [86]. The results of these studies indicate that geftinib could be more effective as a combination drug in chemo-sensitive ovarian malignancies.

(ii) *Erlotinib*: Erlotinib is another TKI tested in advanced-stage ovarian cancer both as single-agent therapy and as combination chemotherapy. In a phase I trial, a combination of erlotinib and carboplatin was tested in chemo-naive metastatic ovarian tumor patients. The drug combination showed a good response in terms of ORR, which was 52% [87]. Another phase II study tested the effectiveness of the combination medications (erlotinb + carboplatin) in both chemo-sensitive and chemo-resistant ovarian cancer patients. Chemo-sensitive patients were more responsive as compared to the chemo-resistant group in terms of the ORR (57% in sensitive group vs 7% in resistant group) [88]. The results of this study pointed to the fact that chemo-sensitive patients are more responsive to the therapy.

Monoclonal antibodies such as cetuximab and matuzumab were tested in advanced-stage ovarian cancer in phase II trials both as single-agent therapy and in combination with other drugs such as carboplatin and paclitaxel. The studies found that the medications were well tolerated but showed significant clinical activity only when they were used as combination therapy in chemo-sensitive patients [89–91].

## Targeting the estrogen-signaling pathways

Ovarian tumors cells carry significant number of estrogen receptors (ERs) including ER alpha and beta, and the action of high estrogen levels on these cells results in tumor formation and growth [92]. Estrogen displays nongenomic actions such as activation of PI3K and MAPK pathways, interaction with ErbB2 and IGF-1R receptors and binding to GPR30 receptors, which ultimately result in cancer chemoresistance [93, 94]. Studies have reported that estrogen plays a role in chemoresistance to various gynecologic cancers. Estrogen is thought to be involved in phosphorylation of AKT-ASK1 complex, which leads to chemoresistance of ovarian tumors to paclitaxel therapy [95]. So, hormonal therapy could be effective in overcoming chemoresistance. An overview of various targeted chemotherapy drugs is shown in Fig. 2. Some of the drugs targeting estrogen receptors include.

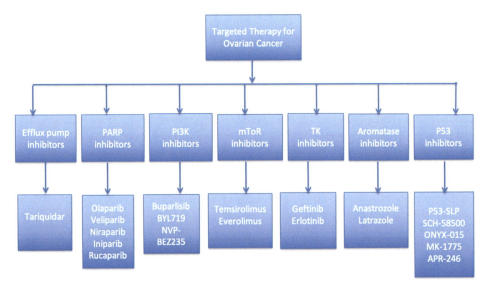

FIG. 2   An overview of various targeted chemotherapy drugs in the management of ovarian cancer. *Abbreviations:* PARP, poly-ADP-ribose polymerase; PI3K, phosphatidylinositol 3-kinase, mTOR, mammalian target of rapamycin; TK, tyrosine kinase.

## Aromatase inhibitors

(i) *Anastrozole*: Anastrozole, a nonsteroidal agent, was tested in a phase II trial in 53 asymptomatic patients against gynecological cancers including 43 people with recurrent ovarian cancer and remaining 10 people with tubal and peritoneal cancers. Anastrozole helped in achieving SD in about 68% of the patients [96]. Another phase II trial consisting of a combination of anastrozole and geftinib was conducted by the same research team, where 35 patients with gynecologic cancers were evaluated. The combination showed only minimal response, however [97].

(ii) *Letrozole*: Letrozole was tested in couple of phase II trials in ovarian cancer patients where it showed modest response clinically. Both of these trials had contradicting results in terms of assessment of letrozole efficiency in estrogen receptor positive patients [98, 99]. In a different phase II trial, patients with ovarian cancer were selected based on the presence of estrogen receptors as a prerequisite and treated with letrozole. Interestingly, letrozole showed promising results in impeding the tumor growth and spread [100]. Later, letrozole was tested in ovarian cancer patients who were resistant to platinum and taxane drugs and had positive estrogen receptor status. Letrozole failed to show clinical response in chemoresistance cancers [101]. This indicated that letrozole efficiency is inversely proportional to cancer chemoresistance. Exemestane is another novel aromatase inhibitor that showed potential in stopping tumor progression in ovarian cancer patients during phase II clinical studies [102].

## Estrogen receptor antagonists

Drugs such as Fulvestrant were tested in phase II clinical trials in ovarian cancer patients as a single-agent therapy where it showed potential in stopping tumor progression [103]. Interestingly, in vitro studies of toremifene in combination with doxorubicin showed positive results in ovarian cancer cell lines. Toremifene increased the efficiency of doxorubicin in chemo-resistant cells [104]. This study provided promising results for the development of further clinical trials.

## Targeting p53

P53 is a tumor suppressor tetramer protein (coded by *TP53* gene) that displays antitumor properties through various mechanisms including repairing the DNA and cell cycle arrest and inducing cell death in tumor cells (Fig. 2). In normal cells, p53 is present in very low levels but gets activated and upregulated through various signals such as damage to DNA strands, hypoxia, and activation of tumor-producing genes. Studies have shown that p53 levels are commonly elevated in >90% of the ovarian cancers [105]. Mutations in *TP53* gene result in overexpression of missense mutant and wild-type (WT) p53 proteins. These mutations also play major roles in chemoresistance to radiation therapy and drugs like paclitaxel through mechanisms such as epithelial-mesenchymal transition (EMT) and changes in the PI3K pathway resulting in tumor metastases [106]. P53 mutations are also associated with chemoresistance to platinum drugs via upregulation of BCL-2 protein. The multitude effects of p53 in tumor progression and acquired chemoresistance make it a target for clinical trials focused on overcoming chemoresistance in ovarian cancer cells. Few trials are being conducted targeting the *TP53* gene (gene therapy) with an aim of producing p53 proteins that has the capability to induce apoptosis in cancer cells. Some of the important drugs in this area include the following:

(i) *P53-SLP*: A synthetic protein, P53-SLP, analogous to middle portion of p53 protein, was developed with the aim of boosting the immune system which further helps in eliciting cytotoxicity against cancer cells that display p53 overexpression. In a phase II trial consisting of 20 patients with ovarian cancer, p53-SLP was able to generate cytotoxic cellular responses [107]. However, it was only able to produce minimal clinical response. In a different phase II study, the protein was administered in patients who also received secondary chemotherapy later. Pretreatment with P53-SLP was unable to produce any significant clinical benefits [108]. However, in a later trial where it was administered in ovarian cancer patients who already received cyclophosphamide treatment, the combination was able to significantly elevate T-cells when compared to the peptide alone [109]. Additional research is required to evaluate the efficacy of the drug in combination therapy.

(ii) *SCH-58500*: It is a synthetic adenovirus producing WT p53 peptides. The efficiency of the SCH-58500 was tested in a phase I/II trial consisting of 24 people with advanced-stage and relapsing ovarian cancer. During the trial, patients were treated with a combination of platinum chemotherapy and varying doses of SCH-58500, and the results were promising. The therapy resulted in p53 reexpression and a 50% decrease in tumor marker CA-125 [110]. Moreover, the combination was able to show dose-dependent impact on median survival in the patients after prolonged use (5 months with single dose and 13 months with multiple dosages) [111].

(iii) *ONYX-015*: Another genetically engineered adenovirus, ONYX-015, which has tumor suppressor properties through replication in mutant p53 cells and further promoting cell death [112]. ONYX-015 has shown clinical benefits in combination therapy against various chemo-resistant malignancies such as gastrointestinal, neck, and head tumors [113, 114]. However, additional research is necessary to evaluate the drug in ovarian malignancies.

(iv) *MK-1775*: This is another molecule that helps in overcoming chemoresistance through inhibition of WEE1, a G2 checkpoint nuclear protein kinase. Since chemo-resistant ovarian cancer cells with p53 mutations lose the G1 checkpoint, targeting G2 checkpoint might help regain sensitivity in such cancer cells [115]. MK-1775 was tested in few phase I/II trials in combination with platinum compounds and taxanes in chemo-resistant ovarian tumors. A phase II study was conducted on 121 patients with platinum-sensitive tumors where MK-1775 was given along with carboplatin and paclitaxel. The combination showed good clinical responses in terms of the ORR, which was 81% in combination therapy vs 74% in standard chemotherapy alone group [116]. Another phase II trial in platinum-resistant ovarian cancer patients has shown positive results with the addition of MK-1775 [117]. A phase II trial is currently ongoing where a combination of gemcitabine and MK-1775 is being tested in advanced-stage ovarian cancer [118].

**(v)** *APR-246*: It is a small compound that has the capability to repair mutated p53 to its functional form thus reactivating cell apoptosis. Studies have shown that APR-246 when used in combination with platinum compounds and doxorubicin in chemo-resistant ovarian cancers cells has demonstrated good response [119, 120]. A phase I/II trial is ongoing with the aim of testing the efficacy of the compound in combination therapy in resistant cancers [121].

Other potential targets, which are under investigation, include the prostate apoptosis response 4 (PAR-4) protein. The PAR-4 has shown selective apoptosis of tumor cells in various in vitro experiments [122]. A recent study has shown that PAR-4 can increase the tumor cell death in ovarian tumor cells pretreated with paclitaxel [32]. However, more focused studies are needed at this avenue for better understanding.

## Conclusions and future perspectives

In summary, chemo-resistant ovarian cancer has been increasing in recent years, and propitiously a lot of translational research about various mechanisms of resistance, both intrinsic and acquired and targeted therapy against resistant cells is expanding rapidly as well. Targeted therapy is designed to selectively target the specific molecules and mechanisms that cause chemoresistance and eliminate them so that chemotherapeutic drugs can work to their full potential of antitumor activity. However, the current clinical trials of various targeted therapies working alone and/or in combination with standard chemotherapy are yet to show the perfect response in impeding the chemo-resistant pathways. A better understanding of tumor genetics and further studies with molecular profiling might help in producing drugs that could effectively target the pathways causing chemoresistance thus ultimately killing the cancer cells and improving the patient survival and quality of life for patients with ovarian cancer.

## Author contributions

Deepika Sarvepalli and Mamoon Ur Rashid primarily conceived the idea, and subsequently, all the coauthors have diligently contributed to the development and preparation of this research chapter, including the literature search, concept organization, data interpretation, and writings. All the authors have read and approved the final draft for publication.

## Financial disclosures

None to disclose.

## References

[1] American Cancer Society. Key statistics for ovarian cancer, https://www.cancer.org/cancer/ovarian-cancer/about/key-statistics.html; 2020. [Accessed 5 September 2020].
[2] Canadian Cancer Society's Advisory Committee on Cancer Statistics. Canadian cancer statistics 2015. Toronto, ON: Canadian Cancer Society; 2015.

[3] Siegel R, Naishadham D, Jemal A. Cancer statistics, 2013. CA Cancer J Clin 2013;63:11–30.

[4] Casciato DA, Territo MC. Manual of clinical oncology. Lippincott Williams & Wilkins; 2009.

[5] Howlader NN, Noone AM, Krapcho M, Miller D, Bishop K, Altekruse SF, Kosary CL, Yu M, Ruhl J, Tatalovich Z, Mariotto A. SEER cancer statistics review, 1975–2013. Bethesda, MD: National Cancer Institute; 2016 April 8. p. 19.

[6] Lengyel E. Ovarian cancer development and metastasis. Am J Pathol 2010;177(3):1053–64. https://doi.org/10.2353/ajpath.2010.100105.

[7] Brasseur K, Gévry N, Asselin E. Chemoresistance and targeted therapies in ovarian and endometrial cancers. Oncotarget 2017;8(3):4008–42. https://doi.org/10.18632/oncotarget.14021.

[8] American Cancer Society. Endometrial (Uterine) cancer detailed guide; 2016.

[9] American Cancer Society. Ovarian cancer detailed guide; 2016.

[10] Sjoquist KM, Martyn J, Edmondson RJ, Friedlander ML. The role of hormonal therapy in gynecological cancers-current status and future directions. Int J Gynecol Cancer 2011;21(7):1328–33. https://doi.org/10.1097/IGC.0b013e31821d6021.

[11] Garrett A, Quinn MA. Hormonal therapies and gynaecological cancers. Best Pract Res Clin Obstet Gynaecol 2008;22:407–21. https://doi.org/10.1016/j.bpobgyn.2007.08.003.

[12] Plataniotis G, Castiglione M, ESMO Guidelines Working Group. Endometrial cancer: ESMO clinical practice guidelines for diagnosis, treatment and follow-up. Ann Oncol 2010;21(Suppl. 5):v41–5. https://doi.org/10.1093/annonc/mdq245.

[13] Lalwani N, Prasad SR, Vikram R, Shanbhogue AK, Huettner PC, Fasih N. Histologic, molecular, and cytogenetic features of ovarian cancers: implications for diagnosis and treatment. Radiographics 2011;31(3):625–46. https://doi.org/10.1148/rg.313105066.

[14] Katano K, Kondo A, Safaei R, et al. Acquisition of resistance to cisplatin is accompanied by changes in the cellular pharmacology of copper. Cancer Res 2002;62(22):6559–65.

[15] Nakayama K, Kanzaki A, Ogawa K, Miyazaki K, Neamati N, Takebayashi Y. Copper-transporting P-type adenosine triphosphatase (ATP7B) as a cisplatin based chemoresistance marker in ovarian carcinoma: comparative analysis with expression of MDR1, MRP1, MRP2, LRP and BCRP. Int J Cancer 2002;101(5):488–95. https://doi.org/10.1002/ijc.10608.

[16] Komatsu M, Sumizawa T, Mutoh M, et al. Copper-transporting P-type adenosine triphosphatase (ATP7B) is associated with cisplatin resistance. Cancer Res 2000;60(5):1312–6.

[17] Samimi G, Safaei R, Katano K, et al. Increased expression of the copper efflux transporter ATP7A mediates resistance to cisplatin, carboplatin, and oxaliplatin in ovarian cancer cells. Clin Cancer Res 2004;10(14):4661–9. https://doi.org/10.1158/1078-0432.CCR-04-0137.

[18] Holzmayer TA, Hilsenbeck S, Von Hoff DD, Roninson IB. Clinical correlates of MDR1 (P-glycoprotein) gene expression in ovarian and small-cell lung carcinomas. J Natl Cancer Inst 1992;84(19):1486–91. https://doi.org/10.1093/jnci/84.19.1486.

[19] Kamazawa S, Kigawa J, Kanamori Y, et al. Multidrug resistance gene-1 is a useful predictor of paclitaxel-based chemotherapy for patients with ovarian cancer. Gynecol Oncol 2002;86(2):171–6. https://doi.org/10.1006/gyno.2002.6738.

[20] Selvakumaran M, Pisarcik DA, Bao R, Yeung AT, Hamilton TC. Enhanced cisplatin cytotoxicity by disturbing the nucleotide excision repair pathway in ovarian cancer cell lines. Cancer Res 2003;63(6):1311–6.

[21] Tan DS, Rothermundt C, Thomas K, et al. "BRCAness" syndrome in ovarian cancer: a case-control study describing the clinical features and outcome of patients with epithelial ovarian cancer associated with BRCA1 and BRCA2 mutations. J Clin Oncol 2008;26(34):5530–6. https://doi.org/10.1200/JCO.2008.16.1703.

[22] Samimi G, Fink D, Varki NM, et al. Analysis of MLH1 and MSH2 expression in ovarian cancer before and after platinum drug-based chemotherapy. Clin Cancer Res 2000;6(4):1415–21.

[23] Lee S, Choi EJ, Jin C, Kim DH. Activation of PI3K/Akt pathway by PTEN reduction and PIK3CA mRNA amplification contributes to cisplatin resistance in an ovarian cancer cell line. Gynecol Oncol 2005;97(1):26–34. https://doi.org/10.1016/j.ygyno.2004.11.051.

[24] Yuan ZQ, Feldman RI, Sussman GE, Coppola D, Nicosia SV, Cheng JQ. AKT2 inhibition of cisplatin-induced JNK/p38 and Bax activation by phosphorylation of ASK1. Implication of AKT2 in chemoresistance [retraction of: J Biol Chem. 2003;278(26):23432–40]. J Biol Chem 2016;291(43):22847. https://doi.org/10.1074/jbc.A116.302674.

[25] Mitsuuchi Y, Johnson SW, Selvakumaran M, Williams SJ, Hamilton TC, Testa JR. The phosphatidylinositol 3-kinase/AKT signal transduction pathway plays a critical role in the expression of p21WAF1/CIP1/SDI1 induced by cisplatin and paclitaxel. Cancer Res 2000;60(19):5390–4.

[26] Asselin E, Mills GB, Tsang BK. XIAP regulates Akt activity and caspase-3-dependent cleavage during cisplatin-induced apoptosis in human ovarian epithelial cancer cells. Cancer Res 2001;61(5):1862–8.

[27] Cheng JQ, Jiang X, Fraser M, et al. Role of X-linked inhibitor of apoptosis protein in chemoresistance in ovarian cancer: possible involvement of the phosphoinositide-3 kinase/Akt pathway. Drug Resist Updat 2002;5(3–4):131–46. https://doi.org/10.1016/s1368-7646(02)00003-1.

[28] Chan DW, Liu VW, Tsao GS, et al. Loss of MKP3 mediated by oxidative stress enhances tumorigenicity and chemoresistance of ovarian cancer cells. Carcinogenesis 2008;29(9):1742–50. https://doi.org/10.1093/carcin/bgn167.

[29] Buttitta F, Marchetti A, Gadducci A, et al. p53 alterations are predictive of chemoresistance and aggressiveness in ovarian carcinomas: a molecular and immunohistochemical study. Br J Cancer 1997;75(2):230–5. https://doi.org/10.1038/bjc.1997.38.

[30] Perego P, Righetti SC, Supino R, et al. Role of apoptosis and apoptosis-related proteins in the cisplatin-resistant phenotype of human tumor cell lines. Apoptosis 1997;2(6):540–8. https://doi.org/10.1023/a:1026442716000.

[31] Hengstler JG, Lange J, Kett A, et al. Contribution of c-erbB-2 and topoisomerase IIalpha to chemoresistance in ovarian cancer. Cancer Res 1999;59(13):3206–14.

[32] Meynier S, Kramer M, Ribaux P, et al. Role of PAR-4 in ovarian cancer. Oncotarget 2015;6(26):22641–52. https://doi.org/10.18632/oncotarget.4010.

[33] Qiu L, Di W, Jiang Q, et al. Targeted inhibition of transient activation of the EGFR-mediated cell survival pathway enhances paclitaxel-induced ovarian cancer cell death. Int J Oncol 2005;27(5):1441–8.

[34] Wang D, Lippard SJ. Cellular processing of platinum anticancer drugs. Nat Rev Drug Discov 2005;4(4):307–20. https://doi.org/10.1038/nrd1691.

[35] Orr GA, Verdier-Pinard P, McDaid H, Horwitz SB. Mechanisms of taxol resistance related to microtubules. Oncogene 2003;22(47):7280–95. https://doi.org/10.1038/sj.onc.1206934.

[36] Jordan MA, Wilson L. Microtubules as a target for anticancer drugs. Nat Rev Cancer 2004;4(4):253–65. https://doi.org/10.1038/nrc1317.

[37] Thorn CF, Oshiro C, Marsh S, et al. Doxorubicin pathways: pharmacodynamics and adverse effects. Pharmacogenet Genomics 2011;21(7):440–6. https://doi.org/10.1097/FPC.0b013e32833ffb56.

[38] Boere IA, van der Burg ME. Review of dose-intense platinum and/or paclitaxel containing chemotherapy in advanced and recurrent epithelial ovarian cancer. Curr Pharm Des 2012;18(25):3741–53. https://doi.org/10.2174/138161212802002634.

[39] Yap TA, Carden CP, Kaye SB. Beyond chemotherapy: targeted therapies in ovarian cancer. Nat Rev Cancer 2009;9(3):167–81. https://doi.org/10.1038/nrc2583.

[40] Longley DB, Johnston PG. Molecular mechanisms of drug resistance. J Pathol 2005;205(2):275–92. https://doi.org/10.1002/path.1706.

[41] Mistry P, Stewart AJ, Dangerfield W, et al. In vitro and in vivo reversal of P-glycoprotein-mediated multidrug resistance by a novel potent modulator, XR9576. Cancer Res 2001;61(2):749–58.

[42] Tutt AN, Lord CJ, McCabe N, et al. Exploiting the DNA repair defect in BRCA mutant cells in the design of new therapeutic strategies for cancer. Cold Spring Harb Symp Quant Biol 2005;70:139–48. https://doi.org/10.1101/sqb.2005.70.012.

[43] Audeh MW, Carmichael J, Penson RT, et al. Oral poly(ADP-ribose) polymerase inhibitor olaparib in patients with BRCA1 or BRCA2 mutations and recurrent ovarian cancer: a proof-of-concept trial. Lancet 2010;376(9737):245–51. https://doi.org/10.1016/S0140-6736(10)60893-8.

[44] Ledermann J, Harter P, Gourley C, et al. Olaparib maintenance therapy in platinum-sensitive relapsed ovarian cancer. N Engl J Med 2012;366(15):1382–92. https://doi.org/10.1056/NEJMoa1105535.

[45] Gelmon KA, Tischkowitz M, Mackay H, et al. Olaparib in patients with recurrent high-grade serous or poorly differentiated ovarian carcinoma or triple-negative breast cancer: a phase 2, multicentre, open-label, non-randomised study. Lancet Oncol 2011;12(9):852–61. https://doi.org/10.1016/S1470-2045(11)70214-5.

[46] Kaufman B, Shapira-Frommer R, Schmutzler RK, et al. Olaparib monotherapy in patients with advanced cancer and a germline BRCA1/2 mutation. J Clin Oncol 2015;33(3):244–50. https://doi.org/10.1200/JCO.2014.56.2728.

[47] Oza AM, Cibula D, Benzaquen AO, et al. Olaparib combined with chemotherapy for recurrent platinum-sensitive ovarian cancer: a randomised phase 2 trial [published correction appears in Lancet Oncol. 2015

Feb;16(2):e55; published correction appears in Lancet Oncol. 2015 Jan;16(1):e6]. Lancet Oncol 2015;16(1): 87–97. https://doi.org/10.1016/S1470-2045(14)71135-0.

[48] Liu JF, Barry WT, Birrer M, et al. Combination cediranib and olaparib versus olaparib alone for women with recurrent platinum-sensitive ovarian cancer: a randomised phase 2 study. Lancet Oncol 2014;15(11):1207–14. https://doi.org/10.1016/S1470-2045(14)70391-2.

[49] Coleman RL, Sill MW, Bell-McGuinn K, et al. A phase II evaluation of the potent, highly selective PARP inhibitor veliparib in the treatment of persistent or recurrent epithelial ovarian, fallopian tube, or primary peritoneal cancer in patients who carry a germline BRCA1 or BRCA2 mutation—an NRG oncology/gynecologic oncology group study. Gynecol Oncol 2015;137(3):386–91. https://doi.org/10.1016/j.ygyno.2015.03.042.

[50] Kummar S, Oza AM, Fleming GF, et al. Randomized trial of oral cyclophosphamide and veliparib in high-grade serous ovarian, primary peritoneal, or fallopian tube cancers, or BRCA-mutant ovarian cancer. Clin Cancer Res 2015;21(7):1574–82. https://doi.org/10.1158/1078-0432.CCR-14-2565.

[51] Sandhu SK, Schelman WR, Wilding G, Moreno V, Baird RD, Miranda S, Hylands L, Riisnaes R, Forster M, Omlin A. The poly (ADP-ribose) polymerase inhibitor niraparib (MK4827) in BRCA mutation carriers and patients with sporadic cancer: a phase 1 dose-escalation trial. Lancet Oncol 2013;14:882–92.

[52] Penson RT, Whalen C, Lasonde B, Krasner CN, Konstantinopoulos P, Stallings TE, Bradley CR, Birrer MJ, Matulonis U. A phase II trial of iniparib (BSI-201) in combination with gemcitabine/carboplatin (GC) in patients with platinum-sensitive recurrent ovarian cancer. J Clin Oncol 2011;29(15_suppl):5004.

[53] Aghajanian C, Sill MW, Secord AA, Powell MA, Steinhoff M. Iniparib plus paclitaxel and carboplatin as initial treatment of advanced or recurrent uterine carcinosarcoma: a gynecologic oncology group study. Gynecol Oncol 2012;126(3):424–7. https://doi.org/10.1016/j.ygyno.2012.05.024.

[54] Ihnen M, Zu Eulenburg C, Kolarova T, et al. Therapeutic potential of the poly(ADP-ribose) polymerase inhibitor rucaparib for the treatment of sporadic human ovarian cancer. Mol Cancer Ther 2013;12(6):1002–15. https://doi.org/10.1158/1535-7163.MCT-12-0813.

[55] Kristeleit RS, Shapiro G, LoRusso P, Infante JR, Flynn M, Patel MR, Tolaney SM, Hilton JF, Calvert AH, Giordano H. A phase I dose-escalation and PK study of continuous oral rucaparib in patients with advanced solid tumors. In: ASCO annual meeting proceedings; 2013. p. 2585.

[56] Drew Y, Ledermann J, Hall G, et al. Phase 2 multicentre trial investigating intermittent and continuous dosing schedules of the poly(ADP-ribose) polymerase inhibitor rucaparib in germline BRCA mutation carriers with advanced ovarian and breast cancer [published correction appears in Br J Cancer. 2016 Jun 14;114(12):e21]. Br J Cancer 2016;114(7):723–30. https://doi.org/10.1038/bjc.2016.41.

[57] Kunkel TA, Erie DA. DNA mismatch repair. Annu Rev Biochem 2005;74:681–710. https://doi.org/10.1146/annurev.biochem.74.082803.133243.

[58] Helleman J, van Staveren IL, Dinjens WN, et al. Mismatch repair and treatment resistance in ovarian cancer. BMC Cancer 2006;6:201 [Published July 31] https://doi.org/10.1186/1471-2407-6-201.

[59] Mesquita B, Veiga I, Pereira D, et al. No significant role for beta tubulin mutations and mismatch repair defects in ovarian cancer resistance to paclitaxel/cisplatin. BMC Cancer 2005;5:101 [Published August 11] https://doi.org/10.1186/1471-2407-5-101.

[60] Plumb JA, Strathdee G, Sludden J, Kaye SB, Brown R. Reversal of drug resistance in human tumor xenografts by 2′-deoxy-5-azacytidine-induced demethylation of the hMLH1 gene promoter. Cancer Res 2000;60(21):6039–44.

[61] Fang F, Balch C, Schilder J, et al. A phase 1 and pharmacodynamic study of decitabine in combination with carboplatin in patients with recurrent, platinum-resistant, epithelial ovarian cancer. Cancer 2010;116(17):4043–53. https://doi.org/10.1002/cncr.25204.

[62] Matei D, Fang F, Shen C, et al. Epigenetic resensitization to platinum in ovarian cancer. Cancer Res 2012;72(9):2197–205. https://doi.org/10.1158/0008-5472.CAN-11-3909.

[63] Cerami E, Gao J, Dogrusoz U, et al. The cBio cancer genomics portal: an open platform for exploring multidimensional cancer genomics data [published correction appears in Cancer Discov. 2012 Oct;2(10):960]. Cancer Discov 2012;2(5):401–4. https://doi.org/10.1158/2159-8290.CD-12-0095.

[64] Juvekar A, Burga LN, Hu H, et al. Combining a PI3K inhibitor with a PARP inhibitor provides an effective therapy for BRCA1-related breast cancer. Cancer Discov 2012;2(11):1048–63. https://doi.org/10.1158/2159-8290.CD-11-0336.

[65] Matulonis U, Wulf GM, Birrer MJ, Westin SN, Quy P, Bell-McGuinn KM, Lasonde B, Whalen C, Aghajanian C, Solit DB. Phase I study of oral BKM120 and oral olaparib for high-grade serous ovarian cancer (HGSC) or triple-negative breast cancer (TNBC). In: ASCO annual meeting proceedings; 2014. p. 2510.

[66] Juric D, Soria J-C, Sharma S, Banerji U, Azaro A, Desai J, Ringeisen FP, Kaag A, Radhakrishnan R, Hourcade-Potelleret F. A phase 1b dose-escalation study of BYL719 plus binimetinib (MEK162) in patients with selected advanced solid tumors. In: ASCO annual meeting proceedings; 2014. p. 9051.

[67] Meyer LA, Slomovitz BM, Djordjevic B, et al. The search continues: looking for predictive biomarkers for response to mammalian target of rapamycin inhibition in endometrial cancer. Int J Gynecol Cancer 2014;24 (4):713–7. https://doi.org/10.1097/IGC.0000000000000118.

[68] Santiskulvong C, Konecny GE, Fekete M, et al. Dual targeting of phosphoinositide 3-kinase and mammalian target of rapamycin using NVP-BEZ235 as a novel therapeutic approach in human ovarian carcinoma. Clin Cancer Res 2011;17(8):2373–84. https://doi.org/10.1158/1078-0432.CCR-10-2289.

[69] Lin YH, Chen BY, Lai WT, et al. The Akt inhibitor MK-2206 enhances the cytotoxicity of paclitaxel (Taxol) and cisplatin in ovarian cancer cells. Naunyn Schmiedebergs Arch Pharmacol 2015;388(1):19–31. https://doi.org/10.1007/s00210-014-1032-y.

[70] Sun H, Yu T, Li J. Co-administration of perifosine with paclitaxel synergistically induces apoptosis in ovarian cancer cells: more than just AKT inhibition. Cancer Lett 2011;310(1):118–28. https://doi.org/10.1016/j.canlet.2011.06.010.

[71] Behbakht K, Sill MW, Darcy KM, et al. Phase II trial of the mTOR inhibitor, temsirolimus and evaluation of circulating tumor cells and tumor biomarkers in persistent and recurrent epithelial ovarian and primary peritoneal malignancies: a Gynecologic oncology group study. Gynecol Oncol 2011;123(1):19–26. https://doi.org/10.1016/j.ygyno.2011.06.022.

[72] Morgan R, Oza A, Qin R, Laumann K, Mackay H, Strevel E, Welch S, Sullivan D, Wenham R, Chen H. A phase II trial of temsirolimus and bevacizumab in patients with endometrial, ovarian, hepatocellular carcinoma, carcinoid, or islet cell cancer: ovarian cancer (OC) subset—a study of the Princess Margaret, Mayo, Southeast phase II, and California Cancer (CCCP) N01 Consortia NCI# 8233. In: ASCO annual meeting proceedings; 2011. p. 5015.

[73] Mabuchi S, Altomare DA, Connolly DC, et al. RAD001 (Everolimus) delays tumor onset and progression in a transgenic mouse model of ovarian cancer. Cancer Res 2007;67(6):2408–13. https://doi.org/10.1158/0008-5472.CAN-06-4490.

[74] Mabuchi S, Altomare DA, Cheung M, et al. RAD001 inhibits human ovarian cancer cell proliferation, enhances cisplatin-induced apoptosis, and prolongs survival in an ovarian cancer model. Clin Cancer Res 2007;13 (14):4261–70. https://doi.org/10.1158/1078-0432.CCR-06-2770.

[75] Robinson DR, Wu YM, Lin SF. The protein tyrosine kinase family of the human genome. Oncogene 2000 Nov;19 (49):5548–57. https://doi.org/10.1038/sj.onc.1203957.

[76] Schlessinger J. Cell signaling by receptor tyrosine kinases. Cell 2000;103(2):211–25. https://doi.org/10.1016/s0092-8674(00)00114-8.

[77] Olayioye MA, Neve RM, Lane HA, Hynes NE. The ErbB signaling network: receptor heterodimerization in development and cancer. EMBO J 2000;19(13):3159–67. https://doi.org/10.1093/emboj/19.13.3159.

[78] Morishita A, Gong J, Masaki T. Targeting receptor tyrosine kinases in gastric cancer. World J Gastroenterol 2014;20(16):4536–45. https://doi.org/10.3748/wjg.v20.i16.4536.

[79] Hubbard SR, Till JH. Protein tyrosine kinase structure and function. Annu Rev Biochem 2000;69:373–98. https://doi.org/10.1146/annurev.biochem.69.1.373.

[80] Lemmon MA, Schlessinger J. Cell signaling by receptor tyrosine kinases. Cell 2010;141(7):1117–34. https://doi.org/10.1016/j.cell.2010.06.011.

[81] Becker JC, Muller-Tidow C, Serve H, Domschke W, Pohle T. Role of receptor tyrosine kinases in gastric cancer: new targets for a selective therapy. World J Gastroenterol 2006;12(21):3297–305. https://doi.org/10.3748/wjg.v12.i21.3297.

[82] Leslie K, Sill M, Darcy K, Baron A, Wilken J, Godwin A, Cook L, Schilder R, Schilder J, Maihle N. Efficacy and safety of gefitinib and potential prognostic value of soluble EGFR, EGFR mutations, and tumor markers in a gynecologic oncology group phase II trial of persistent or recurrent endometrial cancer. In: ASCO annual meeting proceedings; 2009. p. e16542.

[83] Posadas EM, Liel MS, Kwitkowski V, et al. A phase II and pharmacodynamic study of gefitinib in patients with refractory or recurrent epithelial ovarian cancer. Cancer 2007;109(7):1323–30. https://doi.org/10.1002/cncr.22545.

[84] Wagner U, du Bois A, Pfisterer J, et al. Gefitinib in combination with tamoxifen in patients with ovarian cancer refractory or resistant to platinum-taxane based therapy—a phase II trial of the AGO ovarian cancer study group (AGO-OVAR 2.6). Gynecol Oncol 2007;105(1):132–7. https://doi.org/10.1016/j.ygyno.2006.10.053.

[85] Pautier P, Joly F, Kerbrat P, et al. Phase II study of gefitinib in combination with paclitaxel (P) and carboplatin (C) as second-line therapy for ovarian, tubal or peritoneal adenocarcinoma (1839IL/0074). Gynecol Oncol 2010;116(2):157–62. https://doi.org/10.1016/j.ygyno.2009.10.076.

[86] Mavroudis D, Efstathiou E, Polyzos A, Athanasiadis A, Milaki G, Kastritis E, Kalykaki A, Saridaki Z, Dimopoulos A, Georgoulias V. A phase I-II trial of gefitinib in combination with vinorelbine and oxaliplatin as salvage therapy in women with advanced ovarian cancer (AOC). In: ASCO annual meeting proceedings; 2004. p. 5020.

[87] Vasey PA, Gore M, Wilson R, et al. A phase Ib trial of docetaxel, carboplatin and erlotinib in ovarian, fallopian tube and primary peritoneal cancers. Br J Cancer 2008;98(11):1774–80. https://doi.org/10.1038/sj.bjc.6604371.

[88] Hirte H, Oza A, Swenerton K, et al. A phase II study of erlotinib (OSI-774) given in combination with carboplatin in patients with recurrent epithelial ovarian cancer (NCIC CTG IND.149). Gynecol Oncol 2010;118(3):308–12. https://doi.org/10.1016/j.ygyno.2010.05.005.

[89] Seiden MV, Burris HA, Matulonis U, et al. A phase II trial of EMD72000 (matuzumab), a humanized anti-EGFR monoclonal antibody, in patients with platinum-resistant ovarian and primary peritoneal malignancies. Gynecol Oncol 2007;104(3):727–31. https://doi.org/10.1016/j.ygyno.2006.10.019.

[90] Schilder RJ, Pathak HB, Lokshin AE, et al. Phase II trial of single agent cetuximab in patients with persistent or recurrent epithelial ovarian or primary peritoneal carcinoma with the potential for dose escalation to rash. Gynecol Oncol 2009;113(1):21–7. https://doi.org/10.1016/j.ygyno.2008.12.003.

[91] Konner J, Schilder RJ, DeRosa FA, et al. A phase II study of cetuximab/paclitaxel/carboplatin for the initial treatment of advanced-stage ovarian, primary peritoneal, or fallopian tube cancer. Gynecol Oncol 2008;110 (2):140–5. https://doi.org/10.1016/j.ygyno.2008.04.018.

[92] Pearce ST, Jordan VC. The biological role of estrogen receptors alpha and beta in cancer. Crit Rev Oncol Hematol 2004;50(1):3–22. https://doi.org/10.1016/j.critrevonc.2003.09.003.

[93] Björnström L, Sjöberg M. Mechanisms of estrogen receptor signaling: convergence of genomic and nongenomic actions on target genes. Mol Endocrinol 2005;19(4):833–42. https://doi.org/10.1210/me.2004-0486.

[94] Prossnitz ER, Arterburn JB, Sklar LA. GPR30: a G protein-coupled receptor for estrogen. Mol Cell Endocrinol 2007;265–266:138–42. https://doi.org/10.1016/j.mce.2006.12.010.

[95] Mabuchi S, Ohmichi M, Kimura A, et al. Estrogen inhibits paclitaxel-induced apoptosis via the phosphorylation of apoptosis signal-regulating kinase 1 in human ovarian cancer cell lines. Endocrinology 2004;145(1):49–58. https://doi.org/10.1210/en.2003-0792.

[96] Del Carmen MG, Fuller AF, Matulonis U, et al. Phase II trial of anastrozole in women with asymptomatic müllerian cancer. Gynecol Oncol 2003;91(3):596–602. https://doi.org/10.1016/j.ygyno.2003.08.021.

[97] Krasner C, Debernardo R, Findley M, Penson R, Matulonis U, Atkinson T, Roche M, Seiden M. Phase II trial of anastrazole in combination with gefitinib in women with asymptomatic mullerian cancer. In: ASCO annual meeting proceedings; 2005. p. 5063.

[98] Bowman A, Gabra H, Langdon SP, Lessells A, Stewart M, Young A, Smyth JF. CA125 response is associated with estrogen receptor expression in a phase II trial of letrozole in ovarian cancer: identification of an endocrine-sensitive subgroup. Clin Cancer Res 2002;8(7):2233–9.

[99] Papadimitriou CA, Markaki S, Siapkaras J, et al. Hormonal therapy with letrozole for relapsed epithelial ovarian cancer. Long-term results of a phase II study. Oncology 2004;66(2):112–7. https://doi.org/10.1159/000077436.

[100] Smyth JF, Gourley C, Walker G, et al. Antiestrogen therapy is active in selected ovarian cancer cases: the use of letrozole in estrogen receptor-positive patients. Clin Cancer Res 2007;13(12):3617–22. https://doi.org/10.1158/1078-0432.CCR-06-2878.

[101] Ramirez PT, Schmeler KM, Milam MR, et al. Efficacy of letrozole in the treatment of recurrent platinum- and taxane-resistant high-grade cancer of the ovary or peritoneum. Gynecol Oncol 2008;110(1):56–9. https://doi.org/10.1016/j.ygyno.2008.03.014.

[102] Verma S, Alhayki M, Le T, Baines K, Rambout L, Hopkins L, Fung Kee Fung M. Phase II study of exemestane (E) in refractory ovarian cancer (ROC). In: ASCO annual meeting proceedings; 2006. p. 5026.

[103] Argenta PA, Thomas SG, Judson PL, et al. A phase II study of fulvestrant in the treatment of multiply-recurrent epithelial ovarian cancer. Gynecol Oncol 2009;113(2):205–9. https://doi.org/10.1016/j.ygyno.2009.01.012.

[104] Mäenpää J, Sipilä P, Kangas L, Karnani P, Grönroos M. Chemosensitizing effect of an antiestrogen, toremifene, on ovarian cancer. Gynecol Oncol 1992;46(3):292–7. https://doi.org/10.1016/0090-8258(92)90219-9.

[105] Berchuck A, Kohler MF, Marks JR, Wiseman R, Boyd J, Bast Jr RC. The p53 tumor suppressor gene frequently is altered in gynecologic cancers. Am J Obstet Gynecol 1994;170(1 Pt. 1):246–52. https://doi.org/10.1016/s0002-9378(94)70414-7.

[106] Kurrey NK, Jalgaonkar SP, Joglekar AV, et al. Snail and slug mediate radioresistance and chemoresistance by antagonizing p53-mediated apoptosis and acquiring a stem-like phenotype in ovarian cancer cells. Stem Cells 2009;27(9):2059–68. https://doi.org/10.1002/stem.154.

[107] Leffers N, Lambeck AJ, Gooden MJ, et al. Immunization with a P53 synthetic long peptide vaccine induces P53-specific immune responses in ovarian cancer patients, a phase II trial. Int J Cancer 2009;125(9):2104–13. https://doi.org/10.1002/ijc.24597.

[108] Leffers N, Vermeij R, Hoogeboom BN, et al. Long-term clinical and immunological effects of p53-SLP® vaccine in patients with ovarian cancer. Int J Cancer 2012;130(1):105–12. https://doi.org/10.1002/ijc.25980.

[109] Vermeij R, Leffers N, Hoogeboom BN, et al. Potentiation of a p53-SLP vaccine by cyclophosphamide in ovarian cancer: a single-arm phase II study. Int J Cancer 2012;131(5):E670–80. https://doi.org/10.1002/ijc.27388.

[110] Buller RE, Runnebaum IB, Karlan BY, et al. A phase I/II trial of rAd/p53 (SCH 58500) gene replacement in recurrent ovarian cancer. Cancer Gene Ther 2002;9(7):553–66. https://doi.org/10.1038/sj.cgt.7700472.

[111] Buller RE, Shahin MS, Horowitz JA, et al. Long term follow-up of patients with recurrent ovarian cancer after Ad p53 gene replacement with SCH 58500. Cancer Gene Ther 2002;9(7):567–72. https://doi.org/10.1038/sj.cgt.7700473.

[112] Heise C, Sampson-Johannes A, Williams A, McCormick F, Von Hoff DD, Kirn DH. ONYX-015, an E1B gene-attenuated adenovirus, causes tumor-specific cytolysis and antitumoral efficacy that can be augmented by standard chemotherapeutic agents. Nat Med 1997;3(6):639–45. https://doi.org/10.1038/nm0697-639.

[113] Khuri FR, Nemunaitis J, Ganly I, et al. A controlled trial of intratumoral ONYX-015, a selectively-replicating adenovirus, in combination with cisplatin and 5-fluorouracil in patients with recurrent head and neck cancer. Nat Med 2000;6(8):879–85. https://doi.org/10.1038/78638.

[114] Reid T, Galanis E, Abbruzzese J, et al. Hepatic arterial infusion of a replication-selective oncolytic adenovirus (dl1520): phase II viral, immunologic, and clinical endpoints. Cancer Res 2002;62(21):6070–9.

[115] Hirai H, Iwasawa Y, Okada M, et al. Small-molecule inhibition of Wee1 kinase by MK-1775 selectively sensitizes p53-deficient tumor cells to DNA-damaging agents. Mol Cancer Ther 2009;8(11):2992–3000. https://doi.org/10.1158/1535-7163.MCT-09-0463.

[116] Oza AM, Estevez-Diz M, Grischke EM, et al. A Biomarker-enriched, randomized phase II trial of adavosertib (AZD1775) plus paclitaxel and carboplatin for women with platinum-sensitive *TP53*-mutant ovarian cancer [published online ahead of print, 2020 Jul 1]. Clin Cancer Res 2020. https://doi.org/10.1158/1078-0432.CCR-20-0219.

[117] Leijen S, van Geel RM, Sonke GS, et al. Phase II study of WEE1 inhibitor AZD1775 plus carboplatin in patients with TP53-mutated ovarian Cancer refractory or resistant to first-line therapy within 3 months. J Clin Oncol 2016;34(36):4354–61. https://doi.org/10.1200/JCO.2016.67.5942.

[118] Lheureux S, Weberpals JI, Wahner Hendrickson AE, Fleming GF, Olawaiye A, Brana I, Mackay H, Dhani NC, Wilson MK, Rodriguez-Freixinos V. A randomized, placebo-controlled phase II trial comparing gemcitabine monotherapy to gemcitabine in combination with AZD 1775 (MK 1775) in women with recurrent, platinum-resistant epithelial ovarian, primary peritoneal, or Fallopian tube cancers: trial of Princess Margaret, Mayo, Chicago, and California consortia. In: ASCO annual meeting proceedings; 2015. p. TPS5613.

[119] Kobayashi N, Abedini M, Sakuragi N, Tsang BK. PRIMA-1 increases cisplatin sensitivity in chemoresistant ovarian cancer cells with p53 mutation: a requirement for Akt down-regulation. J Ovarian Res 2013;6:7 [Published January 26] https://doi.org/10.1186/1757-2215-6-7.

[120] Mohell N, Alfredsson J, Fransson Å, et al. APR-246 overcomes resistance to cisplatin and doxorubicin in ovarian cancer cells. Cell Death Dis 2015;6(6). https://doi.org/10.1038/cddis.2015.143, e1794.

[121] Mikael VE, Klas GW, Hani G, James DB, Ignace V, Charlie G, Smith A, Jessica A, Nina M, John G. Phase I/ II study of APR-246, a mutant p53 reactivating compound, in combination with standard chemotherapy in platinum sensitive ovarian cancer. In: 12th international congress on targeted anticancer therapies; 2014.

[122] Burikhanov R, Zhao Y, Goswami A, Qiu S, Schwarze SR, Rangnekar VM. The tumor suppressor Par-4 activates an extrinsic pathway for apoptosis [published correction appears in Cell. 2009 Sep 4;138(5):1032]. Cell 2009;138 (2):377–88. https://doi.org/10.1016/j.cell.2009.05.022.

# CHAPTER

# 13

# Overcoming drug resistance in ovarian cancer: Chemo-sensitizing agents, targeted therapies

*Santoshi Muppala*

**Cleveland Clinic, Cleveland, OH, United States**

## Abstract

Ovarian cancer (OC) is one of the highest death-causing cancers among women around the world. Treatments include platinum-based drugs, for example, cisplatin exerts anticancer effects via multiple mechanisms. Although there are major therapeutic options exist, development of strategies based on chemo-sensitization is still a major goal to improve the clinical outcomes. Existing peer-reviewed literature points out multiple factors associated with the development of drug resistance. This chapter is focused on such factors including mutations, signaling pathways, tumor hypoxia, transcription factors, cancer stem cells, epithelial-mesenchymal transition, and tumor microenvironment, etc., that have more potential to be investigated for the development of chemo-resistant drugs and their role(s) in chemosensitivity. Overcoming the drug resistance could significantly be a beneficial treatment strategy to improve chemosensitization. Taken together, multiple mechanisms that lead to resistance of OC cells to drugs might have a great impact in ovarian cancer management.

## Abbreviations

| | |
|---|---|
| **ADP** | adenosine diphosphate |
| **BRAF** | v-raf murine sarcoma viral oncogene homolog B1 |
| **BRCA** | BReast CAncer gene |
| **CICs** | cancer initiating cells |
| **CSCs** | cancer stem cells |
| **EGFR** | epidermal growth factor receptor |
| **EMT** | epithelial mesenchymal transition |
| **EOC** | epithelial ovarian cancer |
| **ERK** | extracellular signal-regulated kinase |
| **EVs** | extracellular vesicles |
| **HGSOC** | high grade serous ovarian cancer |
| **HIF-1α** | hypoxia inducible factor-1α |
| **HK2** | hexokinase-2 |

R. Basha, S. Ahmad (eds.)
*Overcoming Drug Resistance in Gynecologic Cancers*
ISBN 978-0-12-824299-5
https://doi.org/10.1016/B978-0-12-824299-5.00015-0

**HRD**      homologous recombination deficiency
**KRAS**     kirsten rat sarcoma 2 viral oncogene homolog
**MAPK**     mitogen activated protein kinase
**MDR**      multidrug resistance
**MEK**      mitogen-activated protein kinase kinase
**ncRNA**    noncoding RNA
**OC**       ovarian cancer
**PARP**     poly ADP ribose polymerase
**PTEN**     phosphatase and tensin homolog
**ROS**      reactive oxygen species
**STAT3**    signal transducer and activator of transcription 3
**TGF**      transforming growth factor
**TME**      tumor microenvironment
**TP53**     tumor protein p53
**Wnt**      wingless and Int-1
**XBP1**     X-box-binding protein 1

## Conflict of interest

No potential conflicts of interest were disclosed.

## Introduction

Among the gynecologic malignancies, ovarian cancer (OC) is the seventh most common occurring cancers and eighth most common death-causing cancers around the world. OC is heterogeneous in nature and is a metastatic cancer. OC is known to develop from the ovarian surface epithelium and sometimes from serous intraepithelial carcinoma [1]. Epithelial ovarian cancer (EOC), especially high-grade serous OC (HGSOC), accounts for 75% of the cancers arising from ovaries. Existing peer-reviewed studies indicate that the HGSOC is linked to the acquired chemotherapy-resistant mechanisms. It was shown that tumor protein p53 (TP53) gene mutations are highly associated with the occurrence of HGSOC [2]. Mostly, HGSOC is developed form malignant tubal epithelium, while endometrioid and clear cell OC are developed from endometriosis [3]. OC is related to complete remission postsurgery, but sometimes leads to relapse. The interval between these relapses develops into therapy resistance. The underlying mechanisms involved can be genetic alterations of cancer cells, inactivation of drugs, increased DNA repair mechanisms, epithelial mesenchymal transition (EMT), hypoxia, tumor microenvironment (TME), and deregulation of intracellular pathways [1]. Therefore, detailed understanding of the mechanisms that lead to drug resistance are highly crucial in the drug discovery process and better treatment options for OC.

Newer treatment approaches which include chemo-sensitive drugs are more influential for OC treatments [4]. Studies on multidrug resistance (MDR) demonstrated that the use of platinum, taxanes, and non-$p$-glycoprotein substrate therapies are outlined as alternative critical therapeutic approaches to improve the patient outcomes [5]. Clinical therapeutic application of poly ADP ribose polymerase (PARPs), such as talazoparib, veliparib, niraparib, and rucaparib, etc., is identified to be critical to treat patients with EOC, which showed better clinical efficacy in the different stages of OC disease progression [6]. Treatment options include surgery and cytoreduction, followed by the treatments with cytotoxic platinum- and taxane-based chemotherapy. Despite these advance treatments, some tumors will respond but the tumor recurrence is likely to occur within 2 years. More recent advances in the treatment options for OC now include epigenetic therapies, which can be a key research advancement cancer therapy [7].

## Mechanisms of resistance of ovarian cancer cells to drugs

The DNA repair mechanisms are potential pathways that are involved in drug resistance and are associated with the acquired resistance toward many current therapeutic agents [8]. The development of biomarkers should be focused on the mutations in the genes, duplication, deletions, as well as polymorphisms [9]. Existing peer-reviewed studies indicate that the extracellular vesicles (EVs) play an important role in the development of drug resistance in many cancers including OC. The EVs can be exploited as therapeutic agents by which they are known for their improved efficacy and safety properties [10]. Another conventional targeted strategies include nanoscale drug delivery methods which could improve therapeutic effects [11]. Multiple mechanisms to drive the drug resistance and drug sensitivity include cancer stem cells (CSCs), deregulated noncoding RNA (ncRNA), autophagy, and tumor heterogeneity (Fig. 1). Based on the existing evidence on different mechanisms, there is an urgent need of novel concepts of treatments to overcome the MDR for e.g., platinum or PARP inhibitor resistance [12–14].

Current peer-reviewed literature on EOC for cisplatin resistance and tumor advancement is due to the dysregulation of lncRNA-GAS5 and mitogen-activated protein kinase (MAPK) pathways, which are likely to be involved in playing major roles in driving OC drug sensitivity and progression. EOC is one of the malignant tumors, which is known to develop drug resistance to cisplatin at an early stage of OC development [15]. Another study connecting to the development of acquired cisplatin resistance is due to extracellular signal-regulated kinase (ERK)-mediated autophagy and the process of autophagy is enhanced by the hexokinase 2 (HK2) overexpression. Blocking the mitogen-activated protein kinase kinase (MEK)/ERK pathway using MEK inhibitor UO126 reduced autophagy induced by cisplatin and increased by HK2. High expression of HK2 is downregulated by cisplatin-sensitive cells rather than resistant OC cells [16].

## Mutations

Existing peer-reviewed literature evidence on the HGSOC cases are associated with 15% of germline BReast CAncer (BRCA) gene mutation cancers. More than 50% of the HGSOC cases possess the mutations associated with homologous recombination deficiency (HRD) that include germline, somatic BRCA mutations, and BRCA promoter methylation. Therefore, there is a significant potential in addressing the BRCA mutations for clinical implications of HGSOC [17]. EOCs are very common hereditary cancers linked to the mutations in *BRCA1* and *BRCA2* genes. There are studies that reported the association of *BRCA1* gene for the risk of getting OC at the age of 70 years is 40% and *BRCA2* gene is 18% [18]. Patients with these mutations are recommended for the treatment options that include PARP inhibitors [18].

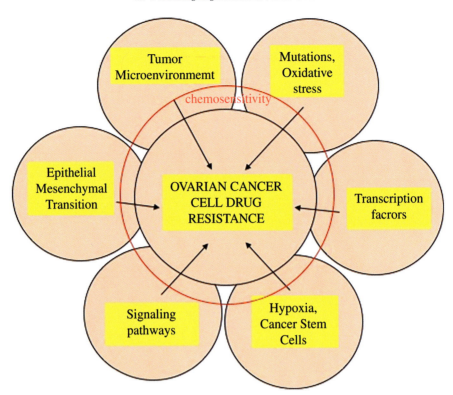

FIG. 1   Mechanisms of drug resistance in ovarian cancer and chemo-sensitizing agents: Multiple mechanisms, such as tumor microenvironment, mutations, oxidative stress, transcription factors, hypoxia, cancer stem cells, signaling pathways, and epithelial-mesenchymal transition, etc., are associated with the induction of drug resistance in ovarian cancer cells—and the need of developing chemo-sensitive drugs is utmost important for the treatment and survival outcomes of ovarian cancer.

Patients under current treatments with PARP inhibitors have become crucial to screen for *BRCA* gene mutations to prioritize the treatment options.

It has been shown that *BRCA* gene mutations are present in most of the EOC, especially accounting for 14% of cases. So *BRCA* gene mutational status plays a major role in the clinical setting to improve treatment options and survival outcomes [19]. Another important aspect to be noted is that the OC patients with homologous recombination deficiencies are observed to have improved responses to treatments that include platinum-based chemotherapies and PARP inhibitors [20]. Based on the hallmark characteristics of OC, they are divided into two stages, i.e., type 1 cancers are characterized by mutations specifically occurring in Kirsten rat sarcoma 2 viral oncogene homolog (KRAS), v-raf murine sarcoma viral oncogene homolog B1 (BRAF), phosphatase and tensin homolog (PTEN), *TP53* genes and are known to be low-grade tumors. Whereas type 2 tumors are very aggressive and high-grade advanced-stage tumors and are characterized by the high occurrence of *TP53* gene mutations [21]. Therefore, treatment options are based on the screening of any associated gene mutations and/or expression.

# Altered signaling pathways

Dysregulated signaling pathways are activated during the progression of EOC and are reported to be related to develop chemoresistance and can be explored to be a potential candidate as a target for chemosensitivity. The wingless and Int-1 (Wnt)/β-catenin signaling pathway is the critical pathway responsible for the EMT transition, and alterations in the Wnt signaling pathway seem to be further investigated for the chemoresistance and chemosensitization [22]. Other known frequently activated pathways that lead to the advancement of the OC include NOTCH signaling, renin-angiotensin system (RAS)/MEK, phosphoinositide 3-kinase (PI3K), and forkhead box protein M1 (FOXM1). Critical understanding of the underlying mechanisms of the dysregulated signaling pathways should be considered for efficient therapies [23].

The protein kinase HIPPO pathway is another important pathway that is altered in many cancer types including OC. Neurofibromin 2 is considered as an upstream regulator of the HIPPO signaling pathway that is inactivated and its biological function is altered [24]. The signal transducer and activator of transcription 3 (STAT3) signaling pathway inhibition leads to suppressed OC progression, metastasis, and chemoresistance. STAT3 deletion in tumor mice models inhibited the cell proliferation, migration, and tumor growth, and further suppressed the important genes that are involved in the EMT progression, E2F (a group of genes that codifies a family of transcription factors in higher eukaryotes) signaling and dysregulated expression of stemness markers [25]. Current studies on cisplatin proved that it is one of the major drugs used as a first-line treatment drug, good initial response, and slowly leads to the development of chemoresistance. Despite the behavior of cisplatin as a key candidate in the treatment of OC, more attention has to be focused on the strategies related to chemosensitization [26].

# Tumor hypoxia

Hypoxia is one of the major regulating factors for driving ovarian tumorigenesis. The interactions between the tumor cells and host immune cells either directly or indirectly respond to the hypoxic environment, which results in modulating the TME. Hypoxic environment leads to the alterations of many oncogenic signaling pathways and oncogenes, which are critical to advance OC progression [27]. Existing evidence from peer-reviewed reports suggest that hypoxia leads to drug resistance in OC. Exosomes derived from the OC cells are able to exert chemoresistance against cisplatin. Exosomes showed the efflux of the cisplatin when the OC cells were treated with cisplatin under hypoxic conditions [24]. This study demonstrated that the exosomes derived from the OC cells conferred resistance against the chemotherapeutic drugs like cisplatin [28].

Hypoxia is the most critical environment that is favorable for many oncogenic processes to be regulated. Hypoxia is also known to regulate the OC cell proliferation, invasion, and adhesion by altering the ErbB signaling pathway, via hypoxia-inducible factor (HIF)-1α-transforming growth factor (TGF)α-epidermal growth factor receptor (EGFR)-ErbB2-MYC axis. Understanding the key processes during hypoxia is critical to develop better therapeutic options in OC [29]. Another study supported the role of exosomes derived from the EOC cells which promote cell proliferation and migration under hypoxic conditions. These OC cell-derived exosomes are rich in microRNAs and determining the role of miRs associated with exosomes may serve as an important diagnostic marker for EOC [30]. Hypoxic conditions also lead to the development of chemoresistance by adapting the cancer cells to the microenvironment and supporting their growth through the regulation of mitochondrial respiratory processes in OC. Therefore, mitochondrial pathways may be important to study for the better treatment outcomes for patients with OC [31].

## Oxidative stress

Oxidative stress is shown to induce OC progression rapidly, and more importantly develops chemoresistance. Existing peer-reviewed literature supports that the oxidative stress is associated with modulating many cancers' cellular processes and also contributes to the progression of endometriosis-associated OC. Understanding the critical genes and microRNAs involved in the pathophysiology of OC under oxidative stress has to be better understood for potential treatments [32]. Higher oxidative stress levels are known to induce higher mitochondrial respiration mechanisms including promyelocytic leukemia (PML)-oxoglutarate carrier (OGC)-1α. Promotion of the mitochondrial respiration especially, oxidative phosphorylation (OXPHOS) metabolism, can in turn promote chemosensitivity in OC [33]. Studies on the use of PARP inhibitors could be a potential strategy to induce oxidative stress in OC cells. PARP inhibitors are able to induce the reactive oxygen species (ROS) which enhances the antitumor effects of OC [34]. Patients with advanced stage of ovarian adenocarcinoma showed increased levels of ROS and these can act as driving force and signaling molecules, which can contribute to the OC disease intensity and progression. All the above-noted studies refer to the importance of oxidative stress, which actually is the generator of the tumor processes and correlates to the advancement of OC [35].

## Resistance of cancer stem cells

Cancer stem cells (CSCs) activation results in the progression of OC and is frequently related to the development of resistance to therapies. Rapid changes in the development and incitement of CSCs steadily impact the mechanisms involved in EMT. The activation of EMT in turn regulates the advancement of OC [1]. CSCs are capable of self-renewal and are able to make the differentiated and undifferentiated cells to grow. They are known to adapt to new tumor microenvironmental conditions and are responsible for tumor recurrence. Based on their unique characteristics, CSCs could be a therapeutic candidate for the treatment of OC [36]. CSCs are summarized to be more important candidates for the occurrence of tumor relapse and drug resistance. Therefore, there is an urgent need for understanding the underlying mechanisms involved in developing the drug-resistant mechanisms/targets, which may be crucial for improving the survival rate of OC patients. CSCs relate to the development of tumor relapse and the best treatment options have yet to be improved, especially for patients with advanced stage of OC [37]. Targeting the CSCs, which play major role(s) in the DNA damage and cell metabolism mechanisms, will therefore be a novel therapeutic advancements in the treatment of OC [38]. CSCs are known to reside at both primary and metastatic areas and effect the progression of OC. Complete or partial killing of the drug-resistant CSCs will impact the overall survival rate of the OC patients [39].

## Transcription factors

A large body of peer-reviewed literature suggests the role of transcription factors in the progression of OC, especially the role of X-box-binding protein (XBP1) in serous OC. For example, XBP1 regulated the inhibition of phosphorylation of mitogen-activated protein kinase (MAPK), p38 pathway, and thereby showing its protective role in SOC cells. XBP1 is also shown to be the critical regulator of ROS in the treatment of SOC. [40] A number of transcription factors linked to Smad family, Forkhead transcription family, and HIFs are known to be associated with the control of many cellular processes such as ovarian cell proliferation, follicular development, and also mediating the effects of hormones. Thus, determining the functions of transcription factors in the regulation of mechanisms of OC growth would be more attractive for novel therapy developments [41]. Other transcription factor named BTB and CNC hololology1 (BACH1) is associated with the EOC progression and metastasis. It was shown that BACH1 transcription factor activated phosphorylated-Akt (p-AKT) and phospho-p70 S6 kinase (p-p70S6K) and increased the expression of cyclinD1, thereby promoting the EMT, tumor growth, and metastasis [42]. In serous borderline ovarian tumors and type 1 ovarian cancers, the EMT transcription factors play pivotal role(s) in the tumor invasiveness and progression. There are differences in the expression patterns of transcription factors including Snail, Slug, TWIST, and ZEB, which imply for their pivotal roles in inducing EMT. Therefore, these transcription factors are highly desirable to be exploited for potential newer treatment strategies/targets for OC [43].

## Epithelial mesenchymal transition

Most of the benign tumors transform into malignant tumors because of the potential mechanisms lead by EMT process. The murine double minute 2 homolog (MDM2) gene is an interacting protein of p53 family, which is considered to be an activator of Smad pathway, thereby promoting EMT and metastasis, which is correlated to a relatively poor prognosis and a high-risk promoting factor for OC [44]. Cancer initiating cells (CICs) could promote tumor aggressiveness and progression. Among the ovarian CICs, one of the differentially expressed genes, CD73 is known to promote the EMT and genes associated with EMT process [45]. Current peer-reviewed literature supports the importance of EMT in the progression and metastasis of OC and in developing chemoresistance. EMT can also be targeted to explore the therapeutic implications for the treatment of OC [46]. Another study on hematopoietic pre-B-cell leukemia transcription factor (PBX) interacting protein (HPIP) is a known scaffold protein, which is highly expressed in high-grade primary ovarian tumors. It plays role in the promotion of metastatic events such as migration, invasion, and EMT which is shown to serve as a potential therapeutic target for cisplatin resistance in ovarian tumors [47]. Overall, it is again emphasized that the EMT is an important mechanism to be considered for developing therapies for OC.

## Tumor microenvironment

Tumor microenvironment (TME) plays important role(s) in the progression and metastasis of OC. Current immunotherapy strategies are focused on the inhibitory networks within the TME. A large body of peer-reviewed literature points out to the influence of TME in the progression of OC and could be a clinical target for the treatment and prevention of OC [48]. Extracellular vesicles, cancer-associated fibroblasts, macrophages, and immune cells have emerging roles in making up the TME favorable for the communication between the normal cells and tumor cells that leads to the accelerated OC progression [49]. A study on the use of losartan as an anticancer agent for OC seems to be modulating the TME and induced the antifibrotic genes expression via its antifibrotic capacity and reduced the tumor burden. Combined treatments of losartan with paclitaxel showed a promising supporting data for the OC patients [50]. Tumor infiltrating lymphocytes (TILs) within the TME are proved to be correlated with the overall survival of OC and are linked to its poor prognosis [51]. TME is associated with the activation of oncogenes, signaling pathways, inactivation of tumor suppressor genes, alteration of normal biological processes, and the initiation and progression of metastatic events, and develop chemoresistance to the therapeutic drugs. Therefore, it again emphasized that TME is highly desirable to be exploited for addressing the challenges of the prevention of immunosuppression against OC [51].

## Conclusions

Ovarian cancer is a deadly disease, which is rapidly invasive and metastatic. Initially, the OC tumors are chemo-sensitive and specific mechanisms that lead to chemoresistance must be focused for therapeutic purposes. Further research needs to be constrained on the development of the combination therapies that are likely to address tumor hypoxia, TME, EMT, and cancer stem cells that could be justified as an effective preventive tool against OC. The drug-resistant therapies should be potentially developed with low toxicity and target to significantly decrease the tumor burden. Furthermore, elucidating the factors that contribute to the development of drug resistance needs to be better understood with a goal to improve the quality of life and survival outcomes of patients with ovarian cancer.

## Highlights

- This chapter highlights the effects of chemo-sensitizing agents and overcoming drug resistance associated with OC progression.
- It reveals the role of mutations and altered signaling pathways in the regulation of OC.

- The role of resistance of cancer stem cells, tumor hypoxia, transcription factors, and TME in connection to OC is discussed.
- It ponders on how different mechanisms are strongly connected in influencing/manipulating the tumor microenvironment and renders drug resistance to therapeutics.

# References

[1] Ottevanger PB. Ovarian cancer stem cells more questions than answers. Semin Cancer Biol 2017;44:67–71. https://doi.org/10.1016/j.semcancer.2017.04.009.

[2] Christie EL, Bowtell DDL. Acquired chemotherapy resistance in ovarian cancer. Ann Oncol 2017;28: viii13–5. https://doi.org/10.1093/annonc/mdx446.

[3] Kujawa KA, Lisowska KM. Ovarian cancer—from biology to clinic. Postepy Hig Med Dosw (Online) 2015;69:1275–90. https://doi.org/10.5604/17322693.1184451.

[4] Norouzi-Barough L, Sarookhani MR, Sharifi M, et al. Molecular mechanisms of drug resistance in ovarian cancer. J Cell Physiol 2018;233:4546–62. https://doi.org/10.1002/jcp.26289.

[5] Freimund AE, Beach JA, Christie EL, et al. Mechanisms of drug resistance in high-grade serous ovarian cancer. Hematol Oncol Clin North Am 2018;32:983–96. https://doi.org/10.1016/j.hoc.2018.07.007.

[6] Mittica G, Ghisoni E, Giannone G, et al. PARP inhibitors in ovarian cancer. Recent Pat Anticancer Drug Discov 2018;13:392–410. https://doi.org/10.2174/1574892813666180305165256.

[7] Moufarrij S, Dandapani M, Arthofer E, et al. Epigenetic therapy for ovarian cancer: promise and progress. Clin Epigenetics 2019;11:7. https://doi.org/10.1186/s13148-018-0602-0.

[8] Cornelison R, Llaneza DC, Landen CN. Emerging therapeutics to overcome chemoresistance in epithelial ovarian cancer: a mini-review. Int J Mol Sci 2017;18. https://doi.org/10.3390/ijms18102171.

[9] Damia G, Broggini M. Platinum resistance in ovarian cancer: role of DNA repair. Cancers (Basel) 2019;11. https://doi.org/10.3390/cancers11010119.

[10] Namee NM, O'Driscoll L. Extracellular vesicles and anti-cancer drug resistance. Biochim Biophys Acta Rev Cancer 2018;1870:123–36. https://doi.org/10.1016/j.bbcan.2018.07.003.

[11] Gupta S, Pathak Y, Gupta MK, et al. Nanoscale drug delivery strategies for therapy of ovarian cancer: conventional vs targeted. Artif Cells Nanomed Biotechnol 2019;47:4066–88. https://doi.org/10.1080/21691401.2019.1677680.

[12] Ren F, Shen J, Shi H, et al. Novel mechanisms and approaches to overcome multidrug resistance in the treatment of ovarian cancer. Biochim Biophys Acta 2016;1866:266–75. https://doi.org/10.1016/j.bbcan.2016.10.001.

[13] Pujade-Lauraine E, Banerjee S, Pignata S. Management of platinum-resistant, relapsed epithelial ovarian cancer and new drug perspectives. J Clin Oncol 2019;37:2437–48. https://doi.org/10.1200/JCO.19.00194.

[14] McMullen M, Karakasis K, Madariaga A, et al. Overcoming platinum and parp-inhibitor resistance in ovarian cancer. Cancers (Basel) 2020;12. https://doi.org/10.3390/cancers12061607.

[15] Long X, Song K, Hu H, et al. Long non-coding RNA GAS5 inhibits DDP-resistance and tumor progression of epithelial ovarian cancer via GAS5-E2F4-PARP1-MAPK axis. J Exp Clin Cancer Res 2019;38:345. https://doi.org/10.1186/s13046-019-1329-2.

[16] Zhang XY, Zhang M, Cong Q, et al. Hexokinase 2 confers resistance to cisplatin in ovarian cancer cells by enhancing cisplatin-induced autophagy. Int J Biochem Cell Biol 2018;95:9–16. https://doi.org/10.1016/j.biocel.2017.12.010.

[17] Moschetta M, George A, Kaye SB, et al. BRCA somatic mutations and epigenetic BRCA modifications in serous ovarian cancer. Ann Oncol 2016;27:1449–55. https://doi.org/10.1093/annonc/mdw142.

[18] Andrews L, Mutch DG. Hereditary ovarian cancer and risk reduction. Best Pract Res Clin Obstet Gynaecol 2017;41:31–48. https://doi.org/10.1016/j.bpobgyn.2016.10.017.

[19] Gadducci A, Guarneri V, Peccatori FA, et al. Current strategies for the targeted treatment of high-grade serous epithelial ovarian cancer and relevance of BRCA mutational status. J Ovarian Res 2019;12:9. https://doi.org/10.1186/s13048-019-0484-6.

[20] da Cunha Colombo Bonadio RR, Fogace RN, Miranda VC, et al. Homologous recombination deficiency in ovarian cancer: a review of its epidemiology and management. Clinics (Sao Paulo) 2018;73. https://doi.org/10.6061/clinics/2018/e450s, e450s.

[21] Kurman RJ, Shih IM. Molecular pathogenesis and extraovarian origin of epithelial ovarian cancer- -shifting the paradigm. Hum Pathol 2011;42:918–31. https://doi.org/10.1016/j.humpath.2011.03.003.

[22] Arend RC, Londono-Joshi AI, Straughn Jr JM, et al. The Wnt/beta-catenin pathway in ovarian cancer: a review. Gynecol Oncol 2013;131:772–9. https://doi.org/10.1016/j.ygyno.2013.09.034.

[23] Testa U, Petrucci E, Pasquini L, et al. Ovarian cancers: genetic abnormalities, tumor heterogeneity and progression, clonal evolution and cancer stem cells. Medicines (Basel) 2018;5. https://doi.org/10.3390/medicines5010016.

[24] Harvey KF, Zhang X, Thomas DM. The Hippo pathway and human cancer. Nat Rev Cancer 2013;13:246–57. https://doi.org/10.1038/nrc3458.

[25] Lu T, Bankhead 3rd A, Ljungman M, et al. Multi-omics profiling reveals key signaling pathways in ovarian cancer controlled by STAT3. Theranostics 2019;9:5478–96. https://doi.org/10.7150/thno.33444.

[26] Galluzzi L, Senovilla L, Vitale I, et al. Molecular mechanisms of cisplatin resistance. Oncogene 2012;31:1869–83. https://doi.org/10.1038/onc.2011.384.

[27] Casey SC, Amedei A, Aquilano K, et al. Cancer prevention and therapy through the modulation of the tumor microenvironment. Semin Cancer Biol 2015;35(Suppl):S199–223. https://doi.org/10.1016/j.semcancer.2015.02.007.

[28] Dorayappan KDP, Wanner R, Wallbillich JJ, et al. Hypoxia-induced exosomes contribute to a more aggressive and chemoresistant ovarian cancer phenotype: a novel mechanism linking STAT3/Rab proteins. Oncogene 2018;37:3806–21. https://doi.org/10.1038/s41388-018-0189-0.

[29] Zhang K, Kong X, Feng G, et al. Investigation of hypoxia networks in ovarian cancer via bioinformatics analysis. J Ovarian Res 2018;11:16. https://doi.org/10.1186/s13048-018-0388-x.

[30] Chen X, Zhou J, Li X, et al. Exosomes derived from hypoxic epithelial ovarian cancer cells deliver microRNAs to macrophages and elicit a tumor-promoted phenotype. Cancer Lett 2018;435:80–91. https://doi.org/10.1016/j.canlet.2018.08.001.

[31] Emmings E, Mullany S, Chang Z, et al. Targeting mitochondria for treatment of chemoresistant ovarian cancer. Int J Mol Sci 2019;20. https://doi.org/10.3390/ijms20010229.

[32] Mari-Alexandre J, Carcelen AP, Agababyan C, et al. Interplay between MicroRNAs and oxidative stress in ovarian conditions with a focus on ovarian cancer and endometriosis. Int J Mol Sci 2019;20. https://doi.org/10.3390/ijms20215322.

[33] Gentric G, Kieffer Y, Mieulet V, et al. PML-regulated mitochondrial metabolism enhances chemosensitivity in human ovarian cancers. Cell Metab 2019;29:156–173 e110. https://doi.org/10.1016/j.cmet.2018.09.002.

[34] Hou D, Liu Z, Xu X, et al. Increased oxidative stress mediates the antitumor effect of PARP inhibition in ovarian cancer. Redox Biol 2018;17:99–111. https://doi.org/10.1016/j.redox.2018.03.016.

[35] Trifanescu O, Gruia MI, Gales L, et al. Tumor is an oxidative stress factor in ovarian cancer patients. Chirurgia (Bucur) 2018;113:687–94. https://doi.org/10.21614/chirurgia.113.5.687.

[36] Al-Alem LF, Pandya UM, Baker AT, et al. Ovarian cancer stem cells: what progress have we made? Int J Biochem Cell Biol 2019;107:92–103. https://doi.org/10.1016/j.biocel.2018.12.010.

[37] Keyvani V, Farshchian M, Esmaeili SA, et al. Ovarian cancer stem cells and targeted therapy. J Ovarian Res 2019;12:120. https://doi.org/10.1186/s13048-019-0588-z.

[38] Zamarin D. Novel therapeutics: response and resistance in ovarian cancer. Int J Gynecol Cancer 2019;29: s16–21. https://doi.org/10.1136/ijgc-2019-000456.

[39] Ahmed N, Escalona R, Leung D, et al. Tumour microenvironment and metabolic plasticity in cancer and cancer stem cells: perspectives on metabolic and immune regulatory signatures in chemoresistant ovarian cancer stem cells. Semin Cancer Biol 2018;53:265–81. https://doi.org/10.1016/j.semcancer.2018.10.002.

[40] Zhang GH, Kai JY, Chen MM, et al. Downregulation of XBP1 decreases serous ovarian cancer cell viability and enhances sensitivity to oxidative stress by increasing intracellular ROS levels. Oncol Lett 2019;18: 4194–202. https://doi.org/10.3892/ol.2019.10772.

[41] Sirotkin AV. Transcription factors and ovarian functions. J Cell Physiol 2010;225:20–6. https://doi.org/10.1002/jcp.22248.

[42] Han W, Zhang Y, Niu C, et al. BTB and CNC homology 1 (Bach1) promotes human ovarian cancer cell metastasis by HMGA2-mediated epithelial-mesenchymal transition. Cancer Lett 2019;445:45–56. https://doi.org/10.1016/j.canlet.2019.01.003.

[43] Crijns AP, Fehrmann RS, de Jong S, et al. Survival-related profile, pathways, and transcription factors in ovarian cancer. PLoS Med 2009;6. https://doi.org/10.1371/journal.pmed.1000024, e24.

[44] Chen Y, Wang DD, Wu YP, et al. MDM2 promotes epithelial-mesenchymal transition and metastasis of ovarian cancer SKOV3 cells. Br J Cancer 2017;117:1192–201. https://doi.org/10.1038/bjc.2017.265.

[45] Lupia M, Angiolini F, Bertalot G, et al. CD73 regulates stemness and epithelial-mesenchymal transition in ovarian cancer-initiating cells. Stem Cell Rep 2018;10:1412–25. https://doi.org/10.1016/j.stemcr.2018.02.009.

[46] Deng J, Wang L, Chen H, et al. Targeting epithelial-mesenchymal transition and cancer stem cells for chemoresistant ovarian cancer. Oncotarget 2016;7:55771–88. https://doi.org/10.18632/oncotarget.9908.

[47] Bugide S, Gonugunta VK, Penugurti V, et al. HPIP promotes epithelial-mesenchymal transition and cisplatin resistance in ovarian cancer cells through PI3K/AKT pathway activation. Cell Oncol (Dordr) 2017;40:133–44. https://doi.org/10.1007/s13402-016-0308-2.

[48] Odunsi K. Immunotherapy in ovarian cancer. Ann Oncol 2017;28:viii1–7. https://doi.org/10.1093/annonc/mdx444.

[49] Feng W, Dean DC, Hornicek FJ, et al. Exosomes promote pre-metastatic niche formation in ovarian cancer. Mol Cancer 2019;18:124. https://doi.org/10.1186/s12943-019-1049-4.

[50] Zhao Y, Cao J, Melamed A, et al. Losartan treatment enhances chemotherapy efficacy and reduces ascites in ovarian cancer models by normalizing the tumor stroma. Proc Natl Acad Sci U S A 2019;116:2210–9. https://doi.org/10.1073/pnas.1818357116.

[51] Ghisoni E, Imbimbo M, Zimmermann S, et al. Ovarian cancer immunotherapy: turning up the heat. Int J Mol Sci 2019;20. https://doi.org/10.3390/ijms20122927.

# Resistance to chemotherapy among ethnic and racial groups: Health disparities perspective in gynecologic cancers

*Begum Dariya[a] and Ganji Purnachandra Nagaraju[b]*

[a]Department of Bioscience and Biotechnology, Banasthali University, Vanasthali, Rajasthan, India
[b]Department of Hematology and Medical Oncology, Winship Cancer Institute, Emory University, Atlanta, GA, United States

## Abstract

The gynecologic cancers originate from any part of female reproductive organs and are commonly diagnosed cancers in women. The health and racial/ethnic group disparities have been associated with poor outcomes in the incidence and mortality rates of cancers, including gynecologic cancers. The etiologic evidences for the disparities detected have also been associated with the levels of education, socioeconomic status, health insurance, and genetic/lifestyle factors. Additionally, the high costs associated with the diagnosis and subsequent treatments are principally cited with the differences in the incidence and mortality rates. The genetic or mutational variations like EGFR, KRAS, PIK3CA, and NF-κB in various racial/ethnic groups have also been the chief causes for developing resistance and altering in responses to the chemotherapeutic drugs. These gene variations result in tumor recurrence and promote metastasis and are with very poor prognosis. Similarly, the health disparities are preventable that showed differences in the diseases and have social disadvantage experienced in few populations as compared with the whole population. In this chapter, we have focused on the racial/ethnical disparities associated with the development of resistance against chemotherapy in major gynecologic cancers (mainly ovarian, endometrial/uterine, and cervical cancers). Additionally, we have also focused on the health disparities with a comprehensive understanding of social determinants that are addressed as the contributory cause for disparities within the races.

## Abbreviations

| | |
|---|---|
| **AACES** | African American Cancer Epidemiology Study |
| **ASR** | age standard rate |

R. Basha, S. Ahmad (eds.)
*Overcoming Drug Resistance in Gynecologic Cancers*
ISBN 978-0-12-824299-5
https://doi.org/10.1016/B978-0-12-824299-5.00016-2

| BMI | body mass index |
| BRCA | BReast CAncer gene |
| CI | confidence interval |
| CNV-H | copy-number high |
| CNV-L | copy-number low |
| EGFR | epidermal growth factor |
| EMT | epithelial mesenchymal transition |
| GOG | Gynecologic Oncology Group |
| HER2 | human epidermal growth factor receptor 2 |
| HLA | human leucocyte antigen |
| hrHPV | high-risk human papilloma virus |
| IACR | International Agency for Cancer Research |
| KNPC | Kaiser Permanent Northern California |
| KRAS | Kristen rat sarcoma viral oncogene |
| NCCN | National Comprehensive Cancer Network |
| NCDB | National Cancer Database |
| NF-κB | nuclear factor kappa B |
| NHANES | National Health and Nutrition Examination Survey |
| OCAC | Ovarian Cancer Association Consortium |
| OR | odds ratio |
| PFS | progression-free survival |
| PIK3CA | phosphotidylinositol-4,5-bisphophate 3-kinase catalytic subunit alpha |
| POLE | polymerase epsilon |
| PTEN | phosphatase and tensin homolog |
| SEER | surveillance epidemiology and end results |
| TCGA | The Cancer Genome Atlas |
| TNF | tumor necrosis factor |
| VEGF | vascular endothelial growth factor |

## Conflict of interest

No potential conflicts of interest were disclosed.

## Introduction

Gynecologic cancers are serious malignancies most commonly detected diseases in women and are considered as a serious public health issue globally. These cancers are the medical conditions that originate from the reproductive organs system of women. Gynecologic cancers include cervical, endometrial/uterine, ovarian (collectively for the cancers of ovary, Fallopian tube, and peritoneal cavity), vaginal, and vulvar cancers. The prevalence of vaginal and vulvar cancers is relatively very low. As estimated for 2020, women diagnosed with gynecologic cancers were about 113,520 and the estimated deaths were about 33,620 [1]. The overall incidence for these gynecologic cancers has drastically increased over the years. Earlier studies on gynecologic cancers largely included white women and therefore, there was a racial/ethnic disparity in the risk for gynecologic cancers that have been deputized. However, from the past decades, researchers have made their efforts to evaluate among the racial/ethnic groups for the risk of gynecologic cancers. The epidemiological research evidences suggest that the variation in the mortality and prognosis are correlated with various disparities like racial/ethnicity, geographic, and health insurance status disparities [2]. For instance, the risk for mortality was high in black women (16%) than in white women as compared among over 200,000 cancer patients [3, 4]. This is because the black women patients are unlikely to receive therapeutic consistently and have shown with worse survival outcomes irrespective of the disease stage. Additionally, the etiological studies also evidenced disparities in cancer therapy and their outcomes are interlinked with the cultural, racial, ethnic, literacy, and socioeconomic factors.

From the past, the major professional/scientific organizations like the American Cancer Society (ACS), the National Cancer Institute (NCI), the American Society of Clinical Oncology (ASCO), the Centre to Reduce Cancer Health Disparities (CRCHD), and the Society of Gynecologic Oncology (SGO) were committed to eradicate disparities in the cancer-related moralities and survival outcomes [5–7].

Lack of standardized/uniform care contributes to the occurrence of gynecologic cancer and creates a disparity among various racial/ethnic groups. Additionally, the less interest toward timely surgery, chemotherapies, lack of participation in clinical trials, genetic counseling, and experimental treatment by the minority groups are other contributing factors for poor quality of life and survival outcome. Furthermore, development of chemoresistance in the cancer cells and cancer recurrence also results in lower survival rate. Therefore, molecular-level studies are also very much essential for the researchers and clinicians to determine the efficacy and outcomes of the drug when used on diverse group of patients. The cancer recurrence is due to the mutations and overexpression of genes like *p53*, *PTEN*, and *HER2* that eventually develops resistance against chemotherapeutic drugs and are with poor histological grade [8–10]. For instance, overexpression of mutated *p53* is detected in black women found to have poor survival rate with diagnosed recurrence stage I level of cancer and are associated with resistance against therapy. Thus, in this chapter, we included all crucial underlying issues/concerns to develop chemoresistance that contribute to the racial and ethnic disparities in the management of various gynecologic cancers (mainly ovarian, endometrial/uterine, and cervical cancers).

## Ovarian cancer

Ovarian cancer (cancers of ovary, Fallopian tube, and peritoneal cavity) is a heterogeneous and fatal disease. Epithelial ovarian carcinomas (EOC, ~90% of ovarian cancers) and ovarian germ cell tumor are the common type of ovarian cancers. This cancer is relatively less prevalent and is the eighth most common type of cancer in females. However, it is the fifth leading cause for cancer-related mortalities in the United States [1]. Researchers and clinicians are now focusing on evaluating the risk for ovarian cancer occurrence in relation with the racial/ethnic disparities that is facilitated via varied consortium formation, and using the pooled data from epidemiological studies provided [11]. Therefore, the data collected from various national/international databases and the associated consortiums like Ovarian Cancer Association Consortium (OCAC) [12] and the multisite African American Cancer Epidemiology Study (AACES) [13] are involved in evaluating the racial/ethnic groups in relation with the risk factors for the cause of ovarian cancer and survival outcomes.

As per the US NCI's surveillance, epidemiology, and end results (SEER), the estimated new cases of ovarian cancer for the year 2020 are 21,750 (1.2%) and estimated mortalities are 13,940 (2.3%). The overall 5-year relative survival rate detected was 48.6% [14, 15]. The lifetime risk developed in women with ovarian cancer is 1 in 75 with the increased chances of mortality having 1 in 100 [16]. The higher mortality rate is due to the asymptomatic nature of ovarian cancer, lack of optimal screening techniques, and delayed diagnosis until their advanced stages [17–19]. Therefore, the disease is often known as "silent killer" [15, 20].

The other risk factors for the cause of ovarian cancer include mutation in BRCA1 and BRCA2 genes, Lynch II syndrome, family history having cancer, and late menopause [21]. Additionally, the genetic and epigenetic factors play crucial roles in promoting cancers. While 10%–15% of the familial ovarian cancer resulted from the breast cancer-mutated genes (BRCA1 and BRCA2), 60%–80% of familial and sporadic ovarian cancers are due to the mutations in Tp53 gene expression [22]. The mutations in BRCA1 and BRCA2 genes are found alike in all racial/ethnic groups. For instance, 1%–4% prevalence is seen in the population of Asians, whites, Africans, and Hispanics. However, Ashkenazi Jews were with ~10-fold higher occurrence of mutations in BRCA1 and BRCA2 genes [23]. The prevalence of mutations in BRCA1 and BRCA2 genes, as per the cross-sectional analyses, showed 5% among Asian, ~7% for white European women, 1% African American, and ~5% Latin-American women [23]. Thus, lower the prevalence of BRCA1 and BRCA2 gene mutations, higher will be the 5-year survival for the carriers and would unlikely to be a risk factor associated for the incidence and survival in the ethnic/racial disparities. These dysregulated oncogenes cause overactivation of the cancer progression signaling pathways causing pathogenicity.

The incidence of ovarian cancer varies significantly in respect to the race/ethical groups. The white women were detected with higher incidence of 12.5% new cases, followed by Hispanic 10.6%, American Indian/Alaska Native of about 10.4, black having 9.6%, and Asian/Pacific Islander had comparatively lower incidence of about 9.3% cases per 100,000 women [17]. The incidence rate is however found to decline considerably with annual percentage change of about −1.8%, for the period of 2005–2014, which is due to the use of oral contraceptives that prevents risk for ovarian cancer. Over time, the incidence rate also declined but vary with race/ethnicity, white women showed a decline of −1.9% of annual percentage change while non-Hispanic whites showed −0.5% percentage decline [24]. However, the cancer is found more fatal for the black women with 5-year survival is detected at 31% and while it is 42% for whites [25]. This disparity is notified as per the tumor phase/stage and age group of the patient of varied racial/ethnic groups. The survival of the patient would be higher if ovarian cancer is diagnosed at the localized stage, and is lower among women diagnosed after the spread of cancer to lymph nodes and other organ systems. The survival rate consistently decreased among black patients due to the potential delays in diagnosis. The other factors for poor survival outcomes include socioeconomic disadvantages and underlying comorbidities. Similarly, the mortality rate also varies markedly via race/ethnicity. The highest mortality rate is detected in non-Hispanic white women with 7.9 deaths per 100,000 women, while black women have 6.4 deaths, followed by Hispanic (5.4 deaths), Alaska Natives/American Indian (4.5 deaths), and Asian/Pacific Islanders of about 4.4 deaths per 100,000 women [24]. Thus, from the analysis, high incidence and mortality is detected among non-Hispanic women and black women while substantially lower detection is seen among Asian/Pacific Islanders.

The therapeutic options for ovarian cancer include chemotherapy while the advanced therapeutic options include targeted therapy, hormone therapy, immunotherapy, and combination thereof. As per the guidelines of the National Comprehensive Cancer Network (NCCN), primary treatment for ovarian cancer includes debulking surgery followed by chemotherapy [26]. The frequently prescribed chemotherapeutic drugs include platinum-based agents like cisplatin and carboplatin, taxane family drugs like paclitaxel and docetaxel [26]. Thus, as analyzed by the NCCN, the survival rate must have been increased for the patient who ever received the treatment as per the guidelines [27, 28]. But the data from case studies showed that only about 35% of white women and 30% of black women have received the therapy as per the stage of the disease with appropriate surgical and chemotherapeutic interventions [29]. Researchers from the meta-analysis data determined that the differences in survival is related to the racial disparity irrespective of the commencement of the NCCN guidelines-based therapy and care [30]; however, the reason for variations is still indistinct. Researchers have utilized data from the registry of cancer patients that includes data for tumor histology [31, 32], comorbidity [31, 33], history of smoking, and health insurance status [32, 33] all together related with racial/ethnic disparities. In respect to this multivariate model, another research study also detected racial disparity regarding the treatment regimen for surgery followed by chemotherapy in ovarian cancer patients with stage II–IV disease [34]. The study included comparison of white women with other races for odds ratio (OR) and confidence interval (CI) like Asians showed 95% CI having 0.73–0.88 and OR 0.80, Hispanic having OR 0.74, 95% CI 0.69–0.79, and black women had OR 0.7 while 95% of CI 0.66–0.74 are all together less likely to receive surgery, followed by chemotherapy in case of stage II-IV ovarian cancer patients [34]. The patients also experienced varied treatment quality with different socioeconomic status. For instance, patients with lower socioeconomic status less probably received therapy as per the NCCN guidelines [35]. The medically underserved communities have less access to best-quality healthcare facilities [35, 36]. Hispanic and African American received lower facilities than Caucasian patients for ovarian cancer patients as they unlikely received surgery followed by chemotherapy [35–37].

There are other studies/organizations that explained variations for the treatment basing on the data analyses available from the racial/ethnic disparities. For instance, Kaiser Permanent Northern California (KNPC) data analysis studied that African American women are likely to receive reduction in chemotherapy doses, and are found to have very poor survival than other races [38]. Notably, while all the racial groups received same health insurance and care support as per the KNPC criteria, but the differences in the survival might have been due to the other inherent risk factors. In addition to the socioeconomic status and healthcare facilities, the efficacy of chemotherapy is also affected due to the underlying comorbid conditions such as obesity, diabetes, and hypertension, etc. The data from the KNPC study reported that the obese women were given decreased doses of chemotherapy drugs, and this dose reduction as per the body weigh/BMI was correlated with the worse survival of the patients [39].

Data analysis from the National Health and Nutrition Examination Survey (NHANES) compared women of different races >20 years of age. As per their estimations, 56.9% of non-Hispanic black women were found to be obese with BMI of $\geq 30 \, kg/m^2$ as compared with other races like non-Hispanic white (35.5%), non-Hispanic Asian women (11.5%), and Hispanic women (45.7%) [40]. The other comorbid conditions like hypertension and diabetes are highly detected in Hispanic and non-Hispanic black women than the white and Asian women [41]. Thus, these differences in age, socioeconomic status, comorbidities, and healthcare factors substantially attribute to the differential response toward the efficacy of chemotherapy in different racial/ethnic groups.

Chemotherapy drug resistance developed by the cancer cells is always an obstacle in cancer therapy. The ovarian cancer cells develop resistance against cisplatin and carboplatin chemotherapeutic drugs, and are often replaced by such agents as doxorubicin, bevacizumab, and gemcitabine [42, 43] as an advanced and combination therapeutic options. Researchers are now identifying the root cause for the development of chemoresistance in cancer cells. Chemoresistance developed during the chemotherapy could be of two types: intrinsic chemoresistance and acquired chemoresistance [44]. The intrinsic resistance is inherent resistant developed to the drug that develop various modifications including inhibition of apoptosis, inhibition of drug uptake and increased detoxification, and drug efflux [45]. The acquired resistance would be due to genetic and epigenetic alterations [45]. Therefore, the knowledge about the mutated genes and dysregulated pathways via employing genetic counseling to the individual will give a better picture for health condition/care of the patients by the clinicians. Unfortunately, the treatment efficacy is also diminished due to less likely participation of the races in clinical trials and experimental therapies. As per the NCCN guidelines, every ovarian cancer patient must receive the genetic counseling. The genetic counseling and genetic testing are essential, as these approaches provide the healthcare professionals a better understanding of the patient's disease condition, including tumor stage, etc. [46, 47].

The Gynecologic Oncology Group (GOG) performed a data analysis for 400 clinical trials done during the years 1985–2013 and evaluated that Caucasian (83%) have actively participated rather than the African American women (8%) and Hispanics (2.2%) [35, 48, 49]. The poor participation by these minority groups in the experimental treatments and clinical trials is due to the lack of understanding about the benefits of clinical trials, limited access to research centers, poor transportation facilities, and illiteracy. Thus, they are unaware of these treatments that differ from the previous therapies and have lesser efficacy. Now since the patients were identified with the *BRCA1/2* gene mutations, they would result in dysregulation of homologous recombination mechanism in the DNA repair mechanism. This, eventually would result in the mutations of downstream signaling genes. Therefore, studying about these downstream-mutated genes would contribute to targeted therapies and help clinicians with informed decisions for the care of patients. However, poor socioeconomic status and relatively less access to genetic counseling facilities provide lesser information on the basic therapy and that essentially requires patient's involvement for her own care [36, 50].

Previous research studies determined that mutations in *BRCA1/2* are also unlikely found to develop resistance against chemotherapy and are the major factors for the increase in incidence and relatively poor survival of the patient. The reviews about the prevalence of *BRCA1/2* genes and the risk for ovarian cancer occurrence found that the mutations in *BRCA1* are most commonly detected in patients with the family history of having both ovarian cancer and breast cancer. White women had a deleterious *BRCA1* mutations (25%–40%) than the other races like African Americans (16%), Hispanics (16%–23%), and Asians (12%) [51]. While the *BRCA2* mutations are less prevalent when compared to *BRCA1* mutations, the *BRCA1* prevalence commonly detected in Asian Americans (13%), 11% in African Americans, and ranging from 6% to 15% in white women and 6% to 8% in Hispanic. Thus, genetic counseling and genetic testing are very much essential for the therapy commencement in patients with ovarian cancer [51].

The in vitro evidence also suggests that the ovarian cancer cells deficient of *BRCA* gene improved sensitivity of the cytotoxic agents or chemotherapeutic drugs. This determines that inhibiting BRCA1 protein in a chemo-resistant ovarian cancer cell lines restore sensitivity to cisplatin [52, 53]. Furthermore, there are various proteins that interact with BRCA, which includes RAD51, RAD50, TP53, E2F, c-myc, and STAT1 proteins, eventually get dysregulated and promote resistance to drugs and inhibit apoptosis. The patients with mutations in *BRCA1/2* most commonly harbor mutations in *TP53* and prevent *p21* expression and inhibits apoptosis [54, 55]. The dysregulated *BRCA1/2* inhibit the cell cycle arrest in G2 phase and also lack transcription-coupled repair [56]. Thus, a complete genetic counseling would potentiate the targeted therapy for better efficacy and outcomes. Additionally, the resistance developed could be due to the overexpression of antiapoptotic proteins like Bcl-2 [57], survivin [58], tyrosine-protein kinase Met (c-Met) [59, 60], and VEGF [61] in ovarian cancer. The inhibitors of apoptotic proteins are correlated with intrinsic and extrinsic signaling cascades. Furthermore, these apoptotic inhibitor proteins develop resistance against chemotherapy and promote proliferation and angiogenesis. Therefore, targeting transcription factors like specific proteins (Sp) [61] and mutated genes like *BRCA1/2* would enhance the sensitivity to chemotherapy drugs. Thus, following the guidelines developed by the scientific/professional organizations would enhance the quality of healthcare for all women irrespective of the racial/ethnic groups and would reduce disparities within ultimately for better efficacy and survival outcomes.

## Endometrial cancer

Endometrial cancer or uterine cancer is the most common gynecologic cancers detected in western population. It is the fourth most common diagnosed and sixth most common cause for cancer-related mortalities in US women [62]. As per the estimation from the ACS for the year 2020, about 65,620 new cases diagnosed while 12,590 will die due to uteruine or endometrial cancer. The main risk factors associated with the occurrence of corpus uteri are obesity, early menarche, anovulation, late menopause, and endometrial hyperplasia [63, 64]. Endometrial cancer is the most commonly detected disease in the United States that affects 1 in 37 women with a continuous increase in the incidence over the years [65, 66]. The uterine cancer can be diagnosed at an early stage based on the symptoms, including postmenopausal status with abnormal vaginal bleeding. The most widely used diagnostic tests include hysteroscopy and endometrial biopsy.

The 5-year survival if considered from the year 2009 to 2015 varies with the time of diagnosis as per the SEER data analysis. So, the estimated 5-year survival could be 95% if diagnosed in its localized stage, 69% if regional, 17% if diagnosed in distant or metastatic stage. However, the racial disparities are rather more pronounced within the incidence, diagnosis, and survival categories. For instance, as estimated by the ACS, 62% of 5-year survival rate is detected in black women with poor survival, while 83% in white women [66, 67]. Similarly, the incidence of endometrial cancer even though higher in white women, black women are more likely to have increased mortality rate [68]. The highest incidence rate with this disease was identified in North America with an age standard rate (ASR) of 2.5 and lowest in South-Central Asia having of 2.5 ASR per 100,000 as reported by the International Agency for Cancer Research (IACR). Additionally, the analyses done by the US Census Bureau project also mentioned about the increased incidence in the black women and are particularly with aggressive histological subtypes [69].

Similarly, in another analysis, the African women showed the highest incidence rate of about 6.3 and low incidence of 4.5 in South Asian women as compared in 100,000 population [70]. The peer-reviewed literatures from the past also demonstrated that the incidence of endometrial cancer is highly prevalent in African American women due to the commonly occurring risk factors like obesity, higher BMI, and poor lifestyle management, and is less prevalent in European American women who are less obese as compared to African Americans [71, 72]. Furthermore, the mortality rate has significantly increased than the incidence rate due to the apparent delay in diagnosis of endometrial cancer as mentioned in few research publications [73]. The black women showed 55% higher mortality rate than white women [66]. If compared with the European Americans, the African Americans showed 2-fold higher mortality rate [66, 74, 75]. Furthermore, the mortality gap within the racial/ethnic groups varies with various other studies and healthcare settings.

Multiple factors including the lack of proper healthcare coverage, socioeconomic status, comorbidities, histology, and treatment inequality are some of the other reasons for the endometrial cancer disparities seen in racial/ethnic groups [2, 76]. The socioeconomic status inequalities are detected over all the incidence, survival, and mortality in various racial/ethnic groups. The factors for socioeconomic status that showed variations include income [77–79], education [80, 81], and health insurance [81, 82]. Topics like cultural differentiations, health education, and awareness of symptoms are the potential influences in endometrial cancer racial disparities [81, 83, 84]. Higher the socioeconomic status, better is the health outcome of individuals, while less socioeconomic status is determined due to the lack of healthcare. Lower socioeconomic status in the races showed up with decreased survival and late diagnosis [85]. For instance, black women having low socioeconomic status and lack of health insurance are very less likely to receive the much needed therapeutic regimens including radiation/chemotherapies. Thus, the variation detected among black women was with 2.5-fold higher mortality as compared to white women that had substantially high socioeconomic status [78]. Additionally, the lower socioeconomic status among black women also resulted in poor prognosis regardless of the stage/grade of endometrial cancer and are also less likely to receive timely treatment [74].

The additional factor that potentiates the socioeconomic status in affecting the racial groups for the incidence and survival of endometrial cancer is education and immigration status. A population of European women ($n = 5500$) was evaluated that are influenced by the immigration status and education level [86]. In this case, higher incidence rate was detected in the patients that had lower education, while the immigrant women having lower education level showed lower incidence rate, which could be due to the geographical influences. The lower survival rate was also attributed with the increased BMI, smoking, obesity, and underlying comorbidities, in addition to the socioeconomic status [87]. The underlying comorbidities like hypertension and diabetes have modest impact on the mortality rate in endometrial cancer in relation with therapy and had racial differences. Black women with comorbidity showed an increase in mortality [88]. While from the white/black women cancer survival study, black women are highly diabetic, obese, and hypertensive than the white women, and are thus, more likely to have more risk for endometrial cancer occurrence [89]. However, the disease specificity, overall survival, and occurrence of cancer showed no correlation [88, 90–92].

The general treatment of patients with endometrial cancer includes hysterectomy, lymph node sampling, and bilateral salpingo-oopherectomy [93]. The surgery is followed with adjuvant treatments including chemotherapy and/or radiation therapy that are based on the tumor differentiation and tumor invasion into the wall of endometrium (i.e., histopathological assessments). The most common chemotherapy drugs offered paclitaxel, carboplatin, doxorubicin, cisplatin, and docetaxel. Moreover, the aggressive tumor differentiation and invasion are widely detected in black women than in white women. There is no difference detected in receiving the adjuvant therapy among the races [79, 94, 95]. However, in gynecologic oncology arena, disparities in racial/ethnic groups exist even in receiving guidelines-based care and are very less likely to be presented in the clinical trials enrollment. For instance, black women very rarely receive guidelines-based therapies and very less likely to participate in clinical trials [7, 49]. As observed through the GOG clinical trials, the enrollment of black women is over ten times lower in participation for clinical trials [49]. Thus, a 3-fold lower enrollment variation is detected between black (6%) and white (21%) women as analyzed from gynecologic oncology trials.

A hypothesis from the National Cancer Database (NCDB) was assessed for the odds in receiving guideline concordant treatment among the women that have had endometrial cancer with endometrioid histology as per the racial and ethnic disparities. The hypothesis showed that non-Hispanic white women (75.3%) are more likely to receive guidelines-concordant treatment than the other non-Hispanic black (70.1%) and Hispanic (71.0%) women. On the other hand, the Asian/Pacific Islander women (72.5%) more prevalently received treatment as per the guidelines.

The overall survival of the patient is associated with the guidelines-based treatment and showed up with worst survival for the racial/ethnic group that did not receive guidelines-based treatment [96]. The evaluation of data by John et al. [97] from clinical trials conducted by the GOG-9 study showed 26% of higher mortality in black women than the white women. The response they detected for Adriamycin and platinum-based chemotherapy in blacks was 35% and 43% for white. The study determined an early discontinuation of chemotherapeutic cycles in both black (55.2%) and white (52.5%) women that did not have any racial disparity but was discontinued due to the range of the disease progression both in black and white women. Additionally, the chemotherapy drug-related toxicities were other reasons to discontinue the therapy. Notably, black women experienced about 82% increased adverse effects like anemia, less frequency of leukopenia, and increased gastrointestinal toxicity than the white patients. However, the racial/ethnic group differences were detected in endometrial cancer patients that received radiation therapy. For instance, taking such factors into account as age, comorbidity, diagnostic stage, and status of lymphadenectomy, Hispanic women (33.4%) and Asian (34.6%) women received radiation therapy less likely than the black (38.3%) and white women (38.7%) [34].

Differences in treatment over the races are another chief contributor for the disparity in endometrial cancer. The data from the NCDB analyzed 50,000 subjects that determined 9% of black women did not receive any kind of treatment as compared to white women that had 4% who did not receive the treatment. While the women who ever received treatment were black women and have received radiation therapy as the primary treatment [78, 90], however, these subjects did not receive surgery as the tumor progression and histology were more likely controlled in all the stages [98]. Previous research studies also cited cultural barriers in some communities that were also responsible for delay in the medical care and surgery even though they were with symptoms, and thus, were presented only in their advanced tumor stages [76, 78]. Furthermore, the analysis of data from the SEER database for not having surgery (e.g., hysterectomy) is because, as the surgery was not recommended and other reasons like lack of uniformity in the data recording at the healthcare centers. Therefore, further analysis on population-based data was very much difficult to process. The healthcare data obtained from the SEER analysis that evaluated 711 women showed that there was no difference or disparity identified in black or white women for the primary surgery, postoperative radiation therapy, and adjuvant therapy [78, 98, 99]. Other research studies also demonstrated that difference in the racial groups was also detected as per the response they show toward chemotherapy and hysterectomy. Furthermore, a meta-analysis done by the GOG identified that response rate decreased in patients and showed up with recurrent metastatic stages who have already received chemotherapy. In this regard, the analysis ascertained that black women showed response rate of 34.9% while white women showed 43.2% [100].

Furthermore, the genetic variations are the other key features or factors for endometrial cancer occurrence influenced by the racial disparity. There are multiple studies that have evaluated molecular variations in black and white women having endometrial cancer that were identified to be the reason for poor prognosis [74, 101]. The studied research analysis widely detected mutations in *p53*, *PTEN*, and overexpression of HER2/neu expression. Thus, in addition to the treatment variations and the socioeconomic status in different racial/ethnic groups, the recurrence of the disease is widely seen in the patients due to the dysregulation of tumor-promoting genes that develops resistance against drugs used for the therapy, as illustrated in Table 1. Mutations in *p53* and *PIK3CA* genes are most widely detected in black women and showed type II endometrial cancer most prevalently, and are with unfavorable prognosis [74, 101]. Mutations in *p53* gene are widely associated with advanced stage,

TABLE 1    Ethnicity variations showing recurrently mutated genes in endometrial cancer.

| Gene | Asian | Caucasian | Black or African American |
| --- | --- | --- | --- |
| PTEN | 85% | 63.26% | 41% |
| RPL22 | 15% | 11.33% | 7.55% |
| PIK3CA | 13% | 50% | 41% |
| ARID5B | 30% | 13.81% | 11.32% |
| ARID1A | 45% | 43.92% | 28.3% |
| FBXW7 | 15% | 17.68% | 22.64% |
| TP53 | 25% | 31.77% | 49.06% |
| PIK3R1 | 30% | 19.61% | 16.98% |
| CTNNB1 | 30% | 27.07% | 16.04% |
| KRAS | 25% | 21% | 14.15% |
| CTCF | 25% | 23.76% | 18.87% |

high-grade tumor, and nonendometroid histology with poor survival outcomes [8]. Black women were more likely to overexpress *p53* mutations in their stage 1 endometrial cancer and have higher likelihood for the disease recurrence [9]. Mutations in *PTEN* and *PIK3CA* genes are most commonly detected in white women in their early stages of endometrial cancer, and however have favorable prognosis [101]. The HER2 upregulation is most commonly detected in black women (70%) than white women (24%) that is correlated with the development of resistance and poor survival of the patient. The white women were estimated to have 16% microsatellite instability (MSI) and 22% of mutations in *PTEN* and are more detected than black women that have 5% *PTEN* and 13% MSI [74, 101]; however, they had better overall survival than the black women. The Caucasian women were also reported with higher prevalence of mutations in *PTEN*, and are with better prognosis [102, 103]. Similarly, MSI is most prevalent in the tumors detected in Caucasian women as compared to black or African American women. The mutations in *PIK3CA* and *KRAS* genes are highly detected in black or African American women than in Caucasian women having low-grade endometrial cancer with endometrioid histology [11]. Unlike the information regarding mutations in relation with races, relatively less is known about ethnic disparities, which is the subject of future investigations.

The Cancer Genome Atlas (TCGA) Research Network has reported genomic analysis of endometrial carcinoma (Table 2). This analysis included transcriptome sequence analysis, whole exome sequence analysis, microsatellite stability testing, methylation testing, and genomic copy-number analysis. These analyses identified four endometrial carcinoma subtypes that are characterized as per the pathologic, distinct clinical and molecular nature. They include: polymerase epsilon (POLE), MSI, copy-number low (CNV-L) and copy-number high

TABLE 2  Genomic analysis of endometrial cancer as reported by the TCGA.

| Molecular alterations in TCGA data | Characterized as | Endometrial carcinoma | Mutated genes | Black | White | Reference (s) |
|---|---|---|---|---|---|---|
| POLE (ultra-mutated 7%) | Somatic mutation in POLE DNA | Low-grade and high-grade endometrioid carcinoma | *TP53* | 6.0% | 4.8% | [104, 105] |
| MSI (hyper mutated 28%) | Deficiency of DNA MMR system, decreased MLH1 mRNA expression. | Low- to high-grade endometrioid tumor | *ARID5B*, *PTEN*, *PIK3CA*, *PIK3R1* | 29% | 14.3% | [106] |
| CNV-low (microsatellite stable 39%) | Microsatellite stable | Low-grade endometrioid tumor | Progesterone receptor, *RAD50* | 41.5% | 19.1% | [107–109] |
| CNV-high (serous like 26%) | Microsatellite stable | Includes all serous carcinoma | *CCNE1*, *PIK3CA*, *CDKN2A*, *MYC*, *TP53*, *Akt* | 23.5% | 16.9% | [107–109] |

TCGA: The Cancer Genome Atlas; MSI: Microsatellite instability; POLE: Polymerase epsilon; CNV: Copy number;

(CNV-H). Among the four subtypes, three were found to have more aggressive molecular alterations in the black patients that are correlated with very poor progression-free survival (PFS) than in the white women. Mitotic subtypes are the strongly associated subtypes for poor survival in black patients with endometrial cancer.

The evidence from previous studies identified that aberrant alternating splicing are closely associated with the cancer progression, development of resistance, and thus, can be targeted at biomarkers level. More recently identified splicing factor is YT521 that was found to promote vascular endothelial growth factor (VEGF)-A alternative splicing while the upregulation of variant VEGF-165 is associated with invasion of endometrial cancer [104]. Targeting alternative splicing would be a promising approach for the targeted therapy of the disease [105, 106]. Other alternative splicing includes HER-2, FBXW7, p53, PTEN, and PIK3CA. Thus, the racial disparity that show variations in developing resistance against the drugs can be offered personalized therapy via targeting these alternate splicing markers.

Similarly, from the immunohistochemistry of endometrial adenocarcinoma, p-mTOR and p-4EBP1 are found to be active [107]. Subsequently, mTORC2 was also found to be upregulated in endometrial cancer that is evidenced from the overexpression of p-mTOR, p-Akt, VEGF-A isoform, and PLD1 [108]. Researchers have also shown that mTOR pathway plays crucial role in endometrial tumorigenesis and was found to be insensitive to rapamycin. Furthermore, overexpression of Akt is also found to be associated with resistance to radiation therapy/chemotherapy [109]. The molecular subtype CNV-H is widely detected in black (23.5%) than the white women (16.9%) that is associated with the dysregulation of Akt and TP53. Black women show increased overexpression of Akt, and thus, induce resistance against chemotherapy drugs. Therefore, inhibiting the PI3K/Akt pathway in endometrial cancer would be a promising approach in enhancing the efficacy of paclitaxel and cisplatin. Additionally, the studies from the GOG have included patients ($n = 1151$) and tested for response toward doxorubicin and platinum-based chemotherapy. Among the patients of different races, black women were very much less likely to have responded to platinum and doxorubicin drugs [97]. Thus, altogether black women were determined to have various odd in receiving adjuvant chemotherapy than the other races which could be due to the socioeconomic status and dysregulation of various genes in their body. Thus, focusing on prospective studies, patient-provider interactions, and personalized treatment are warranted to improve better understanding of the treatment in patients with endometrial cancer.

## Cervical cancer

Cervical cancer initiates progression from the cell lining of the cervix, present at the lower part of the uterus. Globally, it is the most common cancers diagnosed among women and is the major cause for cancer-related mortalities [110]. As estimated by the ACS for the year 2020, about 13,800 new cases will be diagnosed for invasive cervical cancer in the United States and about 4290 women will die from the disease [110]. The cervical precancer is most often diagnosed than the invasive cervical cancer. The main risk factors associated with cervical cancer progression include high-risk human papilloma virus (hrHPV) infection, oral contraception, smoking, age, childbirth, and diet [111–114]. The incidence rate for cervical cancer occurrence has substantially decreased in the United States during the past decade (thanks due to HPV vaccination); however, the racial health disparities still exist and require significant attention in the disease management [35, 115, 116].

As analyzed, the incidence rate frequency was high among African American women (60% higher) than the Caucasian American women [115–119]. The incidence for cervical cancer is also found higher in both Hispanic and Asian women that have low socioeconomic status. As per the data available from the year 2015, the incidence rate was found to be higher in Hispanic women (13.8) than blacks (11.4) and non-Hispanic whites (8.5) per 100,000 women. However, the mortality rate was found to be higher in blacks (4.9%) than in Hispanics (3.3%) and non-Hispanic whites (2.3%) per 100,00 women [120].

Age plays a significant role in increasing the incidence of cervical cancer across all the racial/ethnic groups. If compared within the races of black, white, Hispanic, and Asian women, over 60 years of age have the increased incidence of disease and is associated with cancer-related mortalities [121]. The age-associated incidence of cervical cancer when calculated during the years 2005 to 2009 from the SEER database, black women were 9.8/100,000 and white women were 8.0/100,000 [122] detected. Similarly, the mortality rate also varied with the age as analyzed 4.3 for black and 2.2 for white women [122]. Additionally, these women have never been for screening prior to their diagnosis [123]. Thus, the incidence of cervical cancers increases with the increase in age across all the racial/ethnic groups.

The mortality rate is also twice higher in African American than Caucasian women [119]. The increasing minority population in the United States, i.e.., the Hispanic/Latina have increased rates of cervical cancer, high mortality rate with worst prognosis. The Hispanic women (9.5 per 100,000) were commonly diagnosed in their advanced stages, and thus, experienced higher mortality than the other non-Hispanic women (7.5 per 100,000) and African American women [124]. However, the 5-year survival rate was found to be higher in Hispanic women (75%) than the non-Hispanic women (71%) [125]. In support to this work, another study data analysis showed that the white women (69%) have better 5-year survival rate than the black women (56%) when compared between the races [126]. The variations in racial/ethnic groups were also significantly related to the geographical distribution observed in the incidence and mortality rates of cervical cancer [127]. These incidence and mortality rates also vary in rural and urban areas having least response for screening test, and thus, had higher incidence among minority groups [128, 129].

The screening and therapy for any cancers like cervical cancer is a multistep process and is with successful outcomes only if it is with special care and facilities provided at interdisciplinary levels. However, for few underrepresented races or minorities, patients experience difficulties due to the risk for lack of periodic follow-up. Thus, the variations in the screening of cervical cancer are another possible explanation for the observed disparity in the mortality and morbidity rates. The screening analysis done from the national data between the years 2007 and 2011 found that women in economically strong countries, educated, and having private health insurance are more likely to participate in the screening tests for cervical cancer irrespective of the race, whereas the low-economic countries' women less likely have undergone such screening tests [130]. Furthermore, the WHO also determined that the mortality associated with invasive cervical cancer increased up to 80% in low- and middle-income countries, and is estimated to further increase by 30% by the year 2030 [131–134]. In other words, the incidence and mortality rates have been decreased in high-income countries due to the availability of advanced cytology-based screening tests to diagnose precancerous cervical cancer lesions [132, 135].

Simultaneously, efficient low-cost screening techniques are made available in the low-income countries like India [136], Brazil and Argentina [137], Haiti [138], Soviet Union [139], and China [140]. Despite of modified screening facilities for the minority races, the incidence and death rates other than the white women are twice higher in black and Hispanic women [141]. This is due to the lack of periodic follow-ups in screening, medical comorbidities, illiteracy, public health insurance, and socioeconomic status that inherently promote racial disparities in terms of incidence and death rate [142–144]. In higher income countries like the United States, the incidence and mortality rates are determined to be higher in black women than white women [140]. Furthermore, the Centres for Disease Control and Prevention (CDC) also analyzed that 81% of the US women had screening test within the previous 3 years from the time of their diagnosis. Among them, African American women have had actively participated in screening tests than the Asian and Hispanic women [35, 145]. However, black women were more likely to be diagnosed in their later stages of the disease than the Caucasian women [35, 145, 146]. Similarly, Hispanic women were also diagnosed in their advanced stages of the disease and had poorer response to the treatments [121, 147]. Moreover, a community that include Hispanic and black living together are at higher risk for high-grade cervical cancer that could be due to racial-associated social networking attributed within the community [148].

Thus, socioeconomic status would be responsible for the delayed diagnosis and the risk for advanced stage cancer if combined with other races. For cervical cancer, racial disparity also involves in the examination and prevention; however, the prevention is now potential through the introduction of immunization of HPV vaccination. The survey conducted by the US National Immunization Board demonstrated that the HPV vaccination was lower in black and Hispanic adolescent girls than the white counterparts [149]. Additionally, the black girls though received vaccination initially were found to skip the three-dose vaccination scheduled [150, 151].

The differences in the treatment are well documented and play significant role(s) in cervical cancer racial/ethnic disparities. The therapeutic strategies within the American white women and other minority groups have periodically altered and play crucial roles in the outcome within the racial disparities. Research analysis studies showed that the black women less likely receive therapy due to the lack of periodic follow-up [152]. They also less likely receive surgery and more likely radiation therapy than the other races [153]. The minority racial groups including Asians, blacks, and Hispanics very less likely had health insurance, and therefore, had less resources for optimal treatment [154]. The older women received treatment under Medicaid or no health insurance less likely received guidelines-based care [155, 156]. The white women comparatively received higher radical hysterectomy for early-stage cervical cancer than the black women [157] and more likely received intracavity radiation therapy for advanced-stage disease [158]. The data analyzed from the past showed that the subjects who received surgery as the initial therapy included radical hysterectomy for Hispanics (51.7%) and blacks (32.4%) considerably received local surgery as compared with the other racial groups. If considered for the years 2004 to 2007, black women (70.1%) were found to receive surgery or chemoradiation therapy less likely than the other groups. However, with later in the years 2011 to 2014, the proportion of received chemoradiation therapy increased among black women (76.7%). These findings were recorded as per the multivariate logistics regression model taking age, comorbidity, and year of the diagnosis as confounding factors.

The preferred standard treatment for cervical cancer patients includes radiation therapy and/or chemotherapy. The most commonly prescribed chemotherapeutic drugs include cisplatin, carboplatin, paclitaxel, and topotecan. Additionally, researchers and clinicians are often using combination drugs together with 5-fluorouracil (5-FU), irinotecan, docetaxel, and gemcitabine. These combination therapies are standard therapies for recurrence and metastatic cervical cancers. The cervical cancer cells acquire chemoresistance due to the epithelial-mesenchymal transition (EMT), which promotes upregulation of N-cadherin, Twist and Snail, and downregulates E-cadherin [159]. Additionally, EMT also promotes overexpression of efflux transporters ATP-binding cassette including multidrug-resistant proteins (such as MRP1, MRP3, MRP2, and MRP5) [160–162]. The cisplatin-resistant cells showed decreased activity of Bcl-2, caspase-8, caspase-3, caspase-9, and poly (ADP-ribose) polymerase (PARP) cleavage [163, 164]. Moreover, the loss of p53 activity in cervical cancer cells promote drug resistance and impair apoptosis [165]. The exon sequencing done for cancer-related genes taken from the radiation-refractory recurrent tumor showed increased activity of PIK3CA and KRAS [166]. The mutations in *KMT2A*, TET1, and *NLRP1* genes were also detected in resistant tumors [166]. The mutations in cancer-associated genes including *EGFR* and *KRAS* were also found to promote tumorigenesis and had varied racial background. For instance, a somatic-mutated EGFR gene, *S768I*, was detected in East Asian patients that had potential role to develop resistance in EGFR tyrosine kinase inhibitor [167]. The *EGFR* mutations were also found significantly high in whites (14.1%) than African Americans (2.4%) [168]. Similarly, *KRAS* mutations were also found higher in whites (26%) than African Americans (17%) [169].

Furthermore, the in vitro and in vivo studies demonstrated that transcription factors like nuclear factor (NF)-κB inhibit the chemo drug efficacy via inhibiting apoptosis and promote resistance to chemo drugs in cervical cancer cells. Therefore, inhibiting NF-κB expression would sensitize the tumor cells [170]. Additionally, tumor necrosis factor-α gene (*TNFAIP8*) is found to inhibit apoptosis and induce resistance against cisplatin in cervical cancer cells [171].

Drug resistance is the significant cause for cancer-related mortalities that occurs due to recurrence of the disease and alternate response against the chemotherapeutic drugs that affects the racial/ethnic groups variedly. As detected, Caucasian women showed better response to therapy with the downregulation of TNF-α which eventually inhibits NF-κB and chemokines release in breast cancer than African American [172]. Similarly, in cervical cancer cell lines, the Caucasian women may respond better to the treatment than the African American via targeting TNF-α and sensitizing the cancer cells. Additionally, the response against chemo drugs also differs with the racial groups.

The human leukocyte antigen (HLA) system plays crucial role in HPV virus-associated cervical cancer. The *HLA II* genes are expressed on the peptide epitope of CD4$^+$ T-cells with subtype HLA-DR loci. Studies have demonstrated relation between the gene polymorphism of *HLA-DRB1* and cervical cancer in various racial groups [173]. The family alleles of *HLA-DRB1* were also found to influence cervical cancer by developing resistance in the cells. Additionally, the meta-analysis done for the *HLA-DRB1* alleles showed genetic susceptibility in ethical groups, which showed that polymorphism of *HLA-DRB1* family alleles are associated with cervical cancer in the Chinese Uighur group [173]. Moreover, alleles HLA-DRB1*03 and HLA-DRB1*08 are found protective while HLA-DRB1*10 and HLA-DRB1*15 are the risk factors for the occurrence of cancer in Uighur group and no such differentiation is detected in other ethical groups [173]. Therefore, profiling of genetic, epigenetic factors responsible for cervical cancer progression along with molecular understanding is essential for determining the mutated and resistance genes that show variations in the racial/ethnic groups. This approach may ultimately help support to eradicate the development of cervical cancer worldwide.

## Conclusions and future perspectives

The healthcare disparities in the cancers like gynecologic malignancies are persevered worldwide that are worsening and are with relatively poor clinical outcomes. Understanding these disparities is complicated and includes various factors like racial, geographic, comorbidities/age, socioeconomic status, and genetical factors that influence the therapy and associated survival outcomes. The critical factors including cancer prevention, screening, and surgery positively impact the survival of patients. Thus, a focused observance to therapeutic practices that are based on evidences is found to be effective and would diminish racial/ethnic disparities. Efficient, successful policies, ongoing research, and programs' implementations are essential for investigating barriers to provide high-quality care and evidence-based practice that promote better survival outcomes among women with gynecologic cancers. Additionally, the molecular genetic information of an individual if easily accessible along with decreased cost of novel techniques like sequencing of genome for individualized therapy would be a promising approach for gynecologic cancer therapy/management that would reduce or prevent racial/ethnic disparities. Moreover, awareness via improved educational efforts, well-trained oncology clinician workforce and translational research, individual participation in clinical trials, and care are the crucial interdisciplinary factors to help eradicate/minimize the racial disparities in gynecologic cancers. These initiatives would ultimately improve the overall survival of the population and cost-effective quality of care for patients with gynecologic cancers, especially in the era of coronavirus pandemic.

## References

[1] Siegel RL, Miller KD, Jemal A. Cancer statistics, 2019. CA Cancer J Clin 2019;69(1):7–34.
[2] Allard JE, Maxwell GL. Race disparities between black and white women in the incidence, treatment, and prognosis of endometrial cancer. Cancer Control 2009;16(1):53–6.
[3] Howard J, et al. A collaborative study of differences in the survival rates of black patients and white patients with cancer. Cancer 1992;69(9):2349–60.
[4] Bach PB, et al. Survival of blacks and whites after a cancer diagnosis. JAMA 2002;287(16):2106–13.
[5] Goss E, et al. American society of clinical oncology policy statement: disparities in cancer care. J Clin Oncol 2009;27(17):2881–5.

[6] Zeng C, et al. Disparities by race, age, and sex in the improvement of survival for major cancers: results from the National Cancer Institute surveillance, epidemiology, and end results (SEER) program in the United States, 1990 to 2010. JAMA Oncol 2015;1(1):88–96.

[7] Vickers NJ. Animal communication: when i'm calling you, will you answer too? Curr Biol 2017;27(14):R713–5.

[8] Kohler MF, et al. Overexpression and mutation of p53 in endometrial carcinoma. Cancer Res 1992;52(6):1622–7.

[9] Clifford SL, et al. Racial disparity in overexpression of the p53 tumor suppressor gene in stage I endometrial cancer. Am J Obstet Gynecol 1997;176(6):s229–32.

[10] Santin AD, et al. Racial differences in the overexpression of epidermal growth factor type II receptor (HER2/neu): a major prognostic indicator in uterine serous papillary cancer. Am J Obstet Gynecol 2005;192(3):813–8.

[11] Cannioto RA, et al. Ovarian cancer epidemiology in the era of collaborative team science. Cancer Causes Control 2017;28(5):487–95.

[12] Berchuck A, et al. Role of genetic polymorphisms in ovarian cancer susceptibility: development of an international ovarian cancer association consortium. In: Ovarian Cancer. Springer; 2008. p. 53–67.

[13] Schildkraut JM, et al. A multi-center population-based case–control study of ovarian cancer in African-American women: the African American Cancer epidemiology study (AACES). BMC Cancer 2014;14(1):688.

[14] Stewart C, Ralyea C, Lockwood S. Ovarian cancer: an integrated review. In: Seminars in Oncology Nursing. Elsevier; 2019.

[15] Carlson KJ, Singer DE. Screening for ovarian cáncer. Ann Intern Med 1995;122(2):155.

[16] Howlader N. SEER Cancer Statistics Review, 1975–2008. Bethesda, MD: National Cancer Institute; 2011. http://seer.Cancer.Gov/csr/1975_2008/. based on November 2010 SEER data submission, posted to the SEER web site.

[17] Howlader N, et al. SEER Cancer Statistics Review. 2008. National Cancer Institute; 1975.

[18] Beckmeyer-Borowko AB, et al. The effect of time on racial differences in epithelial ovarian cancer (OVCA) diagnosis stage, overall and by histologic subtypes: a study of the National Cancer Database. Cancer Causes Control 2016;27(10):1261–71.

[19] Morris CR, Sands MT, Smith LH. Ovarian cancer: predictors of early-stage diagnosis. Cancer Causes Control 2010;21(8):1203–11.

[20] Jayson GC, et al. Ovarian cancer. The Lancet 2014;384(9951):1376–88.

[21] Holschneider CH, Berek JS. Ovarian cancer: epidemiology, biology, and prognostic factors. In: Seminars in Surgical Oncology. Wiley Online Library; 2000.

[22] Bast RC, Hennessy B, Mills GB. The biology of ovarian cancer: new opportunities for translation. Nat Rev Cancer 2009;9(6):415–28.

[23] Hall MJ, et al. BRCA1 and BRCA2 mutations in women of different ethnicities undergoing testing for hereditary breast-ovarian cancer. Cancer 2009;115(10):2222–33.

[24] Howlader N, Noone A, Krapcho M. Surveillance, Epidemiology, and End Results (SEER) Program. SEER Cancer Statistics Review, 1975–2012 (updated August 20, 2015). Bethesda, MD: National Cancer Institute; 2016. based on November 2014 SEER data submission, posted to the SEER web site, April 2015.

[25] Stewart SL, et al. Disparities in ovarian cancer survival in the United States (2001-2009): findings from the CONCORD-2 study. Cancer 2017;123:5138–59.

[26] Morgan RJ, et al. Ovarian cancer, version 2.2013. J Natl Compr Canc Netw 2013;11(10):1199–209.

[27] Baldwin LA, et al. Ten-year relative survival for epithelial ovarian cancer. Obstet Gynecol 2012;120(3):612–8.

[28] Bristow RE, et al. Analysis of racial disparities in stage IIIC epithelial ovarian cancer care and outcomes in a tertiary gynecologic oncology referral center. Gynecol Oncol 2011;122(2):319–23.

[29] Warren JL, et al. Trends in the receipt of guideline care and survival for women with ovarian cancer: a population-based study. Gynecol Oncol 2017;145(3):486–92.

[30] Terplan M, Smith EJ, Temkin SM. Race in ovarian cancer treatment and survival: a systematic review with meta-analysis. Cancer Causes Control 2009;20(7):1139–50.

[31] Bristow RE, et al. Sociodemographic disparities in advanced ovarian cancer survival and adherence to treatment guidelines. Obstet Gynecol 2015;125(4):833.

[32] Bristow RE, et al. Disparities in ovarian cancer care quality and survival according to race and socioeconomic status. J Natl Cancer Inst 2013;105(11):823–32.

[33] Liu F, et al. Racial disparities and patterns of ovarian cancer surgical care in California. Gynecol Oncol 2014;132(1):221–6.

[34] Rauh-Hain JA, et al. Racial and ethnic disparities over time in the treatment and mortality of women with gynecological malignancies. Gynecol Oncol 2018;149(1):4–11.

[35] Chatterjee S, et al. Disparities in gynecological malignancies. Front Oncol 2016;6:36.

[36] Sakhuja S, et al. Availability of healthcare resources and epithelial ovarian cancer stage of diagnosis and mortality among blacks and whites. J Ovarian Res 2017;10(1):57.

[37] Srivastava SK, et al. Racial health disparities in ovarian cancer: not just black and white. J Ovarian Res 2017; 10(1):1–9.

[38] Bandera EV, et al. Racial/ethnic disparities in ovarian cancer treatment and survival. Clin Cancer Res 2016; 22(23):5909–14.

[39] Bandera EV, et al. Impact of chemotherapy dosing on ovarian cancer survival according to body mass index. JAMA Oncol 2015;1(6):737–45.

[40] Alsop BR, Sharma P. Esophageal cancer. Gastroenterol Clin 2016;45(3):399–412.

[41] Gunning MN, et al. Coronary artery calcification in middle-aged women with premature ovarian insufficiency. Clin Endocrinol (Oxf) 2019;91(2):314–22.

[42] Sehouli J, et al. Pegylated liposomal doxorubicin (CAELYX®) in patients with advanced ovarian cancer: results of a German multicenter observational study. Cancer Chemother Pharmacol 2009;64(3):585–91.

[43] Ferrandina G, et al. Phase III trial of gemcitabine compared with pegylated liposomal doxorubicin in progressive or recurrent ovarian cancer. J Clin Oncol 2008;26(6):890–6.

[44] Rubin SC, et al. Ten-year follow-up of ovarian cancer patients after second-look laparotomy with negative findings. Obstet Gynecol 1999;93(1):21–4.

[45] Herzog TJ. Update on the role of topotecan in the treatment of recurrent ovarian cancer. Oncologist 2002; 7(90005):3–10.

[46] Nelson HD, et al. Risk assessment, genetic counseling, and genetic testing for BRCA-related cancer in women: a systematic review to update the US preventive services task force recommendation. Ann Intern Med 2014; 160(4):255–66.

[47] Kamara D, et al. Cancer counseling of low-income limited English proficient Latina women using medical interpreters: implications for shared decision-making. J Genet Couns 2018;27(1):155–68.

[48] Tan W, Stehman FB, Carter RL. Mortality rates due to gynecologic cancers in New York state by demographic factors and proximity to a gynecologic oncology group member treatment center: 1979–2001. Gynecol Oncol 2009;114(2):346–52.

[49] Scalici J, et al. Minority participation in gynecologic oncology group (GOG) studies. Gynecol Oncol 2015; 138(2):441–4.

[50] Randall TC, Armstrong K. Health care disparities in hereditary ovarian cancer: are we reaching the underserved population? Curr Treat Options Oncol 2016;17(8):39.

[51] Kurian AW. BRCA1 and BRCA2 mutations across race and ethnicity: distribution and clinical implications. Curr Opin Obstet Gynecol 2010;22(1):72–8.

[52] Husain A, et al. BRCA1 up-regulation is associated with repair-mediated resistance to cis-diamminedichloroplatinum (II). Cancer Res 1998;58(6):1120–3.

[53] Bhattacharyya A, et al. The breast cancer susceptibility gene BRCA1 is required for subnuclear assembly of Rad51 and survival following treatment with the DNA cross-linking agent cisplatin. J Biol Chem 2000; 275(31):23899–903.

[54] Welcsh PL, King M-C. BRCA1 and BRCA2 and the genetics of breast and ovarian cancer. Hum Mol Genet 2001;10(7):705–13.

[55] Welcsh PL, Owens KN, King M-C. Insights into the functions of BRCA1 and BRCA2. Trends Genet 2000;16(2): 69–74.

[56] Larson JS, Tonkinson JL, Lai MT. A BRCA1 mutant alters G2-M cell cycle control in human mammary epithelial cells. Cancer Res 1997;57(16):3351–5.

[57] Witham J, et al. The Bcl-2/Bcl-XL family inhibitor ABT-737 sensitizes ovarian cancer cells to carboplatin. Clin Cancer Res 2007;13(23):7191–8.

[58] Yamamoto T, Tanigawa N. The role of survivin as a new target of diagnosis and treatment in human cancer. Med Electron Microsc 2001;34(4):207–12.

[59] Jagadeeswaran R, et al. Functional analysis of c-met/hepatocyte growth factor pathway in malignant pleural mesothelioma. Cancer Res 2006;66(1):352–61.

[60] Kong DS, et al. Prognostic significance of c-met expression in glioblastomas. Cancer 2009;115(1):140–8.

[61] Safe S, Abdelrahim M. Sp transcription factor family and its role in cancer. Eur J Cancer 2005;41(16): 2438–48.

[62] Siegel RL, Miller KD, Jemal A. Cancer statistics, 2020. CA Cancer J Clin 2020;70(1):7–30.

[63] Francies FZ, et al. Genomics and splicing events of type II endometrial cancers in the black population: racial disparity, socioeconomic and geographical differences. Am J Cancer Res 2020;10(10):3061.

[64] Lortet-Tieulent J, et al. International patterns and trends in endometrial cancer incidence, 1978–2013. J Natl Cancer Inst 2018;110(4):354–61.

[65] Kohler BA, et al. Annual report to the nation on the status of cancer, 1975–2011, featuring incidence of breast cancer subtypes by race/ethnicity, poverty, and state. J Natl Cancer Inst 2015;107(6):djv048.

[66] Kost E, et al. The growing burden of endometrial cancer: a major racial disparity affecting hispanic women. Gynecol Oncol 2019;154(1), e11.

[67] In T. Facts & Figures 2019: US Cancer Death Rate Has Dropped 27% in 25 Years; 2019.

[68] Terplan M, et al. Have racial disparities in ovarian cancer increased over time? An analysis of SEER data. Gynecol Oncol 2012;125(1):19–24.

[69] Gaber C, et al. Endometrial cancer trends by race and histology in the USA: projecting the number of new cases from 2015 to 2040. J Racial Ethn Health Disparities 2017;4(5):895–903.

[70] Shirley MH, et al. Incidence of breast and gynaecological cancers by ethnic group in England, 2001–2007: a descriptive study. BMC Cancer 2014;14(1):979.

[71] Ogden CL, et al. Prevalence of childhood and adult obesity in the United States, 2011–2012. JAMA 2014; 311(8):806–14.

[72] Ogden CL, et al. Prevalence of obesity and trends in body mass index among US children and adolescents, 1999–2010. JAMA 2012;307(5):483–90.

[73] Guzha B, et al. Endometrial cancer with bone marrow metastases: a management dilemma. South Afr J Gynecol Oncol 2017;9(2):22–4.

[74] Long B, Liu F, Bristow RE. Disparities in uterine cancer epidemiology, treatment, and survival among African Americans in the United States. Gynecol Oncol 2013;130(3):652–9.

[75] Collins Y, et al. Gynecologic cancer disparities: a report from the health disparities taskforce of the Society of Gynecologic Oncology. Gynecol Oncol 2014;133(2):353–61.

[76] Yap OS, Matthews RP. Racial and ethnic disparities in cancers of the uterine corpus. J Natl Med Assoc 2006; 98(12):1930.

[77] Armstrong K, et al. Racial differences in surgeons and hospitals for endometrial cancer treatment. Med Care 2011;49(2):207.

[78] Madison T, et al. Endometrial cancer: socioeconomic status and racial/ethnic differences in stage at diagnosis, treatment, and survival. Am J Public Health 2004;94(12):2104–11.

[79] Fader AN, et al. Disparities in treatment and survival for women with endometrial cancer: a contemporary national cancer database registry analysis. Gynecol Oncol 2016;143(1):98–104.

[80] Barrett II RJ, et al. Endometrial cancer: stage at diagnosis and associated factors in black and white patients. Am J Obstet Gynecol 1995;173(2):414–23.

[81] Fedewa S, et al. Insurance status and racial differences in uterine cancer survival: a study of patients in the National Cancer Database. Gynecol Oncol 2011;122(1):63–8.

[82] Dolly D, et al. A delay from diagnosis to treatment is associated with a decreased overall survival for patients with endometrial cancer. Front Oncol 2016;6:31.

[83] Rauh-Hain JA, et al. Racial disparities in treatment of high-grade endometrial cancer in the Medicare population. Obstet Gynecol 2015;125(4):843–51.

[84] Ozen A, et al. Effect of race and histology on patterns of failure in women with early stage endometrial cancer treated with high dose rate brachytherapy. Gynecol Oncol 2015;138(2):429–33.

[85] Darin-Mattsson A, Fors S, Kåreholt I. Different indicators of socioeconomic status and their relative importance as determinants of health in old age. Int J Equity Health 2017;16(1):1–11.

[86] Svanvik T, et al. Sociodemographic disparities in stage-specific incidences of endometrial cancer: a registry-based study in West Sweden, 1995–2016. Acta Oncol 2019;58(6):845–51.

[87] Donkers H, et al. Systematic review on socioeconomic deprivation and survival in endometrial cancer. Cancer Causes Control 2019;1–10.

[88] Olson SH, et al. The impact of race and comorbidity on survival in endometrial cancer. Cancer Epidemiol Biomarkers Prev 2012;21(5):753–60.

[89] Hill HA, et al. Racial differences in endometrial cancer survival: the black/white cancer survival study. Obstet Gynecol 1996;88(6):919–26.

[90] Hicks ML, et al. The national cancer data base report on endometrial carcinoma in African-American women. Cancer 1998;83(12):2629–37.

[91] Folsom AR, et al. Diabetes as a risk factor for death following endometrial cancer. Gynecol Oncol 2004;94(3):740–5.

[92] Setiawan VW, et al. Racial/ethnic differences in endometrial cancer risk: the multiethnic cohort study. Am J Epidemiol 2007;165(3):262–70.

[93] Holbeck SL, et al. The National Cancer Institute ALMANAC: a comprehensive screening resource for the detection of anticancer drug pairs with enhanced therapeutic activity. Cancer Res 2017;77(13):3564–76.

[94] Bregar AJ, et al. Disparities in receipt of care for high-grade endometrial cancer: a National Cancer Data Base analysis. Gynecol Oncol 2017;145(1):114–21.

[95] Elshaikh MA, et al. The impact of race on outcomes of patients with early stage uterine endometrioid carcinoma. Gynecol Oncol 2013;128(2):171–4.

[96] Kaspers M, et al. Black and Hispanic women are less likely than white women to receive guideline-concordant endometrial cancer treatment. Am J Obstet Gynecol 2020.

[97] Farley JH, et al. Chemotherapy intensity and toxicity among black and white women with advanced and recurrent endometrial cancer: a gynecologic oncology group study. Cancer 2010;116(2):355–61.

[98] Randall TC, Armstrong K. Differences in treatment and outcome between African-American and white women with endometrial cancer. J Clin Oncol 2003;21(22):4200–6.

[99] Trimble EL, et al. Pre-operative imaging, surgery and adjuvant therapy for women diagnosed with cancer of the corpus uteri in community practice in the United States. Gynecol Oncol 2005;96(3):741–8.

[100] Plaxe SC, Saltzstein SL. Impact of ethnicity on the incidence of high-risk endometrial carcinoma. Gynecol Oncol 1997;65(1):8–12.

[101] Althubiti MA. Mutation frequencies in endometrial Cancer patients of different ethnicities and tumor grades: an analytical study. Saudi J Med Med Sci 2019;7(1):16.

[102] Maxwell GL, et al. Racial disparity in the frequency of PTEN mutations, but not microsatellite instability, in advanced endometrial cancers. Clin Cancer Res 2000;6(8):2999–3005.

[103] Risinger JI, et al. PTEN mutation in endometrial cancers is associated with favorable clinical and pathologic characteristics. Clin Cancer Res 1998;4(12):3005–10.

[104] Zhang B, et al. YT521 promotes metastases of endometrial cancer by differential splicing of vascular endothelial growth factor a. Tumour Biol 2016;37(12):15543–9.

[105] Buhtoiarova TN, Brenner CA, Singh M. Endometrial carcinoma: role of current and emerging biomarkers in resolving persistent clinical dilemmas. Am J Clin Pathol 2016;145(1):8–21.

[106] Eritja N, et al. Endometrial carcinoma: specific targeted pathways. In: Molecular Genetics of Endometrial Carcinoma. Springer; 2017. p. 149–207.

[107] Darb-Esfahani S, et al. Phospho-mTOR and phospho-4EBP1 in endometrial adenocarcinoma: association with stage and grade in vivo and link with response to rapamycin treatment in vitro. J Cancer Res Clin Oncol 2009;135(7):933.

[108] Shen Q, et al. Morphoproteomic analysis reveals an overexpressed and constitutively activated phospholipase D1-mTORC2 pathway in endometrial carcinoma. Int J Clin Exp Pathol 2011;4(1):13.

[109] Cheng JQ, et al. The Akt/PKB pathway: molecular target for cancer drug discovery. Oncogene 2005;24(50):7482–92.

[110] Siegel RL, et al. Colorectal cancer statistics. CA Cancer J Clin 2020;2020.

[111] Schiffman M, Solomon D. Cervical-cancer screening with human papillomavirus and cytologic cotesting. N Engl J Med 2013;369(24):2324–31.

[112] Gaffney DK, et al. Too many women are dying from cervix cancer: problems and solutions. Gynecol Oncol 2018;151(3):547–54.

[113] Crosbie EJ, et al. Human papillomavirus and cervical cancer. Lancet 2013;382(9895):889–99.

[114] Tsu V, Jerónimo J. Saving the world's women from cervical cancer. N Engl J Med 2016;374(26):2509–11.

[115] Owusu GA, et al. Race and ethnic disparities in cervical cancer screening in a safety-net system. Matern Child Health J 2005;9(3):285–95.

[116] Campbell CMP, et al. Prevention of invasive cervical cancer in the United States: past, present, and future. Cancer Epidemiol Biomarkers Prev 2012;21(9):1402–8.

[117] Heintzman J, et al. Role of race/ethnicity, language, and insurance in use of cervical cancer prevention services among low-income Hispanic women, 2009–2013. Prev Chronic Dis 2018;15:E25.

[118] Myint ZW, et al. Disparities in prostate cancer survival in Appalachian Kentucky: a population-based study. Rural Remote Health 2019;19(2).

[119] Maguire R, et al. Disparities in cervical cancer incidence and mortality: can epigenetics contribute to eliminating disparities? In: Advances in Cancer Research. Elsevier; 2017. p. 129–56.

[120] Harper S, Lynch J. Methods for Measuring Cancer Disparities: Using Data Relevant to Healthy People 2010 Cancer-Related Objectives. DC: National Cancer Institute Washington; 2006.

[121] Eng TY, et al. Persistent disparities in Hispanics with cervical cancer in a major city. J Racial Ethn Health Disparities 2017;4(2):165–8.

[122] Howlader N, et al. SEER Cancer Statistics Review, 1975–2009 (Vintage 2009 Populations). Bethesda, MD: National Cancer Institute; 2012.

[123] Khan HM, et al. Disparities in cervical cancer characteristics and survival between white Hispanics and white non-Hispanic women. J Womens Health 2016;25(10):1052–8.

[124] Moore de Peralta A, Holaday B, Hadoto IM. Cues to cervical cancer screening among US Hispanic women. Hisp Health Care Int 2017;15(1):5–12.

[125] Bandi P, et al. Cancer Facts & Figures for Hispancis/Latinos. 2014. American Cancer Society; 2012.

[126] Facts C. Cancer Facts and Figures (2017). Atlanta: American Cancer Society; 2013.

[127] Rauh-Hain JA, et al. Racial disparities in cervical cancer survival over time. Cancer 2013;119(20):3644–52.

[128] Coker AL, et al. Ethnic disparities in cervical cancer survival among Medicare eligible women in a multiethnic population. Int J Gynecol Cancer 2009;19(1).

[129] Farley JH, et al. Equal care ensures equal survival for African-American women with cervical carcinoma. Cancer 2001;91(4):869–73.

[130] Walsh B, O'Neill C. Socioeconomic disparities across ethnicities: an application to cervical cancer screening. Am J Manag Care 2015;21(9):e527–36.

[131] Bychkovsky BL, et al. Cervical cancer control in Latin America: a call to action. Cancer 2016;122(4):502–14.

[132] Torre LA, et al. Global cancer incidence and mortality rates and trends—an update. Cancer Epidemiol Biomarkers Prev 2016;25(1):16–27.

[133] Pimple S, Mishra G, Shastri S. Global strategies for cervical cancer prevention. Curr Opin Obstet Gynecol 2016;28(1):4–10.

[134] Mathers CD, Loncar D. Projections of global mortality and burden of disease from 2002 to 2030. PLoS Med 2006;3(11), e442.

[135] Hariri S, et al. Human papillomavirus genotypes in high-grade cervical lesions in the United States. J Infect Dis 2012;206(12):1878–86.

[136] Di J-L, et al. Knowledge of cervical cancer screening among health care workers providing services across different socio-economic regions of China. Asian Pac J Cancer Prev 2016;17(6):2965–72.

[137] Syrjänen S, et al. Clearance of high-risk human papillomavirus (HPV) DNA and PAP smear abnormalities in a cohort of women subjected to HPV screening in the new independent states of the former Soviet Union (the NIS cohort study). Eur J Obstet Gynecol Reprod Biol 2005;119(2):219–27.

[138] Longatto-Filho A, et al. Performance characteristics of pap test, VIA, VILI, HR-HPV testing, cervicography, and colposcopy in diagnosis of significant cervical pathology. Virchows Arch 2012;460(6):577–85.

[139] Musselwhite LW, et al. Racial/ethnic disparities in cervical cancer screening and outcomes. Acta Cytol 2016;60(6):518–26.

[140] Walmer DK, et al. Human papillomavirus prevalence in a population of women living in Port-Au-Prince and Leogane, Haiti. PLoS One 2013;8(10), e76110.

[141] McDougall JA, et al. Racial and ethnic disparities in cervical cancer incidence rates in the United States, 1992–2003. Cancer Causes Control 2007;18(10):1175–86.

[142] Downs LS, et al. The disparity of cervical cancer in diverse populations. Gynecol Oncol 2008;109(2):S22–30.

[143] Barnholtz-Sloan J, et al. Incidence trends of invasive cervical cancer in the United States by combined race and ethnicity. Cancer Causes Control 2009;20(7):1129–38.

[144] Singh GK, et al. Socioeconomic, rural-urban, and racial inequalities in US cancer mortality: part I—all cancers and lung cancer and part II—colorectal, prostate, breast, and cervical cancers. J Cancer Epidemiol 2011;2011.

[145] Khan HMR, et al. Health Disparities Between Black Hispanic and Black Non-Hispanic Cervical Cancer Cases in the USA; 2014.

[146] Sheppard CS, et al. Assessment of mediators of racial disparities in cervical cancer survival in the United States. Int J Cancer 2016;138(11):2622–30.

[147] Cohen TF, Legg JS. Factors associated with HPV vaccine use among Hispanic college students. J Allied Health 2014;43(4):241–6.

[148] Gakidou E, Nordhagen S, Obermeyer Z. Coverage of cervical cancer screening in 57 countries: low average levels and large inequalities. PLoS Med 2008;5(6), e132.

[149] Niccolai LM, Mehta NR, Hadler JL. Racial/ethnic and poverty disparities in human papillomavirus vaccination completion. Am J Prev Med 2011;41(4):428–33.

[150] Widdice LE, et al. Adherence to the HPV vaccine dosing intervals and factors associated with completion of 3 doses. Pediatrics 2011;127(1):77–84.

[151] Curtis CR, et al. National and state vaccination coverage among adolescents aged 13–17 years—United States, 2012. MMWR Morb Mortal Wkly Rep 2013;62(34):685.

[152] Akers AY, Newmann SJ, Smith JS. Factors underlying disparities in cervical cancer incidence, screening, and treatment in the United States. Curr Probl Cancer 2007;3(31):157–81.

[153] Calle EE, et al. Demographic predictors of mammography and pap smear screening in US women. Am J Public Health 1993;83(1):53–60.

[154] Charlot M, et al. Impact of patient and navigator race and language concordance on care after cancer screening abnormalities. Cancer 2015;121(9):1477–83.

[155] Uppal S, et al. Association of hospital volume with racial and ethnic disparities in locally advanced cervical cancer treatment. Obstet Gynecol 2017;129(2):295–304.

[156] Uppal S, et al. Variation in care in concurrent chemotherapy administration during radiation for locally advanced cervical cancer. Gynecol Oncol 2016;142(2):286–92.

[157] Del Carmen MG, et al. Ethnic differences in patterns of care of stage 1A1 and stage 1A2 cervical cancer: a SEER database study. Gynecol Oncol 1999;75(1):113–7.

[158] Mundt AJ, et al. Race and clinical outcome in patients with carcinoma of the uterine cervix treated with radiation therapy. Gynecol Oncol 1998;71(2):151–8.

[159] Lamouille S, Xu J, Derynck R. Molecular mechanisms of epithelial–mesenchymal transition. Nat Rev Mol Cell Biol 2014;15(3):178–96.

[160] Jiang Z-S, et al. Epithelial-mesenchymal transition: potential regulator of ABC transporters in tumor progression. J Cancer 2017;8(12):2319.

[161] Saxena M, et al. Transcription factors that mediate epithelial–mesenchymal transition lead to multidrug resistance by upregulating ABC transporters. Cell Death Dis 2011;2(7):e179.

[162] Takano M, et al. Analysis of TGF-β1-and drug-induced epithelial–mesenchymal transition in cultured alveolar epithelial cell line RLE/Abca3. Drug Metab Pharmacokinet 2015;30(1):111–8.

[163] Venkatraman M, et al. Biological and chemical inhibitors of NF-κB sensitize SiHa cells to cisplatin-induced apoptosis. Mol Carcinog 2005;44(1):51–9. Published in cooperation with the University of Texas MD Anderson Cancer Center.

[164] Brozovic A, et al. Long-term activation of SAPK/JNK, p38 kinase and fas-L expression by cisplatin is attenuated in human carcinoma cells that acquired drug resistance. Int J Cancer 2004;112(6):974–85.

[165] Siddik ZH. Cisplatin: mode of cytotoxic action and molecular basis of resistance. Oncogene 2003;22(47):7265–79.

[166] Nuryadi E, et al. Mutational analysis of uterine cervical cancer that survived multiple rounds of radiotherapy. Oncotarget 2018;9(66):32642.

[167] Chen Y, et al. Distinctive activation patterns in constitutively active and gefitinib-sensitive EGFR mutants. Oncogene 2006;25(8):1205–15.

[168] Yang SH, et al. Mutations in the Tyrosine Kinase Domain of the Epidermal Growth Factor Receptor in Non-Small Cell Lung Cancer. Clin Cancer Res 2005;11(6):2106–10.

[169] Bower J, Ylagan L. Frequency of EGFR and Kras mutation in lung adenocarcinomas and its metastases: an institutional analysis. J Am Soc Cytopath 2012;1(1):S84–5.

[170] Zheng X, et al. Synergistic effect of pyrrolidine dithiocarbamate and cisplatin in human cervical carcinoma. Reprod Sci 2014;21(10):1319–25.

[171] Shi T-Y, et al. Functional variants in TNFAIP8 associated with cervical cancer susceptibility and clinical outcomes. Carcinogenesis 2013;34(4):770–8.

[172] Mendonca P, et al. The inhibitory effects of butein on cell proliferation and TNF-α-induced CCL2 release in racially different triple negative breast cancer cells. PLoS One 2019;14(10), e0215269.

[173] Wei L-Z, et al. Meta-analysis on the relationship between HLA-DRBl gene polymorphism and cervical cancer in Chinese population. PLoS One 2014;9(2), e88439.

C H A P T E R

# 15

# Future perspectives and new directions in chemosensitizing activities to reverse drug resistance in gynecologic cancers: Emphasis on challenges and opportunities

Ishna Sharma[a], Nathan Hannay[a], Swathi Sridhar[a], Sarfraz Ahmad[b], and Riyaz Basha[a]

[a]Department of Pediatrics and Women's Health, Texas College of Osteopathic Medicine, University of North Texas Health Science Center, Fort Worth, TX, United States [b]Gynecologic Oncology, AdventHealth Cancer Institute, Florida State University, University of Central Florida, Orlando, FL, United States

## Abstract

Resistance to treatment is an unmet challenge in cancer treatment. Gynecologic cancers (cervical, endometrial/uterine, ovarian, vaginal, and vulvar cancers) originate in the female reproductive system tissues. Drug resistance is well established in many of these cancers. Gynecologic cancers are the fourth most common tumors diagnosed in women; as such, it is imperative to understand and address resistance to treatment. During the past decades, the gynecologic oncology field has evolved from radical en-bloc surgical procedures to more limited resections with systematic lymphadenectomy procedures. Cancer treatment has historically been developed through clinical trials; treatment is now determined based on tumor histology and demographic factors, leading to percentages of response for various groups of patients. With advent of the human genome project, tumors can now be categorized on genomic and proteomic signatures while drugs and treatments can be developed/tailored based on specific targets from these bioinformatics analyses (also referred to as personalized medicine approach). Disrupted apoptotic pathways, increased efflux of therapeutic agents, and modulation of the expression of tumor suppressors are major causes for drug resistance in gynecologic cancers. Current and upcoming research is focused to strategize and counteract the mechanisms of drug resistance. The heterogenic nature of tumors, genetic variability, and mutations are major setbacks in addressing the resistance to gynecologic cancers treatment. Information on critical biomarkers, signaling cascades, genes, and transcription factors associated with gynecologic cancers that are potentially involved in acquiring tumor resistance to

R. Basha, S. Ahmad (eds.)
*Overcoming Drug Resistance in Gynecologic Cancers*
ISBN 978-0-12-824299-5
https://doi.org/10.1016/B978-0-12-824299-5.00017-4

better standard chemotherapeutic is emphasized. While several strategies are under testing to reverse resistance to chemotherapy, the major focus of this chapter is on chemosensitizers. The ongoing development of immunotherapy in the treatment of gynecologic cancers is promising due to its ability to manipulate the patients' own immune system as a therapy against cancer. Oncolytic virus therapy is another promising approach of evading the tumor immune suppressive environment for heavily pretreated patients with gynecologic malignancies. Chemosensitization-focused research includes the prospects of using phytochemicals and nanomedicine coupled immunotherapy in gynecological cancer treatment; preclinical/translational and clinical trials of these techniques are highlighted herein.

## Abbreviations

| | |
|---|---|
| **AMPK** | 5-monophosphate-activated protein kinase |
| **ASCO** | American Society of Clinical Oncology |
| **BRCA** | BReast CAncer |
| **BRC1** | BRCA1 interacting protein C-terminal helicase 1 |
| **BRD1** | BRCA1-associated RING domain 1 |
| **CA** | cancer antigen |
| **CDH1** | cadherin-1 |
| **CDNK2A** | cyclin-dependent kinase inhibitor 2A |
| **CHEK2** | checkpoint kinase 2 |
| **CT** | computerized or computed tomography |
| **DNA** | deoxyribonucleic acid |
| **EPCAM** | epithelial cell adhesion molecule |
| **EPR** | enhanced permeability and retention |
| **HE-4** | human epididymis protein 4 |
| **HPV** | human papilloma virus |
| **IP** | intraperitoneal |
| **KMT2D** | lysine methyltransferase 2D |
| **LAP** | latency-associated peptide |
| **LDFRT** | low-dose fraction radiation therapy |
| **MLH1** | MutL homolog 1 |
| **miRNA** | micro ribonucleic acid |
| **MRI** | magnetic resonance imaging |
| **MSH2** | MutS homolog 2 |
| **NBN** | nibrin |
| **Nrf2** | nuclear factor, erythroid 2 like 2 |
| **OCT1** | organic cation transporter |
| **PALB1** | parvalbumin |
| **PARP** | poly(ADP-ribose) polymerase 1 |
| **PCDA-PEG-Biotin** | poly(curcumin-dithiodipropionic acid)-*b*-poly(ethylene glycol)-biotin |
| **PET** | positron emission tomography |
| **PIK3CA** | phosphatidylinositol-4,5-bisphosphate 3-kinase catalytic subunit alpha |
| **PMS2** | PMS1 homolog 2, mismatch repair system component |
| **PTEN** | phosphatase and tensin homolog |
| **RAD51C** | RAD51 paralog C |
| **SCC** | squamous cell carcinoma |
| **SLC22A2** | solute carrier family 22 member 2 |
| **STK11** | serine/threonine kinase 11 |

## Conflict of interest

No potential conflicts of interest were disclosed.

# Introduction

The ability of cancerous cells to develop resistance to oncologic treatment is an increasing problem in the world of cancer therapy [1]. One area in which resistance to chemotherapeutic agents is being explored is in the treatment of gynecologic cancers. Briefly, gynecologic cancers include those malignant tumors that originate from tissues involved in the female reproductive system, with some of the most common being cervical, endometrial (uterine), ovarian, vaginal, and vulvar cancers [2–4]. Many of these cancers share common gene mutations, symptoms, and clinical presentations. However, key differences between them are important to note as they provide diagnostic clues, indications of prognosis, and help select possible treatment therapies. Because many of these gynecologic cancers are highly associated with the inheritance of various genetic mutations, a thorough family history is recommended during diagnostic work-up [5, 6]. Biomarkers may also be used in the early detection of various gynecologic cancers [7], although more translational research and developmental strategies need to be explored in this complex disease area [5, 6].

Globally, cervical cancer is the most common type of gynecologic cancer. It usually occurs in women between the ages of 30 and 55 years with the average age at diagnosis being 50 years (it rarely develops in women younger than 20 years). The disease originates from cervical tissue located at the junction between the vagina and the uterus [2–4]. Common symptoms associated with cervical cancer include pelvic pain and pelvic pressure [2–4]. The most frequent mutations present in cervical tumor cells involve the BReast CAncer (BRCA) genes, *BRCA1* and *BRCA2*. Other common mutations that are also present in various autosomal dominant cancer syndromes include Li-Fraumeni syndrome [Partner and localizer of BRCA2 (*PALB1*), tumor protein *p53* (*TP53*)], Cowden Syndrome [Phosphatase and tensin homolog (*PTEN*)], Diffuse Gastric and Lobular Breast Cancer Syndrome [Cadherin-1 (*CDH1*)], and Peutz-Jeghers Syndrome [Serine/threonine kinase 11 (*STK11*)] [5, 6]. In terms of biomarkers that are used during early detection and diagnosis, squamous cell carcinoma (SCC) antigen is often targeted as it is elevated in 28%–85% of cervical squamous cell carcinomas [7, 8]. Prevention includes regular Pap smears and avoiding human papillomavirus virus (HPV) infections via safe sex practices and from the HPV vaccine [9–11].

Endometrial (uterine) cancer most commonly occurs in postmenopausal women, with Caucasian women affected most frequently and African American women having higher fatality risk. The most common type of uterine cancer originates in the endometrium, or the inner epithelial layer of the uterus [2–4]. However, it should be noted that endometrial/uterine cancers may originate from other tissue layers of the uterus [2–4]. Associated symptoms include pain or pressure in the pelvis as well as abnormal vaginal discharge or bleeding [2–4]. Lynch Syndrome, or hereditary nonpolyposis colorectal cancer, has been associated with endometrial cancer as well as other cancers like colorectal [5, 6]. Lynch Syndrome involves the mismatch repair genes MutL homolog 1 (*MLH1*), MutS homolog 2 (*MSH2*), MutS homolog 2 (*MSH6*), PMS1 homolog 2, mismatch repair system component, *PMS2*, and epithelial cell adhesion molecule (*EPCAM*) [5, 6]. Other common mutations involve such genes as checkpoint kinase 2 (*CHEK2*), Adenomatous polyposis coli (*APC*), serine-protein kinase ATM, BRCA1-associated RING domain 1 (*BARD1*), BRCA1, BRCA2, BRCA1-interacting protein C-terminal helicase 1 (*BRIP1*), Nibrin (*NBN*), *PTEN*, and RAD51 paralog C (*RAD51C*) [5, 6]. As of now, no reliable and independent biomarkers have been used in the early detection of uterine cancers [7]. Surgical management/staging coupled with adequate lymphadenectomy (particularly sentinel lymph node mapping) is the hallmark of survival outcomes and quality of life for patients with endometrial cancer. Minimally invasive surgery approach (particularly robotic-assisted laparoscopic hysterectomy) is now widely used for successfully managing patients with endometrial/uterine cancer.

Ovarian cancer (collectively known for cancers of the ovary, Fallopian tube, and peritoneal cavity) is the fifth most common female cancer. It originates from ovarian tissue, which is responsible for producing female gonads and reproductive hormones [2–4]. Although there are many different symptoms associated with ovarian cancer, these symptoms may be vague and vary from patient to patient. Potential symptoms may include pain or pressure in the pelvic area, abnormal vaginal discharge or bleeding, back or abdominal pain, bloating, and/or variations in bathroom habits [2]. Ovarian cancer, although not as common as endometrial cancer and cervical cancer (globally), may also be associated with Lynch Syndrome [2–4]. The biomarker cancer antigen (CA)-125 can be used for early detection and is elevated in over 50% of the patients with early-stage ovarian tumors [2–4]. In patients who are at higher risk for ovarian cancer, specifically those carrying the *BRCA1* and *BRCA2* gene mutations, surgical removal of the ovaries is an option to reduce risk of cancer development [5, 6]. Many studies also suggest that regular use of oral contraceptives can reduce the risk of ovarian cancer [5, 6]. Intraperitoneal (IP) chemotherapy approach, which allows delivery of a larger dose directly to the tumor location, is gaining momentum for patients with advanced-stage disease. This approach may have short-term drug-related toxicity, but is associated with long-term survival outcomes of patients with ovarian cancer.

Gynecologic cancers also include vaginal cancer, which is one of the less common types of gynecologic cancers and is considered to have a favorable prognosis [2–4]. Pressure or pain in the pelvis and variations in bathroom habits are common presentations of vaginal cancer [2]. The most common genetic mutations found in vaginal cancers include the genes *TP53*, cyclin-dependent kinase inhibitor 2A (*CDKN2A*), andphosphatidylinositol-4,5-bisphosphate 3-kinase catalytic subunit alpha(*PIK3CA*). Other mutations of the genes *NOTCH1* and lysine methyltransferase 2D (*KMT2D*) are less commonly seen [6]. Because the HPV virus is a common cause of vaginal cancer, practicing safe sex and receiving the HPV vaccine substantially decreases the likelihood of vaginal cancer [2–4].

Vulvar cancers occur on the female external genitalia, and usually present on the skin [2–4]. As vulvar cancers represent only about 5% of all cancers originating from the female reproductive system, it is considered relatively rare, but shares similar genetic mutations as those found in vaginal cancers [2–4, 12]. Serum biomarkers, including cyclooxygenase-2 and caspase-3, levels may be elevated in patients with such tumors [13]. Regular exams are required to prevent the progression of vulvar cancer, and women should monitor the vulvar area for burning or any skin abnormalities, including rashes, warts, or lesions [2–4].

Together, these types of gynecologic cancers are the fourth most common tumors diagnosed in women, and their resistance to current chemotherapeutic agents is a major issue that needs further and rapid addressing [1]. The methods of resistance by gynecologic tumor cells include increasing the efflux of the agents, upregulating the expression of tumor suppressors, and disrupting apoptotic pathways [1, 14]. As such, various state-of-the-art methods and strategies have been introduced in an attempt to counteract these effects.

Chemosensitizers, in particular, are a group of compounds designed to overcome the resistance to chemotherapeutic agents by overriding the resistance mechanisms of tumor cells.

Agents such as calcium channel blockers, sodium channel blockers, and steroids are used to help reduce the efflux of the agents from inside the rapidly growing cells [1, 14]. This helps to increase the intracellular concentration of the agent, thus increasing its efficacy. Other compounds, known as bioactive phytochemicals, are being used to target the dysregulation of tumor-suppressive pathways [15]. One major pathway that is targeted is the PI3K/Akt, which is commonly overexpressed in various different types of cancers and promotes irregular metabolism, proliferation, and angiogenesis of tumor cells [15]. The radiation therapy treatment, low-dose fraction radiation therapy (LDFRT), is also being explored as an option to decrease resistance by regulating the expression of oncogenes [16]. All of these methods and more are being explored in the treatment of gynecologic cancers in order to reduce chemoresistance, increase agent's therapeutic efficacy, and improve the outcomes of gynecologic cancer treatment.

Unfortunately, numerous challenges exist in chemosensitization research for gynecologic cancer treatments. It is well known that genetic variations play a significant role in determining the efficacy of chemotherapeutic agents. Genetic variations have also been found to play a vital role in the expression of chemosensitizers' targets as well as in the metabolism of these drugs. Further characterization of "which" and "how" polymorphisms affects the response variability, and side-effects of chemosensitizers are very much warranted. In addition to the genetic factors, lifestyle behaviors have also been found to affect chemosensitizer efficacy. Such behaviors include diet, activity levels, and other environmental factors that influence microbiota in patients with gynecologic cancers [17]. Lifestyle behaviors necessarily add a layer of complexity to chemosensitization research by limiting the sampling pool available for study and increasing requirements for close monitoring of participants to ensure accurate results. Another important hurdle is the investigation of synergism that exists among chemosensitizers and other nondrug therapies, which is limited by current technology and understanding.

Although these challenges put a strain on the progress of gynecologic cancer research, there are many opportunities for future development in this area of oncologic treatment. Some of the many potential areas of improvement include investigating the use of nanomedicine to enhance diagnosis and treatment involving chemosensitizers. Further, the use of genotyping to identify the presence of chemosensitizer targets in tumor cells can improve the specificity and efficacy of chemosensitizers-based treatments. There are also opportunities to broaden what constitute a chemosensitizer, such as using immunotherapy and naturally occurring agents. In addition, while there is a shift in clinical trial designs for gynecologic cancer treatments, further focus on trials investigating chemosensitizer-based combinations holds great promise.

## Current challenges in chemosensitization research for gynecologic cancers

### Population-dependent variations affect chemosensitization response and side-effects

A major hurdle faced by research in the field of gynecologic cancer chemosensitization stems from the genetic and lifestyle variations between populations. In particular, these variations may significantly affect the response of some chemosensitization methods as they provide alternative resistance pathways and/or eliminate the target of chemosensitizers [18].

Genetic polymorphisms between ethnic and/or racial groups have been implicated as a cause for the wide activity variations seen with chemosensitizers. A particularly important instance of polymorphisms affecting therapeutic response for gynecologic patients involves the mainstay chemotherapeutic agent cisplatin. Up to 75% of ovarian cancer patients develop drug resistance to cisplatin, and thus, patients often also take the chemosensitizer metformin with cisplatin. Metformin enhances the agent's chemotherapeutic effects through the 5-monophosphate-activated protein kinase (AMPK) pathway [19, 20]. Current work in genetic polymorphisms show that metformin's renal clearance and secretion are significantly greater for Caucasians and African Americans that have a variant genotype as compared to those with the wild-type genotype of the solute carrier family 22 member 2 (SLC22A2) gene coding for organic cation transporter 1 (OCT1) [21]. As a result of this polymorphism, the efficacy of the cisplatin-metformin therapy may be reduced in Caucasians and African Americans due to rapid excretion of metformin from the body. On the other hand, ethnic and/or racial groups that do not have such a genetic variation of their OCT1 may experience greater untoward effects than Caucasians and African Americans. This is due to increasing metformin drug concentrations and its increased influence on cisplatin. Besides affecting the clearance rate of metformin, variations among the SLC22A2 gene also increase efflux of the chemosensitizers from the tumor cells, thereby reducing the interactions of the drug with its target [21, 22] (Fig. 1).

Additionally, the use of phytochemicals, or naturally occurring and plant-derived compounds, has been actively investigated over the past decades, with high hopes that such chemosensitizers will improve the chemotherapeutic agents' response(s) without inducing side-effects of their own [15]. Phytochemicals largely act on the key transcription factor

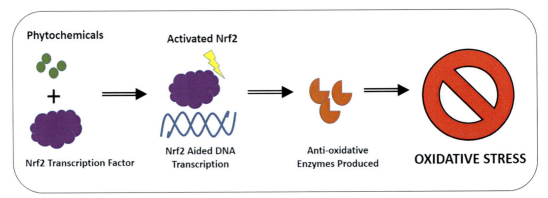

FIG. 1    Illustration on the mechanism(s) by which phytochemicals lead to a reduction in oxidative stress. Phytochemicals are a class of chemosensitizers that are naturally occurring or plant-derived compounds. Phytochemicals act on transcription factor Nrf2, which regulates hundreds of genes involved in decreasing cellular oxidative stress. When Nrf2 is activated by these chemosensitizers, transcription of genes coding for antioxidative enzymes is increased. These enzymes, through various mechanisms, work in concert to decrease the oxidative stress, which often contributes to the development of malignant cells [23–26]. Abbreviation: NRF2=nuclear factor erythroid 2-related factor 2.

Nrf2, which regulates over 600 genes, many of which are involved in cellular protection against oxidative stress [23]. It has previously been shown that increased oxidative stress contributes to malignant cell development in gynecologic cancers. Phytochemicals combat this by increasing the activation of Nrf2 and thus, antioxidative enzymes production [23, 24]. Like metformin, however, phytochemicals are victims to genetic variations of Nrf2, which causes variations in the response of diabetic medications and possibly chemosensitization [25, 26].

While a few key instances through which genetic polymorphism can affect chemosensitizer response have been discussed earlier, there are other such interactions noted in the peer-reviewed literature. However, because of the multiple genes that impact the chemosensitizers and the polymorphisms that exist for most of these genes, further research in this area is necessarily laborious and extensive. Yet, such research would enable selection of those treatment combinations that are most effective for the ethnic and/or racial makeup of the patient (Fig. 2).

Further, while engagement in certain lifestyle behaviors has been shown to predispose certain women to gynecologic cancer, it also appears to influence the treatment efficacy of chemotherapeutic agents and chemosensitizers. Diet, and its effects on the uterine microbiome, is heavily scrutinized and well-studied. It has been found that diets which encourage obesity and diabetes mellitus type 2 may drive progression of gynecologic cancers through pro-inflammatory conditions caused by dysbiosis [17, 27, 28]. Interestingly, these changes in gut and tumor microbiota also modify efficacy of agents and chemosensitizers via interactions of the microbiota with the immune system, with checkpoint inhibitors, and with epithelial barriers, among others [17, 29–32]. Other environmental factors that modify the uterine homeostasis, such as acidity, have also been shown to alter microbiota such that

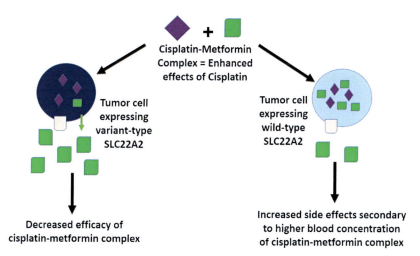

FIG. 2   Metformin is a chemosensitizer used to enhance the effect of cisplatin. Polymorphic variations in the clearance and secretion of metformin can change the effects of the cisplatin-metformin complex. In Caucasians and African Americans with the variant type of the *SLC22A2* gene, metformin is more rapidly excreted, reducing the efficacy of the drug complex. Patients with the wild-type *SCL22A2* gene retain more of the drug complex, which can lead to more untoward effects of cisplatin [21, 22].

there is increased resistance to chemosensitizers [33]. Given the multitude of environmental factors that affect chemosensitization in gynecologic cancers and the significant variations in patient lifestyle behaviors, research in this area may be limited to a small sample pool and certainly requires further investigation into the mechanisms behind microbiota' effects on chemosensitizers.

## Shift in clinical trial landscape for gynecologic cancers

In recent decades, there has been a shift in the landscape of clinical trials for gynecologic cancers to include a phase II/III design [34]. In such instances, the trial design involves a largely phase III trial with an interim phase II assessment of the experimental treatment. As such, a phase II/III design includes smaller sample sizes and separate endpoints for the phase II portion and the phase III portion, thereby improving efficiency [28–30, 34, 35].

Such trial designs, however, present challenges for research investigating the functionality of chemosensitizers. These challenges include reduced flexibility in adjusting phase III trial designs based on results of the phase II portion, earlier commitment to phase III, and an increase in the perceived time for complete accrual as compared to the trial designs that are confined to a single phase [34–36]. As such, clinical trial research for gynecologic cancer chemosensitizers must be designed keeping these constraints in mind. Notably, the 2020 virtual American Society of Clinical Oncology (ASCO) annual meeting reflected on continued progress and outcomes in such areas as PARP inhibitors, improved secondary cytoreductive surgery, and evaluation of novel immunotherapeutic agents for patients with gynecologic malignancies would benefit from these trials [37].

## Synergy of chemosensitized chemotherapy and nondrug therapies

To overcome chemotherapy resistance in gynecologic cancers, chemosensitizers have been used in conjunction with other methods. For example, for ovarian cancer, one of the most common and aggressive presentations of gynecologic cancers, a synergistic effect has been noted between chemosensitizers and surgical cytoreduction [38]. As expected, surgical cytoreduction would reduce the tumor burden of a patient. Because dosing of chemosensitizers is limited by their therapeutic upper limit, reduction in the tumor burden may enable a greater amount of chemosensitizer drug for a given dose to reach each tumor cell. Unfortunately, adequate surgical cytoreduction is limited by current imaging precision, which generally requires at least 30 to 40 cell doublings before the tumor appears on imaging. Thus, undetectable microscopic lesions may result in remaining tumor nodules (also referred to as residual disease), premature withdrawal of chemosensitized treatments, and resistance to both the chemotherapeutic agent and the chemosensitizer.

Additionally, effectiveness of chemosensitizers is limited by efficiency in drug delivery to the gynecologic tumor cells [39]. Conventionally, chemosensitizers are delivered nonspecifically. Depending on the physiochemical properties of the drug, this nonspecific delivery may cause an increase in sensitization of both the malignant and normal tissues to cytotoxic anticancer drugs [40, 41]. In fact, it has been noted that the chemosensitizer wortmannin is often used with cisplatin to synergistically improve treatment response in ovarian cancer [40, 42]. However, unlike some chemosensitizers, wortmannin affects both the tumor cells and normal tissues, resulting in greater toxicity and treatment side-effects [40].

Chemosensitization research would be wise to study the influence of nondrug therapies in gynecologic cancer treatment and particularly as to how such therapies may provide synergistic effects when used in combination with chemosensitizers. Investigations into more sensitive imaging techniques and drug delivery options require a multidisciplinary approach into the identification of current and future limitations of such therapies (Table 1).

TABLE 1   Gene mutations and biomarkers associated with gynecologic cancers.

| Cancer type | Biomarker for early detection [4, 5] | Common gene mutations [3] | Genetic syndromes associated with mutations [3] |
|---|---|---|---|
| Cervical | Squamous cell carcinoma antigen HPV and cytologic screening | • *BRCA1, BRCA2* <br> • *PALB1, TP53* <br> • *PTEN* <br> • *CDH1* <br> • *STK11* | • Li-Fraumeni syndrome <br> • Cowden Syndrome <br> • Diffuse Gastric and Lobular Breast Cancer Syndrome <br> • Peutz-Jeghers Syndrome |
| Endometrial (uterine) | No reliable markers available (hence the detection is mostly based on clinico-pathologic features coupled with precision assessments of gene mutations and risk factors) | • *MLH1, MSH2, MSH6, PMS2,* and *EPCAM* <br> • *CHEK2, APC, ATM, BARD1, BRCA1, BRCA2, BRIP1, NBN, PTEN,* and *RAD51C* | • Lynch syndrome |
| Ovarian | CA-125 <br> HE-4[a] <br> CA-72.4[a] <br> CA-15.3[a] <br> LAP[a] | • *BRCA1, BRCA2* <br> • *MLH1, MSH2, MSH6, PMS2,* and *EPCAM* | • Lynch syndrome |
| Vaginal | None—but routine pelvic exams and Pap tests detect precancerous conditions | • *TP53, CDKN2A, PIK3CA* <br> • *NOTCH1, KMT2D* | – |
| Vulvar | • Cyclooxygenase-2 <br> • Caspase-3 | Similar to vaginal cancers | – |

A compilation of critical mutations and biomarkers associated with gynecologic cancers.
Abbreviations: *CA,* cancer antigen; *HE-4,* human epididymis protein 4; *HPV,* human papilloma virus; *LAP,* latency-associated peptide.
[a] *Worked better in combination with CA-125 and helps in differential diagnosis [74].*

## Future development in chemosensitization research for gynecologic cancers

### Addressing current challenges in chemosensitization research: Elucidating interactions

As mentioned earlier, many chemosensitization methods exploit certain genetic mechanisms to increase the efficacy of chemotherapeutic agents. Genetic variations between populations, however, prevent uniform improvement in chemotherapy efficacy since such variations provide alternative resistance pathways and/or eliminate the target of chemosensitizers [18]. To overcome this hurdle, chemosensitization research is increasingly borrowing from chemotherapeutic research strategies.

One such strategy involves genotyping of tumor cells to better understand gynecologic tumor characteristics and as to how the chemosensitizers interact with the phenotype of the cancers [43]. Certain gynecologic cancers may express a mutated *BRCA* gene, such that the cancer phenotype is *BRCA*-deficient. Such cancer phenotypes respond extremely well to poly-ADP-ribose polymerase (PARP) inhibitors, which inhibit a DNA repair pathway protein [44–46]. Indeed, when Olaparib (a PARP inhibitor, recently FDA-approved for some gynecologic indications) is given to patients expressing a *BRCA*-deficient ovarian cell line, the chemosensitization effect enables up to a 30% enhancement in treatment [42]. This benefit is not seen in those ovarian cancers that do not express this gene mutation. Further understanding of how certain onco-genetic and/or chemo-resistant polymorphisms are more susceptible to certain chemosensitization methods holds great promise for increasing the overall success of chemotherapy-reliant treatments and improved outcomes.

Additionally, while current works on chemosensitizers identify the synergistic effect of these molecules with chemotherapeutic drugs, a better understanding of how the synergy arises would enable selection of the most potent drug combinations for the cancer phenotype. Current studies have emphasized the positive/favorable use of metformin and phytochemicals, among others, in gynecologic cancers while other studies have noted that curcumin and vitamins interfere with certain agents used with these cancers [47]. Understanding how metformin and phytochemicals interact with anticancer drugs, like cisplatin, would enable efficient selection of a chemosensitizer-chemotherapeutic agent combination to tackle the tumor growth.

Further, chemosensitizers used for other cancers, such as prostate and pancreatic cancers, that share similar phenotypes with gynecologic cancers may also induce a positive response when combined with current gynecologic cancer chemotherapeutic agents. Androgen receptor expression is an important marker for aggressive development and progression in prostate cancer [48]. Similarly, androgen receptor signaling affects the growth and risk level for ovarian cancer [49]. Current work in the area of androgen receptors has found that inactivation of androgen receptors in tumor cells create chemosensitization effects in prostate cancer and potentially in ovarian cancer [49]. Future research should explore whether "prostate-specific" chemosensitizers that target androgen receptor transcription, such as Quercetin, may also act as chemosensitizers for ovarian cancers expressing a similar phenotype [50]. This principle may be extended to a variety of cancer phenotypes beyond prostate cancer and androgen receptors.

Future investigations into how chemosensitizers interact with the gynecologic tumor cells and with coadministered drugs provide an exciting prospect in the field. This information, combined with clinical trials of an appropriate trial design, may lead to enhancement in tumor treatment and discovery of other positive effects on prognosis.

## Addressing current challenges in chemosensitization research: Nanomedicine

Chemosensitizers themselves may be combined with nondrug therapies to further improve their effectiveness in gynecologic cancers. A nondrug therapy that holds great opportunities for further investigation includes nanomedicine and its use in precise imaging and drug delivery.

A synergistic effect has been noted between chemosensitizers and surgical cytoreduction for ovarian cancer, stemming in part from a reduced cancer burden [38]. Current diagnostic and prognostic imaging includes the use of photoacoustics, computerized or computed tomography (CT) scans, magnetic resonance imaging (MRI) scans, and positron emission tomography (PET) scans. These imaging methods may be relatively slow, expensive, increase patient exposure to radiation, and/or have relatively poor resolution, requiring at least 30 doubling cycles before tumor cells are detected. Instead, the use of nanostructured probes has been proposed to improve the imaging resolution of tumor cells and identification of biomarkers [38, 51]. In particular, nanoparticles are often used as the probes since they offer attractive imaging features with their flexible functional properties and their small scale. For example, magnetic nanoparticles may be used with T2 contrast to provide a signal enhancement of 58% while conjugated nanoparticle formulations may be used to identify biomarkers exclusive to the cancer cells, even when those cells are hidden from current imaging modalities [51].

Indeed, current utilization of nanoprobes in optical and electrochemical immunoassays for CA-125 marker has limits of detection at 1.3 and 40 U/mL, respectively [52]. Biomarker detection may occur in vivo using a combination of magnetic nanoparticles with surface conjugation and/or may occur in vitro using conjugated beads in assay form [53]. Thus, the use of nanotechnology in cancer care may allow for better evaluation of a patient's tumor burden, identify situations requiring synergistic reduction of the tumor burden by surgery, and identification of situations in which chemotherapeutic agents and chemosensitization should not be prematurely withdrawn.

Nanomedicine is also playing an important role in personalized medicine via nanotheranostics, which combines the diagnostic functionalities of nanotechnologies, as previously described, with their therapeutic functionalities into a single system. In other words, nanoparticles may be used to enhance diagnostic imaging of current modalities and may also serve as a means for precise drug delivery to the target tumor cells [51]. Current work is focused on the use of such drug delivery nanosystems to enhance chemotherapeutic efficacy on tumor cells, and thereby the nanoparticles themselves act as chemosensitizers. These drug delivery strategies include targeting specific receptors unique to gynecologic cancers, preventing drug efflux by drug-resistant cancer cells, and reducing the cytotoxicity of health tissues [54–56]. As an example, the commonly used anticancer drug, cisplatin, may be loaded into a telodendrimer-styled nanoparticle that has surface conjugation targeted to SKOV3 ovarian cancer cells. The result is prolonged drug delivery to systemic circulation, targeted delivery of cisplatin to tumor cells at higher dose concentrations, and a reduction in renal toxicity [57]. Other agents and nanoparticle formulations may be used to further tailor drug delivery and chemosensitization effects.

Opportunities exist in chemosensitization research to similarly incorporate these strategies into drug delivery options for chemosensitizers. Motivation for using this strategy with chemosensitizers stems from their reduced in vivo effects due to poor accumulation in tumor cells and, as noted previously, cytotoxicity that extends to normal tissue [58]. Nanoparticles copolymerized with curcumin and poly(curcumin-dithiodipropionic acid)-*b*-poly(ethylene glycol)-biotin (PCDA-PEG-Biotin) enable high loading capacity of the chemosensitizer curcumin and responsive release of the chemosensitizer at target tumor cells [58]. Further, specific delivery targeting is achieved via biotin and the enhanced permeability and retention (EPR) effect [58]. It was found that this method prevented the need to increase cisplatin drug dose by 1 to 2 orders of magnitude to achieve therapeutic response levels of predrug resistance [58].

With promise of nanomedicine to enhance the precision of diagnostic and treatment capabilities, further investigation and innovative works on different nanomedicine formulations may be capitalized on to improve chemosensitization efficacy.

## Nonconventional chemosensitizers

Chemosensitizers are normally thought of as molecules that interact with chemotherapy and/or tumor cells to enhance anticancer drug response. Chemosensitizers, however, may take the form of nanoparticles as either immunotherapy and/or epigenetic modifiers, among others. As such, future development in chemosensitization research should explore these "nonconventional" means for chemosensitizing gynecologic tumor cells.

A normal-behaving immune system inherently prevents the development of malignant cells. Once these natural systems of checks and balances fail, however, there is growth of cancerous cells since tumor cells are able to evade tumor-specific T-cell-mediated immune response by releasing cytokines [59]. The goal of immunotherapy is to resensitize the immune system to tumor-associated antigens and thereby restore the body's natural ability to prevent cancer development.

Interestingly, immunotherapy may be used as a chemosensitizer itself. Indeed, it has been found that with nongynecologic cancers, the use of immunotherapy in conjunction with chemotherapeutic agents enhances the sensitivity of the tumor cells to the agents [60]. Vaccination of mice with anti-MDR antibodies augmented the sensitivity of leukemia cells to doxorubicin, both in vivo and in vitro [61]. In colon cell carcinoma, immunotherapy targeting the interleukin-2 (IL-2) gene has also been shown to confer chemosensitization [62]. Further, it has been shown that immunotherapy based on aGal epitope causes enhanced presentation of tumor-specific antigen and thus, the immune system is better able to opsonize the tumor cells. When the aGal epitope immunotherapy is used in combination with chemotherapy, a synergistic effect was noted [63, 64]. Although there is a lack of research with regards to immunotherapy's chemosensitization effects in gynecologic cancers, the synergistic effect noted in other cancers is promising avenue. Investigating the chemosensitization response of the plethora of gynecologic cancer immunotherapies provides an exciting prospect for future chemotherapeutic-based treatment.

Drugs that modify the epigenome of gynecologic tumor cells also hold promise as alternative chemosensitizers. The epigenome comprises molecules and proteins that attach to the deoxyribonucleic acid (DNA) and affects the expression of gene and gene products. Lifestyle factors, such as diet, drug consumption, smoking, and stress, have been found to affect epigenetic mechanisms [65]. Most commonly, these mechanisms include DNA methylation, histone acetylation, and microRNA (miRNA) expression, all of which modify the expression of gene and gene products [65]. Epithelial ovarian cancer is associated with global DNA hypomethylation. Further, increases in satellite DNA hypomethylation are associated with aggressive progression and tumor formation due to overexpression of oncogenes [43]. For BRCA-mutated ovarian tumor cells, there is also a hypermethylation and reduced expression of the BRCA1 gene [43]. In ovarian cancer, the use of epigenetic drugs that modify DNA methylation has been shown to augment chemotherapeutic responses [43]. For example, decitabine can be used with cisplatin to resensitize cisplatin-resistant epithelial ovarian cancer cells [43]. Determining the methylation status of certain genes in the tumor cells may not only provide insight on prognosis, but also on appropriate selection of methylation-modifying chemosensitizers based on genotypic presentation.

Additionally, research into alternative chemosensitization techniques should include the impact of nutrition on epigenetic modulation of noncoding ribonucleic acid (miRNA). The miRNAs are endogenous RNA pieces that regulate gene expression, but do not themselves code for a gene product. Chemosensitizers, including phytochemicals, can influence miRNAs' interactions with translation components to silence or overexpress genetic products [66]. Increase in the miR-93 levels results in a decrease in Nrf2, which regulates the genes involved in cellular protection against oxidative stress [23, 66]. The use of phytochemical vitamin C in ovarian cancers has been found to decrease miR-93 levels, and thus, ultimately increases the oxidative-protective enzymes levels [66]. While only one example of how nutrition affects epigenetic modulation and chemosensitization abilities is discussed here, it is telling of the potency of a combinational approach to cancer treatment. This is an exciting area for future development since the structure-activity relationship and functional consequences between the phytochemicals and miRNA are poorly understood and documented, but the combination of which has proved potent.

# New direction in clinical trials

Biomarker-driven and natural chemosensitizer-focused research are fairly recent developments in gynecologic cancer research arena. Extension of these studies to encompass in vivo clinical trials of chemosensitizers may result in additional avenues that improve chemotherapeutic treatment efficacy in this patient population.

There has been extensive research to determine prognostic biomarkers for gynecologic cancers over the past 40 years [67]. Identifying predictive biomarkers associated with chemosensitization as well as how these biomarkers interact with chemosensitizing molecules are important areas of new development in this field. It has been shown that overexpression of miR-378-based, miR-30d-based, CA-125, let-7g, and other biomarkers in ovarian cancers increases chemosensitization of platinum treatments [66–68]. Notably, however, why the expression of such biomarkers increases effects for only a portion of the known gynecologic chemosensitizers is still largely unanswered. Moreover, within the last decade, there has been a push to utilize nanomedicine to detect these cancer cell biomarkers at an earlier stage of the diagnosis [38, 40]. As discussed earlier, nanoparticles with biomarker-friendly conjugated surfaces provide more sensitive and earlier detection of prognostic and predictive biomarkers [57]. Thus, biomarker-driven clinical trials that utilize a better understanding of interactions between biomarkers, chemosensitizing molecules, and nanomedicine.

In studies conducted in vitro, naturally occurring chemosensitizers have been found to overcome multidrug resistance in gynecologic cancers. Such chemosensitizers include flavonoids that resensitizes the cancer cells by providing antioxidant abilities, leukocyte immobilization, and the like [69–72]. Although many studies have looked at flavonoids in vitro, few have evaluated the in vivo responses of such naturally occurring chemosensitizers when used in combination with chemotherapeutic agents [73]. This observation extends generally to most phytochemicals, although there has been increasing interest during the past several years to investigate their chemosensitization properties.

Moreover, these chemosensitization-based clinical trials may utilize the phase II/III design to evaluate their efficacy. As discussed earlier, clinical trial designs for gynecologic cancers have seen a recent shift toward a combined phase II/III design. Currently, only a handful of clinical trials that include chemosensitizers is underway, with one of the most notable being a randomized study of paclitaxel, carboplatin, and metformin *vs.* paclitaxel, carboplatin, and placebo in endometrial cancer (NCT02065687) [34]. With a better understanding of the mechanisms, interactions, and limitations of the chemosensitization areas listed earlier, future clinical trials would be wiser to test the efficacy of these drugs in combination with current treatments.

## Conclusions

Gynecologic malignancies (cervical, endometrial/uterine, ovarian, vaginal, and vulvar cancers) are relatively complex reproductive tract cancers with varying symptoms, risk factors, and diagnostic/therapeutic responsiveness. Recognizing early symptoms and managing the disease requires multidisciplinary approach with novel therapeutic options. Chemosensitizing molecules have been shown to greatly enhance the capabilities and longevity of chemotherapeutic treatments in gynecologic cancers since they overcome many of the mechanisms that enable drug resistance in tumor cells. While chemosensitization is well known in the field, translational research in this area faces numerous challenges as well as many avenues of promise for future diagnostic/therapeutic development. Genetic and lifestyle variations may reduce the response of the chemosensitizers to lower than what is expected. Genotyping of tumor cells, evaluation of interactions between chemosensitizers and chemotherapeutics, and critical evaluation of lifestyle behaviors provide the possibility to use better tailored chemosensitizers that may fit the tumor phenotype and patient profile. Additionally, the use of chemosensitizers may be limited by current imaging and drug delivery precision. Future developments in nanomedicine, shifting the focus of clinical trials, and evaluating other innovative methods for chemosensitization can potentially overcome these challenges and improve upon the current methods of diagnosis and treatment for patients with gynecologic cancers thereby improving the survival, quality of life, and cost effectiveness.

## Acknowledgments

RB is supported by grants from the National Institute on Minority Health and Health Disparities (#1S21MD012472-01;2U54 MD006882-06) and National Cancer Institute (#P20CA233355-01).

## References

[1] Fracasso PM. Overcoming drug resistance in ovarian carcinoma. Curr Oncol Rep 2001;3(1):19–26.

[2] Dhakal HP, Pradhan M. Histological pattern of gynecological cancers. JNMA J Nepal Med Assoc 2009;48 (176):301–5.

[3] Ledford LRC, Lockwood S. Scope and epidemiology of gynecologic cancers: an overview. Semin Oncol Nurs 2019;35(2):147–50.

[4] Whitcomb BP. Gynecologic malignancies. Surg Clin North Am 2008;88(2):301–17. vi.

[5] Gardiner C. Family history of gynaecological cancers. Obstet Gynaecol Reprod Med 2007;17(12):356–61.

[6] Genetics of breast and gynecologic cancers (PDQ®)–health professional version https://www.cancer.gov/types/breast/hp/breast-ovarian-genetics-pdq#link/_4.

[7] Ueda Y, Enomoto T, Kimura T, Miyatake T, Yoshino K, Fujita M, Kimura T. Serum biomarkers for early detection of gynecologic cancers. Cancers (Basel) 2010;2(2):1312–27.

[8] Massuger LF, Koper NP, Thomas CM, Dom KE, Schijf CP. Improvement of clinical staging in cervical cancer with serum squamous cell carcinoma antigen and CA 125 determinations. Gynecol Oncol 1997;64(3):473–6.

[9] Fowler JR, Jack BW. Cervical cancer. Treasure Island (FL): StatPearls; 2020.

[10] Sundstrom B, Smith E, Delay C, Luque JS, Davila C, Feder B, Paddock V, Poudrier J, Pierce JY, Brandt HM. A reproductive justice approach to understanding women's experiences with HPV and cervical cancer prevention. Soc Sci Med 2019;232:289–97.

[11] Suris JC, Dexeus S, Lopez-Marin L. Epidemiology of preinvasive lesions. Eur J Gynaecol Oncol 1999;20(4):302–5.

[12] Weinberg D, Gomez-Martinez RA. Vulvar cancer. Obstet Gynecol Clin North Am 2019;46(1):125–35.

[13] Fons G, Burger MP, Ten Kate FJ, van der Velden J. Identification of potential prognostic markers for vulvar cancer using immunohistochemical staining of tissue microarrays. Int J Gynecol Pathol 2007;26(2):188–93.

[14] Higgins CF. Multiple molecular mechanisms for multidrug resistance transporters. Nature 2007;446(7137):749–57.

[15] Farrand L, Oh SW, Song YS, Tsang BK. Phytochemicals: a multitargeted approach to gynecologic cancer therapy. Biomed Res Int 2014;2014:890141.

[16] Zhao L, Liu S, Liang D, Jiang T, Yan X, Zhao S, Liu Y, Zhao W, Yu H. Resensitization of cisplatin resistance ovarian cancer cells to cisplatin through pretreatment with low-dose fraction radiation. Cancer Med 2019;8(5):2442–8.

[17] McKenzie ND, Hong H, Ahmad S, Holloway RW. The gut microbiome and cancer immunotherapeutics: a review of emerging data and implications for future gynecologic cancer research. Crit Rev Oncol Hematol 2020;157:103165.

[18] Helm CW, States JC. Enhancing the efficacy of cisplatin in ovarian cancer treatment—could arsenic have a role. J Ovarian Res 2009;2:2.

[19] Dong H, Huang J, Zheng K, Tan D, Chang Q, Gong G, Zhang Q, Tang H, Sun J, Zhang S. Metformin enhances the chemosensitivity of hepatocarcinoma cells to cisplatin through AMPK pathway. Oncol Lett 2017;14(6):7807–12.

[20] Yellepeddi VK, Vangara KK, Kumar A, Palakurthi S. Comparative evaluation of small-molecule chemosensitizers in reversal of cisplatin resistance in ovarian cancer cells. Anticancer Res 2012;32(9):3651–8.

[21] Goswami S, Yee SW, Xu F, Sridhar SB, Mosley JD, Takahashi A, Kubo M, Maeda S, Davis RL, Roden DM, et al. A longitudinal HbA1c model elucidates genes linked to disease progression on metformin. Clin Pharmacol Ther 2016;100(5):537–47.

[22] Kimchi-Sarfaty C, Marple AH, Shinar S, Kimchi AM, Scavo D, Roma MI, Kim IW, Jones A, Arora M, Gribar J, et al. Ethnicity-related polymorphisms and haplotypes in the human ABCB1 gene. Pharmacogenomics 2007; 8(1):29–39.

[23] Eggler AL, Savinov SN. Chemical and biological mechanisms of phytochemical activation of Nrf2 and importance in disease prevention. Recent Adv Phytochem 2013;43:121–55.

[24] Calaf GM, Urzua U, Termini L, Aguayo F. Oxidative stress in female cancers. Oncotarget 2018;9(34):23824–42.

[25] Boettler U, Volz N, Teller N, Haupt LM, Bakuradze T, Eisenbrand G, Bytof G, Lantz I, Griffiths LR, Marko D. Induction of antioxidative Nrf2 gene transcription by coffee in humans: depending on genotype? Mol Biol Rep 2012;39(6):7155–62.

[26] Ishikawa T. Genetic polymorphism in the NRF2 gene as a prognosis marker for cancer chemotherapy. Front Genet 2014;5:383.

[27] Belizário JE, Faintuch J. Microbiome and gut dysbiosis. In: Silvestre R, Torrado E, editors. Metabolic interaction in infection experientia supplementum, vol. 109. Springer; 2018.

[28] Mert I, Walther-Antonio M, Mariani A. Case for a role of the microbiome in gynecologic cancers: clinician's perspective. J Obstet Gynaecol Res 2018;44(9):1693–704.

[29] Bhatt AP, Redinbo MR, Bultman SJ. The role of the microbiome in cancer development and therapy. CA Cancer J Clin 2017;67(4):326–44.

[30] Cogdill AP, Gaudreau PO, Arora R, Gopalakrishnan V, Wargo JA. The impact of intratumoral and gastrointestinal microbiota on systemic cancer therapy. Trends Immunol 2018;39(11):900–20.

[31] Panebianco C, Andriulli A, Pazienza V. Pharmacomicrobiomics: exploiting the drug-microbiota interactions in anticancer therapies. Microbiome 2018;6(1):92.

[32] Rea D, Coppola G, Palma G, Barbieri A, Luciano A, Del Prete P, Rossetti S, Berretta M, Facchini G, Perdona S, et al. Microbiota effects on cancer: from risks to therapies. Oncotarget 2018;9(25):17915–27.

[33] Walther-Antonio MR, Chen J, Multinu F, Hokenstad A, Distad TJ, Cheek EH, Keeney GL, Creedon DJ, Nelson H, Mariani A, et al. Potential contribution of the uterine microbiome in the development of endometrial cancer. Genome Med 2016;8(1):122.

[34] Annunziata CM, Kohn EC. Clinical trials in gynecologic oncology: past, present, and future. Gynecol Oncol 2018;148(2):393–402.

[35] Korn EL, Freidlin B. Adaptive clinical trials: advantages and disadvantages of various adaptive design elements. J Natl Cancer Inst 2017;109(6).

[36] Thall PF. A review of phase 2-3 clinical trial designs. Lifetime Data Anal 2008;14(1):37–53.

[37] Salani R, Liu JF. Meeting report from the 2020 Annual (virtual) Meeting of the American Society of Clinical Oncology. Gynecol Oncol 2020;159(1):13–6.

[38] Di Lorenzo G, Ricci G, Severini GM, Romano F, Biffi S. Imaging and therapy of ovarian cancer: clinical application of nanoparticles and future perspectives. Theranostics 2018;8(16):4279–94.

[39] Zhang RX, Wong HL, Xue HY, Eoh JY, Wu XY. Nanomedicine of synergistic drug combinations for cancer therapy—strategies and perspectives. J Control Release 2016;240:489–503.

[40] Caster JM, Sethi M, Kowalczyk S, Wang E, Tian X, Nabeel Hyder S, Wagner KT, Zhang YA, Kapadia C, Man Au K, et al. Nanoparticle delivery of chemosensitizers improve chemotherapy efficacy without incurring additional toxicity. Nanoscale 2015;7(6):2805–11.

[41] Husain K, Malafa MP. Chapter 4—Role of tocotrienols in chemosensitization of cancer. In: Bharti AC, Aggarwal BB, editors. Cancer sensitizing agents for chemotherapy, role of nutraceuticals in cancer chemosensitization, vol. 2. Academic Press; 2018. p. 77–97.

[42] Zhang M, Hagan CT, Min Y, Foley H, Tian X, Yang F, Mi Y, Au KM, Medik Y, Roche K, et al. Nanoparticle co-delivery of wortmannin and cisplatin synergistically enhances chemoradiotherapy and reverses platinum resistance in ovarian cancer models. Biomaterials 2018;169:1–10.

[43] Krakowsky RH, Tollefsbol TO. Impact of nutrition on non-coding RNA epigenetics in breast and gynecological cancer. Front Nutr 2015;2:16.

[44] Audeh MW, Carmichael J, Penson RT, Friedlander M, Powell B, Bell-McGuinn KM, Scott C, Weitzel JN, Oaknin A, Loman N, et al. Oral poly(ADP-ribose) polymerase inhibitor olaparib in patients with BRCA1 or BRCA2 mutations and recurrent ovarian cancer: a proof-of-concept trial. Lancet 2010;376(9737):245–51.

[45] Chen A. PARP inhibitors: its role in treatment of cancer. Chin J Cancer 2011;30(7):463–71.

[46] Ledermann JA, Pujade-Lauraine E. Olaparib as maintenance treatment for patients with platinum-sensitive relapsed ovarian cancer. Ther Adv Med Oncol 2019;11, 1758835919849753.

[47] Kallifatidis G, Labsch S, Rausch V, Mattern J, Gladkich J, Moldenhauer G, Buchler MW, Salnikov AV, Herr I. Sulforaphane increases drug-mediated cytotoxicity toward cancer stem-like cells of pancreas and prostate. Mol Ther 2011;19(1):188–95.

[48] Fujita K, Nonomura N. Role of androgen receptor in prostate cancer: a review. World J Mens Health 2019;37(3):288–95.

[49] Mizushima T, Miyamoto H. The role of androgen receptor signaling in ovarian cancer. Cells 2019;8(2):176.

[50] Cimino S, Sortino G, Favilla V, Castelli T, Madonia M, Sansalone S, Russo GI, Morgia G. Polyphenols: key issues involved in chemoprevention of prostate cancer. Oxid Med Cell Longev 2012;2012:632959.

[51] Kim TH, Lee S, Chen X. Nanotheranostics for personalized medicine. Expert Rev Mol Diagn 2013;13(3):257–69.

[52] Al-Ogaidi I, Aguilar ZP, Suri S, Gou H, Wu N. Dual detection of cancer biomarker CA125 using absorbance and electrochemical methods. Analyst 2013;138(19):5647–53.

[53] Chinen AB, Guan CM, Ferrer JR, Barnaby SN, Merkel TJ, Mirkin CA. Nanoparticle probes for the detection of cancer biomarkers, cells, and tissues by fluorescence. Chem Rev 2015;115(19):10530–74.

[54] Gupta S, Pathak Y, Gupta MK, Vyas SP. Nanoscale drug delivery strategies for therapy of ovarian cancer: conventional vs targeted. Artif Cells Nanomed Biotechnol 2019;47(1):4066–88.

[55] Venkatas J, Singh M. Cervical cancer: a meta-analysis, therapy and future of nanomedicine. Ecancermedicalscience 2020;14:1111.

[56] Wang H, Agarwal P, Zhao G, Ji G, Jewell CM, Fisher JP, Lu X, He X. Overcoming ovarian cancer drug resistance with a cold responsive nanomaterial. ACS Cent Sci 2018;4(5):567–81.

[57] Cai L, Xu G, Shi C, Guo D, Wang X, Luo J. Telodendrimer nanocarrier for co-delivery of paclitaxel and cisplatin: a synergistic combination nanotherapy for ovarian cancer treatment. Biomaterials 2015;37:456–68.

[58] Guo S, Lv L, Shen Y, Hu Z, He Q, Chen X. A nanoparticulate pre-chemosensitizer for efficacious chemotherapy of multidrug resistant breast cancer. Sci Rep 2016;6:21459.

[59] Chen DS, Mellman I. Oncology meets immunology: the cancer-immunity cycle. Immunity 2013;39(1):1–10.

[60] Curiel TJ. Immunotherapy: a useful strategy to help combat multidrug resistance. Drug Resist Updat 2012;15(1–2):106–13.

[61] Gatouillat G, Odot J, Balasse E, Nicolau C, Tosi PF, Hickman DT, Lopez-Deber MP, Madoulet C. Immunization with liposome-anchored pegylated peptides modulates doxorubicin sensitivity in P-glycoprotein-expressing P388 cells. Cancer Lett 2007;257(2):165–71.

[62] Stein U, Walther W, Shoemaker RH, Schlag PM. IL-2 gene transfer for chemosensitization of multidrug-resistant human colon carcinoma cells. Adv Exp Med Biol 1998;451:145–9.

[63] Mroz P, Szokalska A, Wu MX, Hamblin MR. Photodynamic therapy of tumors can lead to development of systemic antigen-specific immune response. PLoS ONE 2010;5(12), e15194.

[64] Rossi GR, Vahanian NN, Ramsey WJ, Link CJ. HyperAcute vaccines: a novel cancer immunotherapy. In: Prendergast GC, Jaffee EM, editors. Cancer immunotherapy. 2nd ed. Academic Press; 2013. p. 497–516.

[65] Alegria-Torres JA, Baccarelli A, Bollati V. Epigenetics and lifestyle. Epigenomics 2011;3(3):267–77.

[66] Matei DE, Nephew KP. Epigenetic therapies for chemoresensitization of epithelial ovarian cancer. Gynecol Oncol 2010;116(2):195–201.

[67] Pal MK, Jaiswar SP, Dwivedi VN, Tripathi AK, Dwivedi A, Sankhwar P. MicroRNA: a new and promising potential biomarker for diagnosis and prognosis of ovarian cancer. Cancer Biol Med 2015;12(4):328–41.

[68] Biamonte F, Santamaria G, Sacco A, Perrone FM, Di Cello A, Battaglia AM, Salatino A, Di Vito A, Aversa I, Venturella R, et al. MicroRNA let-7g acts as tumor suppressor and predictive biomarker for chemoresistance in human epithelial ovarian cancer. Sci Rep 2019;9(1):5668.

[69] Nijveldt RJ, van Nood E, van Hoorn DE, Boelens PG, van Norren K, van Leeuwen PA. Flavonoids: a review of probable mechanisms of action and potential applications. Am J Clin Nutr 2001;74(4):418–25.

[70] Amawi H, Ashby Jr CR, Tiwari AK. Cancer chemoprevention through dietary flavonoids: what's limiting? Chin J Cancer 2017;36(1):50.

[71] Burkard M, Leischner C, Lauer UM, Busch C, Venturelli S, Frank J. Dietary flavonoids and modulation of natural killer cells: implications in malignant and viral diseases. J Nutr Biochem 2017;46:1–12.

[72] Lin TH, Hsu WH, Tsai PH, Huang YT, Lin CW, Chen KC, Tsai IH, Kandaswami CC, Huang CJ, Chang GD, et al. Dietary flavonoids, luteolin and quercetin, inhibit invasion of cervical cancer by reduction of UBE2S through epithelial-mesenchymal transition signaling. Food Funct 2017;8(4):1558–68.

[73] Ye Q, Liu K, Shen Q, Li Q, Hao J, Han F, Jiang RW. Reversal of multidrug resistance in cancer by multi-functional flavonoids. Front Oncol 2019;9:487.

[74] Granato T, Midulla C, Longo F, Colaprisca B, Frati L, Anastasi E. Role of HE4, CA72.4, and CA125 in monitoring ovarian cancer. Tumour Biol 2012;33(5):1335–9.

# Index

Note: Page numbers followed by *f* indicate figures and *t* indicate tables.